T0327557

Suspension Geometry and Computation

By the same author:

The Shock Absorber Handbook, 2nd edn (Wiley, PEP, SAE)
Tires, Suspension and Handling, 2nd edn (SAE, Arnold).
The High-Performance Two-Stroke Engine (Haynes)

Suspension Geometry and Computation

John C. Dixon, PhD, F.I.Mech.E., F.R.Ae.S.

Senior Lecturer in Engineering Mechanics
The Open University, Great Britain.

A John Wiley and Sons, Ltd, Publication

This edition first published 2009

Registered office
John Wiley & Sons Ltd, The Atrium, Southern Gate, Chichester, West Sussex, PO19 8SQ, United Kingdom

For details of our global editorial offices, for customer services and for information about how to apply for permission to reuse the copyright material in this book please see our website at www.wiley.com.

Library of Congress Cataloging-in-Publication Data

Dixon, John C., 1948-
Suspension geometry and computation / John C. Dixon.
p. cm.
Includes bibliographical references and index.
ISBN 978-0-470-51021-6 (cloth)
1. Automobiles–Springs and suspension–Mathematics.
2. Automobiles–Steering-gear–Mathematics. 3. Automobiles–Stability. 4. Roads–Mathematical models. I. Title.
TL257.D59 2009
629.2'43–dc22

2009035872

ISBN: 9780470510216
A catalogue record for this book is available from the British Library.

Typeset in 9/11 pt Times by Thomson Digital, Noida, India.

This work is dedicated to

Aythe

the beautiful goddess of truth, hence also of science and mathematics, and of good computer programs.

Her holy book is the book of nature.

Contents

Preface

The motor car is over one hundred years old. The suspension is an important part of its design, and there have been many research papers and several books on the topic. However, suspension analysis and design are so complex and bring together so many fields of study that they seem inexhaustible. Certainly, the number of different suspension designs that have been used over the years is considerable, and each one was right for a particular vehicle in that designer's opinion. This breadth of the field, at least, is a justification for another book. It is not that the existing ones are not good, but that new perspectives are possible, and often valuable, and there are new things to say. Here, the focus is on the most fundamental aspect of all, the geometry of the road, the vehicle and the suspension, the basic measurement which is the foundation of all subsequent dynamic analysis.

The process of modern engineering has been deeply affected by the computer, not always for the better because analytical solutions may be neglected with loss of insight to the problem. In this case the solution of complex three-dimensional geometrical problems is greatly facilitated by true coordinate geometry solutions or by iteration, methods which are discussed here in some detail.

In principle, geometry is not really conceptually difficult. Wrestling with actual problems shows otherwise. Analytical geometry, particularly on the computer with its many digits of precision, mercilessly shows up any approximations and errors, and, surprisingly, often reveals incomplete understanding of deep principles.

New material presented in detail here includes relationships between bump, heave and roll coefficients (Table 8.10.1), detailed analysis of linear and non-linear bump steer, design methods for determining wishbone arm lengths and angles, methods of two-dimensional and three-dimensional solutions of suspension-related geometry, and details of numerical iterative methods applied to three-dimensional suspensions, with examples.

As in my previous work, I have tried to present the basic core of theory and practice, so that the book will be of lasting value. I would be delighted to hear from readers who wish to suggest any improvements to presentation or coverage.

John C. Dixon.

1

Introduction and History

1.1 Introduction

To understand vehicle performance and cornering, it is essential to have an in-depth understanding of the basic geometric properties of roads and suspensions, including characteristics such as bump steer, roll steer, the various kinds of roll centre, and the relationships between them.

Of course, the vehicle is mainly a device for moving passengers or other payload from A to B, although in some cases, such as a passenger car tour, a motor race or rally, it is used for the interest of the movement itself. The route depends on the terrain, and is the basic challenge to be overcome. Therefore road characteristics are examined in detail in Chapter 2. This includes the road undulations giving ride quality problems, and road lateral curvature giving handling requirements. These give rise to the need for suspension, and lead to definite requirements for suspension geometry optimisation.

Chapter 3 analyses the geometry of road profiles, essential to the analysis of ride quality and handling on rough roads. Chapter 4 covers suspension geometry as required for ride analysis. Chapter 6 deals with steering geometry. Chapters 6–9 study the geometry of suspensions as required for handling analysis, including bump steer, roll steer, camber, roll centres, compliance steer, etc., in general terms.

Subsequent chapters deal with the properties of the main particular types of suspension, using the methods introduced in the earlier chapters. Then the computational methods required for solution of suspension geometry problems are studied, including two- and three-dimensional coordinate geometry, and numerical iteration.

This chapter gives an overview of suspensions in qualitative terms, with illustrations to show the main types. It is possible to show only a sample of the innumerable designs that have been used.

1.2 Early Steering History

The first common wheeled vehicles were probably single-axle hand carts with the wheels rotating independently on the axle, this being the simplest possible method, allowing variations of direction without any steering mechanism. This is also the basis of the lightweight horse-drawn chariot, already important many thousands of years ago for its military applications. Sporting use also goes back to antiquity, as illustrated in films such as *Ben Hur* with the famous chariot race. Suspension, such as it was, must have been important for use on rough ground, for some degree of comfort, and also to minimise the stress of the structure, and was based on general compliance rather than the inclusion of special spring members. The axle can be made long and allowed to bend vertically and longitudinally to ride the bumps. Another important factor in riding over rough roads is to use large wheels.

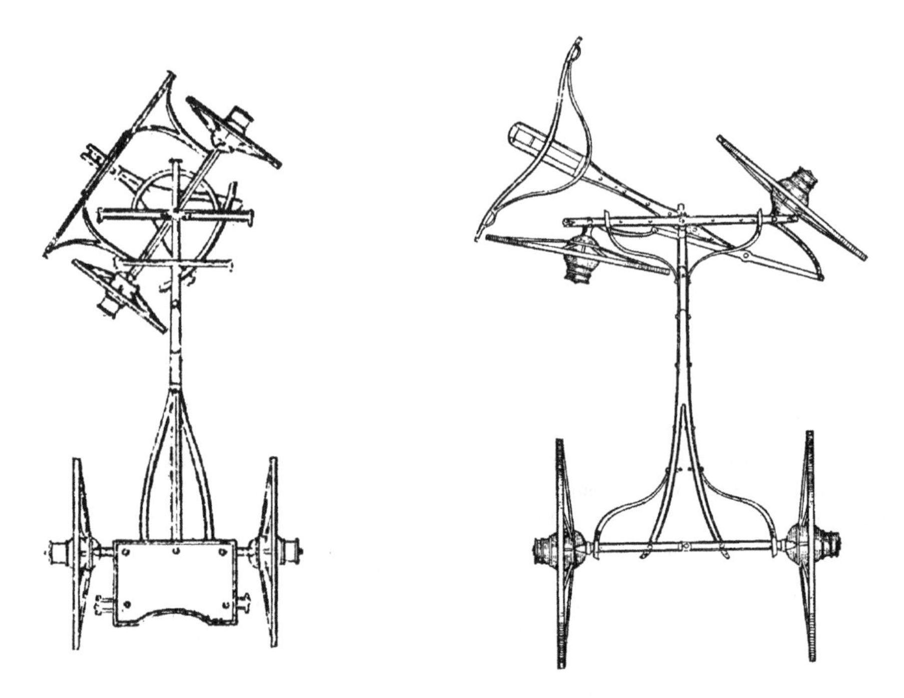

Figure 1.2.1 Steering: (a) basic cart steering by rotating the whole axle; (b) Langensperger's independent steering of 1816.

For more mundane transport of goods, a heavier low-speed two-axle cart was desirable, and this requires some form of steering mechanism. Initially this was achieved by the simple means of allowing the entire front axle to rotate, as shown in Figure 1.2.1(a).

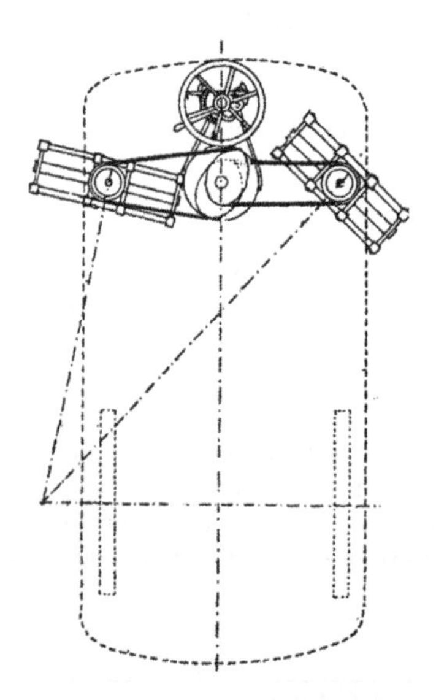

Figure 1.2.2 Ackermann steering effect achieved by two cams on L'Obeissante, designed by Amedée Bollée in 1873.

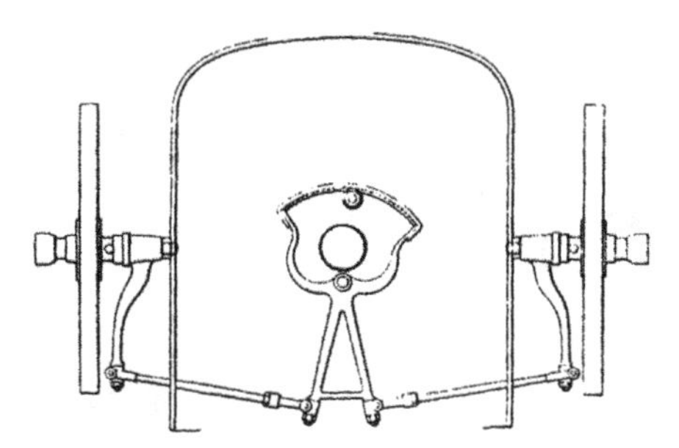

Figure 1.2.3 Ackermann steering effect achieved with parallel steering arms, by using angled drive points at the inner end of the track rods: 'La Mancelle', 1878.

Steering by the movement of the whole axle gives good geometric positioning, with easy low-speed manoeuvring, but the movement of the axle takes up useful space. To overcome this, the next stage was to steer the wheels independently, each turning about an axis close to the wheel. The first steps in this direction were taken by Erasmus Darwin (1731–1802), who had built a carriage for his doctor's practice, allowing larger-diameter wheels of great help on the rough roads. However, if the two wheels are steered through the same angle then they must slip sideways somewhat during cornering, which greatly increases the resistance to motion in tight turns. This is very obvious when a parallel-steered cart is being moved by hand. To solve this, the two wheels must be steered through different angles, as in Figure 1.2.1(b). The origin of this notion may be due to Erasmus Darwin himself in 1758, or to Richard Edgeworth, who produced the earliest known drawing of such a system. Later, in 1816 Langensperger obtained a German patent for such a concept, and in 1817 Rudolf Ackermann, acting as Langensperger's agent, obtained a British patent. The name Ackermann has since then been firmly attached to this steering design. The first application of this steering to a motor vehicle, rather than hand or horse-drawn carts, was by Edward Butler. The simplest way to achieve the desired geometry is to angle the steering arms inwards in the straight-ahead position, and to link them by a tie rod (also known as a track rod), as was done by Langensperger. However, there are certainly other methods, as demonstrated by French engineer Amedée Bollée in 1873, Figure 1.2.2, possibly allowing a greater range of action, that is, a smaller minimum turning circle.

The 'La Mancelle' vehicle of 1878 (the name refers to a person or thing from Le Mans) achieved the required results with parallel steering arms and a central triangular member, Figure 1.2.3. In 1893 Benz obtained a German patent for the same system, Figure 1.2.4. This shows tiller control of the steering, the common method of the time. In 1897 Benz introduced the steering wheel, a much superior system to the tiller, for cars. This was rapidly adopted by all manufacturers. For comparison, it is interesting to note that dinghies use tillers, where it is suitable, being convenient and economic, but ships use a large wheel, and aircraft use a joystick for pitch and roll, although sometimes they have a partial wheel on top of a joystick with only fore–aft stick movement.

1.3 Leaf-Spring Axles

Early stage coaches required suspension of some kind. With the limited technology of the period, simple wrought-iron beam springs were the practical method, and these were made in several layers to obtain the required combination of compliance with strength. These multiple-leaf springs became known simply as leaf springs. To increase the compliance, a pair of leaf springs were mounted back-to-back. They were curved, and so then known, imprecisely, as elliptical springs, or elliptics for short. Single ones were called

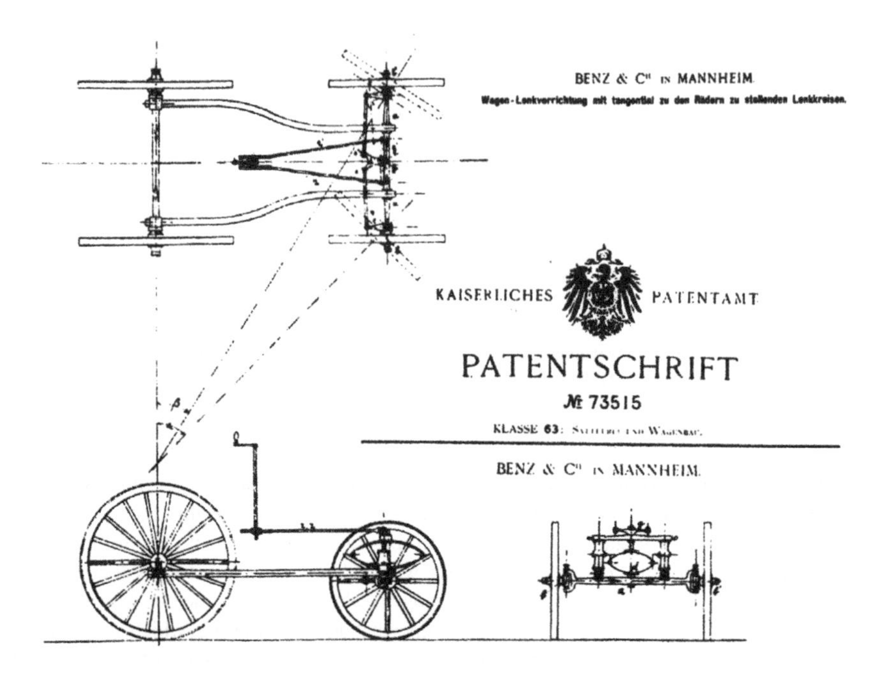

Figure 1.2.4 German patent of 1893 by Benz for a mechanism to achieve the Ackermann steering effect, the same mechanism as La Mancelle.

semi-elliptics. In the very earliest days of motoring, these were carried over from the stage coaches as the one practical form of suspension, as may be seen in Figure 1.3.1.

The leaf spring was developed in numerous variations over the next 50 years, for example as in Figure 1.3.2. With improving quality of steels in the early twentieth century, despite the increasing average

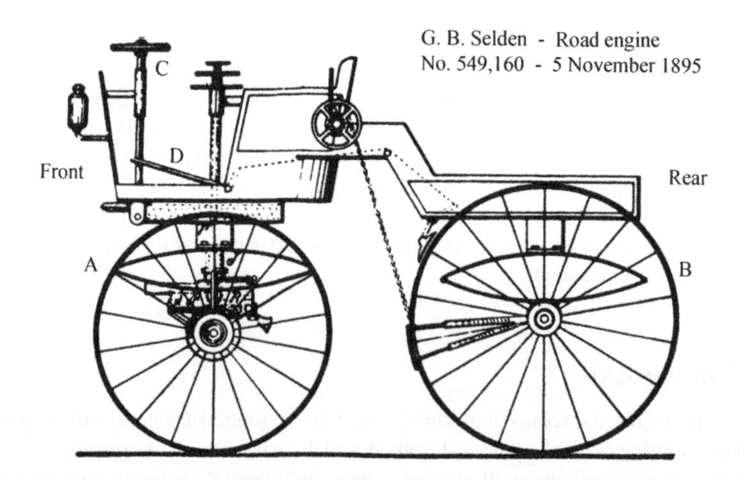

Figure 1.3.1 Selden's 1895 patent showing the use of fully-elliptic leaf springs at the front A and rear B. The steering wheel is C and the foot brake D.

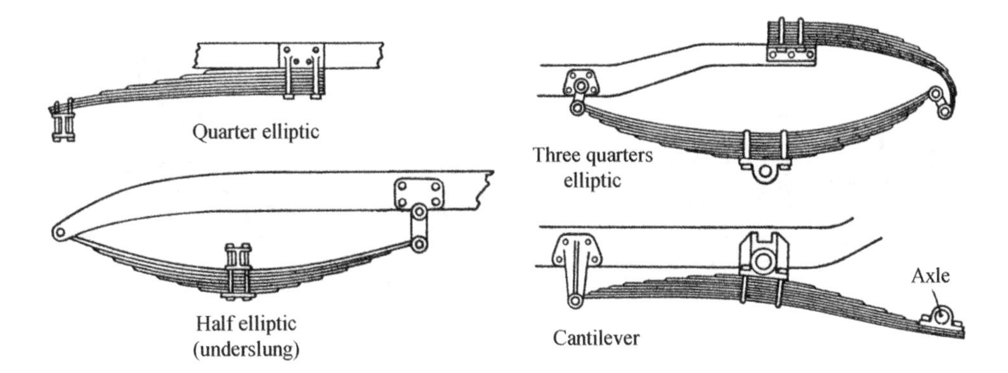

Figure 1.3.2 Some examples of the variation of leaf springs in the early days. As is apparent here, the adjective 'elliptical' is used only loosely.

weight of motor cars, the simpler semi-elliptic leaf springs became sufficient, and became widely standardised in principle, although with many detailed variations, not least in the mounting systems, position of the shackle, which is necessary to permit length variation, and so on. The complete vehicle of Figure 1.3.3 shows representative applications at the front and rear, the front having a single compression shackle, the rear two tension shackles. A very real advantage of the leaf spring in the early days was that the spring provides lateral and longitudinal location of the axle in addition to the springing compliance action. However, as engine power and speeds increased, the poor location geometry of the leaf spring became an increasing problem, particularly at the front, where the steering system caused many problems in bump and roll. To minimise these difficulties, the suspension was made stiff, which caused poor ride quality.

Figures 1.3.4 and 1.3.5 show representative examples of the application of the leaf spring at the rear of normal configuration motor cars of the 1950s and 1960s, using a single compression shackle.

Greatly improved production machinery by the 1930s made possible the mass production of good quality coil springs, which progressively replaced the leaf spring for passenger cars. However, leaf-spring use on passenger cars continued through into the 1970s, and even then it functioned competitively, at the rear at least, Figure 1.3.6. The leaf spring is still widely used for heavily loaded axles on trucks and military vehicles, and has some advantages for use in remote areas where only basic maintenance is possible, so leaf-spring geometry problems are still of real practical interest.

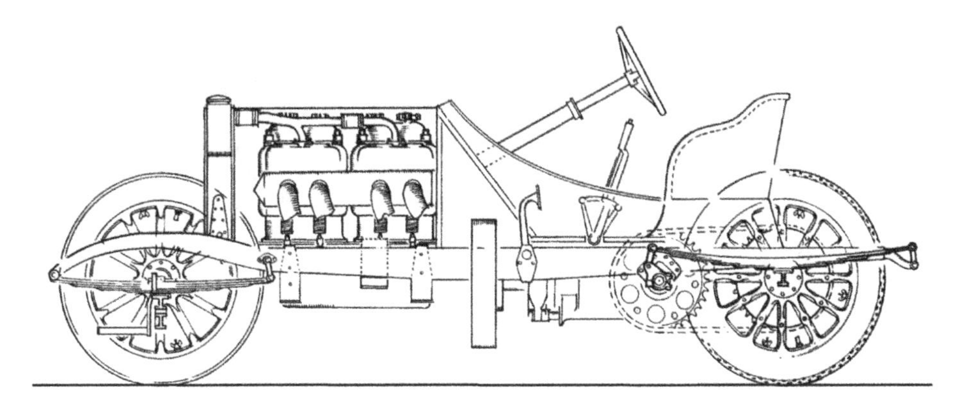

Figure 1.3.3 Grand Prix car of 1908, with application of semi-elliptic leaf springs at the front and rear (Mercedes-Benz).

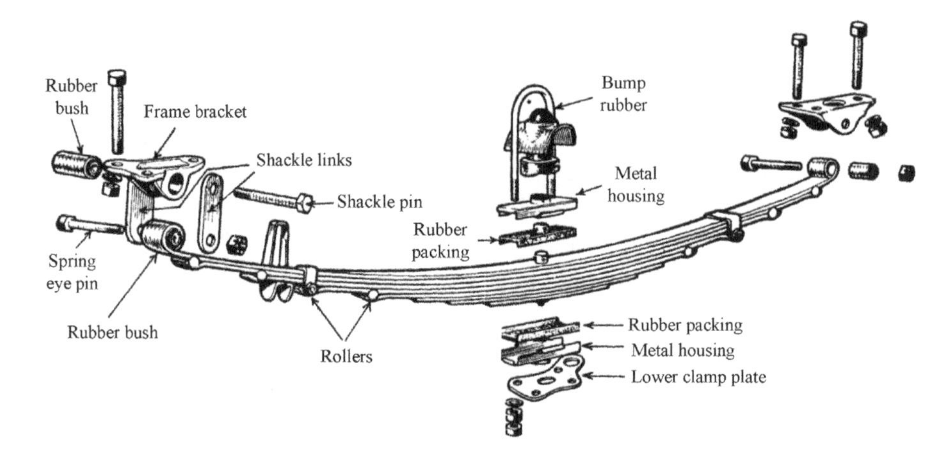

Figure 1.3.4 A representative rear leaf-spring assembly (Vauxhall).

At the front, the leaf spring was much less satisfactory, because of the steering geometry difficulties (bump steer, roll steer, brake wind-up steering effects, and shimmy vibration problems). Figure 1.3.7 shows a representative layout of the typical passenger car rigid-front-axle system up to about 1933. In bump, the axle arc of movement is centred at the front of the spring, but the steering arm arc is centred at

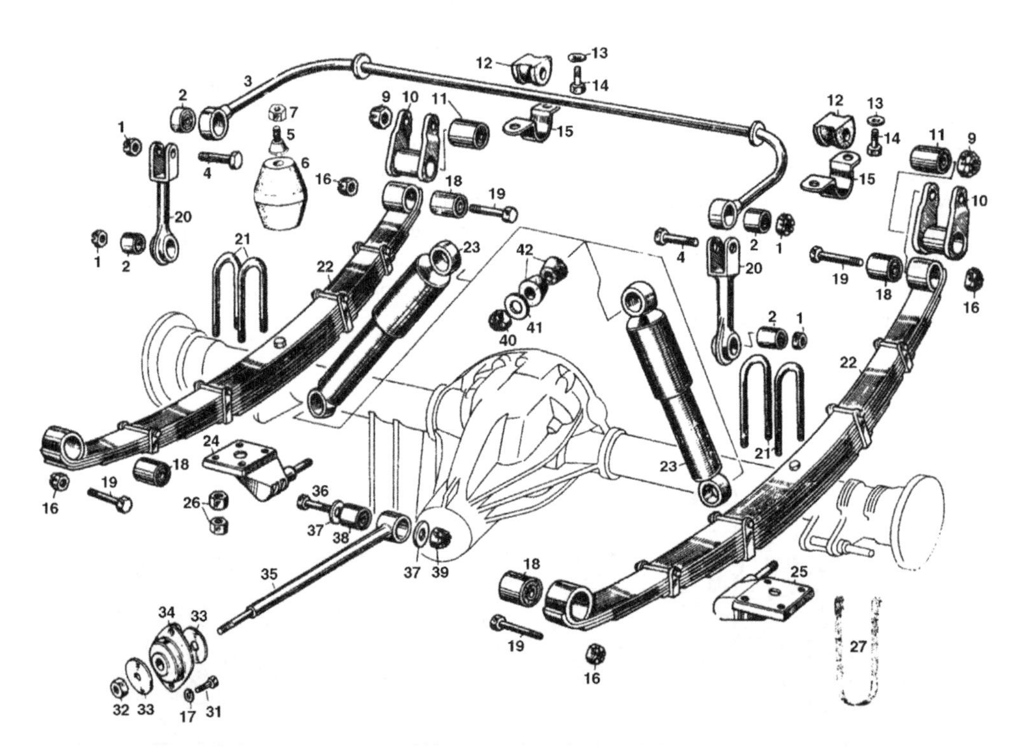

Figure 1.3.5 A 1964 live rigid rear axle with leaf springs, anti-roll bar and telescopic dampers. The axle clamps on top of the springs (Maserati).

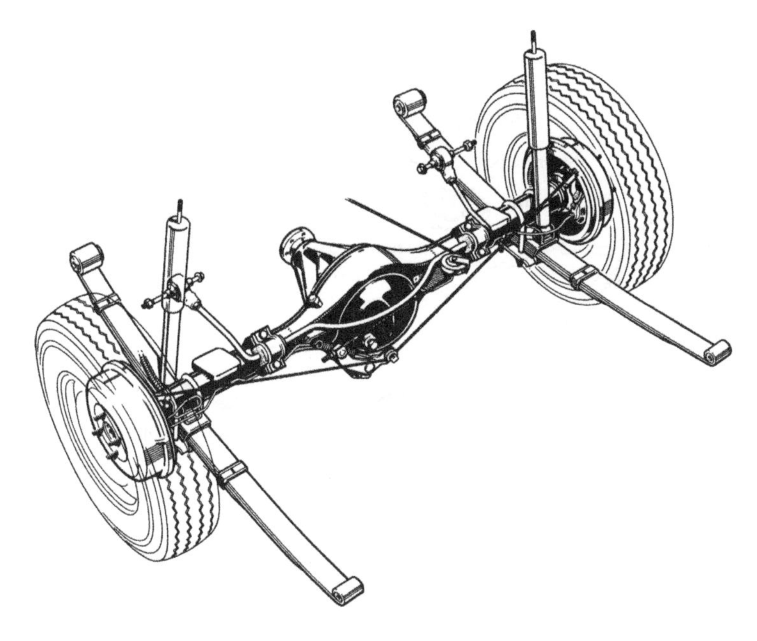

Figure 1.3.6 Amongst the last of the passenger car leaf-spring rear axles used by a major manufacturer was that of the Ford Capri. Road testers at the time found this system in no way inferior to more modern designs.

the steering box. These conflicting arcs give a large and problematic bump steer effect. The large bump steer angle change also contributed to the shimmy problems by causing gyroscopic precession moments on the wheels. Figure 1.3.8 shows an improved system with a transverse connection.

Truck and van steering with a leaf spring generally has the steering box ahead of the axle, to give the maximum payload space, as seen in Figure 1.3.9. In bump, the arc of motion of the steering arm and the axle on the spring are in much better agreement than with the rear box arrangement of Figure 1.3.7, so bump steer is reduced. Also, the springs are likely to be much stiffer, with reduced range of suspension movement, generally reducing the geometric problems.

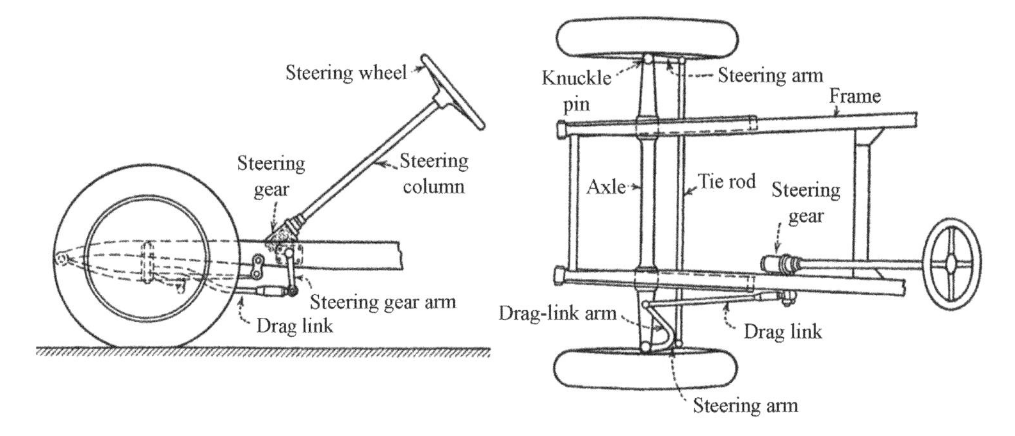

Figure 1.3.7 Classical application of the rigid axle at the front of a passenger car, the normal design up to 1933. Steering geometry was a major problem because of the variability of rigid axle movements.

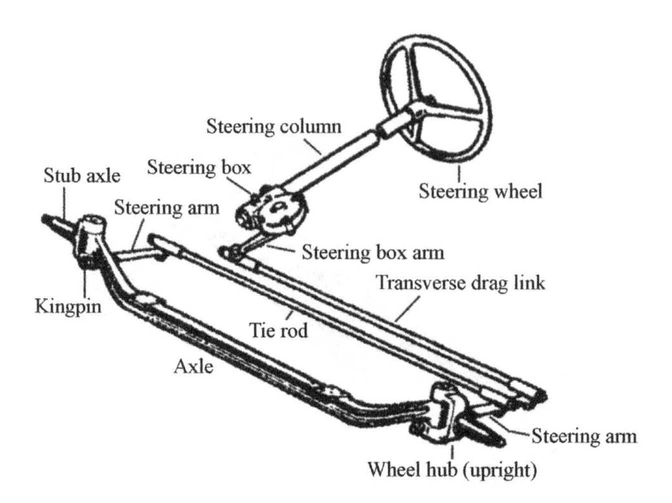

Figure 1.3.8 Alternative application of the rigid axle at the front of a passenger car, with a transverse steering link between the steering box on the sprung mass and the axle, reducing bump steer problems.

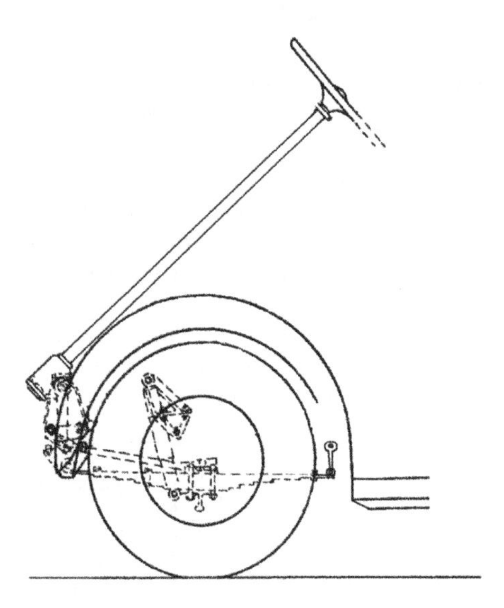

Figure 1.3.9 Van or truck steering typically has a much steeper steering column with a steering box forward of the axle, as here. The steering geometry problems are different in detail, but may be less overall because a stiffer suspension is more acceptable.

1.4 Transverse Leaf Springs

Leaf springs were not used only in longitudinal alignment. There have been many applications with transverse leaf springs. In some cases, these were axles or wheel uprights located by separate links, to overcome the geometry problems, with the leaf spring providing only limited location service, or only the springing action. Some transverse leaf examples are given in Figures 1.4.1–1.4.4

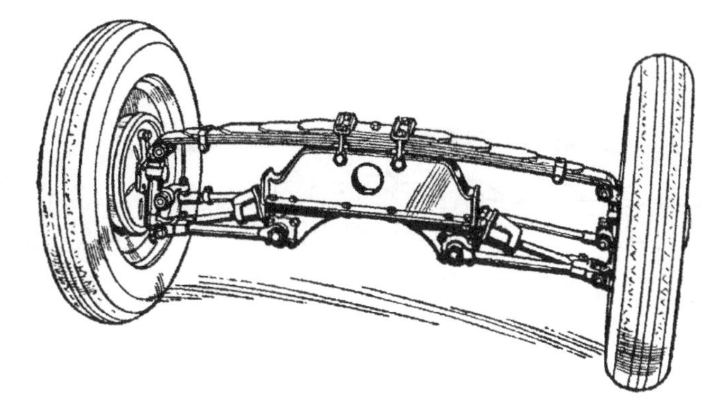

Figure 1.4.1 A transverse leaf spring at the top also provides upper lateral and longitudinal location on this front axle, with a lower wishbone (early BMW).

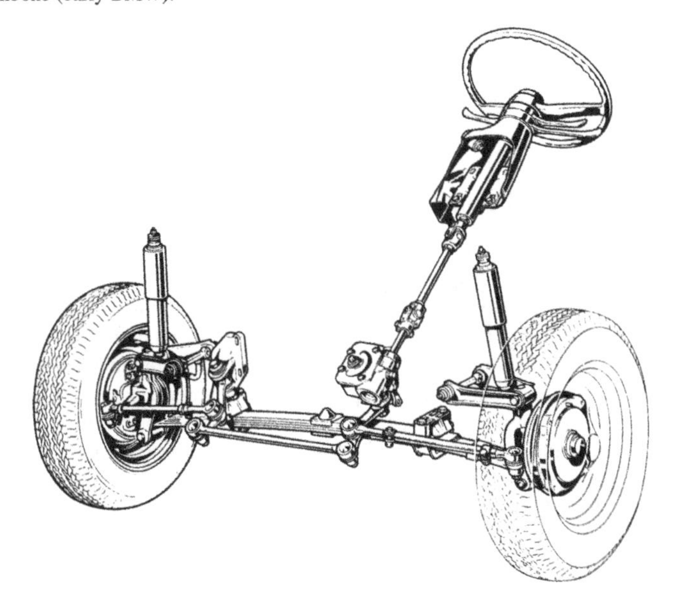

Figure 1.4.2 This more modern small car front suspension has a transverse leaf spring at the bottom with an upper wishbone (Fiat).

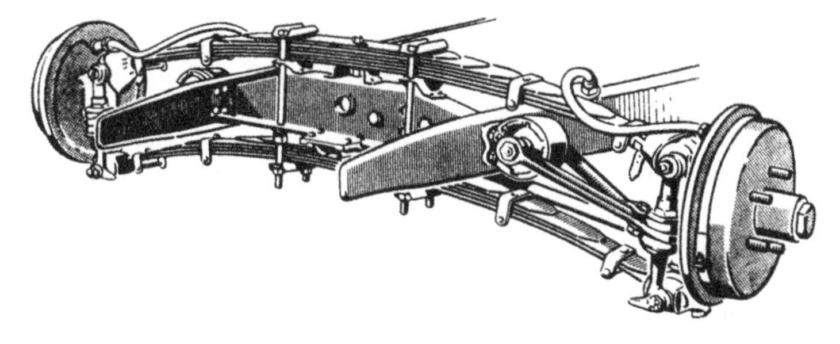

Figure 1.4.3 Two transverse leaf springs providing complete hub location acting as equal-length wishbones without any additional links (1931, Mercedes Benz).

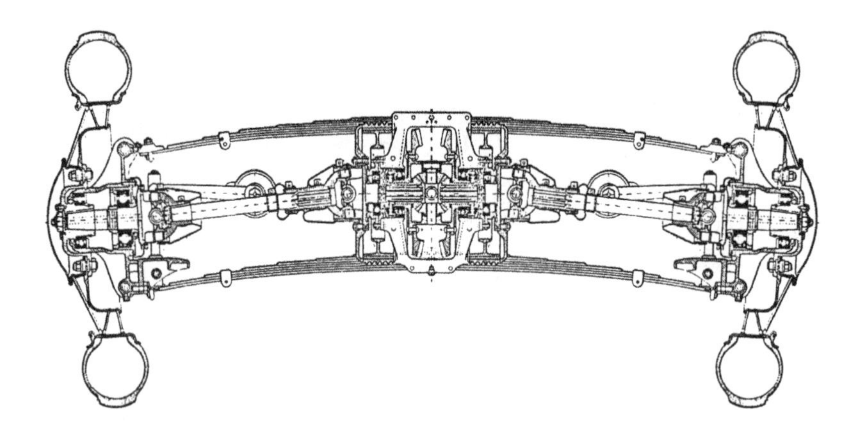

Figure 1.4.4 The Cottin-Desgouttes of 1927 used four leaf springs on the driven rear axle, in a square configuration, also featuring inboard brakes. The parallel pair of springs at the top or bottom acted as equal-length wishbones, with length equal to three-quarters of the cantilever length. The driveshaft length can be chosen to match this length, to minimise the plunge requirement at the splines. The wheels have 2.5° static camber but do not rotate in bump (zero bump camber).

1.5 Early Independent Fronts

Through the 1920s, the rigid axle at the front was increasingly a problem. Despite considerable thought and experimentation by suspension design engineers, no way had been found to make a steering system that worked accurately. In other words, there were major problems with bump steer, roll steer and spring wind-up, particularly during braking. Any one of these problems might be solved, but not all at once. With increasing engine power and vehicle speeds, this was becoming increasingly dangerous, and hard front springs were required to ameliorate the problem, limiting the axle movement, but this caused very poor ride comfort. The answer was to use independent front suspension, for which a consistently accurate steering system could be made, allowing much softer springs and greater comfort. Early independent suspension designs were produced by André Dubonnet in France in the late 1920s, and a little later for Rolls-Royce by Donald Bastow and Maurice Olley in England. These successful applications of independent suspension became known in the USA, and General Motors president Alfred P. Sloan took action, as he describes in his autobiography (Sloan, 1963).

Around 1930, Sloan considered the problem of ride quality as one of the most pressing and most complex in automotive engineering, and the problem was getting worse as car speeds increased. The early solid rubber tyres had been replaced by vented thick rubber, and then by inflated tyres. In the 1920s, tyres became even softer, which introduced increased problems of handling stability and axle vibrations. On a trip to Europe, Sloan met French engineer André Dubonnet who had patented a successful independent suspension, and had him visit the US to make contact with GM engineers. Also, by 1933 Rolls-Royce already had an independent front suspension, which was on cars imported to the USA. Maurice Olley, who had previously worked for Rolls-Royce, was employed by GM, and worked on the introduction of independent suspensions there. In Sloan's autobiography, a letter from Olley describes an early ride meter, which was simply an open-topped container of water, which was weighed after a measured mile at various speeds. Rolls-Royce had been looking carefully at ride dynamics, including measuring body inertia, trying to get a sound scientific understanding of the problem, and Olley introduced this approach at GM. In 1932 they built the K-squared rig (i.e. radius of gyration squared), a test car with various heavy added masses right at the front and rear to alter the pitch inertia in a controlled way. This brought home the realisation that a much superior ride could be achieved by the use of softer front springs, but soft springs caused shimmy problems and bad handling. Two experimental Cadillac cars were built, one using Dubonnet's type of suspension, the other with a double-wishbone (double A-arm) suspension of GM's design. The engineers were pleased with the ride and handling, but shimmy steering vibration was a persistent problem requiring intensive development work. In March 1933 these two experimental cars were demonstrated to GM's top management, along with an automatic transmission. Within a couple of miles, the 'flat ride' suspension was evidently well received.

March 1933 was during the Great Depression, and financial constraints on car manufacturing and retail prices were pressing, but the independent front suspension designs were enthusiastically accepted, and shown to the public in 1934. In 1935 Chevrolet and Pontiac had cars available with Dubonnet suspension, whilst Cadillac, Buick and Oldsmobile offered double-wishbone front suspension, and the rigid front axle was effectively history, for passenger cars at least. A serious concern for production was the ability of the machine tool industry to produce enough suitable centreless grinders to make all the coil springs that would be required. With some practical experience, it became apparent that with development the wishbone suspension was easier and cheaper to manufacture, and also more reliable, and was universally adopted.

Figure 1.5.1 shows the 1934 Cadillac independent suspension system, with double wishbones on each side, in which it may be seen that the basic steering concept is recognisably related to the ones described earlier. As covered in detail in Chapter 6, the track-rod length and angle can be adjusted to give good steering characteristics, controlling bump steer and roll steer. The dampers were the lever-operated double-piston type, incorporated into the upper wishbone arms. Such a system would still be usable today.

Figure 1.5.2 shows the Dubonnet type suspension, used by several other manufacturers, which was unusually compact. The wheels are on leading or trailing arms, with the spring contained in a tube on the

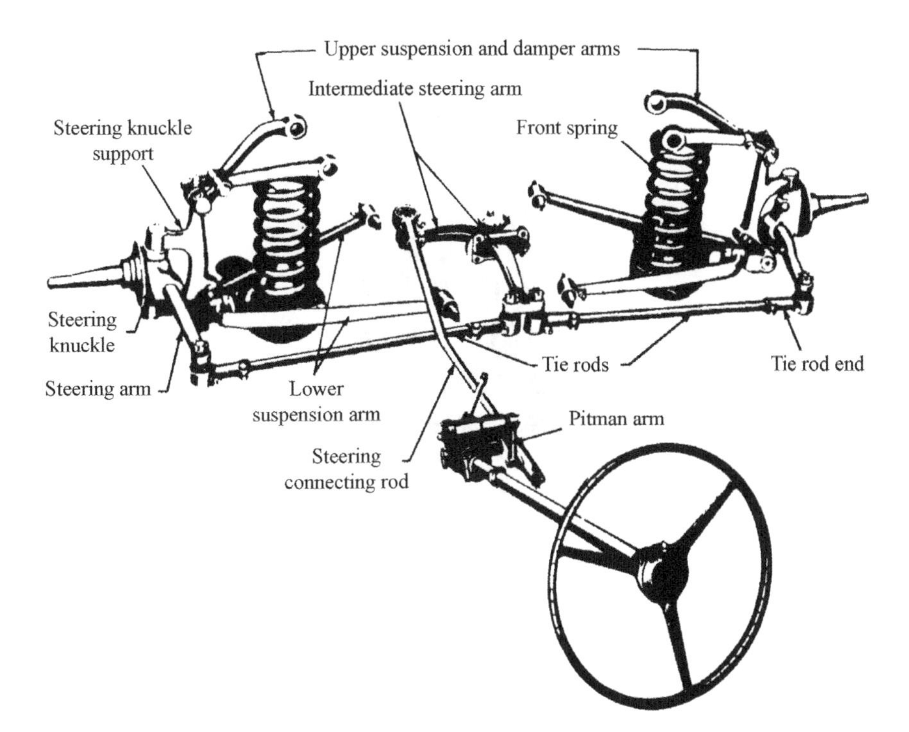

Figure 1.5.1 The new Cadillac steering and independent suspension of 1934.

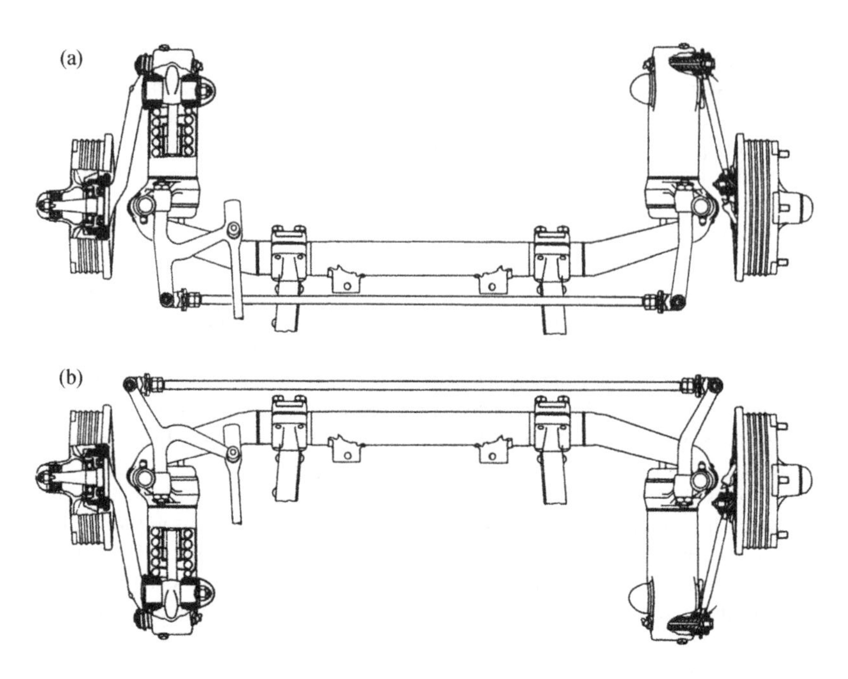

Figure 1.5.2 The Dubonnet type suspension in plan view, front at the top: (a) with trailing links; (b) with leading links (1938 Opel).

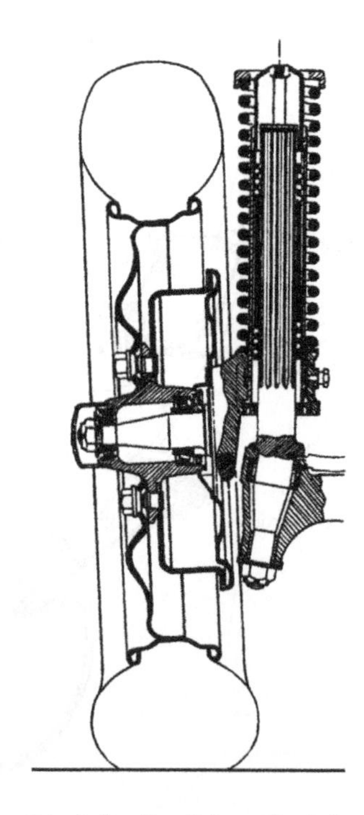

Figure 1.5.3 Broulhiet ball-spline sliding pillar independent suspension.

steerable part of the system. The type shown has a single tie rod with a steering box, as was usual then, but the system is equally adaptable to a steering rack. The Ackermann effect is achieved here by angling the steering arms backwards and inwards in Figure 1.5.2(a) with the trailing arms, or forwards and outwards in Figure 1.5.2(b) with the leading arms. The steering action is entirely on the sprung mass, so there is no question of bump or roll steer due to the steering, and there are no related issues over the length of the steering members. Bump steer effects depend only on the angle of the pivot axis of the arms, in this case simply transverse, with zero bump steer and zero bump camber. Other versions had this axis at various angles. The leading link type at the front of a vehicle gives considerable anti-dive in braking, but is harsh over sharp bumps. The trailing-arm version is better over sharp bumps but has strong pro-dive in braking.

Another early form of independent suspension was that due to Brouhliet in France, who used sliding splines, with ball bearings for low friction, for the suspension action, Figure 1.5.3, again allowing the steering to be entirely sprung, eliminating the steering problems of the rigid front axle.

1.6 Independent Front Suspension

Some independent suspensions have already been shown. Section 1.4 illustrates some with transverse leaf springs. Section 1.5 shows two from the mid 1930s – the Dubonnet, now effectively defunct, and the double wishbone which was the *de facto* standard front suspension for many years, although now that could be perhaps be said instead of the strut and wishbone. Subsequent to the leaf spring, torsion bar suspensions were quite common. However, the modern independent suspension is almost invariably based on the coil spring, with location by two wishbones (A-arms) or by a strut with one wishbone at the bottom.

Figure 1.6.1 shows a sliding pillar suspension, not representative of common modern practice, but this was an early success of some historical interest. With the spring and damper unit enclosed, it was very

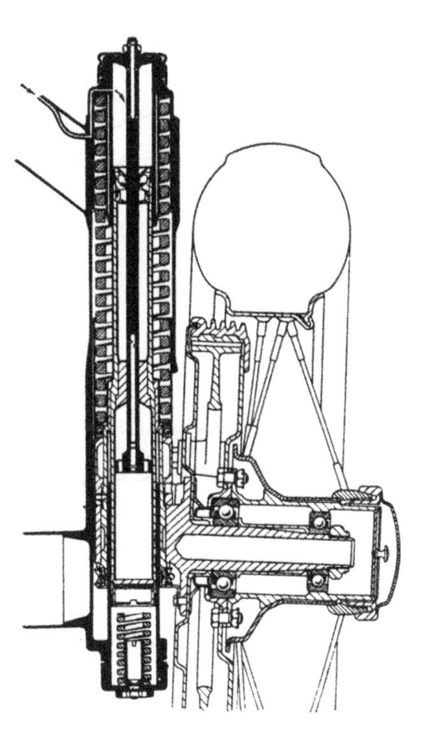

Figure 1.6.1 Sliding pillar front suspension (Lancia Lambda).

reliable, particularly compared with other designs of the early days. When introduced, this was regarded by the manufacturer as the best suspension design regardless of cost.

Figure 1.6.2 shows various versions and views of the twin parallel-trailing-arm suspension, which most often used torsion-bar springing in the cross tubes. On the Gordon-Armstrong this could be supplemented with, or replaced by, coils springs used in compression with draw bars, with double action on the spring. Again, this was a very compact system. The steering can be laid out to give zero bump steer, as in the Aston Martin version of Figure 1.6.2(d), or even with asymmetrical steering as in Figure 1.6.2(a).

A transverse single swing-arm type of front suspension can be used, as in Figure 1.6.3, but with lower body pivot points than the usual driven rear swing axle, giving a lower roll centre. There is a large bump camber effect with this design, such as to effectively eliminate roll camber completely. Steering is by

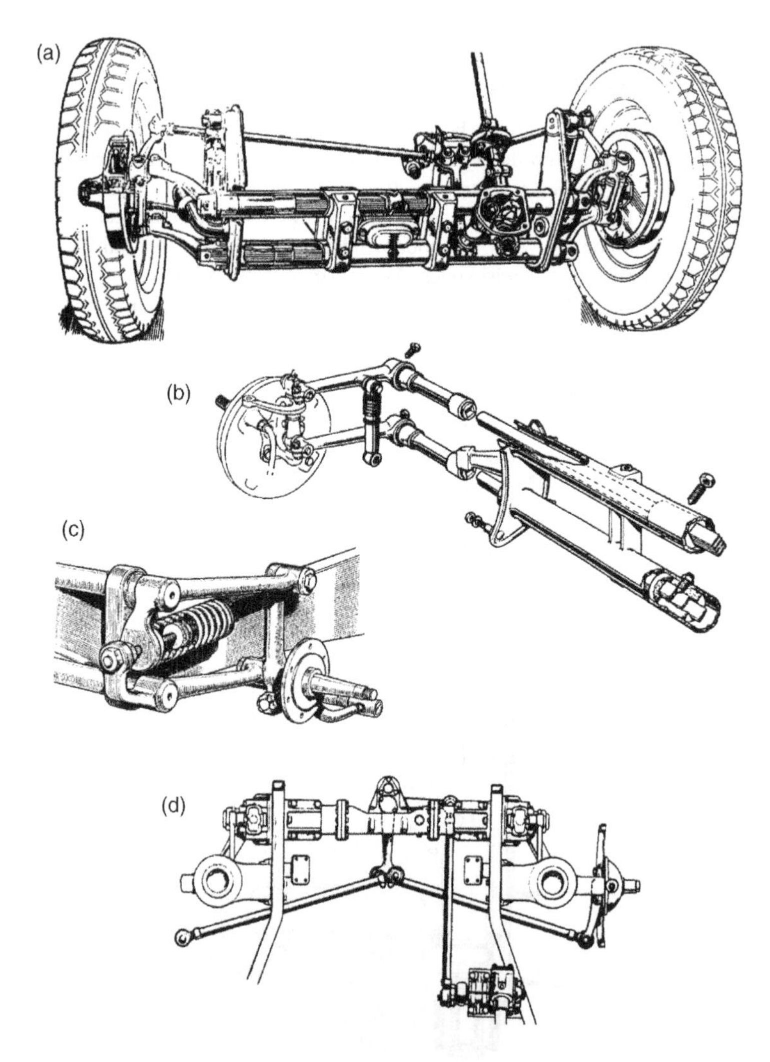

Figure 1.6.2 Parallel-trailing-arm front suspension: (a) general front view (VW); (b) rear three-quarter view of torsion bar type construction; (c) parallel-trailing-arm suspension with laid-down compression coil spring and tension bar (Gordon-Armstrong); (d) steering layout for parallel-trailing-arm system, plan view (early Aston Martin).

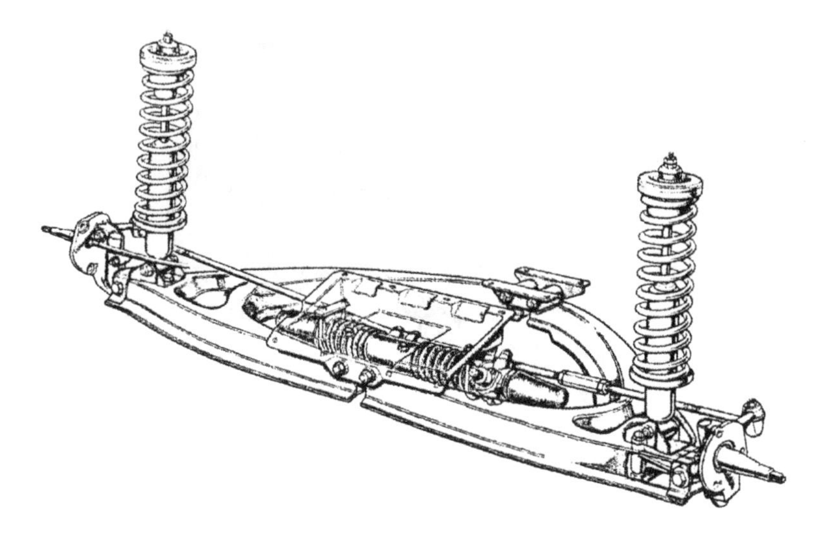

Figure 1.6.3 Single transverse swing-arm independent front suspension with rack-and-pinion steering (1963 Hillman).

rack-and-pinion, with appropriately long track rods, giving no bump steer, this requiring the pick-up points on the rack to be aligned with the arm pivot axes in the straight-ahead position. Unusually, the track-rod connections are on the rear of the rack, which affects only the plan view angle of the track rods, and hence the Ackermann factor.

The Glas Isar had double wishbones, as seen in Figure 1.6.4, but the upper wishbone had its pivot axis transverse, so in front view the geometry was similar to a strut-and-wishbone suspension. The steering system is high up, and asymmetrical. Analysis of bump steer requires a full three-dimensional solution, but with the asymmetrical steering on this design there could be problems unless the track-rod connections to the steering box arm are aligned with the upper wishbone axes.

Some early double-wishbone systems were very short, particularly on racing cars, as in Figure 1.6.5. With the relatively long track rod shown there would have been significant bump steer, which could have been only marginally acceptable by virtue of the stiff suspension and small suspension deflections. This makes an interesting contrast with the very long wishbones on modern racing cars, although in that case the deflections are still small and it is done for different reasons.

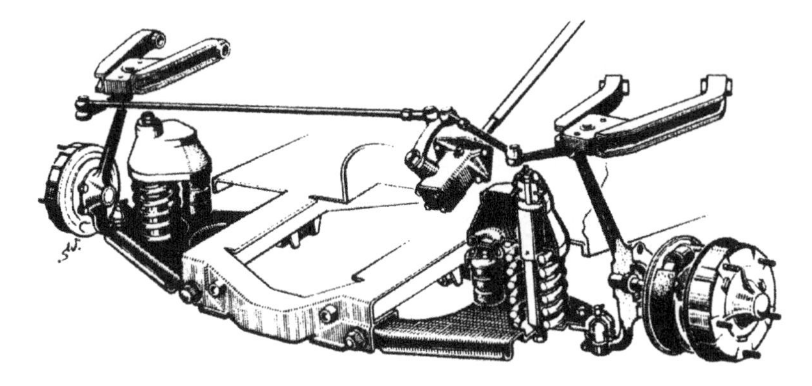

Figure 1.6.4 A double-wishbone suspension in which the wishbone axes are crossed (Glas Isar).

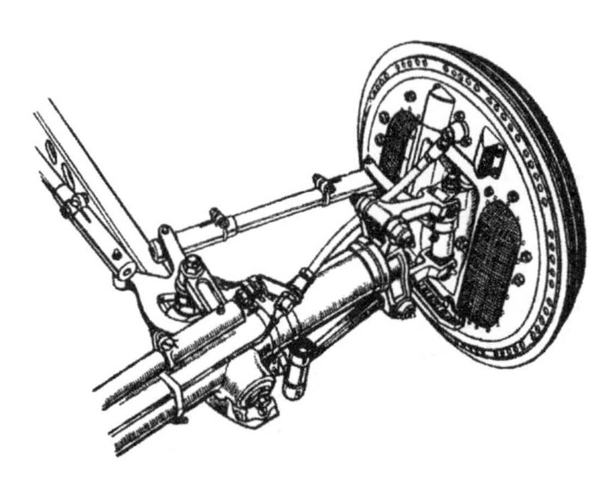

Figure 1.6.5 An early double-wishbone system with very short arms (Mercedes Benz). The suspension spring is inside the transverse horizontal tube.

Figure 1.6.6 shows an engineering section of a fairly representative double-wishbone system, with unequal-length arms, nominally parallel in the static position. As is usual with double wishbones, the spring acts on the lower arm at a motion ratio of about 0.5. The steering axis is defined by ball joints rather than by the old kingpin system, with wide spacing giving lower joint loads.

In Figure 1.6.7, a more recent double-wishbone system, the upper wishbone is partially defined by the rear-mounted anti-roll bar. The steering arms are inclined to the rear as if to give an Ackermann effect, but the track rods are also angled (see Chapter 5). The offset connections on the steering box and idler arm give some Ackermann effect.

The modern double-wishbone system of Figure 1.6.8 is different in that the spring acts on the upper wishbone, with vertical forces transmitted into the body at the top of the wheel arch in the same way as for a strut. However, despite the spring position this is certainly not a strut suspension, which is defined by a

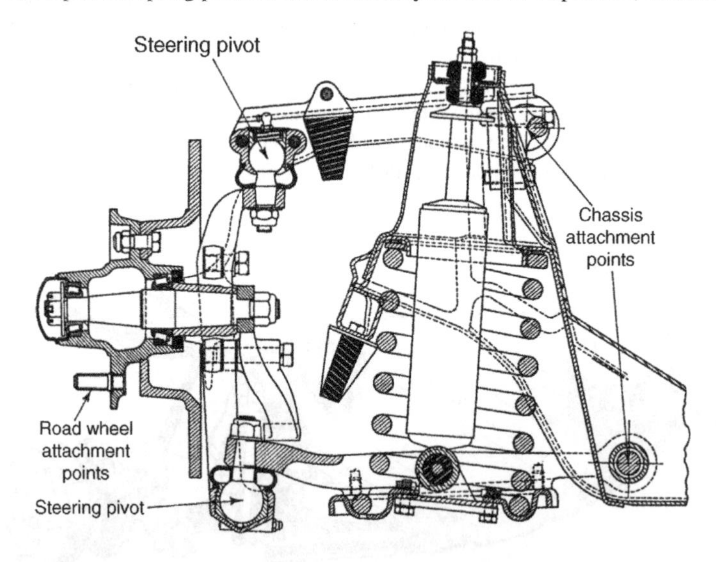

Figure 1.6.6 Traditional configuration of passenger car double-wishbone suspension, with the spring and damper acting on the lower wishbone (Jaguar).

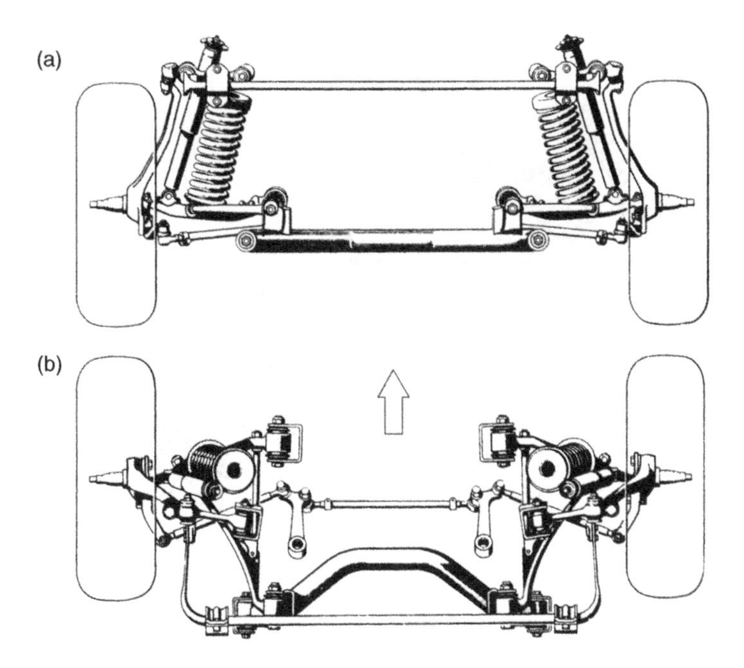

Figure 1.6.7 A passenger car double-wishbone system with a wide-base lower wishbone, and the upper wishbone partially defined by the anti-roll bar (Mercedes Benz).

rigid camber connection between the wheel upright and the strut. The steering connections are to the ends of the rack, to give the correct track-rod length to control bump steer.

The commercial vehicle front suspension of Figure 1.6.9 is a conventional double-wishbone system with a rear-mounted anti-roll bar, and also illustrates the use of a forward steering box and steep steering

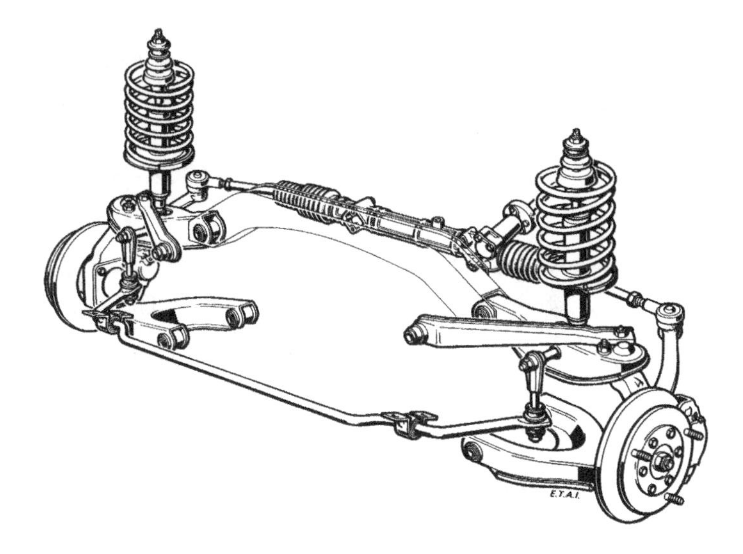

Figure 1.6.8 A representative modern double-wishbone suspension, with spring and damper acting on the upper arm (Renault). This is not a strut suspension.

Figure 1.6.9 A double wishbone system from a light commercial vehicle, also illustrating a forward steering system (VW).

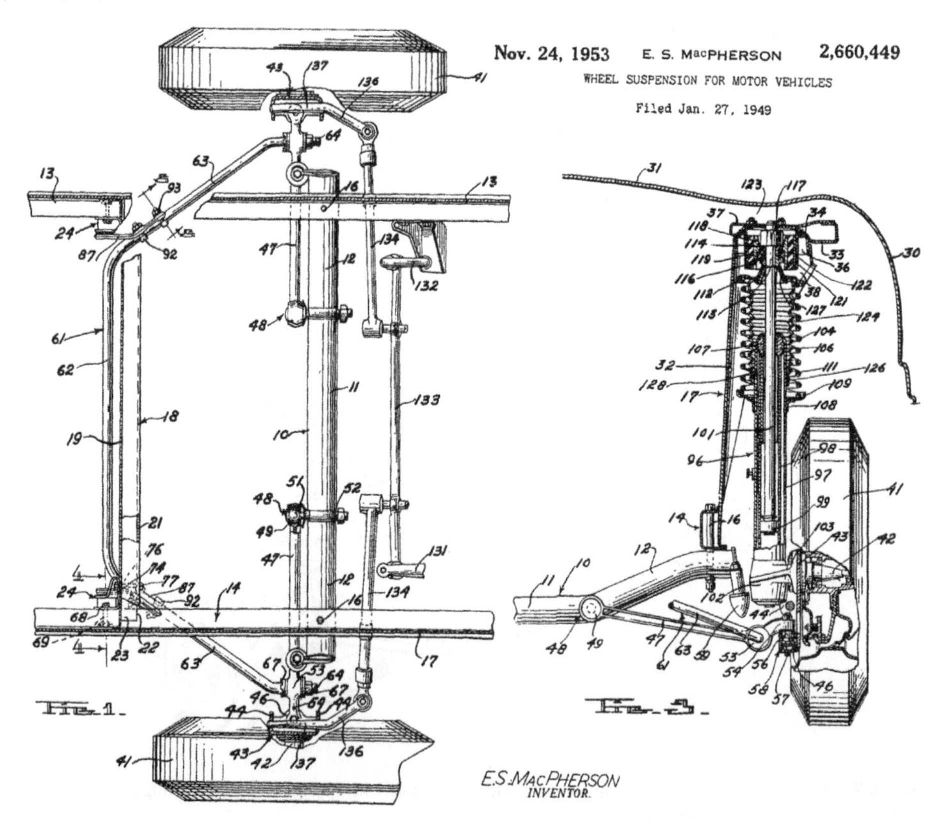

Figure 1.6.10 MacPherson's 1953 US patent for strut suspension (front shown).

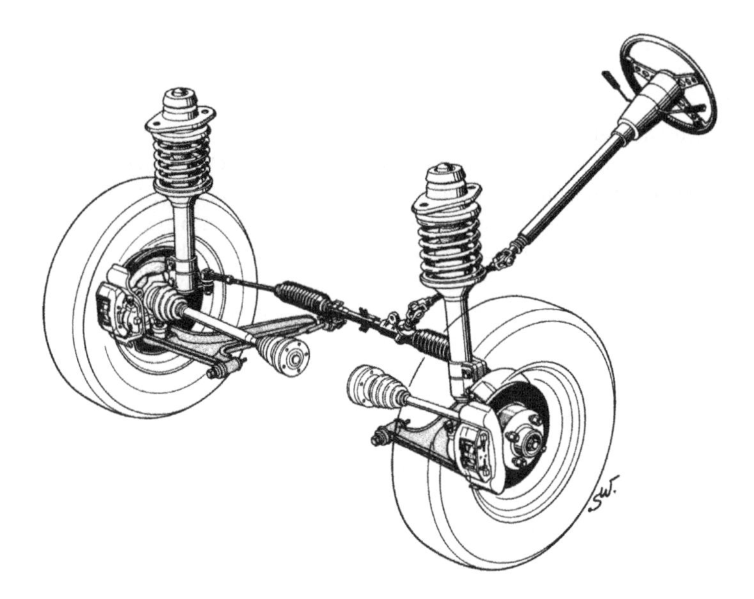

Figure 1.6.11 Passenger car strut suspension with wide-based lower wishbone and low steering (VW).

column on this kind of vehicle. Again, the tie rod and idler arm allow the two track rods to be equal in length and to have correct geometry for the wishbones.

Finally, Figure 1.6.10 shows the MacPherson patent of 1953 for strut suspension, propsed for use at the front and the rear. A strut suspension is one in which the wheel upright (hub) is controlled in camber by a rigid connection to the strut itself. This was popularised for front suspension during the 1950s and 1960s by Ford, with the additional feature that the function of longitudinal location was combined with an anti-roll bar. Strut suspension lacks the adaptability of double-wishbone suspension to desired geometric properties, but can be made acceptable whilst giving other benefits. The load transmission into the body is

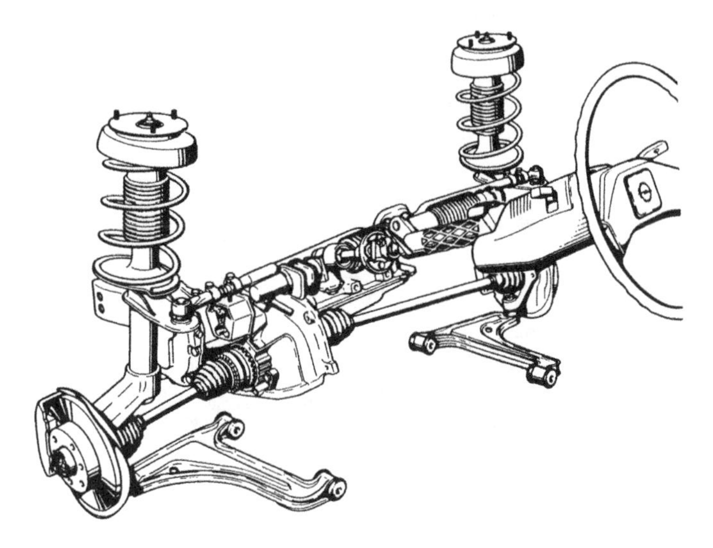

Figure 1.6.12 Passenger car strut suspension with wide-based lower wishbone and high steering (Opel).

widely spread, and pre-assembled units of the suspension can be fitted to the body in an efficient manner on the final assembly line.

Figures 1.6.11 and 1.6.12 give two more modern examples, in fact of strut-and-wishbone suspensions, normally just called strut suspensions, as now so commonly used on small and medium passenger vehicles. The two illustrations show conventional struts. In contrast, a Chapman strut is a strut suspension with a driveshaft, the shaft providing the lateral location of the bottom member, this requiring and allowing no length change of the driveshaft, eliminating a number of problems such as sticking splines under load (see Figure 1.10.7). In Figure 1.6.11 the steering rack is low, close to the level of the wishbone, and the track-rod connections are, correspondingly, at the ends of the rack, to give the correct track-rod length. In Figure 1.6.12 the steering is higher, at the level of the spring seat, so for small bump steer the track rods must be longer, and are connected to the centre of the rack.

1.7 Driven Rigid Axles

The classic driven rigid rear axle, or so-called 'live axle', is supported and located by two leaf springs, in which case it is called a 'Hotchkiss axle', as shown previously in Figures 1.3.5 and 1.3.6. Perhaps due to the many years of manufacturers' experience of detailing this design, sometimes it has been implemented with great success. In other cases, there have been problems, such as axle tramp, particularly when high tractive force is used. To locate the axle more precisely, or more firmly, sometimes additional links are used, such as the longitudinal traction bars above the axle in Figure 1.7.1, opposing pitch rotation. These used to be a well-known aftermarket modification for some cars, but were often of no help. In other cases, the leaf springs have been retained as the sole locating members but with the springing action assisted by coils, as in Figure 1.7.2, giving good load spreading into the body.

However, with the readier availability of coil springs, in due course the rear leaf-spring axle finally disappeared from passenger cars, typically being replaced by the common arrangement of Figure 1.7.3, with four locating links, this system being used by several manufacturers. The two lower widely-spaced parallel links usually also carry the springs, as this costs less boot (trunk) space than placing them directly on the axle. Lateral positioning of the axle is mainly by the convergent upper links, although this gives rather a high roll centre. The action of axle movement may not be strictly kinematic, and may depend to some extent on compliance of the large rubber bushes that are used in each end of the links.

The basic geometry of the four-link system is retained in the T-bar system of Figure 1.7.4, with the cross-arm of the T located between longitudinal ribs on the body, allowing pivoting with the tail of the T, connected to the axle, able to move up and down in an arc in side view. This gives somewhat more precise location than the four-link system, and requires less bush compliance for its action, but again the roll centre is high, satisfactory for passenger cars, but usually replaced by a sliding block system for racing versions of these cars.

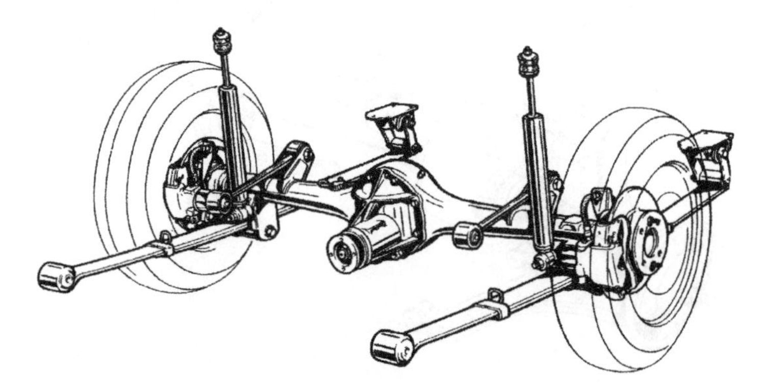

Figure 1.7.1 Leaf-spring axle with the addition of traction bars above the axle (Fiat).

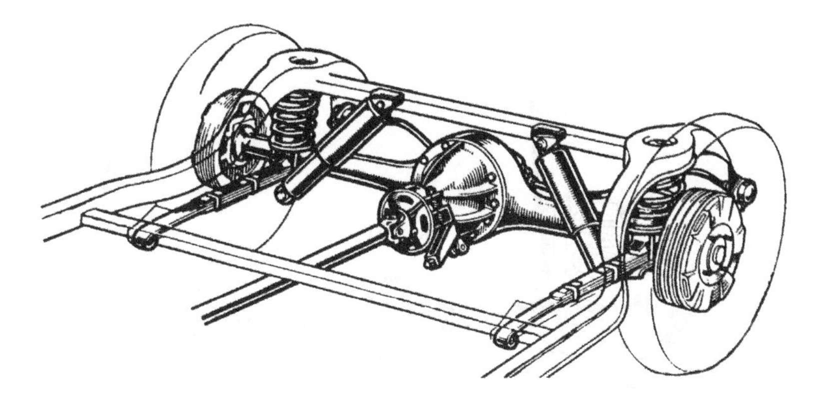

Figure 1.7.2 Leaf-spring axle with additional coil springs (Fiat).

Figure 1.7.3 Widely-used design of four-link location axle (Ford).

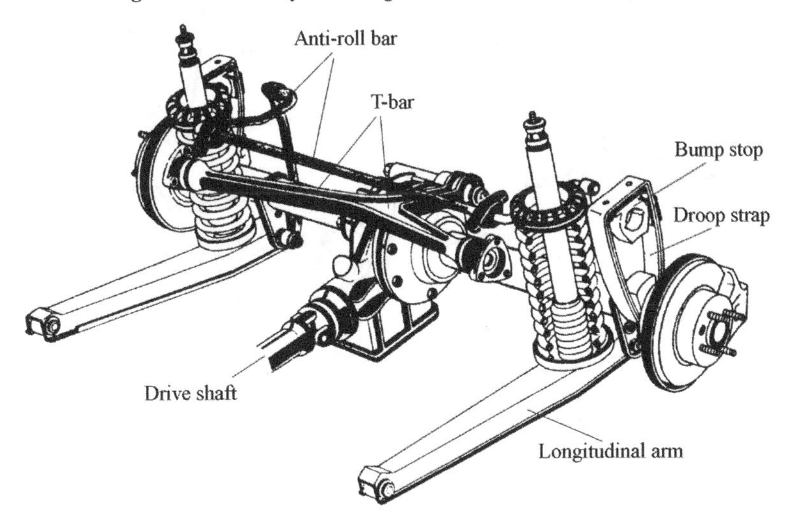

Figure 1.7.4 Alfa-Romeo T-bar upper lateral location.

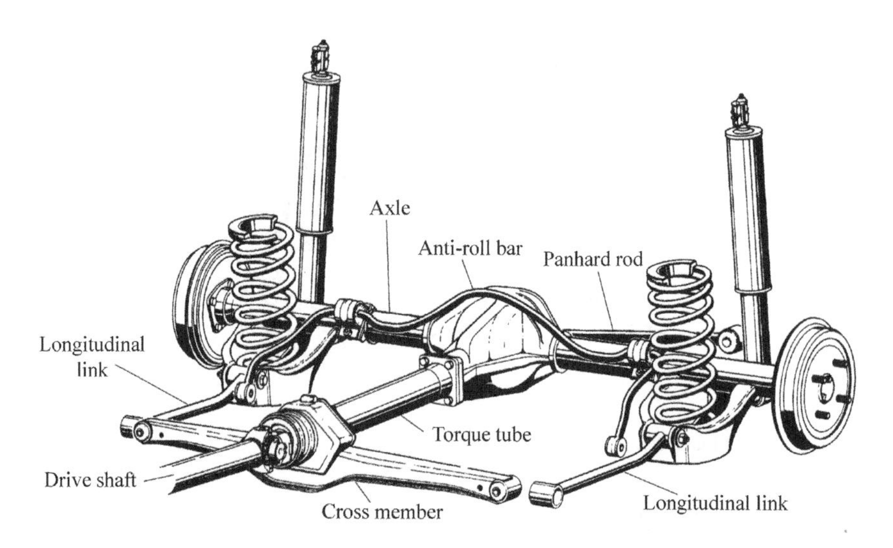

Figure 1.7.5 Rear axle with torque tube and Panhard rod (Opel).

There have, of course, been many other lateral location systems for axles, including the Panhard rod and the Watt's linkage, shown in Figures 1.7.5 and 1.7.6, respectively.

The rigid axle is sometimes fitted with a rigid tube going forward to a cross member in which it can rotate as in a ball joint. This, perhaps confusingly, is called a 'torque tube', presumably because it reacts to the pitch effect of torque in the driveshafts acting on the wheels. It does give very good location of the axle in pitch. Additional lateral location is required at the rear, such as by a Panhard rod as in Figure 1.7.5. In the similar torque tube system of Figure 1.7.6 the rear lateral location is by a Watt's linkage. Generally, the torque tube arrangement is a superior but more costly design used on more expensive vehicles.

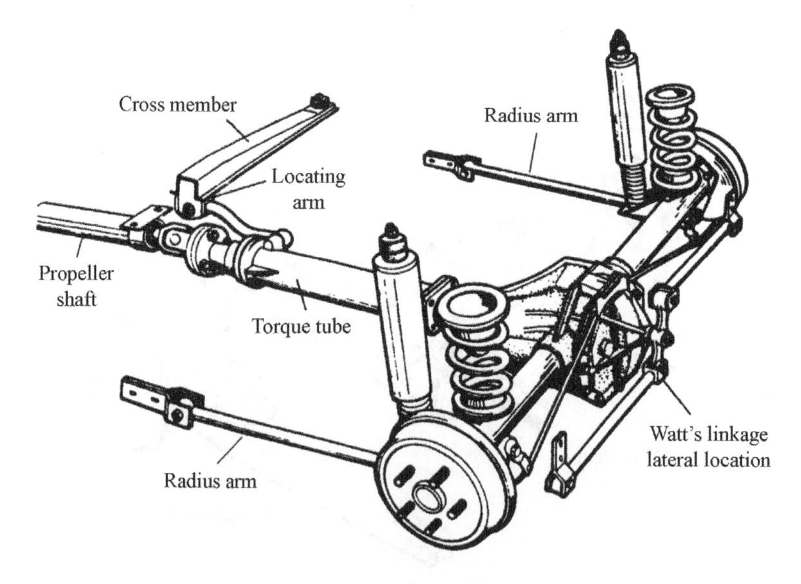

Figure 1.7.6 Axle with torque tube and Watt's linkage (Rover).

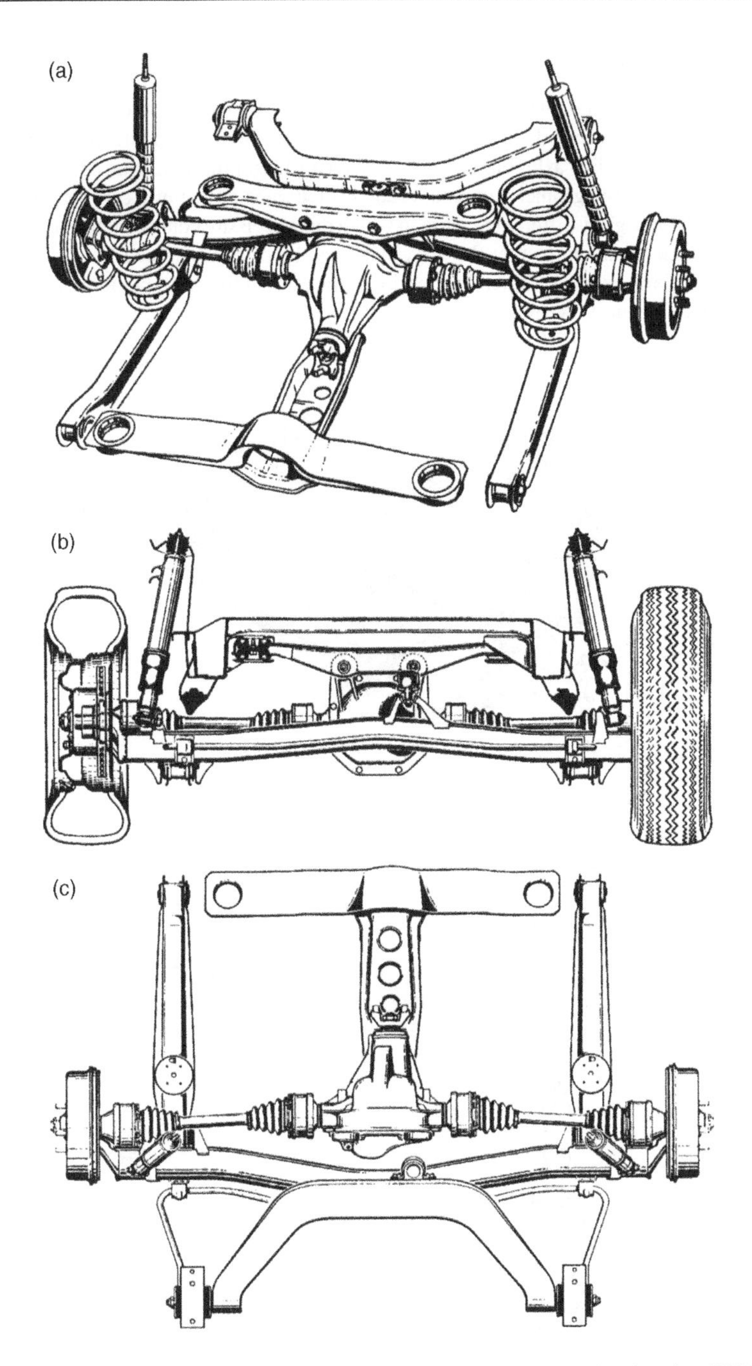

Figure 1.8.1 De Dion axle: (a) front three-quarter view; (b) rear elevation; (c) plan view (1969 Opel).

1.8 De Dion Rigid Axles

The de Dion design is an old one going back to the earliest days of motoring. In this axle, the two wheel hubs are linked rigidly together, but the final drive unit is attached to the body, so the unsprung mass is greatly reduced compared to a conventional live axle, Figure 1.8.1. Driveshaft length must be allowed some variation, for example by splines. The basic geometry of axle location is the same as that of a conventional axle.

Figure 1.8.2 shows a slightly different version in which the wheels are connected by a large sliding tube permitting some track variation, so that the driveshafts can be of constant length.

In general, the de Dion axle is technically superior to the normal live axle, but more costly, and so has been found on more expensive vehicles.

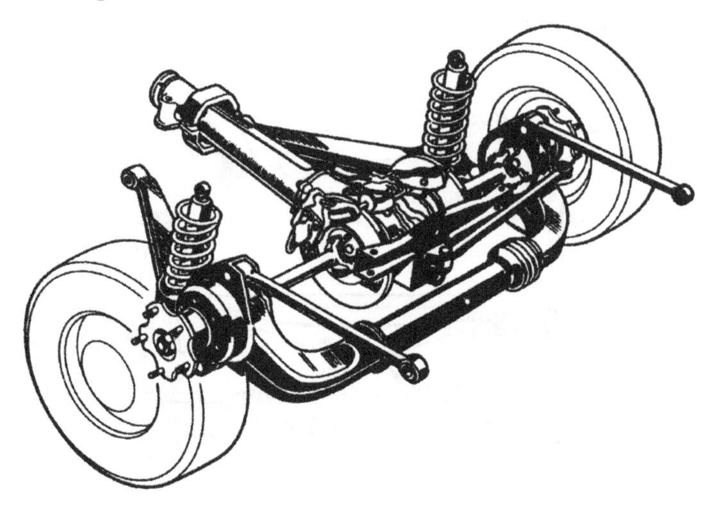

Figure 1.8.2 De Dion axle variation with sliding-tube variable track (1963 Rover).

1.9 Undriven Rigid Axles

Undriven rigid axles, used at the rear of front-drive vehicles, have the same geometric location requirements as live rigid axles, but are not subject to the additional forces and moments of the drive action, and can be made lower in mass. Figure 1.9.1 is a good example, with two lower widely-spaced

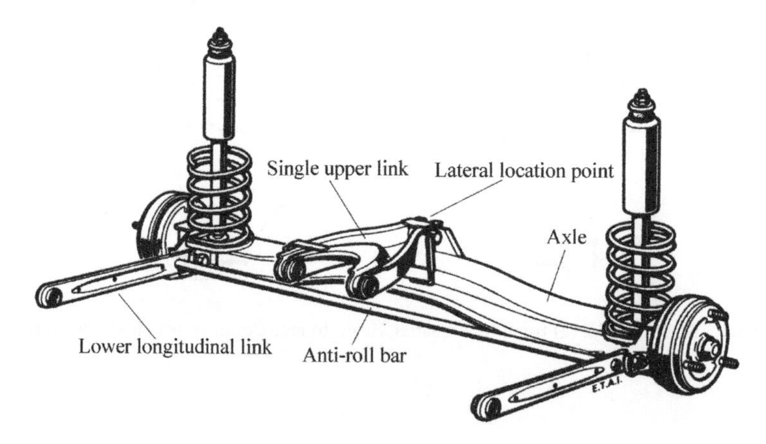

Figure 1.9.1 Simple undriven rigid axle (Renault).

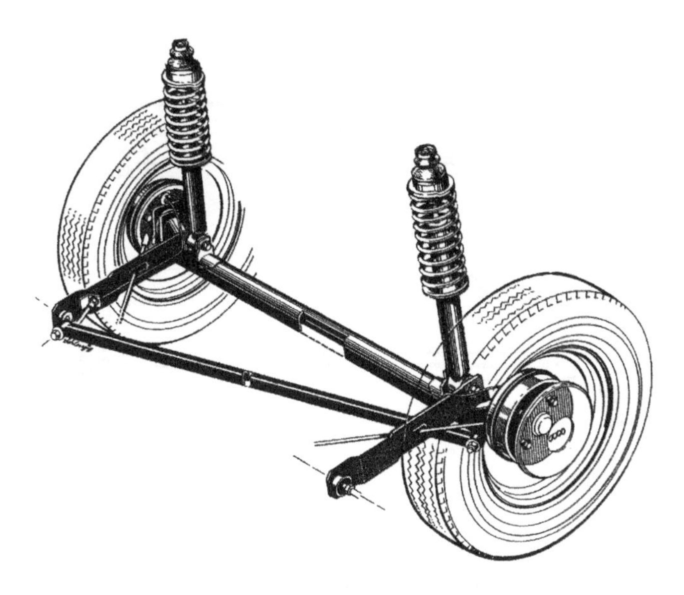

Figure 1.9.2 Undriven rigid axle with diagonal lateral location (Audi).

longitudinal arms and a single upper link for lateral and pitch location. The arms are linked by an anti-roll bar. The roll centre is high.

In Figure 1.9.2, lateral location is by the long diagonal member. This form eliminates the lateral displacements in bump of the Panhard rod. If the longitudinal links are fixed rigidly to the axle then the axle acts in torsion as an anti-roll bar, the system then being a limiting case of a trailing-twist axle.

The undriven axle of Figure 1.9.3 has location at each side by a longitudinal Watt's linkage, giving a truer linear vertical movement to the wheel centre than most systems, and also affecting the pitching angle of the hub, introducing an axle torsion anti-roll bar effect, but in a non-linear way.

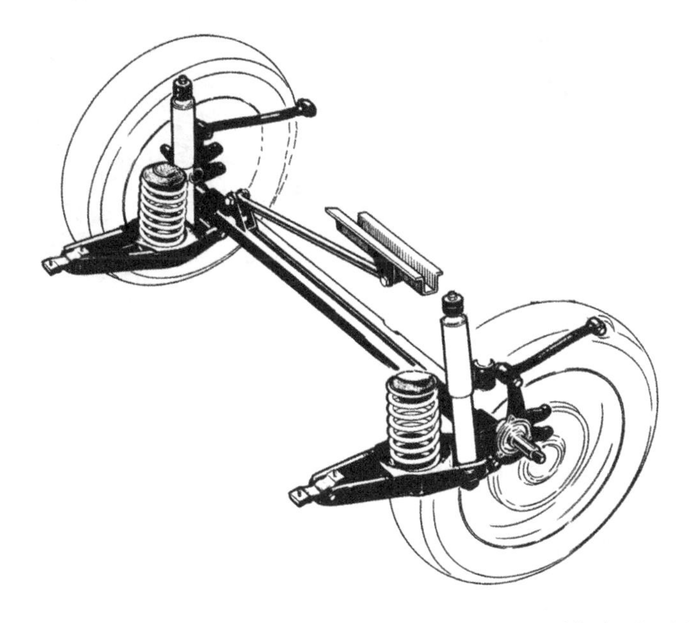

Figure 1.9.3 Undriven rigid axle with longitudinal Watt's linkages and Panhard rod (Saab).

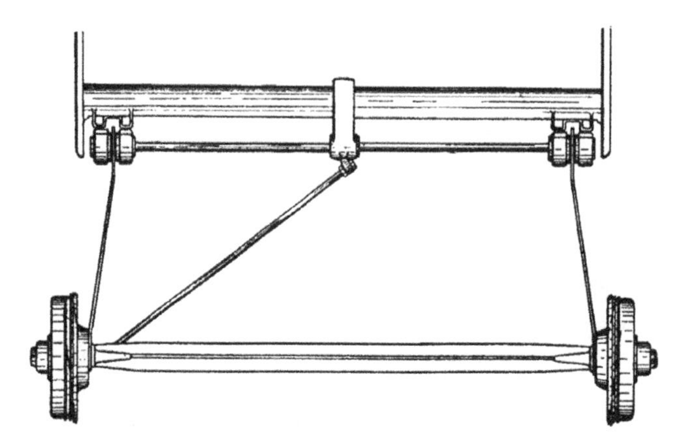

Figure 1.9.4 Rigid axle with diagonal lateral location and torsion bar springing, plan view (nascent trailing-twist axle) (Citroën).

Figure 1.9.4 shows a system with trailing arms operating half-width torsion bars, and with a short diagonal link for lateral location. If the trailing arms are fixed to the axle then in roll the axle will deflect torsionally, giving an anti-roll bar effect, whilst the thin trailing arms deflect relatively freely. This is another example of a limiting case of a trailing-twist axle, with the cross member at the wheel position.

Figure 1.9.5 shows a modern rigid axle with location system designed to give controlled side-force oversteer, the axle being able to yaw slightly about the front location point, according to the stiffness of the bushes in the outer longitudinal members.

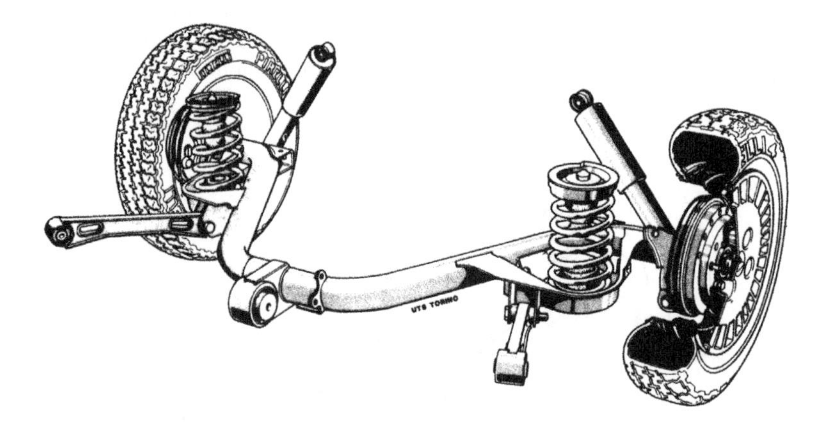

Figure 1.9.5 Tubular-structure undriven rigid axle with forward lateral location point and two longitudinal links (Lancia).

1.10 Independent Rear Driven

In the early days, most road vehicles had a rear drive, using a rigid axle. There were, however, some adventurous designers who tried independent driven suspension, such as on the Cottin-Desgoutes, which was shown in Figure 1.4.4. The most common early independent driven suspension was the simple swing axle, which has the advantage of constant driveshaft lengths, and low unsprung mass. The driveshafts can

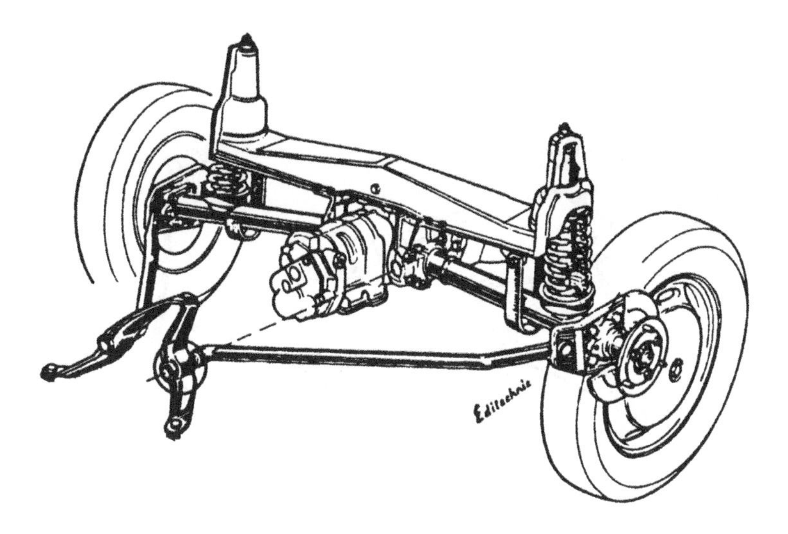

Figure 1.10.1 Swing axle with long leading links for longitudinal location (Renault).

swing forward so they require some extra location. Initially, a simple longitudinal pivot was used. Sometimes the supporting member had pivot points both in front of and behind the driveshaft. Figure 1.10.1 shows one with a single, forward, link. The swing axle has a large bump camber and little roll camber. The roll centre is not as high as with many rigid axles, but it is more of a problem because with a high roll centre on independent suspension there is jacking, which *in extremis* can get out of control with the outer wheel tucking under.

To overcome the problem of the roll centre of the basic swing axle, a low-pivot swing axle may be used, as in Figure 1.10.2, now requiring variable-length driveshafts by splines or doughnuts. The bottom pivots are offset slightly, longitudinally. This is still considered to be a swing axle because the axis of pivot of the axle part is longitudinal.

The obvious alternative to the swing axle is to use simple trailing arms, with the pivot axis perpendicular to the vehicle centre plane and parallel to the driveshafts. Again, this requires allowance for length

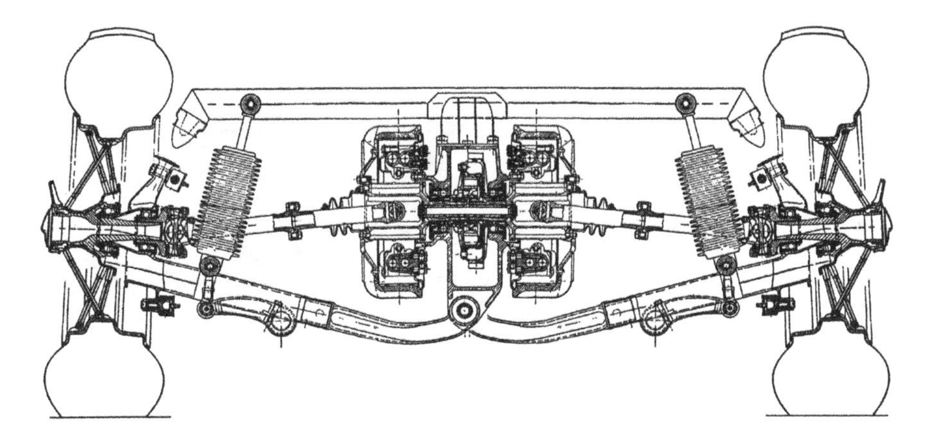

Figure 1.10.2 Low-pivot swing axle with inboard brakes (Mercedes Benz).

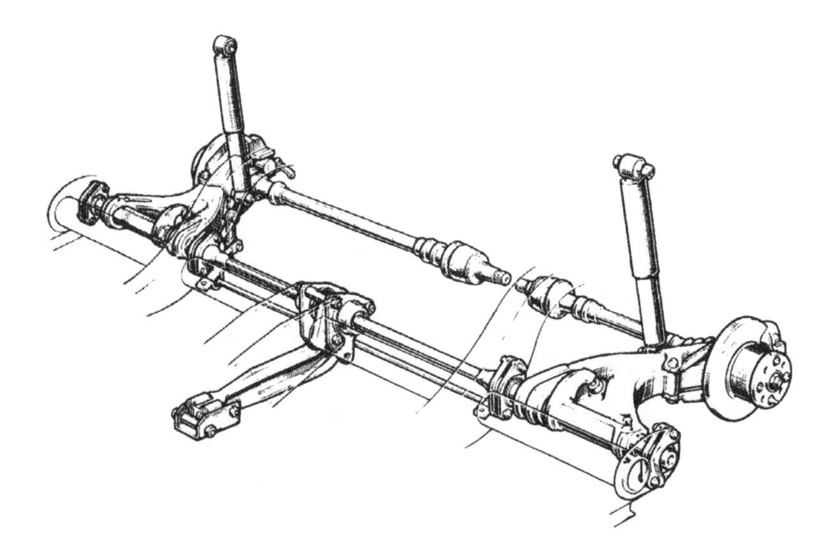

Figure 1.10.3 Plain trailing arms with 90° transverse axis of pivot (Matra Simca).

variation, a significant complication, Figure 1.10.3. In the example shown, the springing is by half-width torsion bars anchored at the vehicle centreline. There is also an anti-roll bar.

The next development, introduced in 1951, was the semi-trailing arm in which the arm pivot axis is a compromise between the swing axle and the plain trailing arm, typically in the range 15° to 25°, as in Figure 1.10.4. A more recent and simpler semi-trailing arm system is shown in Figure 1.10.5. Bump camber is greatly reduced compared with the swing axle.

To control the geometric properties more closely to desired values, a double wishbone system may be used, although this is less compact and on the rear of a passenger car it is detrimental to luggage space, but it is very widely used on sports and racing cars. Figure 1.10.6 shows an example sports car application, where the camber angle and the roll centre height were made adjustable.

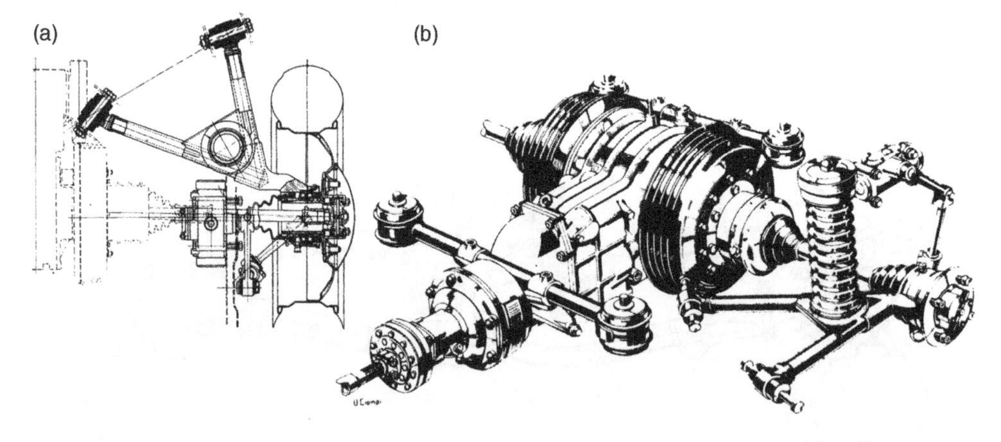

Figure 1.10.4 The first semi-trailing-arm design, also with transaxle and inboard brakes: (a) plan view; (b) front three-quarter view (1951 Lancia).

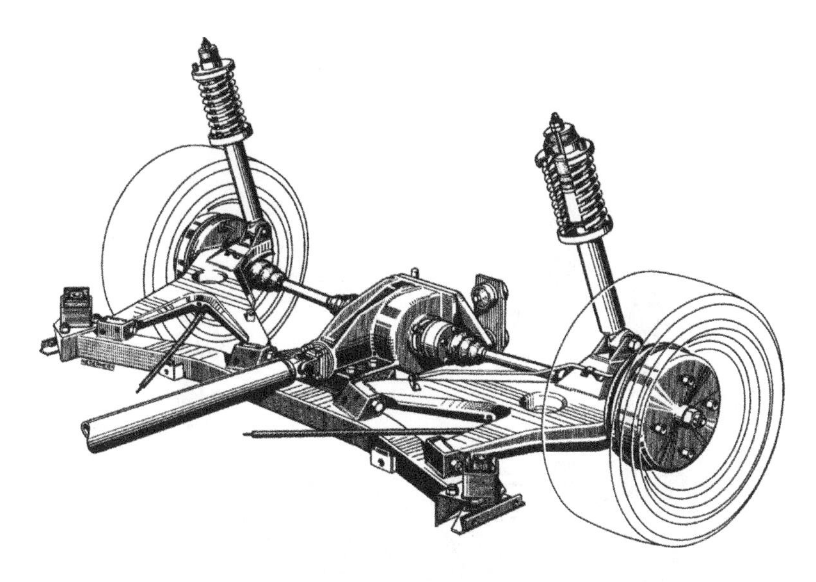

Figure 1.10.5 Semi-trailing arms (BMW).

The Chapman strut is a strut suspension in which the lower lateral location is provided by a fixed-length driveshaft. Figure 1.10.7 gives an example. Lower longitudinal location must also be provided, as seen in the forward diagonal arms which also, here, carry the springs.

Figure 1.10.8 shows the 'Weissach axle', which uses controlled compliance to give some toe-in on braking, or on power lift-off, for better handling.

A relatively recent extension of the wishbone concept is to separate each wishbone into two separate simple links. There are then five links in total, two for each wishbone and one steer angle link. This system

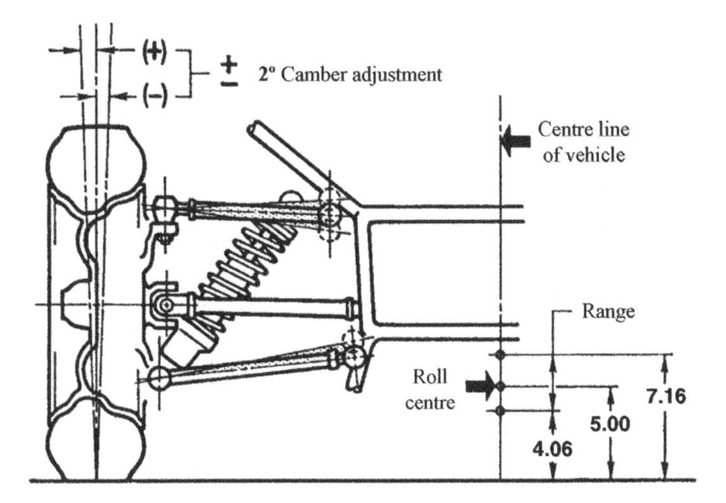

Figure 1.10.6 Double-wishbone sports car suspension with diagonal spring–damper unit, roll centre height dimensions in inches (Ford).

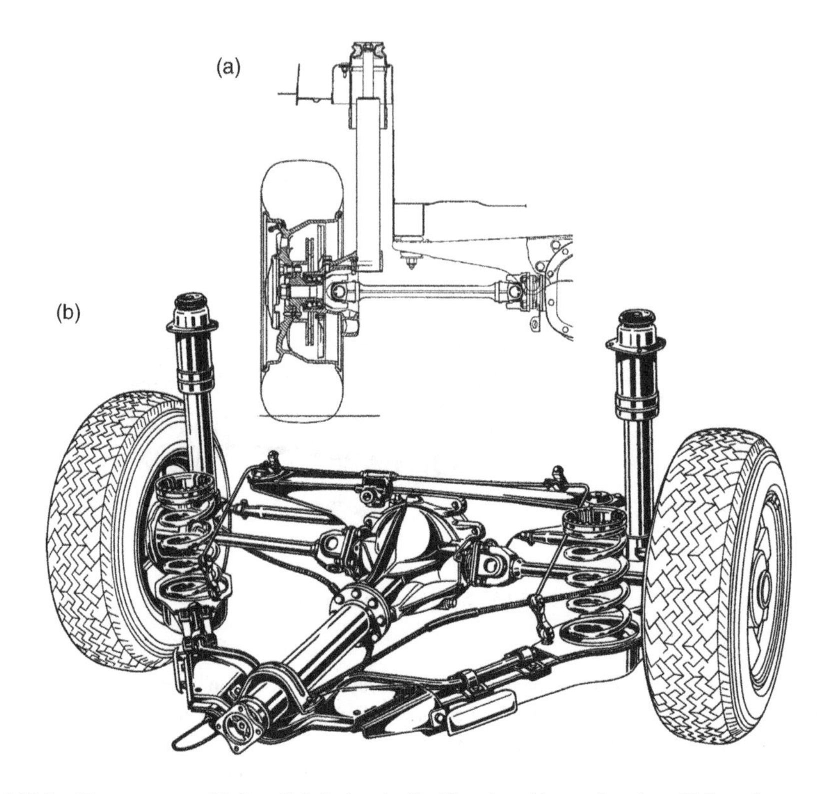

Figure 1.10.7 Chapman strut with front link for longitudinal location: (a) rear elevation; (b) front three-quarter view (Fiat).

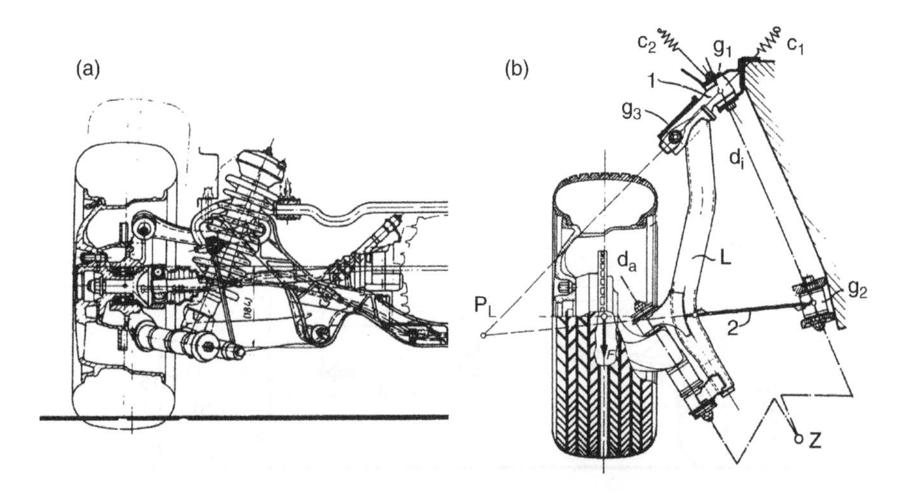

Figure 1.10.8 'Weissach axle' (Porsche).

has been used at the front and the rear, and, with careful design, makes possible better control of the geometric and compliance properties. Figure 1.10.9 shows an example. The advantages seem real for driven rear axles, but undriven ones have not adopted this scheme. The concept has also been used at the front for steered wheels.

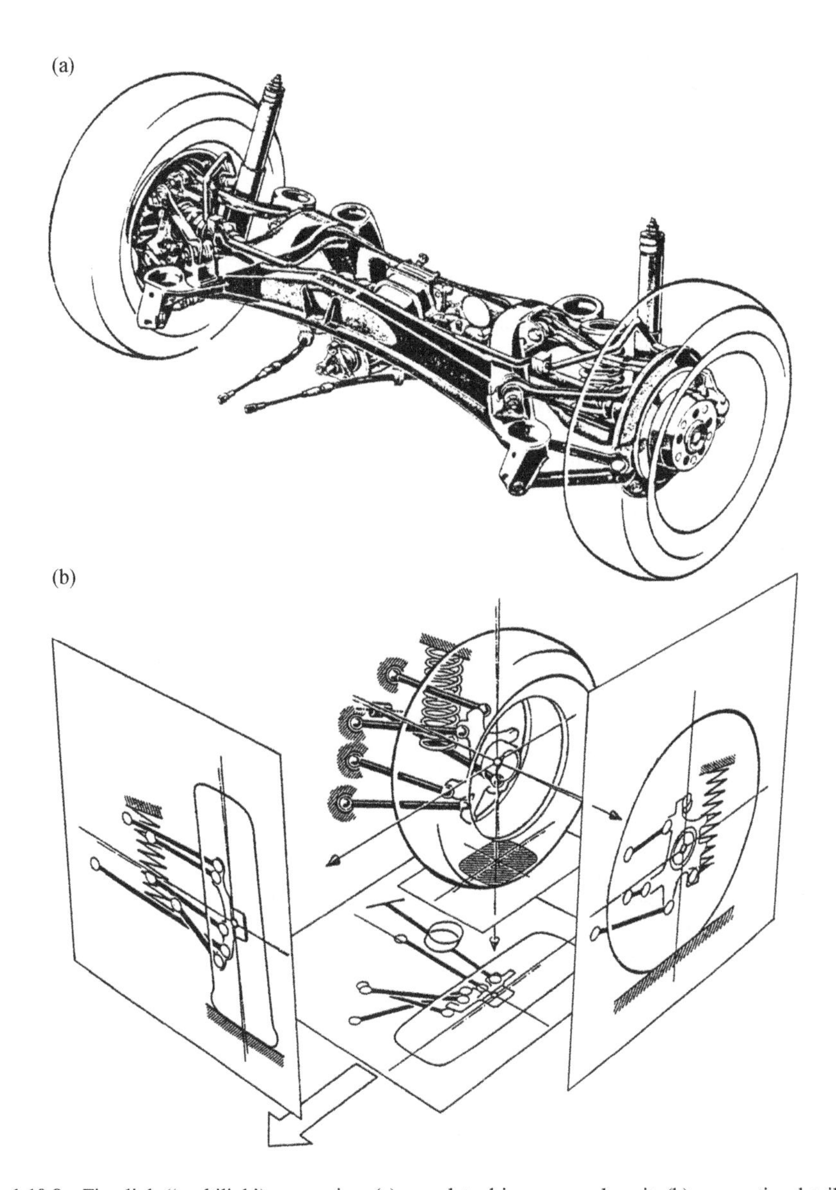

Figure 1.10.9 Five-link ('multilink') suspension: (a) complete driven rear-axle unit; (b) perspective details of one side with plan and front and side elevations (Mercedes Benz).

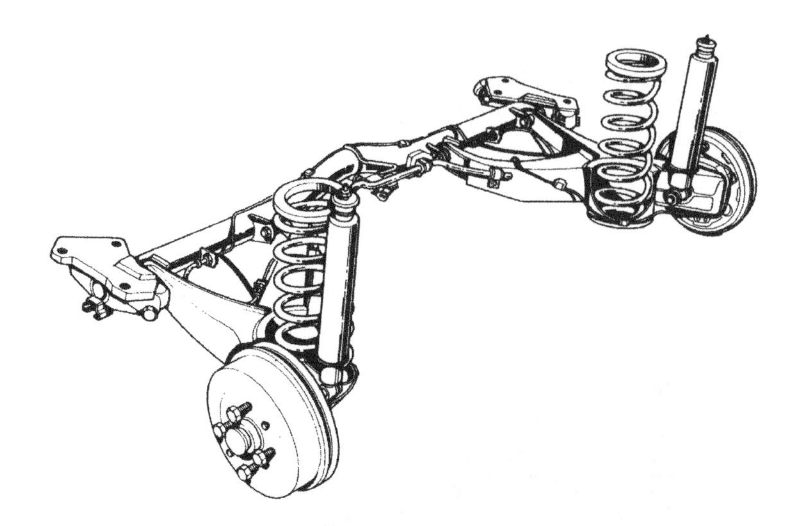

Figure 1.11.1 Plain trailing arms, 90° pivot axis, coil springs (Simca).

1.11 Independent Rear Undriven

At the rear of a front-drive vehicle it seems quite natural and easy to use independent rear suspension. Figures 1.11.1–1.11.4 give some examples.

The plain trailing arm with transverse pivot at 90° to the vehicle centre plane has often been used. The original BMC Mini, on which it was used in conjunction with rubber suspension, was a particularly compact example. A subframe is often used, as seen in Figure 1.11.1. Vertical coil springs detract from the luggage compartment space, so torsion bars are attractive. For symmetry, these have a length of only half of the track (tread), which is less than ideal. Figure 1.11.2 shows an example where slightly offset full-length bars are used. The left and right wheelbases are slightly different, but this does not seem to be of practical detriment.

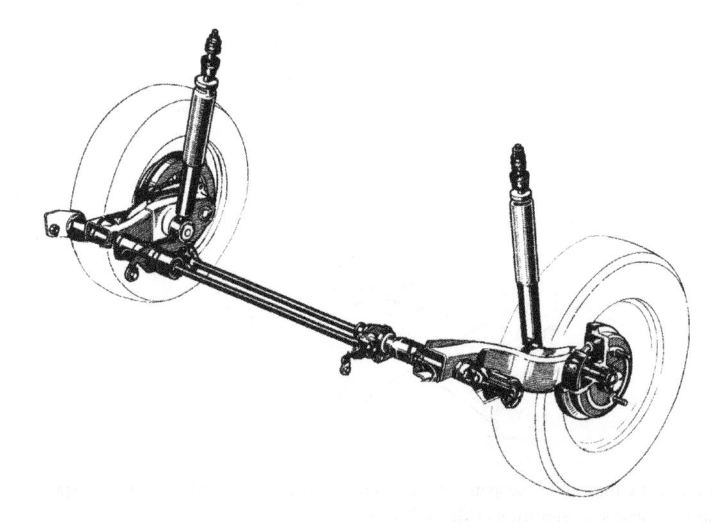

Figure 1.11.2 Plain trailing arms, 90° pivot axis, offset torsion-bar springs, unequal wheelbases (Renault).

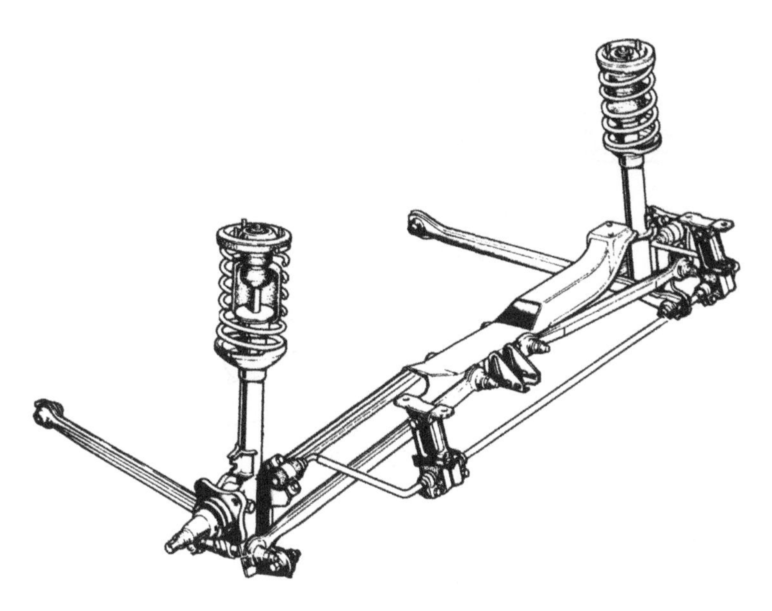

Figure 1.11.3 Strut suspension with long twin lateral/yaw location arms and leading link for longitudinal location (Lancia).

Figure 1.11.3 shows an independent strut rear suspension, the wheel-hub camber and pitch being controlled by the spring–damper unit. Twin lateral arms control scrub (track change) and the steer angle. The anti-roll bar has no effect on the geometry. Figure 1.11.4 shows another strut suspension, with a single wide lateral link controlling the steer angle.

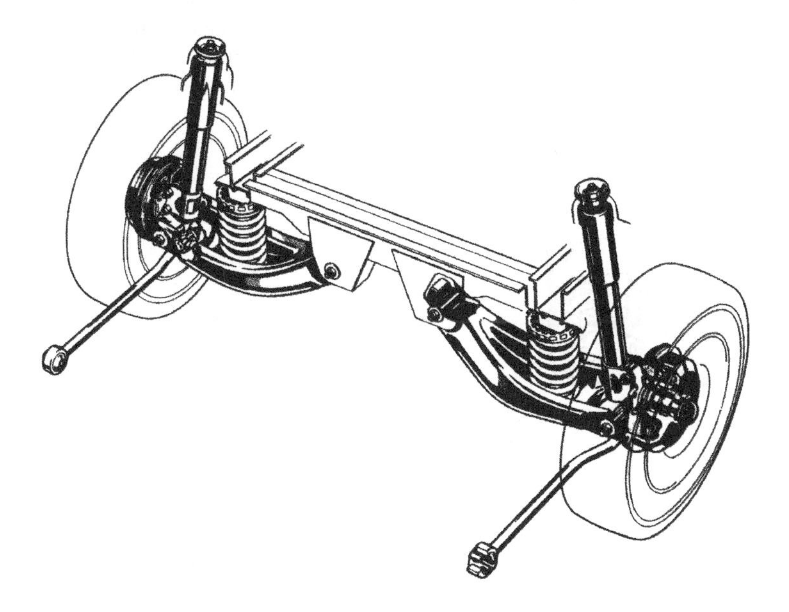

Figure 1.11.4 Strut suspension with wide lateral links and longitudinal links (Ford).

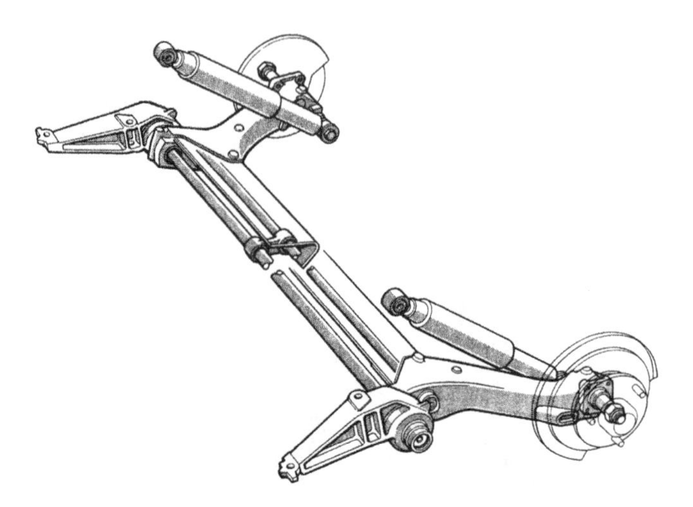

Figure 1.12.1 A compound crank axle using torsion-bar springs (Renault).

1.12 Trailing-Twist Axles

The 'trailing-twist' axle, now often known as the 'compound-crank' axle, is illustrated in Figures 1.12.1–1.12.3. The axle concept is good, but the new name is not an obvious improvement over the old one. This design is a logical development of the fully-independent trailing-arm system. Beginning with a simple pair of trailing arms, it is often desired to add an anti-roll bar. Originally, this was done by a standard U-shaped bar with two mountings on the body locating the bar, but allowing it to twist. Drop links connected the bar to the trailing arms. A disadvantage of this basic system was that the anti-roll bar transmitted extra noise into the passenger compartment, despite being fitted with rubber bushes. This problem was reduced by deleting the connections to the body, instead using two rubber-bushed connections on each trailing arm, so that the bar was still constrained in torsion. This was then simplified

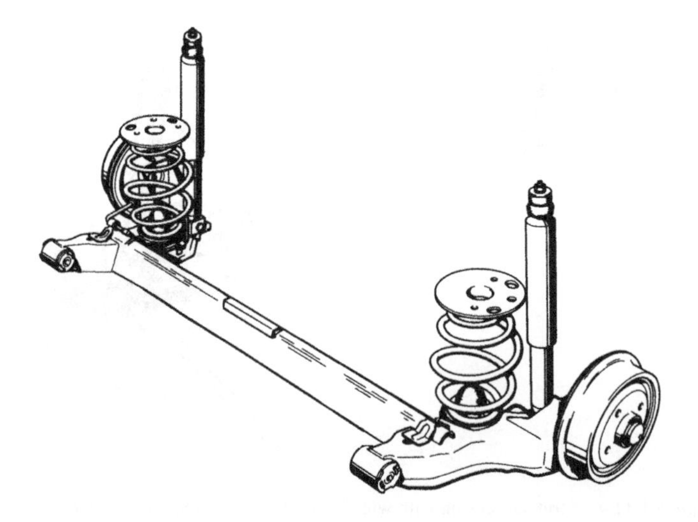

Figure 1.12.2 A compound crank axle with coil springs (Opel).

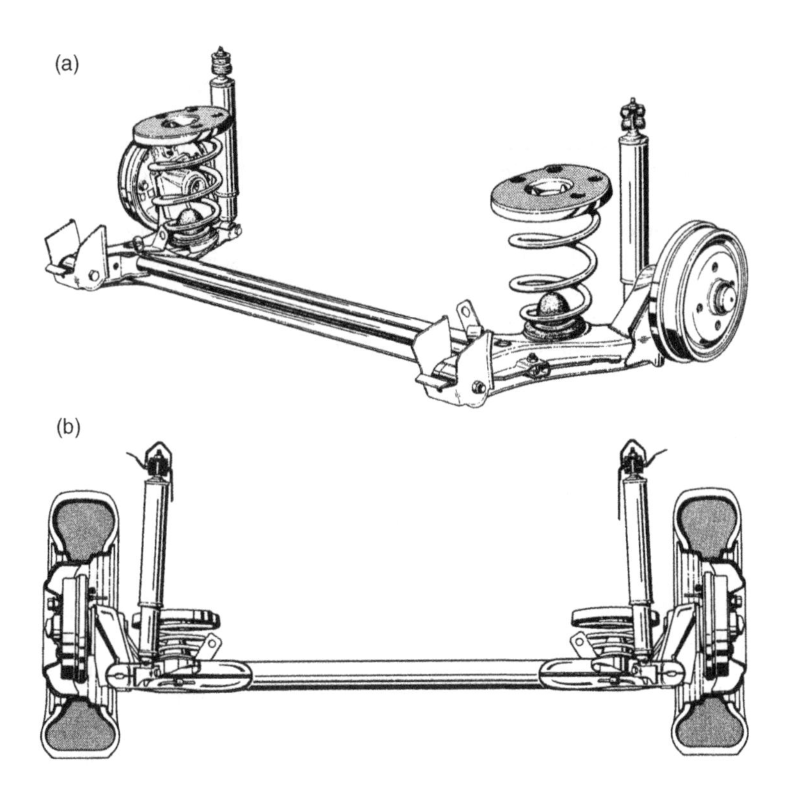

Figure 1.12.3 A compound crank axle similar to that in Figure 1.12.2, also showing in (a) the body fixture brackets, and in (b) the front elevation (Opel).

mechanically by making the two arms and the bar in one piece, requiring the now only semi-independent axle to flex in bending and torsion. This complicated the geometry, but allowed a compact system that was easy to install as a prepared unit at the final assembly stage. As seen in the figures, to facilitate the necessary bending and torsion the cross member of the axle is an open section pressing. The compound crank axle is now almost a standard for small passenger cars.

1.13 Some Unusual Suspensions

Despite the wide range of conventional suspensions already shown, many other strange and unusual suspensions have been proposed, and patented, and some actually used, on experimental vehicles at least, the designers claiming better properties of one kind or another, sometimes with justification. For most vehicles, the extra complication is not justified. Some designs are presented here for interest, and as a stimulus for thought, in approximate chronological order, but with no endorsement that they all work as claimed by the inventors. Proper design requires careful consideration of equilibrium and stability of the mechanism position, the kind of analysis which is usually conspicuously lacking from patents, which are generally presented in vague qualitative and conceptual terms only. As the information about these suspensions is mainly to be found in patent applications, engineering analysis is rarely offered in their support, the claims are only qualitative. The functioning details of actual roll angles and camber angles and the effect on handling behaviour are not discussed in the patents in any significant way.

There are many known designs of suspension with supplementary roll-camber coupling, that is, beyond the basic roll camber normally occurring. So far, however, this has not been a very successful theme. The proposed systems add cost, weight and complexity, and often have packaging difficulties. Also, there does not appear to have been a published analysis of the vehicle-dynamic handling consequences of such systems to provide an adequate basis for design. Typically, where the expected use is on passenger vehicles, the intention is to reduce or eliminate body roll, or even to bank the body into the corner, with claims of improved comfort. Where racing applications are envisaged, the main concern is elimination of adverse camber in roll.

Most of the suspensions discussed are of pure solid mechanical design. Such suspension design tends to add undesirable weight, is usually bulky, and is therefore space-consuming with packaging problems. Camber adjustment accomplished by lateral movement of the lower suspension arms causes adverse scrub effects at the tyre contact patch which is probably not acceptable.

In the Hurley suspension, shown in Figure 1.13.1, a pendulum drives four hydraulic master cylinders in two independent circuits to body roll and camber slave cylinders, the intention being to control both body roll and wheel camber. Either system could be used alone. A serious problem here, as with many other designs, is that an apparently reasonable kinematic design may fail to operate as expected, dynamically. In this case, force and energy are to be provided by the pendulum. Instead, body roll and wheel camber may simply force the pendulum over to the other side, unless it is sufficiently massive. One may also doubt the practicality of the general configuration for normal service. However, the Hurley design is of great historical interest.

In the Bennett 'Fairthorpe TX-1' suspension, Figure 1.13.2, the basically independent suspension has trailing arms, but the wheel hubs have additional camber freedom by rotation on the arms, and there is a crossed intercoupling not unlike cruciate ligaments. Double bump gives zero camber effect. Cornering body roll is said to give wheel camber into the curve. Only in single wheel bump is the camber characteristic considered to be less than ideal. An experimental version in racing was claimed to work well. The characteristics are basically like a rigid axle.

In the Drechsel suspension, shown in Figure 1.13.3, the body is suspended from an intermediate member, a mobile subframe, at each axle. The suspension is connected to the subframe. The body is intended to roll into the corner, pivoting like a pendulum on the subframe, simultaneously adjusting the wheel camber. Energy is provided by the outward motion of the body, which in effect provides a massive

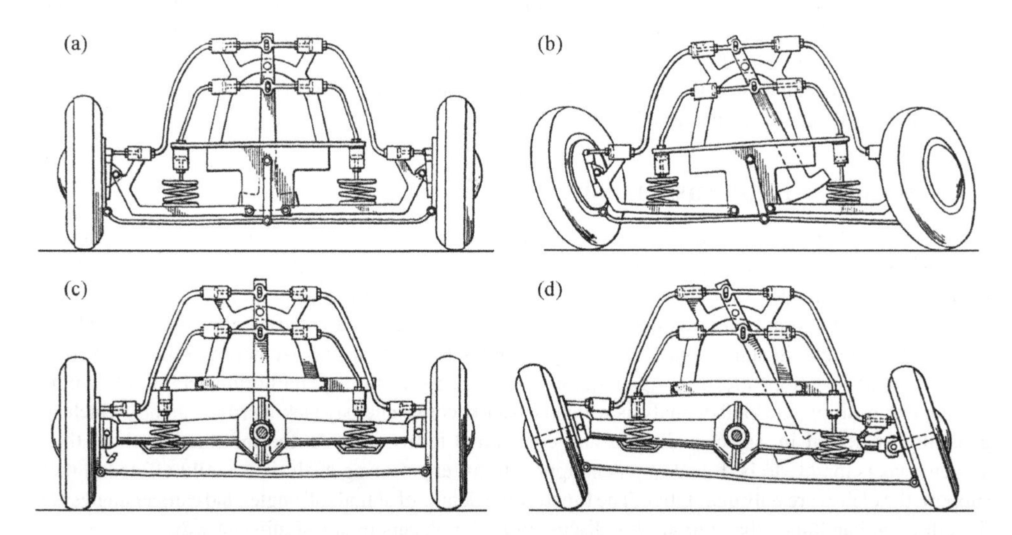

Figure 1.13.1 Hurley suspensions, all in front view: (a) applied to independent suspension; (b) independent in right-hand turn; (c) applied to an axle; (d) axle in right-hand turn (Hurley, 1939).

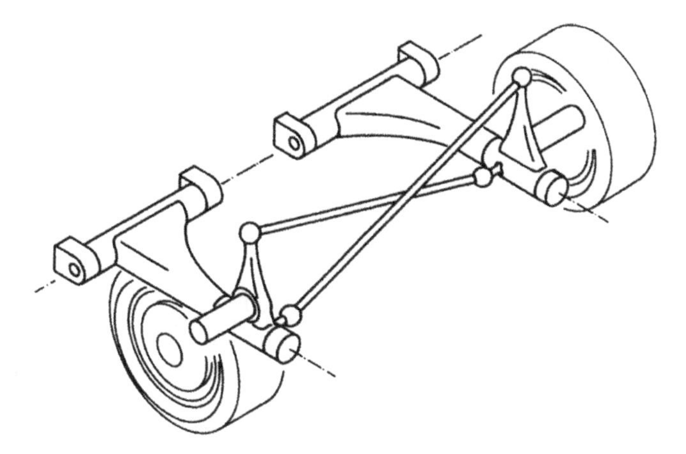

Figure 1.13.2 'Fairthorpe TX-1' suspension (T. Bennett, 1965).

pendulum as an improvement to the Hurley separate-pendulum concept. There are two geometric roll centres, one for the body on the subframe, the other for the suspension on the subframe.

The Trebron double roll centre (DRC) suspension, Figure 1.13.4, due to N. Hamy, is similar in operation to the Drechsel concept, but with changes to the details. An adapted passenger car test vehicle operated successfully, demonstrating negated chassis roll and camber change in cornering.

In the Bolaski suspension, shown in Figure 1.13.5, the operating principle is again similar to that of Drechsel, using the body as a massive pendulum, but in this case the main body rests on compression links, and in cornering is intended to deflect the central triangular member which in turn adjusts the wheel camber by the lower wishbones. Bolaski limits his invention for application to front suspensions by the title of this patent.

In the Parsons system, Figure 1.13.6, each axle has two mobile subframes. The body and the two subframes have a common pivot as shown, but this is not an essential feature. The front suspension design is expected to use struts. Each upper link, on rising in bump, pulls on the opposite lower wishbone, changing the camber angle of that side. On the rear suspension design, with double wishbones, in bump the rising lower link pushes the opposite upper wishbone out, having the same type of camber effect. In the double-wishbone type shown, the spring as shown would give a negative heave stiffness, but this would be used in conjunction with a pair of stiff conventional springs to give an equivalent anti-roll bar effect. Of course, the springs shown are not an essential part of the geometry.

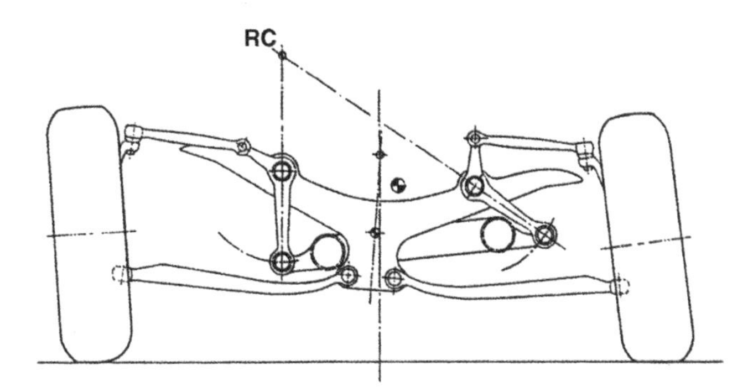

Figure 1.13.3 Drechsel suspension, rear view, left-hand cornering (Drechsel, 1956).

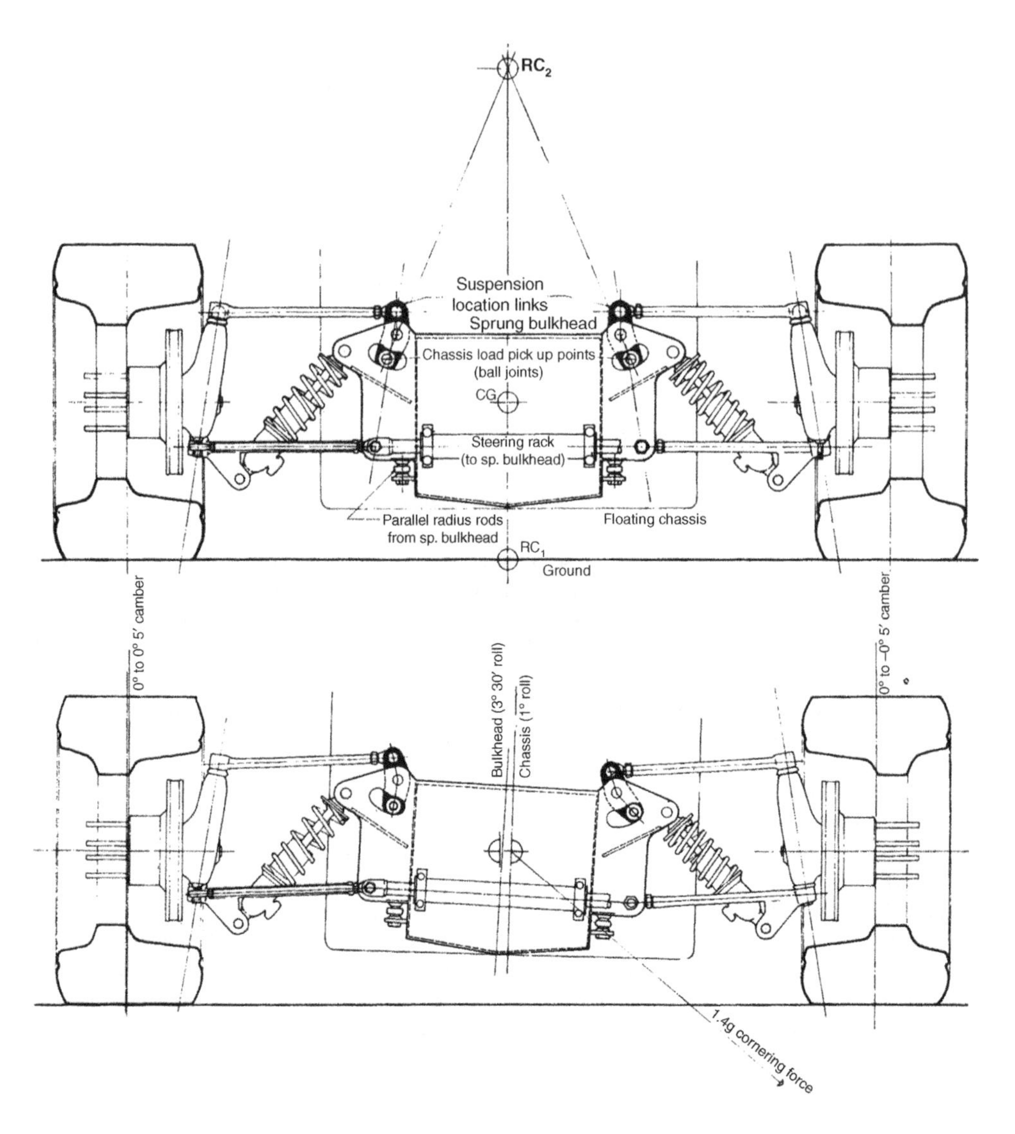

Figure 1.13.4 Trebron DRC suspension rear views: (a) configuration, (b) in turn, (Hamy, 1968).

The Jeandupeux system, shown in Figure 1.13.7, has one mobile subframe at each axle, as a slider, carrying the lower two wishbone inner pivots, giving simultaneous lateral shifting of the lower wishbones. In addition there are extra links connecting to the upper wishbones. In cornering, the wheel lateral forces produce forces in the lower wishbones, which move the slider subframe across, turning the vertical rocker, and pulling the centre of the V across, tending to lift or lower the opposing wheels, hence opposing the body roll and opposing camber change. In other words, there is a direct, basically linear, relationship between lateral tyre force and consequent load transfer through the mechanism, to be compared with the natural value for a simple suspension. With correct design, there should be no roll in cornering. Double bump causes no camber because of the equal parallel wishbone design.

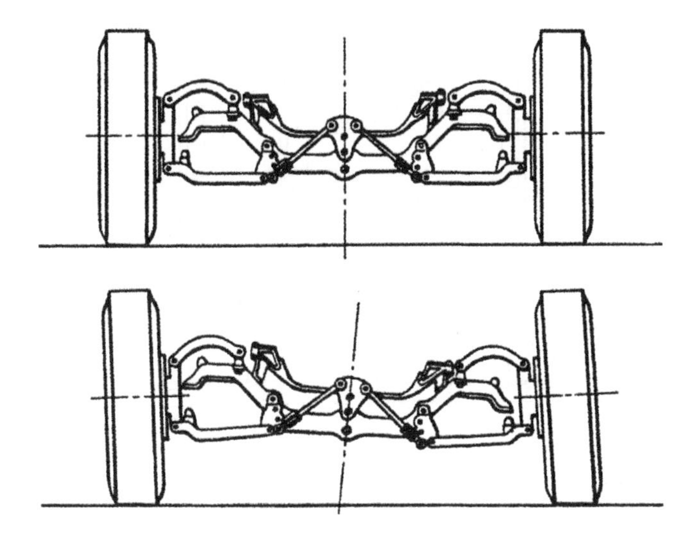

Figure 1.13.5 Bolaski suspension (Bolaski, 1967).

In contrast to the earlier suspensions, all claimed to be passive in action, the Phillippe suspension, Figure 1.13.8, is declared to be an active system with power input from a hydraulic pump, with camber control by hydraulic action on a slider linking the inner pivots of the upper wishbones. This is direct active control of the geometry, quite different from the normal active suspension concept which replaces the usual springs and dampers in the vertical action of the suspension. The light pendulum shown acts only as a sensor.

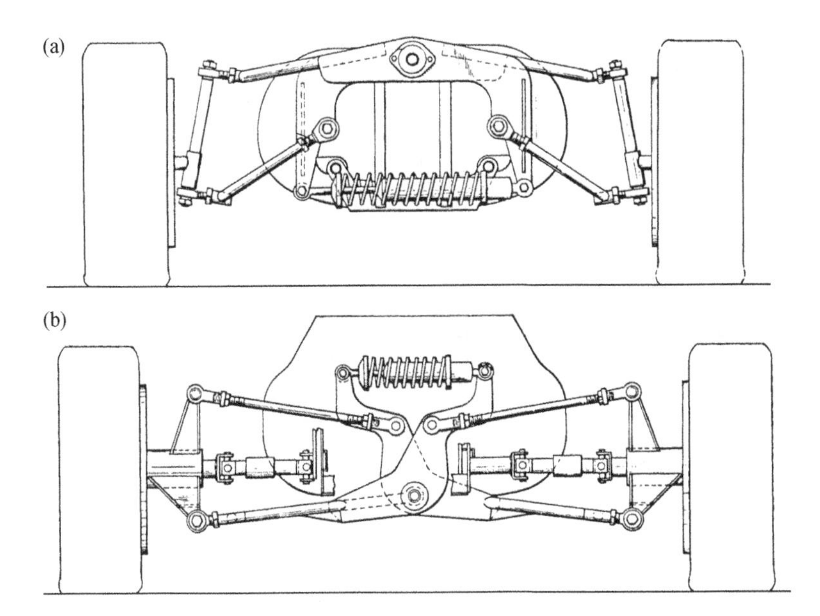

Figure 1.13.6 Parsons suspensions: (a) for the steerable front using struts; (b) for the driven rear using double wishbones. Additional springs and dampers would be used on each wheel (Parsons, 1971).

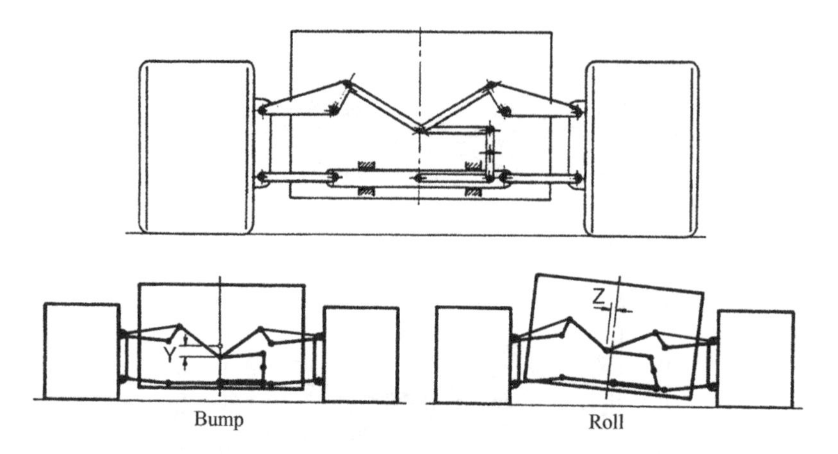

Figure 1.13.7 Jeandupeux suspension: (a) configuration; (b) kinematic action in bump and roll without camber change (Jeandupeux, 1971).

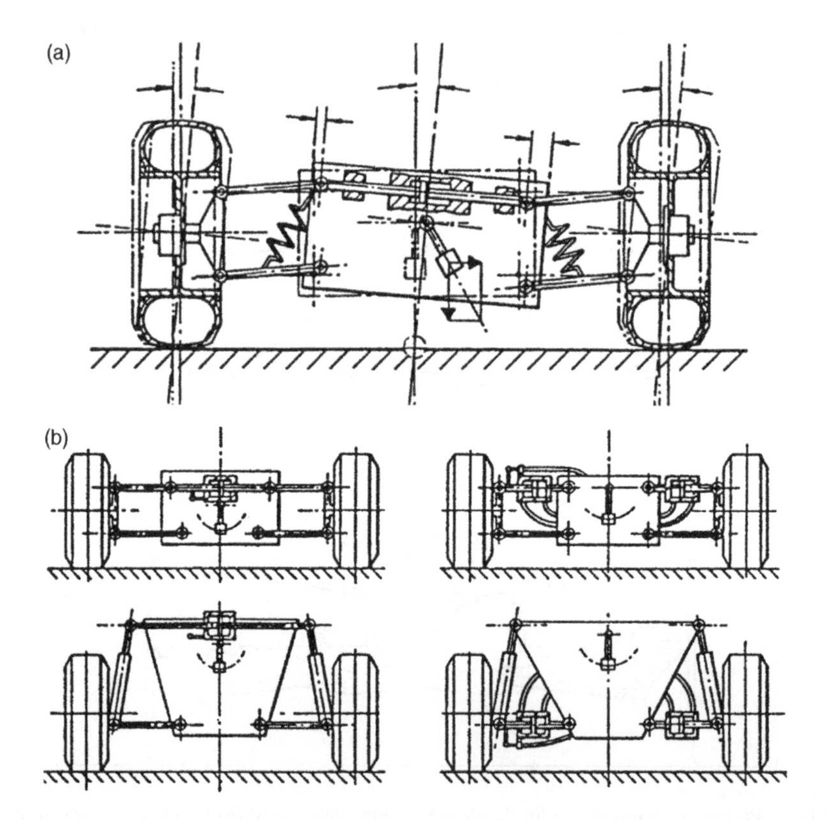

Figure 1.13.8 (a) Phillippe suspension (Phillippe, 1975). (b) Variations of Phillippe suspension.

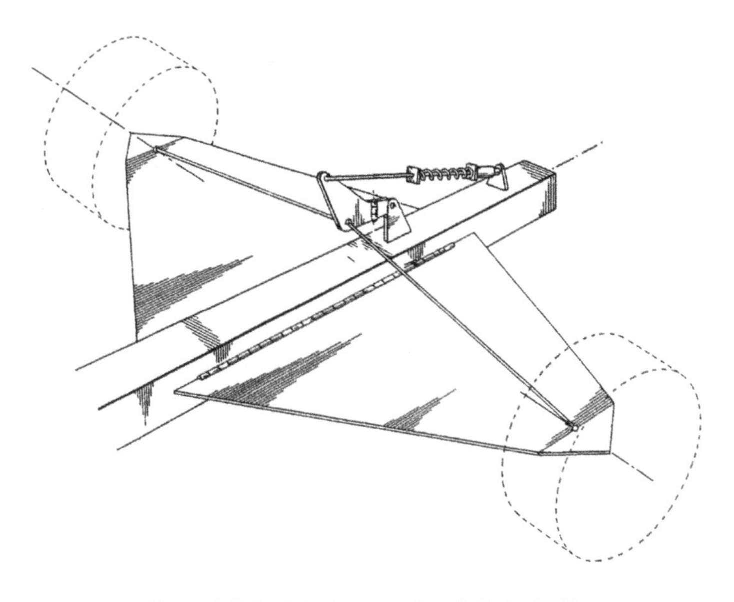

Figure 1.13.9 Pellerin suspension (Pellerin, 1997).

The Pellerin suspension is aimed specifically at formula racing cars which use front suspensions which are very stiff in roll. These have sometimes used a mono-spring–damper suspension, the so-called 'monoshock' system, which is a suspension with a single spring–damper unit active in double bump, with nominally no roll action in the suspension mechanism itself. The complete front roll stiffness then depends mainly on the tyre vertical stiffness with some contribution from suspension compliance. In the Pellerin system, Figure 1.13.9, the actuator plate of the spring–damper unit is vertically hinged, allowing some

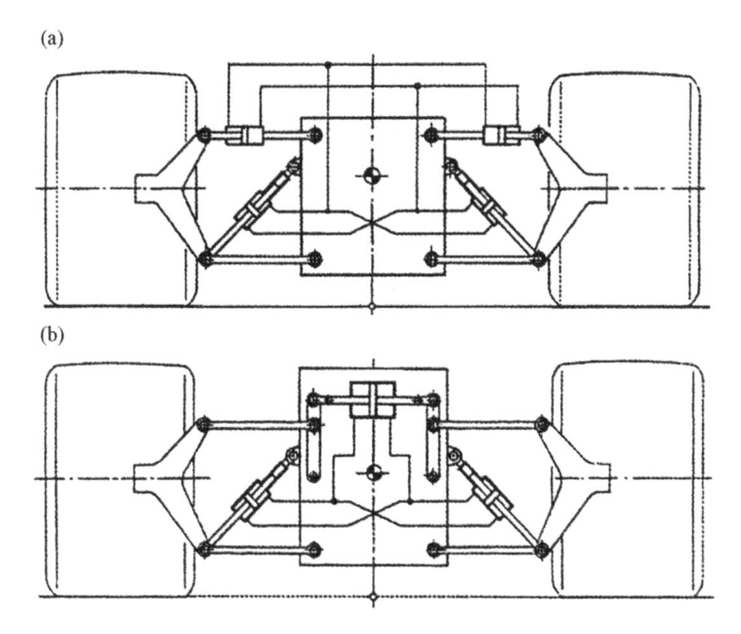

Figure 1.13.10 Weiss PRCC suspension (springs not shown): (a) version 1, camber cylinders in wishbones; (b) version 2, camber cylinder between the two wishbone pivots (Weiss, 1997).

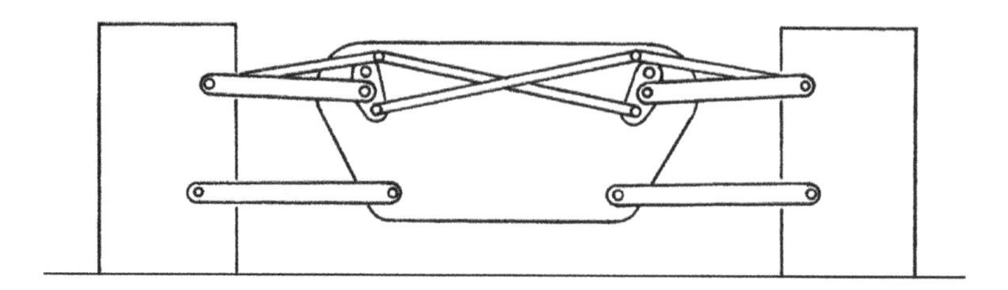

Figure 1.13.11 Walker 'Camber Nectar' suspension (Walker, 2003).

lateral motion of the pushrod connection point. The roll stiffness of the suspension mechanism is then provided by the tension in the hinged link, and is dependent on the basic pushrod compression force, which increases with vehicle speed due to aerodynamic downforce.

The Weiss passive roll-camber coupled (PRCC) system, shown in Figure 1.13.10, uses hydraulic coupling of the wheel camber angle with body roll, in several configurations. Normal springs are retained. Essentially, in cornering the tyre side forces are used to oppose body roll, and body forces oppose adverse camber development, mediated by the hydraulics. Simple double bump moves the diagonal bump cylinders, but these are cross coupled, so in this action they merely interchange fluid.

Finally, Figure 1.13.11 shows the Walker 'Camber Nectar' intercoupling system simply operating between the upper wishbones, each of which is hinged at its inner end on a vertical rocker, the lower end of which is coupled to a drive rod on top of the opposite wishbone. The basic wishbone geometry is such as to give about 50% compensation of roll camber by the basic geometry and 50% by the action of the extra links, so that double bump and roll both give zero camber change.

The above examples give at least a flavour of the many unconventional systems that have been proposed. Many others may be found in the patent literature.

References

Bolaski, M. P. (1967) Front vehicle suspension with automatic camber adjustment. US Patent 3,497,233, filed 3 Nov 1967, patented 24 Feb 1970.

Drechsel A. (1956) Radführungsanordnung für Fahrzeuge. German Patent 12-14-100, filed 18 Dec 1956, patented 10 Nov 1966.

Hamy N. (1968) The Trebron DRC racing chassis. *Motor*, 21 Dec, pp. 20–24.

Hurley L. G. (1939) Automobile chassis. US Patent 2,279,120, filed 5 June 1939, patented 7 April 1942.

Jeandupeux P. A. (1971) Suspension pour véhicules automobiles. Swiss Patent 537 818, filed 4 May 1971, patented 31 July 1973.

MacPherson E. S. (1949) Wheel Suspension System for Motor Vehicles, US Patent 2,660,449, filed 27 Jan 1949, patented 24 Nov 1953.

Parsons C. F. D. (1969) Vehicle Wheel Suspension System, US Patent 3,598,385, filed 27 Mar 1969, Patented 10 Aug 1971.

Pellerin L. (1997) Vehicle suspension system. UK Patent GB2309206, priority date 19 Jan 1996, filed 17 Jan 1997, published 11 Mar 1998.

Phillippe M. P. (1975) Vehicle suspension systems. UK Patent GB1526970, filed 30 Dec 1975, published 4 Oct 1978.

Sloan, A. P. (1963) *My Years with General Motors*. Garden City, NY: Doubleday.

Walker P. (2003) Camber Nectar. US Patent 7,407,173, filed 29 May 2003, published 5 Aug 2008.

Weiss W. (1997) Wheel suspension with automatic camber adjustment. US Patent 6,267,387, filed 12 Sep 1997, published 31 July 2001.

——— // ———

2

Road Geometry

2.1 Introduction

The purpose of a vehicle is to go from A to B, moving the passengers or payload comfortably, safely and expediently. Considering in general any surface of the globe, the terrain to be covered may vary enormously. In fact, most of the Earth's ground surface is difficult terrain for a conventional motor vehicle due to roughness, swampiness, etc. Specialised ground vehicles are required over much of the area, in some cases tracked, in less difficult cases a good all-wheel-drive vehicle. Obviously, to ease the problem the surface has been improved along many frequently-used routes, with a vast capital investment in roads which generally have a broken rock base and are surfaced with concrete in some cases, or with fine gravel held together by very high viscosity tar.

The general form of the ground and roads varies greatly, as may readily be seen by studying a road map. Figure 2.1.1(a) shows the contour map of a small section of Belgian pavé type rough road surface made with stone blocks, which may be contrasted with the relatively smooth nature of good quality roads such a motorways. Figure 2.1.1(b) shows a theoretically generated isotropic surface which is of similar character.

The shape of a road or desired path may be analysed in various ways, largely dependent on the application. In handling analysis, the general geometric form is required, for example the path lateral curvature, which governs the lateral tyre force requirement. On the other hand, in ride quality analysis it is usual to think in terms of a Fourier spectral analysis of the road surface quality. This chapter deals primarily with road geometry in conventional geometric terms, as is of use in performance and handling analysis. The next chapter deals with the finer scale vertical perturbations of importance in comfort (ride quality) analysis.

In general, the ground surface is defined by some sort of coordinate system. The Earth itself is a close approximation to spherical, the sea-level reference surface being a slightly oblate spheroid due to the diurnal rotation. The mean radius is 6371 km (SI units are described in Appendix B). The highest mountain, Everest, is 8848 m above sea level, barely one-thousandth of the Earth's radius. Locally, a simple rectangular coordinate system may be used. The ISO vehicle-dynamics system has X forwards, Y to the left and Z upwards. Consider, then, (X,Y) Earth-fixed coordinates, with the ground altitude Z, relative to some defined reference plane, such as notional local sea level, being a function of X and Y, as in Figure 2.1.2. (The notation used in each chapter is summarised in Appendix A.) In principle, this effectively fully defines the surface shape, excepting the extreme case of overhangs, where Z is multivalued. The ground is notionally a continuum, although really discrete at the molecular level. In any case, the surface would normally be represented in a digital form of $Z(X,Y)$ values. With modern computation, the memory size available is large, and good resolution of ground shape is possible, obtained by satellite global-positioning systems and other means.

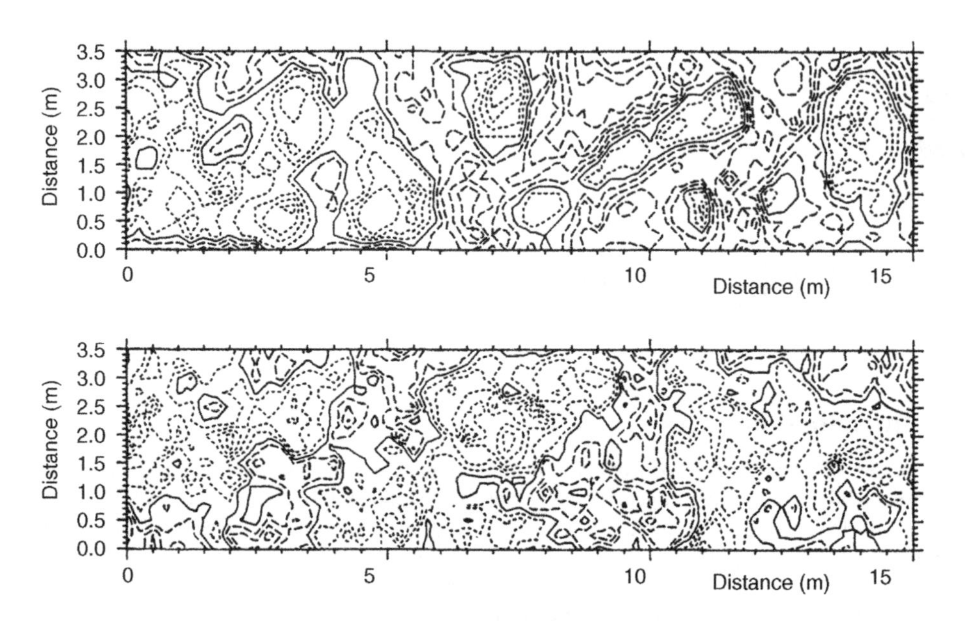

Figure 2.1.1 Road contour maps: (a) MIRA Belgian pavé test track; (b) theoretically generated isotropic surface. Long dashes are contours below datum, short ones above (Cebon and Newland, 1984).

Nevertheless, the quality of the ground surface itself may be considered to be a separate feature from the geographical ground shape, the distinction being at the scale depending on the vehicle, and is analysed for its effect on tyre grip and ride quality rather than handling requirements. Thus the road shape on a small scale, called the macrostructure and microstructure, which affects the tyre grip and aquaplaning, is handled separately, as detailed in the next chapter.

Once a particular route is selected over the terrain, the nature of the representation of the surface changes. In general, the road is defined by a band, of varying width, along the surface of the Earth. The road surface is then a section of the complete surface, its edges being the limits of the usable surface. In a

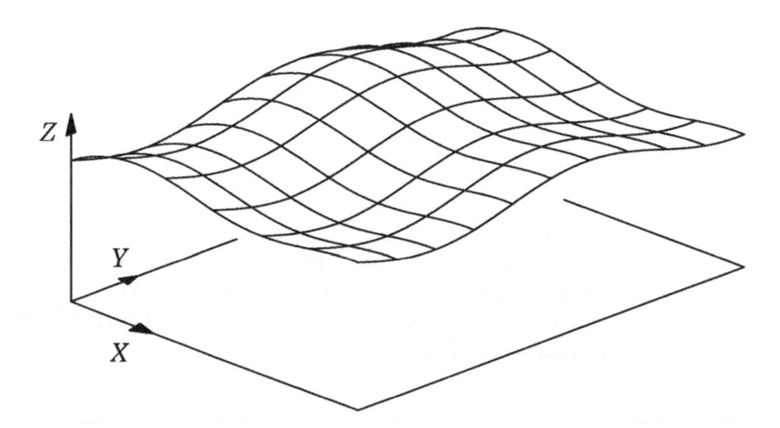

Figure 2.1.2 General ground surface height Z above the datum plane is a function of rectangular coordinates (X,Y).

simpler representation, the road shape then becomes a single path line, essentially a function of one variable rather than two, defined by values as a function of distance along the road, the path length from a reference point.

The actual path of the vehicle centre of mass, or other reference point, projected down into the road surface, gives a specific line. The path is like a bent wire in space, with curvature at any position on its length. In the context of racing, this should be the ideal lap line forming a closed loop, but compromised in reality by driver and vehicle imperfection and inconsistency, and by the requirement to avoid other vehicles.

2.2 The Road

In plan view, the route from A to B is simplistically defined by a line, Figure 2.2.1, specified by a definite relationship between X and Y. In general, the functions $X(Y)$ and $Y(X)$ are not single-valued. Measuring the vehicle's position by the distance travelled from the point A (i.e. the path length from A), then there is a parametric representation of the route position, $X(s)$ and $Y(s)$. For a given path length s (i.e. position along the road), various values may be deduced from the general surface shape, *inter alia* those in Table 2.2.1.

Strictly, the path length includes vertical positional variations, so the path length seen in the horizontal (X,Y) plane is really

$$s_{\mathrm{H}} = \int_{\mathrm{A}}^{\mathrm{B}} \cos\theta_{\mathrm{R}}\, \mathrm{d}s$$

where θ_{R} is the road longitudinal gradient angle.

The longitudinal slope gives a longitudinal weight force component affecting the tractive force required for steady speed. The banking slope similarly introduces a lateral component of the weight force. The longitudinal normal curvature, combined with the speed, gives vertical accelerations which cause variations of the tyre normal forces. The longitudinal lateral curvature creates the cornering force requirement. The banking curvature affects tyre camber angles. The road camber twist (spatial rate of change of bank angle) affects the tyre normal forces (diagonal distribution of normal force). The rate of change of lateral curvature is the path turn-in, discussed later.

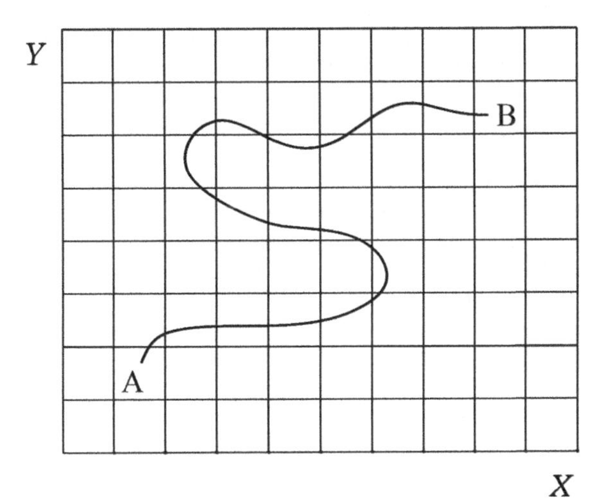

Figure 2.2.1 Route A to B specified by a line in coordinates (X,Y).

Table 2.2.1 The basic geometrical properties of a road, at a path point s

(1) the X coordinate
(2) the Y coordinate
(3) the road height (altitude) Z_R
(4) the usable width
(5) the longitudinal slope, angle θ_R
(6) the banking (lateral) slope, angle ϕ_R
(7) the longitudinal normal curvature, κ_N
(8) the longitudinal lateral curvature, κ_L
(9) the banking (camber) curvature, κ_B
(10) the road camber twist
(11) rate of change of lateral curvature
(12) etc.

The basic geometric properties in Table 2.2.1 require careful definition because the values obtained depend in general on the resolution of the data, or of the wavelength filtering applied, which must be chosen to suit the application. As an example, a pot-hole in a road may have an extreme slope at an edge, but this would not generally be interpreted as a road slope in the general sense, but as a local disturbance.

In principle, the geometric shape of the road is continuous with wavelength, and if subject to Fourier analysis may be represented in a form of spectral analysis, with no special significance for any particular wavelengths. This is correct geometrically, but not true dynamically. The vehicle itself has a definite size, as do the wheels and tyres, and the vehicle on its suspension has natural frequencies, and a speed of travel, from which it is apparent that certain wavelengths may be significant. In particular, there is a dividing frequency, not extremely well defined, which is some fraction of the natural heave frequency of the vehicle. This dividing frequency separates road shapes which are dynamic in effect from those which are essentially static in effect. In other words, in general, driving over a hill is not the same as driving over a bump. For a passenger car, the basic natural frequency of the vehicle in heave, the basic ride frequency, is typically about 1.4 Hz (8.8 rad/s). At a forward velocity of 20 m/s, this corresponds to a resonant wavelength of 14 m. For frequencies less than about 0.2 times the natural ride frequency there is no significant dynamic effect, so wavelengths greater than about $5 \times 14\,\mathrm{m} = 70\,\mathrm{m}$ do not stimulate any dynamic response. Therefore they can be separated out and dealt with by different methods. A hump with a wavelength of 100 m is a small hill, not a ride or handling issue.

From this consideration, then, the geometric properties of the road are, in practice, separated out by wavelength into static and dynamic factors, although this is not inherent in the basic geometric specification of the road.

The altitude graph $Z(s)$ is the longitudinal profile of the road as a function of distance along the road, Figure 2.2.2. This may be derived conceptually from Figures 2.1.2 and 2.2.1. Of course, in many cases this would actually be obtained directly, for example by driving the route with instrumentation (a logger, GPS, etc.), rather than by actual calculation from the general topological form $Z_R(X,Y)$.

The local road angles are defined as follows, as seen in Figure 2.2.3. For a given point on the path, there is a plane tangential to the road surface at that point, generally not horizontal. The path lies locally in that plane; draw the normal and tangent of the path which lie in that plane. Consider the longitudinal vertical plane containing the tangent line. This is also simply the vertical plane tangent to the path at the point of analysis. In this longitudinal vertical plane, the tangent line is at angle θ_R to the horizontal. This is the road pitch angle (gradient angle) θ_R considered positive for an upward (climbing) gradient. The longitudinal road gradient is actually

$$G = \sin\theta_R = \frac{dZ_R}{ds}$$

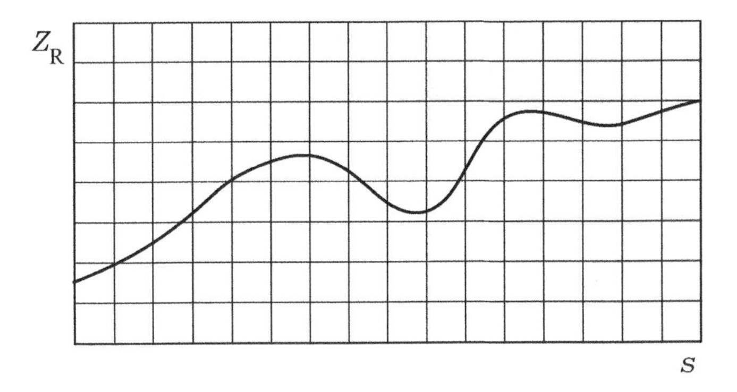

Figure 2.2.2 The road profile $Z_R(s)$ shows the path height as a function of path position.

Now consider the line normal to the path, in the road tangent plane. By rotating this normal line about an axis along the path tangent line, the former may be brought to a horizontal position, that is, into the horizontal plane. The angular displacement required to achieve this is the road bank angle. Note that the bank angle is strictly measured in a plane perpendicular to the tangent line, not in a vertical plane. Expressed in another way, the road bank rotation is about the tangent line axis, in general not about a horizontal line.

This relates to the positional Euler angles of a vehicle body, which use yawing, pitching and rolling angular displacements in that sequence about the vehicle-fixed (x,y,z) axes.

In practice, for cars, the path longitudinal slope is fairly small. One-in-seven is a considerable slope by normal road standards (8.2°), and one-in-five is extreme (11.5°), so the road bank angle can normally be approximately measured in the vertical plane. However, the equations of road geometry are strictly correct for the formal definition of the bank angle, not the approximation in the vertical plane. This is one of the problems of combined gradient and banking. Considering verticals dropped from the path to a true

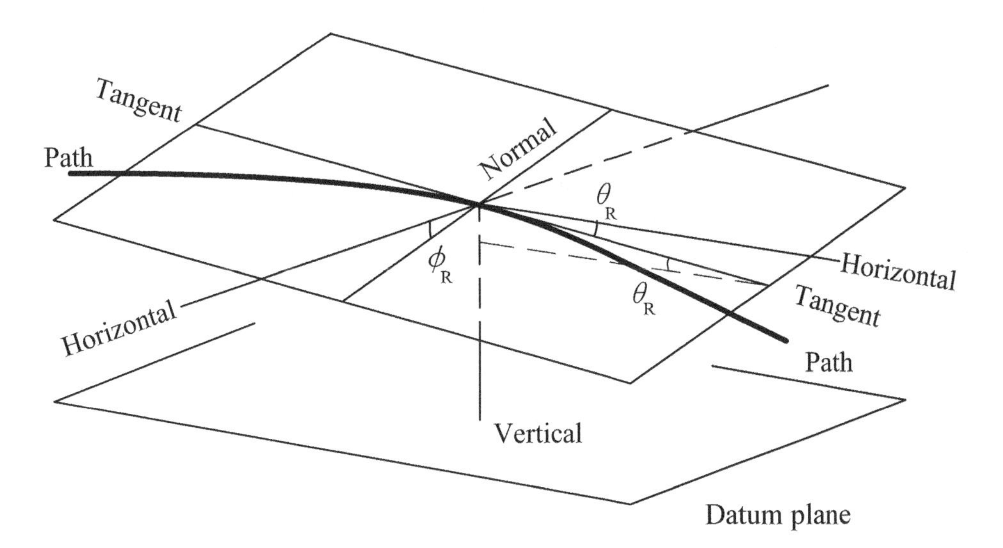

Figure 2.2.3 Road angle geometry. The plane is that of the road surface at the point, containing the path line at the point.

horizontal plane, at the datum level, the result can be seen to be that the path elevation Z_R may be plotted against path position, giving $Z_R(s_H)$. The path slope is normally sufficiently small for this to be a minor distinction, locally at least (e.g. cos 8° = 0.99027). Mathematically, the gradient is

$$G = \frac{dZ_R}{ds}$$

whereas numerically the mean gradient between two data points is

$$G_r = \frac{Z_{R,r} - Z_{R,r-1}}{s_r - s_{r-1}}$$

It may be that the above expression gives G_{r-1}. The gradient belongs to the space between the two points, but may be attributed to one or other point. This is just a matter of definition.

2.3 Road Curvatures

Any path curvature, κ (lower case Greek kappa, units m^{-1}) is the reciprocal of the local radius of curvature, R (m). (Use of the Greek alphabet is described in Appendix C.)

For a flat horizontal road with path radius R there is therefore simply a lateral curvature

$$\kappa_L = \frac{1}{R}$$

For a 'straight' road, in the sense of no corners or banking, but with undulations of road elevation, the road gradient is

$$G = \frac{dZ_R}{ds}$$

and the vertical curvature is defined as

$$\kappa_V = \frac{dG}{ds} = \frac{d^2 Z_R}{ds^2}$$

with corresponding numerical definitions. In this case, with no road bank angle, and small G, the vertical curvature is the same as the normal curvature.

In the general case, the vehicle path line has lateral (cornering) curvature on a banked road which also has longitudinal gradient. It is then necessary to consider five curvatures and two angles (see Figure 2.3.1 and Table 2.3.1).

Table 2.3.1 Path curvatures and angles, Figure 2.3.1

(1) κ, the total path curvature
(2) κ_N, the normal curvature
(3) κ_L, the lateral curvature
(4) κ_V, the vertical curvature
(5) κ_H, the horizontal curvature
(6) ϕ_R, the road bank angle
(7) ϕ_κ, the curvature vector angle from horizontal.

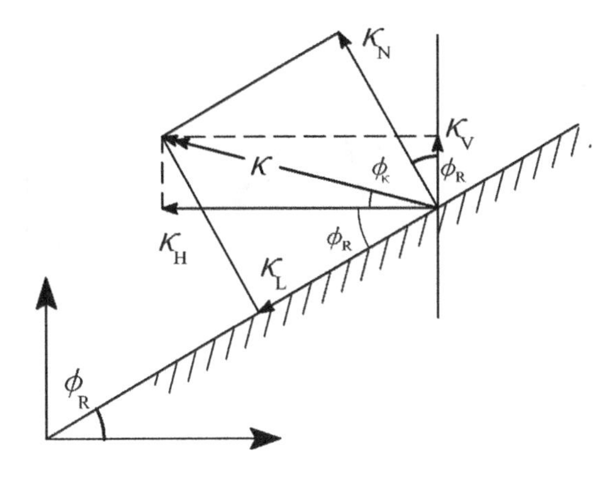

Figure 2.3.1 Path curvature seen in rear view.

From Figure 2.3.1, the total curvature κ can be resolved into perpendicular components. There are two principal ways to do this. In the Earth-fixed datum coordinates, the horizontal and vertical components are

$$\kappa_H = \kappa \cos \phi_\kappa$$
$$\kappa_V = \kappa \sin \phi_\kappa$$

In the road-surface-aligned coordinates, the lateral and normal components are

$$\kappa_L = \kappa \cos (\phi_\kappa + \phi_R)$$
$$\kappa_N = \kappa \sin (\phi_\kappa + \phi_R)$$

The horizontal curvature κ_H is the curvature appearing in a road map (i.e. in plan view) with the particular path drawn in.

The vertical curvature κ_V is the curvature derived from the path elevation graph $Z_R(s)$.

The normal curvature κ_N, the curvature normal to the track surface, is the one which affects the symmetrical suspension deflections and the tyre vertical forces.

The lateral curvature κ_L, parallel to the local road surface, is the curvature requiring lateral tyre forces for the car to follow the road, and causing the vehicle roll angles.

In view of the above, dynamic analysis is most conveniently performed in coordinates aligned with the banked track surface, and with the (κ_N, κ_L) curvature pair, that is, with the normal and lateral curvatures.

For simulation purposes, it is necessary to know the road bank angle ϕ_R, and the curvature vector (i.e. κ and ϕ_κ, or other combination of data allowing κ and ϕ_κ to be evaluated). Methods of determining this information are considered later.

The dynamic behaviour of the vehicle is normally analysed in road surface-aligned coordinates. These are distinct from the Frenet–Serret coordinates generally used by mathematicians to analyse a curved line in space (see Appendix E).

2.4 Pitch Gradient and Curvature

Consider a track with no bank angle, but with longitudinal gradient angle θ_R, as in Figure 2.4.1. The angle is considered constant, that is, the pitch curvature is temporarily zero.

Analysis proceeds in road-aligned coordinates. The weight force W needs to be replaced by its components in those coordinates, as shown. The consequence of the normal component change is a reduction in the tyre vertical forces, although this is usually fairly small. The longitudinal component exerts an effective resistance on the vehicle for positive θ_R and, because it acts at the centre of mass, above

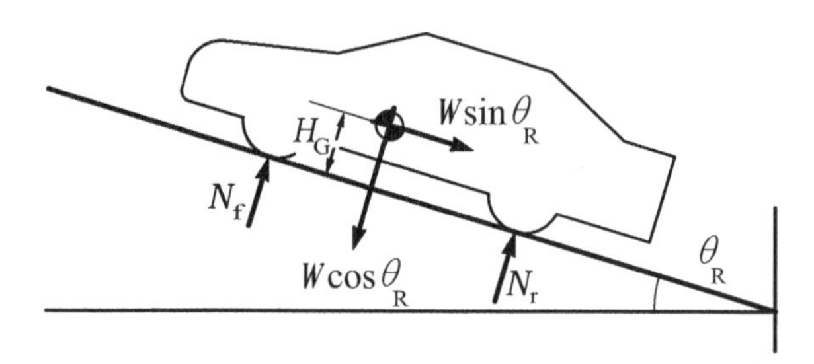

Figure 2.4.1 Vehicle on longitudinal gradient.

ground level, in combination with the required opposing tractive force it gives a force couple in pitch and a resulting longitudinal load transfer.

Figure 2.4.2 shows a vehicle in a trough of pitch radius R_N (and positive pitch curvature κ_N), which, with zero road bank angle, is

$$\kappa_V \equiv \kappa_N$$

The normal curvature is

$$\kappa_N = \frac{1}{R_N}$$

The vehicle is shown in a position at which the road normal is vertical, in order to temporarily separate out the pitch curvature and pitch angle effects. The bank angle is still zero. A trough is considered to have positive curvature, whereas a crest has negative curvature. Also shown, incidentally, is the angle θ_{LN}

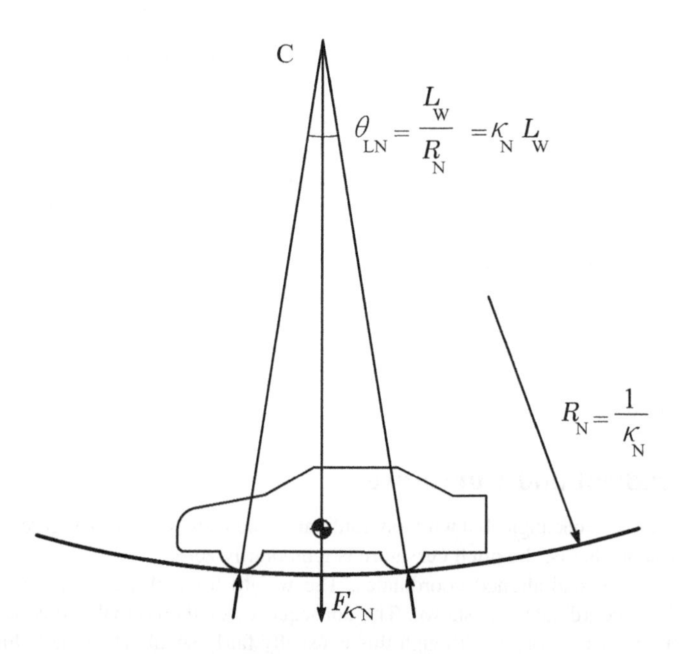

Figure 2.4.2 Road pitch curvature.

subtended by the wheelbase. Considering an effective steady state, to follow the path curvature the vehicle must have a normal acceleration

$$A_N = \kappa_N V^2$$

For a full simulation, with varying κ_N, the vehicle ride position on the suspension must be time-stepped, and, allowing for suspension deflections, the vehicle centre of mass will not have exactly the same normal path curvature as the road. However, consider a suspension which reaches equilibrium quickly, in a well-damped manner, with displacement small relative to the radius, so that the vehicle centre of mass has the same path normal curvature as the known track normal curvature, at least for longer wavelengths (low frequencies). Then, for analysis in vehicle-fixed coordinates to determine the axle and tyre vertical forces, it is necessary to add a normal pseudo-force at G:

$$F_{\kappa N} = -m\kappa_N V^2$$

In a trough, which has positive curvature, this pseudo-force acts along the downward normal, increasing the 'apparent weight' and therefore increasing the tyre vertical forces.

To study the suspension response, this correction force can be applied separately to sprung and unsprung masses. For this, really a time-stepped Earth-fixed coordinate analysis should then be used, for clarity, so that pseudo-forces are not needed.

With pitch curvature, the road normal direction at the two axles is different, by the wheelbase normal-curvature angle

$$\theta_{LN} = \frac{L_W}{R_N} = \kappa_N L_W$$

This has a small effect on the tyre-road normal force magnitude and direction, but is probably better neglected as of little importance for normal roads. It may be significant when the pitch curvature is large, arising for off-road applications.

2.5 Road Bank Angle

Figure 2.5.1 shows a rear view of a vehicle on a straight road with a road bank angle ϕ_R, but without any longitudinal gradient or cornering, and having zero lateral acceleration.

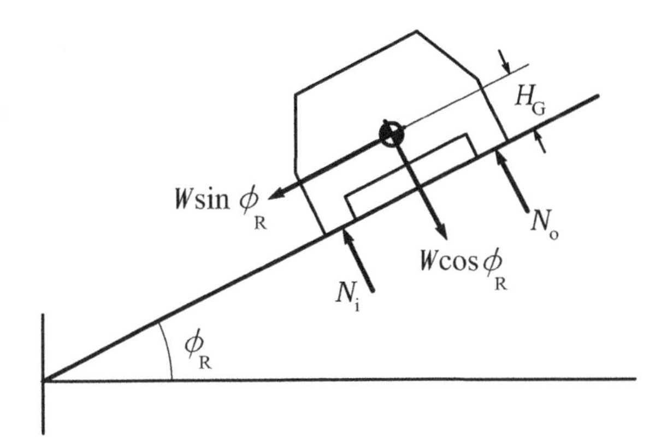

Figure 2.5.1 Vehicle on a straight banked road.

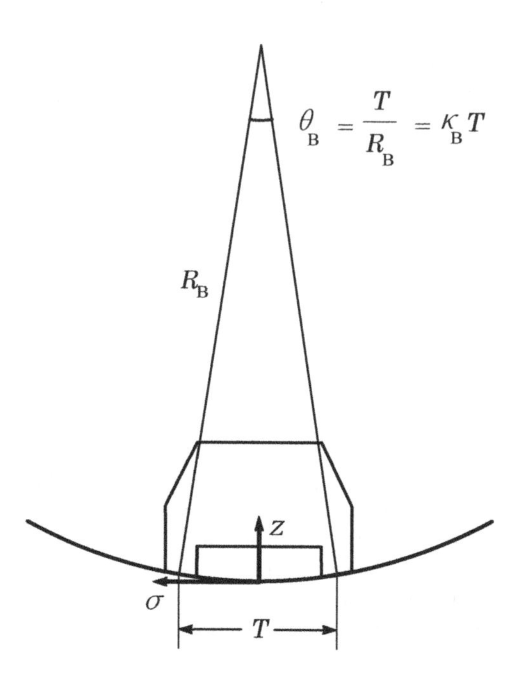

Figure 2.5.2 Road camber curvature (bank curvature).

To see the effects, analysis is performed in road-aligned coordinates, so the weight force is resolved into components normal to the road and tangential, down the slope. The result of the reduction of normal component is a reduction of tyre normal forces. The result of the lateral component is a lateral load transfer. There is also some resulting body and axle roll, and the front-to-rear distribution of the lateral load transfer should be considered. Lateral tyre forces are needed to maintain height on the bank. The complete analysis is therefore similar to cornering with lateral acceleration $g \sin \phi_R$, but the tyre lateral forces do not produce an actual acceleration, they act up the slope to oppose the weight force component.

In general, the vehicle will be cornering when on the banking, and the banking itself will have some horizontal component of curvature; these must be distinguished. It is possible in principle for a vehicle to corner on a straight-banked road, and possible, but very unlikely in practice, for the vehicle to travel straight (zero horizontal curvature of the path) on a road with a curved bank, by changing its height.

At the position of the vehicle, the road has certain properties, in the present context the local bank angle (zero gradient assumed at present). The driver's control inputs will give the vehicle some cornering curvature in the plane of the road, through control of the tyre lateral forces. The relationship between the various curvature values was considered earlier, Figure 2.3.1.

As zero gradient is currently assumed, the vehicle is in a special case of cornering at constant elevation on the banking. For a constant bank angle, zero cornering force is required for a horizontal road curvature given by

$$\frac{\kappa_H V^2}{g} = \tan \phi_R$$

High-speed test tracks are designed to allow this condition to be obtained at various speeds by using a track with bank curvature making various ϕ_R values available. This is done to minimise the tyre drag to allow maximum sustained speed. Despite the elimination of cornering drag in this way, there is still a considerable increase of tyre vertical forces, and hence of tyre rolling resistance and tyre power dissipation, so the tyre temperatures are still increased.

If the vehicle on the banking is not in this special condition of balanced constant-height cornering then there will in general be a change of longitudinal road pitch angle (θ_R) and additional longitudinal curvature (i.e. κ_V) and a more general analysis is required.

The road bank curvature (road camber curvature) κ_B implies that the road bank angle ϕ_B varies with lateral position, where

$$\kappa_B = \frac{d\phi_B}{d\sigma}$$

in which σ (sigma) is the lateral position on the road from the centreline (analogous to longitudinal path position s). Actual differences of ϕ_B arise only once a non-zero width of road is considered. Bank curvature may be considered relevant for some forms of corner analysis where the actual path is not predetermined but remains to be optimised. It also results in different directions for the road normals at the two wheels of an axle, having an effect on tyre camber forces. The actual angle difference between the normals is $\kappa_B T$, where T is the axle track width (tread).

2.6 Combined Gradient and Banking

Considering the foregoing special case analyses, it is apparent that the vehicle will in general be on a curved-banked road, and that it is necessary to account at least for the simultaneous effects of

(1) road pitch angle,
(2) road pitch curvature,
(3) road bank angle.

The most satisfactory approach, and one in accord with the definition of angles in Section 2.2, is to consider the longitudinal road profile first. Hence the weight force is resolved into $F_{WX} = W \sin\theta_R$ to the rear, and $W \cos\theta_R$ normally. Because of the road bank angle, the $W \cos\theta_R$ component is further resolved into

$$F_{WY} = W \cos\theta_R \sin\phi_R$$

acting laterally parallel to the banked road surface, and

$$F_{WZ} = W \cos\theta_R \cos\phi_R$$

acting normal to the local road surface.

The longitudinal profile curvature is no longer the correct curvature for evaluating the normal (compensation) pseudo-force, rather the normal curvature κ_N must be used to give $\kappa_N m V^2$ acting at G perpendicular to the local road surface.

The particular curvature components required are to be evaluated from Figure 2.3.1, depending upon the form of the data. If κ_N and κ_L are known, along with ϕ_R, then these are used directly. However, if the actual path elevations are given, then κ_V will be deduced from these, and if the corner plan shape is deduced from a map analysis, then this will give κ_H. The road-aligned curvatures will then need to be calculated for each point.

2.7 Path Analysis

A path position is defined in coordinate geometry by the path length s from an initial point on the path. The path shape is defined by the parametric three-dimensional form

$$X \equiv X(s); \quad Y \equiv Y(s); \quad Z \equiv Z(s)$$

where (X,Y,Z) are Earth-fixed axes. Much vehicle dynamic analysis is performed in two dimensions, effectively on a flat plane (X,Y). In these two dimensions (road plan view), it is often convenient to express the path shape by $Y \equiv Y(X)$, that is, Y as a function of X. The path radius of curvature is also dependent on the path position,

$$R_P \equiv R_P(s)$$

and the path curvature itself is

$$\kappa_P \equiv \kappa_P(s) = \frac{1}{R_P} \tag{2.7.1}$$

The path angle ν (nu) is defined in Figure 2.7.1. Hence,

$$\frac{dX}{ds} = \cos\nu; \quad \frac{dY}{ds} = \sin\nu; \quad \frac{dY}{dX} = \tan\nu$$

Also

$$\nu = \operatorname{atan}\left(\frac{dY}{dX}\right)$$

but this must be evaluated into the correct quadrant (BASIC Angle(dx,dy), FORTRAN Atan2(dy,dx) functions, not the common single-argument Atan(dy/dx) functions).

The path curvature and angle are related by

$$\kappa_P = \frac{1}{R_P} = \frac{d\nu}{ds} \tag{2.7.2}$$

that is, the path curvature is the first spatial derivative of path angle (with the angle expressed in radians, of course). The rate of change of path curvature with path length is the path turn-in τ_P, expressed by

$$\tau_P \equiv \frac{d\kappa_P}{ds} \equiv \frac{d}{ds}\left(\frac{1}{R_P}\right) \equiv \frac{d^2\nu}{ds^2} \tag{2.7.3}$$

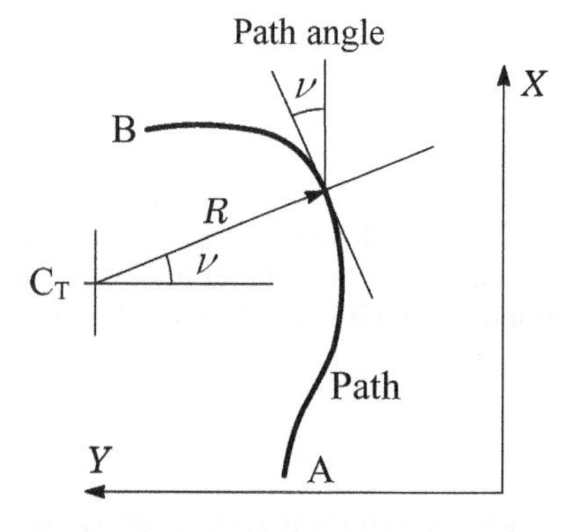

Figure 2.7.1 Path angle and path radius.

Hence the local path parameters include path angle ν, path curvature ρ_P and path turn-in τ_P, with

$$\begin{aligned}\kappa_P &= \frac{d\nu}{ds} \\ \tau_P &= \frac{d\kappa_P}{ds} = \frac{d^2\nu}{ds^2}\end{aligned} \tag{2.7.4}$$

The units are m^{-1} or rad/m for curvature and m^{-2} or rad/m^2 for turn-in (spatial rate of change of curvature). In practical applications, caution must be exercised with the sign of path turn-in. In a left turn, curvature is positive, path angle increasing. Positive path turn-in then corresponds to driver-perceived turn-in. Negative path turn-in is a release from the turn. However, in a right-hand turn, path curvature is negative, and positive path turn-in corresponds to release from the corner, with driver-perceived turn-in corresponding to negative $d^2\nu/ds^2$. The driver-perceived turn-in, which is positive when the absolute radius of curvature is decreasing, is therefore

$$\tau_{P,D} = \text{sign}(\kappa)\tau_P \tag{2.7.5}$$

The spatial derivative of path radius is

$$\frac{dR}{ds} = \frac{d}{ds}\left(\frac{1}{\kappa_P}\right) = -\frac{\tau_P}{\kappa_P^2}$$

$$\frac{dR}{ds} = -\tau_P R^2 \tag{2.7.6}$$

Also

$$\frac{d}{ds}\left(\frac{1}{R}\right) = \frac{d\kappa_P}{ds} = \tau_P \tag{2.7.7}$$

In terms of rectangular coordinate geometry, in two dimensions the path curvature is given by the standard expression

$$\kappa_P = \frac{d^2y/dx^2}{[1 + (dy/dx)^2]^{3/2}}$$

and in terms of first and second time derivatives as

$$\begin{aligned}\kappa_P &= \frac{\dot{x}\ddot{y} - \ddot{x}\dot{y}}{(\dot{x}^2 + \dot{y}^2)^{3/2}} \\ &= \frac{\dot{x}\ddot{y} - \ddot{x}\dot{y}}{\dot{s}^3} \\ &= \frac{\dot{x}\ddot{y} - \ddot{x}\dot{y}}{V^3}\end{aligned}$$

2.8 Particle-Vehicle Analysis

Consider a vehicle of no spatial extent (i.e. a particle) at a point on the path at (X,Y). Coincident with the vehicle at this instant is the origin of another set of Earth-fixed axes, here denoted (x,y). These vehicle-coincident path-aligned axes have the longitudinal axis x tangential to the path, the y axis transverse, along the local path radius. The vehicle passes at this instant through the origin of these axes, which are stationary, with speed V and absolute longitudinal acceleration A_x. These axes may be thought of as local path axes. In the more general case of a non-particulate vehicle, which has yaw, roll and pitch angles relative to the path, vehicle-fixed coordinates must be distinguished from local path-fixed coordinates, and both of these from datum Earth-fixed coordinates.

In two dimensions, the speed is V at path angle ν, with the velocity vector $\boldsymbol{V}$ having components

$$V_X = V\cos\nu$$
$$V_Y = V\sin\nu$$

The speed is

$$V = \frac{\mathrm{d}s}{\mathrm{d}t} \tag{2.8.1}$$

In path-aligned coordinates (i.e. tangential x and normal y):

$$V_x = V$$
$$V_y = 0$$

The longitudinal (tangential) acceleration A_x is determined essentially independently of the path by the engine and brakes, etc. The lateral acceleration depends on the path curvature:

$$A_y = \kappa_P V^2 = \frac{V^2}{R_P} \tag{2.8.2}$$

This is the standard result for circular motion, but remains applicable when the radius of curvature is varying.

The accelerations in the original Earth-fixed coordinates are

$$A_X = A_x\cos\nu - A_y\sin\nu$$
$$A_Y = A_x\sin\nu + A_y\cos\nu$$

The rate of change of lateral acceleration, the lateral jerk (m/s^3), is

$$\begin{aligned} J_y &= \frac{\mathrm{d}}{\mathrm{d}t}(A_y) \\ &= \frac{\mathrm{d}}{\mathrm{d}t}(\kappa_P V^2) \\ &= V^2\frac{\mathrm{d}}{\mathrm{d}t}(\kappa_P) + \kappa_P\frac{\mathrm{d}}{\mathrm{d}t}(V^2) \\ &= V^2\frac{\mathrm{d}s}{\mathrm{d}t}\frac{\mathrm{d}}{\mathrm{d}s}(\kappa_P) + \kappa_P\frac{\mathrm{d}V}{\mathrm{d}t}\frac{\mathrm{d}}{\mathrm{d}V}(V^2) \end{aligned}$$

so

$$J_y = \tau_P V^3 + 2\kappa_P V A_x \tag{2.8.3}$$

The longitudinal jerk is simply

$$J_x = \frac{\mathrm{d}A_x}{\mathrm{d}t} \tag{2.8.4}$$

2.9 Two-Axle-Vehicle Analysis

Consider now a vehicle with extent in length, having two axles – that is, a wheelbase L and a vehicle yaw position. The vehicle heading angle is ψ in absolute coordinates (X,Y), as shown in Figure 2.9.1. The path angle is ν, and the difference is the attitude angle β. The attitude angle is the angle of the vehicle longitudinal axis relative to the direction of motion, measured at the centre of mass, to be distinguished carefully from the heading angle. In general, the sign convention adopted is

$$\psi = \nu - \beta$$

However, consider here the tyres to have zero slip angle, no roll steer etc., so that the vehicle moves fixed rigidly (for lateral motion) to the tangent of the single track path, with zero yaw angle. This is an approximation to having the centre of each axle follow the path. In this model the vehicle behaves rather like a simplified train on a track, with a rigid lateral constraint at each end. However, the wheelbase is still considered to be small relative to the path radius. The vehicle has zero attitude angle β (sideslip angle), so in this special case the heading angle equals the path angle:

$$\psi = \nu - \beta = \nu \tag{2.9.1}$$

that is, the vehicle remains simply tangential to the path. Hence also

$$\begin{aligned}\dot{\psi} &= \dot{\nu}\\ \ddot{\psi} &= \ddot{\nu}\end{aligned} \tag{2.9.2}$$

where time variations of path angle refer of course to the point at the moving position of the vehicle, taken at the centre of the wheelbase or the centre of mass.

More specifically, the vehicle yaw angular velocity (rad/s) is in this case

$$\dot{\psi} = \dot{\nu} = \frac{d\nu}{dt} = \frac{d\nu}{ds} \cdot \frac{ds}{dt}$$

$$\dot{\psi} = \kappa_P V \tag{2.9.3}$$

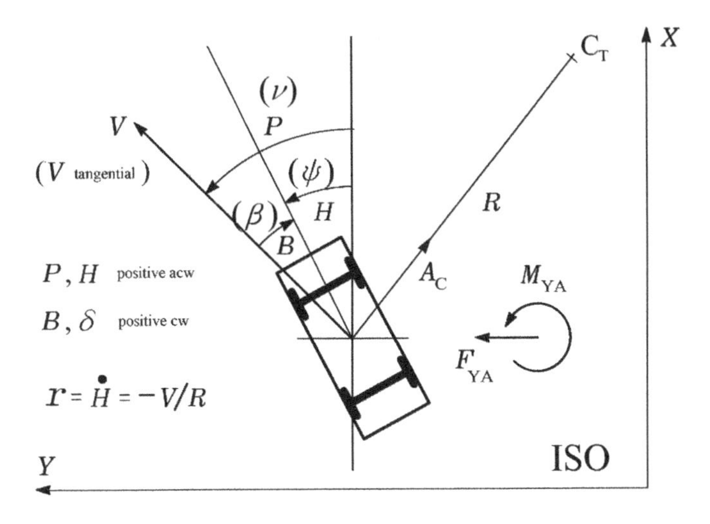

Figure 2.9.1 Path, heading and attitude angles.

The vehicle angular acceleration (rad/s^2) – again for $\beta = 0$, zero attitude angle – is

$$\ddot{\psi} = \ddot{\nu}$$

so

$$\begin{aligned} \ddot{\nu} &= \frac{\mathrm{d}\dot{\psi}}{\mathrm{d}t} = \frac{\mathrm{d}}{\mathrm{d}t}(\kappa_\mathrm{P} V) \\ &= \kappa_\mathrm{P}\frac{\mathrm{d}V}{\mathrm{d}t} + V\frac{\mathrm{d}\kappa_\mathrm{P}}{\mathrm{d}t} \\ &= \kappa_\mathrm{P} A_\mathrm{x} + V\frac{\mathrm{d}\kappa_\mathrm{P}}{\mathrm{d}s}\cdot\frac{\mathrm{d}s}{\mathrm{d}t} \end{aligned}$$

giving

$$\begin{aligned} \ddot{\nu} &= \kappa_\mathrm{P} A_\mathrm{x} + \tau_\mathrm{P} V^2 \\ &= \nu' A_\mathrm{x} + \nu'' V^2 \end{aligned} \tag{2.9.4}$$

where a prime indicates differentiation with respect to path length. So, for a vehicle without yaw freedom relative to the track (i.e. $\beta = 0$),

$$\ddot{\psi} = \ddot{\nu} = \kappa_\mathrm{P} A_\mathrm{x} + \tau_\mathrm{P} V^2 \tag{2.9.5}$$

Hence the yaw angular acceleration required arises from two terms. The first term, $\kappa_\mathrm{P} A_\mathrm{x}$, is the result of a given path curvature with a changing speed. The second term, $\tau_\mathrm{P} V^2$, arises from the change of path curvature (the path turn-in) with existing speed. In particular, the time rate of change of the vehicle-position path curvature is

$$\frac{\mathrm{d}\kappa_\mathrm{P}}{\mathrm{d}t} = \frac{\mathrm{d}\kappa_\mathrm{P}}{\mathrm{d}s}\cdot\frac{\mathrm{d}s}{\mathrm{d}t} = \tau_\mathrm{P} V \tag{2.9.6}$$

Note, however, that for a fixed point $\mathrm{d}\kappa_\mathrm{P}/\mathrm{d}t = 0$, the path itself is fixed in Earth-fixed coordinates.

Possibly also useful, the spatial rate of change of path radius is

$$\begin{aligned} \frac{\mathrm{d}R}{\mathrm{d}s} &= \frac{\mathrm{d}}{\mathrm{d}s}\left(\frac{1}{\kappa_\mathrm{P}}\right) \\ &= \frac{\mathrm{d}\kappa_\mathrm{P}}{\mathrm{d}s}\frac{\mathrm{d}}{\mathrm{d}\kappa_\mathrm{P}}\left(\frac{1}{\kappa_\mathrm{P}}\right) \end{aligned}$$

Hence

$$\frac{\mathrm{d}R}{\mathrm{d}s} = -\frac{\tau_\mathrm{P}}{\kappa_\mathrm{P}^2} = -\tau_\mathrm{P} R^2 \tag{2.9.7}$$

Also

$$\frac{\mathrm{d}R}{\mathrm{d}t} = \frac{\mathrm{d}R}{\mathrm{d}s}\frac{\mathrm{d}s}{\mathrm{d}t} = -\tau_\mathrm{P} R^2 V \tag{2.9.8}$$

and

$$\frac{\mathrm{d}}{\mathrm{d}t}\left(\frac{1}{R}\right) = \frac{\mathrm{d}\kappa_P}{\mathrm{d}t} = \tau_P V = -\frac{\dot{R}}{R^2} \tag{2.9.9}$$

$$\frac{\mathrm{d}}{\mathrm{d}s}\left(\frac{1}{R}\right) = \frac{\mathrm{d}\kappa_P}{\mathrm{d}s} = \tau_P \tag{2.9.10}$$

Associated with the path turn-in τ_P (units m^{-2}),

$$\tau_P = \frac{\mathrm{d}\kappa_P}{\mathrm{d}s}$$

we may define a car turn-in τ_C for a given velocity V:

$$\begin{aligned}\tau_C &= \frac{\mathrm{d}\kappa_P}{\mathrm{d}t} \\ &= \frac{\mathrm{d}\kappa_P}{\mathrm{d}s}\cdot\frac{\mathrm{d}s}{\mathrm{d}t} \\ &= \tau_P V\end{aligned} \tag{2.9.11}$$

This has units $\mathrm{m}^{-1}\,\mathrm{s}^{-1}$. The subscript C for car rather than V for vehicle is used because V is used elsewhere as a subscript for vertical.

2.10 Road Cross-Sectional Shape

Figure 2.10.1 shows the general case where the effective road bank angle is ϕ_R, measured between the tyre contact points. Note that for a road with non-straight (curved or piecewise linear) cross-section, the road bank angle must be defined by the line joining the centres of the contact patches. Because the road cross-section is not straight, the road angles measured from the ϕ_R position – the road camber angles, at the tyres, left and right – are γ_{RL} and γ_{RR}, defined positive as illustrated. These angles give corresponding changes of the tyre camber angles, influencing the tyre camber forces.

The road bank angles at the tyre positions are then

$$\phi_{RL} = \phi_R - \gamma_{RL}$$
$$\phi_{RR} = \phi_R + \gamma_{RR}$$

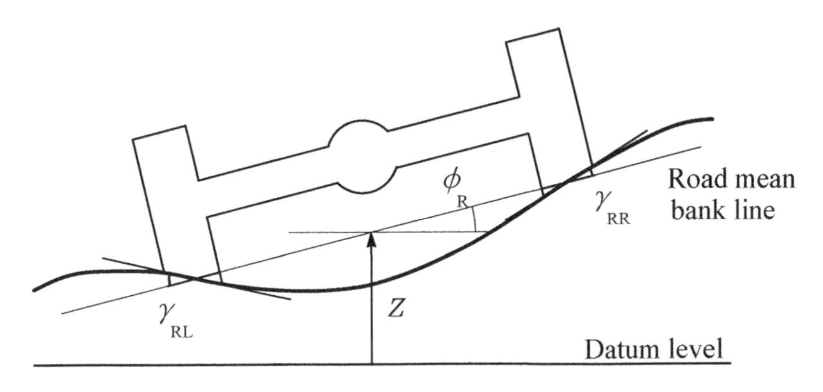

Figure 2.10.1 Road bank and camber geometry (track cross-sectional shape).

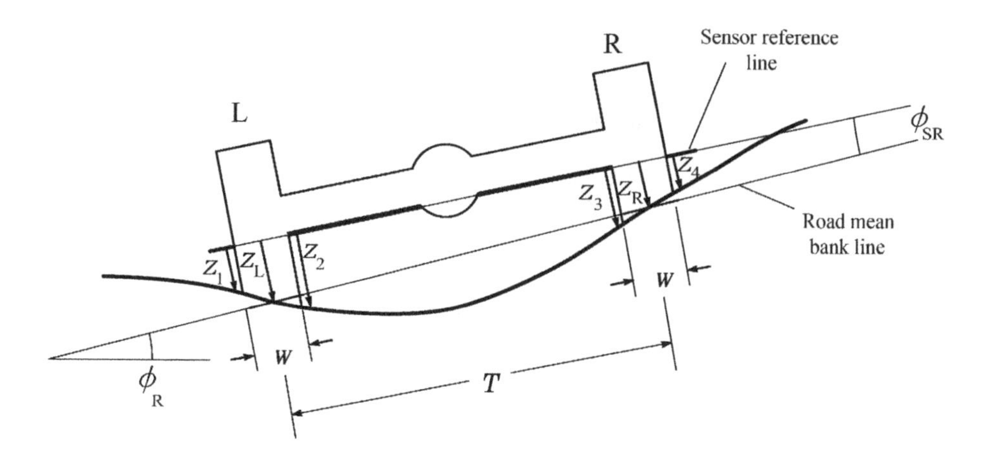

Figure 2.10.2 Use of four ride-height sensors for logger evaluation of road camber angles (rear view).

To analyse this in a simulation, the data set for the road needs, in principle, four extra data items for each point of the path, these being the road camber angles, one for each of the four tyres. It would normally be acceptable to use only two road camber angles on the basis that the front and rear tracks are nearly the same.

The road camber angles, as defined in Figure 2.10.1, cannot be deduced directly from normal logger data, but are easily obtained with a special set-up. The equipment required is a set of four body ride-height meters, for example lasers, placed along a transverse line (perpendicular to the centreline), as in Figure 2.10.2. For best results, the spacing of the tyre-area sectional mid-points should equal the track (tread). This is assumed in the following analysis, but is not absolutely essential.

The spacing width between sensors across the tyre area is w. The road bank angle ϕ_R is not known, and cannot be deduced here, but is not required. The four measured ride-height values are $z_1, \ldots, z_4$. The values at the centre of the left and right tyres (at track spacing) are therefore

$$z_L = \frac{1}{2}(z_1 + z_2)$$

$$z_R = \frac{1}{2}(z_3 + z_4)$$

The sensor reference line is rolled from the road mean bank line by

$$\begin{aligned}\phi_{\text{Sen}} &= \arcsin\left(\frac{z_L - z_R}{T}\right)\\ &= \arcsin\left(\frac{z_1 + z_2 - z_3 - z_4}{2T}\right)\end{aligned}$$

Relative to the sensor reference line, the two road angles are

$$\phi_{RL} = \arcsin\left(\frac{z_2 - z_1}{w}\right)$$

$$\phi_{RR} = \arcsin\left(\frac{z_3 - z_4}{w}\right)$$

The actual road camber angles are therefore

$$\gamma_{RL} = \phi_{RL} + \phi_{Sen}$$
$$\gamma_{RR} = \phi_{RR} - \phi_{Sen}$$

Explicitly,

$$\gamma_{RL} = \arcsin\left(\frac{z_2 - z_1}{w}\right) + \arcsin\left(\frac{z_L - z_R}{T}\right)$$

$$\gamma_{RR} = \arcsin\left(\frac{z_3 - z_4}{w}\right) - \arcsin\left(\frac{z_L - z_R}{T}\right)$$

2.11 Road Torsion

The road bank angle varies with longitudinal path position s, and hence the bank angles are different at the two axles, according to the local road torsion $k_{\phi RS}$ (rad/m), defined as

$$k_{\phi RS} = \frac{d\phi_R}{ds}$$

The difference of road bank angles between the two axles depends on the road torsion. Considering the road torsion to changes only slowly along the road, approximately

$$\phi_{Rf} - \phi_{Rr} = k_{\phi RS} L_W$$

where L_W is the wheelbase, and the road torsion is taken at the centre of the vehicle.

For vehicles with stiff suspensions (e.g. Formula 1 or equivalent) this may have a significant effect on vehicle behaviour at some points of some circuits. In a few special cases, severe road torsion occurs, for example, when moving onto or off the apron on a banked oval track, or over chicane edges in Formula 1. Considering that, in Formula 1 for example, the maximum roll angle is about 1 degree, then a road torsion of 0.1 degree or more over the wheelbase length could be expected to have significant effect. This is a road torsion of less than 0.05 deg/m (0.9 mrad/m). Obviously, racing tracks frequently exceed this level of torsion, even in the absence of the special cases mentioned.

Off-road and agricultural vehicles may, of course, be subject to large track torsion angles. For passenger cars on normal roads the effect is small because the suspension is more compliant.

2.12 Logger Data Analysis

When a vehicle path is to be specified, the path data requirement may arise in two ways. The first is the invented path, for a specific test, for example a slalom test or a lane change, when the path or road, with width, can be specified geometrically *ab initio*. The second case is when an existing road is to be specified geometrically. A detailed survey may be made, particularly for test tracks. However, this may be impractical, or uneconomical, and it may be desired to obtain the path geometry by direct means, by driving a vehicle over the road and in some way recording the path shape. Naturally, this often arises in racing, when data loggers are in common use, and the track shape is required for use in simulations for subsequent vehicle optimisation, of handling set-up, gear ratios, etc.

For each point of the path it is desired to evaluate, *inter alia*, the parameters in Table 2.12.1. In actual cases, some more detail may be appropriate, but these are the basic geometrical factors.

Table 2.12.1 Primary path parameters for evaluation

(1) θ_R, the road pitch angle
(2) ϕ_R, the road bank angle
(3) κ_L, the path lateral curvature
(4) κ_N, the path normal curvature.

Figure 2.12.1 shows the logger-carrying vehicle in rear view. Accelerometers on the body will give A_{By} and A_{Bz}. These must be corrected for the body roll angle to give the road-aligned accelerations. In the case of a vehicle with a rigid axle, in principle the accelerometers could be fixed to the axle, reducing the roll angle corrections, although the general vibration environment is worse. The analysis here assumes body-fixed accelerometers.

On a level road, for given known vertical and horizontal accelerations A_V and A_H, the observed (accelerometer indicated) body-coordinate accelerations are given by

$$A_{Bz} = A_V \cos \phi_B - A_H \sin \phi_B$$

$$A_{By} = A_V \sin \phi_B + A_H \cos \phi_B$$

These two simultaneous equations may be solved in the usual way to give the body-aligned accelerations, eliminating A_V and A_H in turn. The result is

$$A_H = A_{By} \cos \phi_B - A_{Bz} \sin \phi_B$$

$$A_V = A_{By} \sin \phi_B + A_{Bz} \cos \phi_B$$

The accelerometers readings will include the effect of gravity acting to negative V, so the actual vertical acceleration is $A_V - g$.

The speed must also be known. The curvatures are then given simply by $\kappa = A/V^2$. The accuracy of this depends on the quality of the value for the body roll angle ϕ_B, which is the sum of the suspension and axle roll angles:

$$\phi_B = \phi_S + \phi_A$$

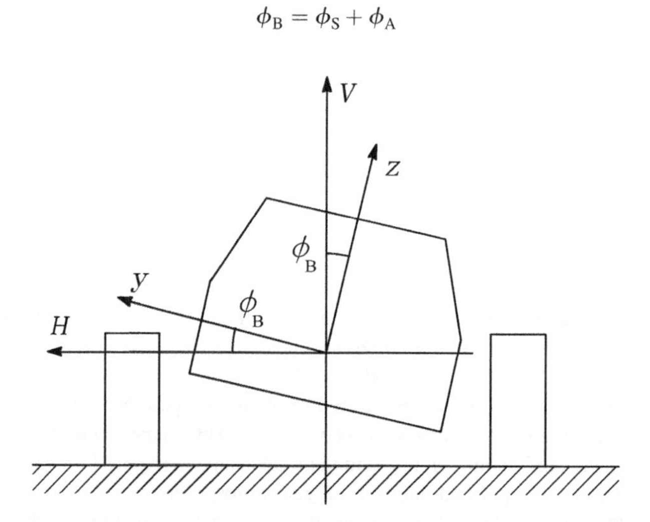

Figure 2.12.1 Accelerometer-carrying vehicle in rolled position.

The suspension roll at each axle can be deduced accurately from the individual suspension bump deflections. For a body that is not ideally torsionally rigid, interpolation of intermediate effective suspension roll angles requires knowledge of the longitudinal distribution of torsional compliance. It is probably better to locate the accelerometers over an axle.

The axle roll angle is due to the tyre normal deflections. For a passenger car, the axle roll is smaller than the suspension roll by a factor of 10 or so, but for a racing car they may be almost equal in value, but fortunately, then, small. Tyre normal deflections can be calculated accurately only with some difficulty because the normal stiffness is increased by speed (centrifugal carcase stiffening and peripheral momentum impact force) and reduced by tyre lateral forces. Where possible, it may be better to measure body roll directly by lasers, or to use both methods.

If the road itself has a bank angle, this cannot be detected by the vehicle. The direction of gravity in the vehicle-fixed axes is changed, altering the equations. Hence some independent means is necessary to obtain the road angles. Some possibilities include:

(1) gyros on the car giving angular velocities, integrated to angles;
(2) inclinometers placed on the road;
(3) inclinometers in a car driven very slowly along the road (corrected for body roll and pitch angles).

In principle it may be possible to obtain the road pitch angle from analysis of the car with additional other instrumentation. For example, from the driveshaft torques and knowledge of the aerodynamics and tyre drag, the longitudinal acceleration expected may be deduced and compared with that derived by considering the wheel angular speeds. Any discrepancy is then due to a component of the weight force in the longitudinal direction (i.e. a road pitch angle). However, it is doubtful that this would form the basis of a practical method because of inaccuracies.

In analysing data to obtain the normal curvature, similar problems will arise as in derivation of the cornering curvature – that is, the double differentiation amplifies any noise in the base data, and careful filtering of the data may be required.

The use of gyro sensors to provide angular velocity data, which may be integrated to give angular position (with various transformations of axes), appears to be an ideal solution. However, very high-quality gyro data are needed for this purpose, to minimise drift which gives a cumulative angular position error.

Lack of correct zeroing of the accelerometers will, after integration, give a cumulative lateral velocity, equivalent to a false road curvature $\kappa_0 = A_{Y0}/V^2$. The presence of lateral road slope without correction gives a similar effect. The accumulation of these is significant. This becomes apparent when the logger data are used to reconstruct the shape of a racing track when it is found that the two end positions and angles may be far from agreement. It is possible to deduce an accelerometer zero correction to minimise the end discrepancy. Fortunately, when the logger-data reconstructed circuit is used for subsequent vehicle optimisation in a simulation, there is no need for the track to join up correctly in position and angle, so this is mainly an aesthetic problem for the resulting track shape display.

References

Cebon, D and Newland, D. E. (1984) The artificial generation of road surface topography by the inverse FFT method. In *8th IAVSD Conference on Dynamics of Vehicles on Roads and Tracks*, pp. 29–42. Lisse: Swets and Zeitlinger.

———— // ————

3

Road Profiles

3.1 Introduction

Any particular track along a road has a longitudinal profile, expressed as a function of the whole path length $Z(s)$, or as a function of the plan-view (horizontal) path length $Z(s_H)$, representing its vertical sectional shape for the purpose of ride quality analysis. This neglects lateral curvature (cornering). For advanced analysis, two parallel tracks may be considered, for left and right wheels, or even four road tracks if the vehicle axle tracks (US treads) are different. In physical testing and simulations, the type of road may classified mainly as

(1) isolated bump,
(2) sinusoidal,
(3) fixed waveform,
(4) stochastic, independent of speed.

An isolated bump is a local disturbance on a nominally smooth basic road, in effect a solitary wave. These are modelled as idealised theoretical bumps with simple geometrical properties. Such 'bumps' may be further classified as ramps or true bumps. A ramp is a disturbance which results in a residual change of height. A non-ramp bump returns to its original elevation. Such solitary bumps are quite easily arranged experimentally, whereas sinusoidal roads are difficult to make physically, although easy analytically or numerically, and stochastic roads are also hard to generate to a given specification in reality, although again easy enough in theory. Idealised bumps are therefore useful tests, and also have the merit of being close to practical problems, for example the common depressed drainage grid at the side of a road and the 'sleeping policeman' hump for traffic calming.

3.2 Isolated Ramps

The ramp is a disturbance with a residual change of road height. The main types of ramp, in its broadest sense, are listed in Table 3.2.1 and Figure 3.2.1.

The simple step has a height specification, but no length dimension. The height may be positive (step up) or negative (step down). Notionally, it is perfectly sharp at the edges. The nearest equivalent in reality is driving up onto a kerb, or down off one, with a height of about 100 mm. If this is done at any speed it is a severe shock to the tyre and suspension, with the tyre suffering a severe initial deflection and local negative curvature of the tyre tread on the corner of the step, so when tested in reality the step height is fairly limited.

Table 3.2.1 Isolated ramps

(1) Simple step, zero length, height up or down.
(2) True linear ramp, with length, and height up or down.
(3) Haversine ramp, with length, and height up or down.

The maximum positive deflection from static allowed by a passenger car tyre is about 50 mm, compared with a static deflection of about 14 mm, i.e. a total deflection of about 64 mm, varying with the section aspect ratio. When the 64 mm positive deflection is reached, a rim impact occurs, with two layers of tyre pinched against the rim.

The true linear ramp is a softened version of the step, being specified by the height H and length L, where L is the plan-view length rather than the true path length. It has the merit that it can be produced physically fairly easily, and is a realistic representation of a problem occurring in reality, such as temporary road works or poor quality resurfacing edges. The gradient of the ramp is simply

$$G = \frac{H}{L}$$

The ramp angle is $\theta_R = \text{atan}(H/L)$. The gradient is discontinuous at the ends of the ramp. Realistic test values would be a height change of 50 mm over a length of 1 m, which is an angle of 2.86°. At practical vehicle speeds the rise time of the ramp is short (e.g. 50 ms), so realistic ramp lengths are almost as bad for maximum tyre deflection as a step, although the local tyre distortion at the ramp corners is much reduced.

The haversine ramp is a more smoothed version, lacking the gradient discontinuities of the linear ramp. The haversine function is so named because it is one half of the versine, which in turn is the complement of the cosine, so the haversine function is defined as

$$\text{hav}(\theta) = \tfrac{1}{2}\{1 - \cos(\theta)\}$$

It has a sinusoidal form, varying from 0 to 1, being 0 at $\theta = 2N\pi$ and 1 at $\theta = 2(N-1)\pi$, as shown in Figure 3.2.2.

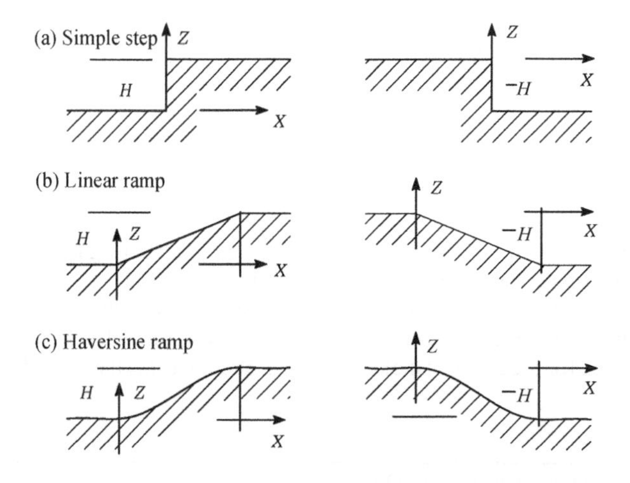

Figure 3.2.1 Isolated ramps up and down: (a) simple step; (b) true linear ramp; (c) haversine ramp.

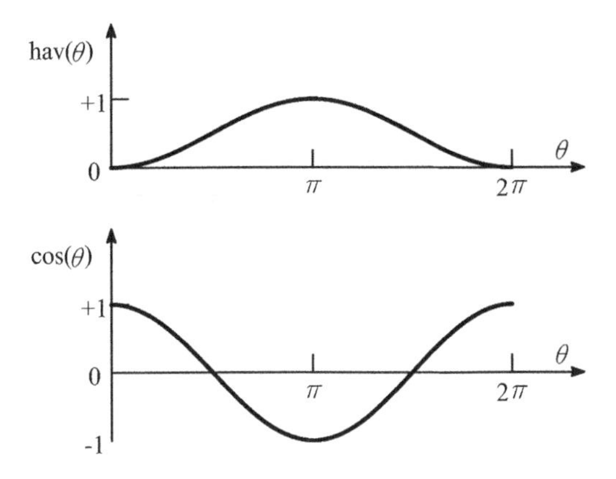

Figure 3.2.2 Haversine and cosine curves compared.

The haversine ramp, which is half a wavelength long (i.e. $\lambda = 2L$), may therefore be expressed as

$$Z = H\,\mathrm{hav}\left(\frac{\pi X}{L}\right) = \tfrac{1}{2}H\left(1 - \cos\left(\frac{\pi X}{L}\right)\right); \quad 0 \le X \le L$$

The gradient of the haversine ramp is given by the derivative of Z,

$$G = \frac{\pi H}{2L}\sin\left(\frac{\pi X}{L}\right) \le \frac{\pi H}{2L}$$

with a maximum value of $\pi H/2L$. This maximum gradient is therefore more severe than the corresponding linear ramp by a factor $\pi/2$.

3.3 Isolated Bumps

The main types of model isolated bump are listed in Table 3.3.1 and Figure 3.3.1. In all cases, the height may be negative, giving a trough rather than a bump. In some cases this is a realistic problem – for example, the trapezoidal trough (or negative trapezoidal bump) is a good representation of the common badly-levelled gutter drainage grid.

The simple step bump is rectangular. The triangular bump is two ramps back-to-back. The trapezoidal bump is just a flattened triangle. The sine half-wave bump is defined by

$$Z = H\sin(\pi X/L); \quad 0 \le X \le L$$

Table 3.3.1 Isolated bumps

(1) Simple step, various lengths, height up or down.
(2) Triangular, with length and height.
(3) Trapezoidal, with total length, plateau length, and height.
(4) Sine half-wave, with length and height.
(5) Haversine bump, with length, and height.

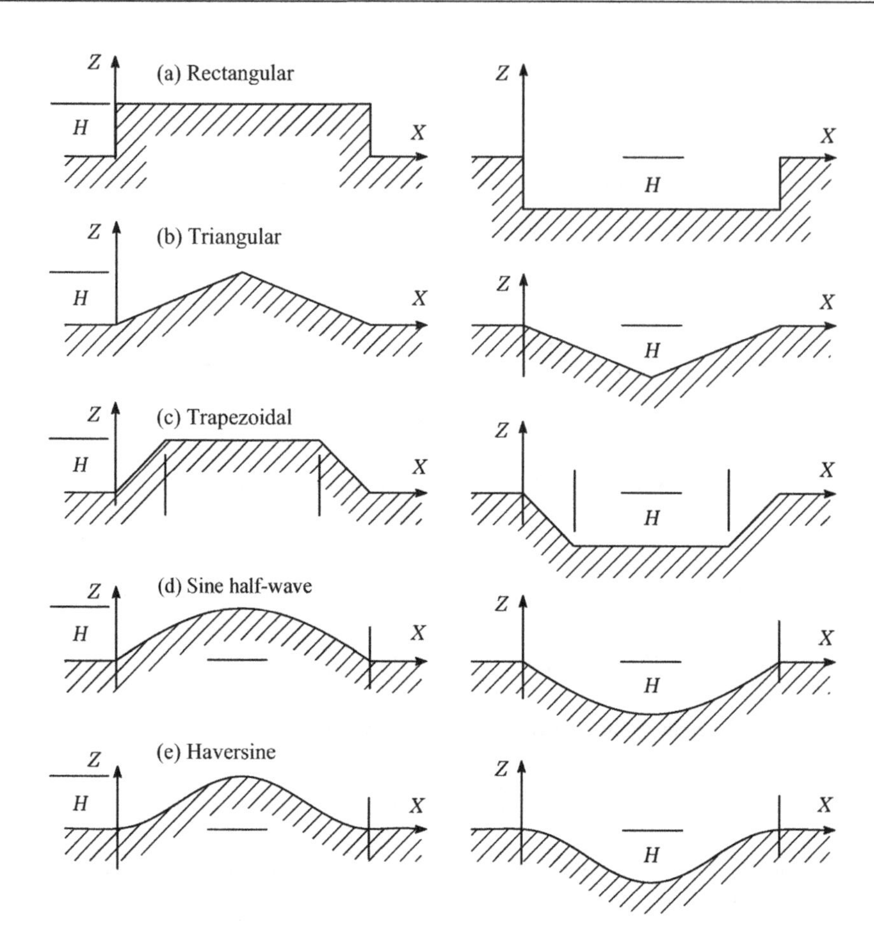

Figure 3.3.1 Isolated bumps and troughs up and down: (a) simple step; (b) triangular; (c) trapezoidal; (d) sine half-wave; (e) haversine.

The gradient, by differentiation, is

$$G = \left(\frac{\pi H}{L}\right)\cos\left(\frac{\pi X}{L}\right)$$

and is discontinuous at the ends. This is a typical shape used for a deliberately made traffic-slowing bump, the 'sleeping policeman', with length from 1 m to 3 m.

The full haversine bump is defined by

$$Z = H\,\mathrm{hav}\left(\frac{2\pi X}{L}\right); \quad 0 \le X \le L$$

The gradient is

$$G = \left(\frac{\pi H}{L}\right)\sin\left(\frac{2\pi X}{L}\right)$$

and is continuous, with a maximum value equal to that of the sine half-wave. (The sine wave actually has twice the amplitude, but only one half is used.)

3.4 Sinusoidal Single Paths

In simple analytical ride studies the path or road is considered to have only one spatial frequency of bumps at a time. This is known as a sinusoidal single-path road, Figure 3.4.1. All frequencies may be considered, but only one at a time. It is also known in simulation studies as a single-track road, but this is potentially confusing because in general motoring a single-track road is one with the operating width of a single vehicle.

Considering a sinusoidal road shape of wavelength λ_R, Figure 3.4.2, the spatial frequency of the road n_{SR} (cycles per metre, c/m) is

$$n_{SR} = \frac{1}{\lambda_R} \tag{3.4.1}$$

The radian spatial frequency of the road, ω_{SR} (rad/m), is

$$\omega_{SR} = 2\pi n_{SR} = \frac{2\pi}{\lambda_R} \tag{3.4.2}$$

At a vehicle longitudinal speed V, the observed frequency of road vibration stimulus is f_R (Hz) with corresponding radian frequency ω_R (rad/s) and period T_R (in seconds):

$$T_R = \frac{\lambda_R}{V} \tag{3.4.3}$$

$$f_R = \frac{1}{T_R} = \frac{V}{\lambda_R} = V n_{SR} \tag{3.4.4}$$

$$\omega_R = V \omega_{SR} \tag{3.4.5}$$

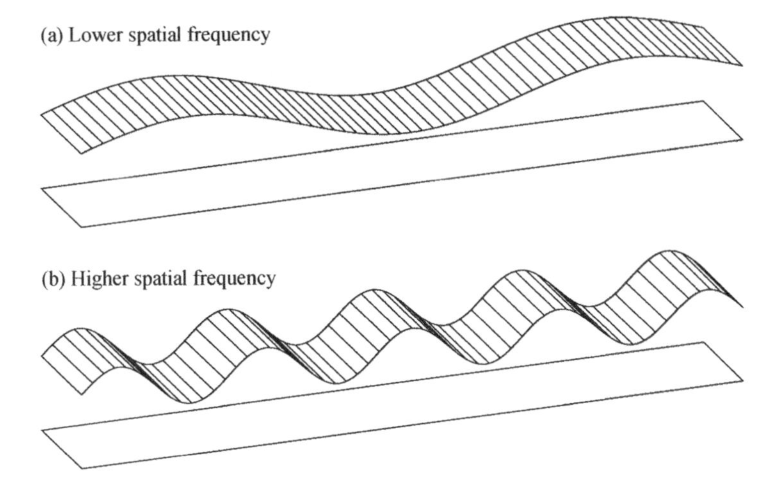

Figure 3.4.1 Sinusoidal single paths of lower and higher spatial frequencies.

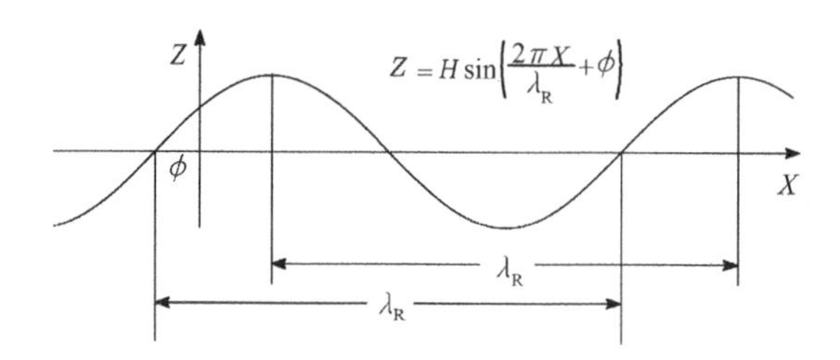

Figure 3.4.2 Sinusoidal wavelength and phase angle.

It is therefore easy to find the vehicle speed or road wavelength which will stimulate vehicle resonance at a natural frequency f_N:

$$V = \lambda f_N \tag{3.4.6}$$

A heave or pitch resonance at 1.4 Hz occurs for a wavelength of 10 m at a speed of $10 \times 1.4 = 14$ m/s. At a typical speed of 20 m/s and $f_N = 1.4$ Hz the resonant wavelength is 14 m. A wheel hop resonance at 10 Hz will be stimulated at a speed of 20 m/s by a wavelength of $20/10 = 2.0$ m. At a wavelength of 0.1 m, as found with cobblestones, wheel hop is stimulated at a speed of around 1 m/s. A wavelength of about 0.160 m equals the tyre contact patch length, and stimulates longitudinal vibrations at the wheel, independent of speed.

In Figure 3.4.2, when a purely sinusoidal road is being considered, the phase angle is normally zero. However, when the sinusoids are the result of a Fourier analysis, the phase must be considered. In general, for a given single wavelength λ_R and spatial radian frequency ω_{SR} there are sine and cosine components, so

$$Z = S\sin(\omega_{SR}X) + C\cos(\omega_{SR}X) \tag{3.4.7}$$

These may be combined into a single sine wave of amplitude Z_0 with a phase angle ϕ:

$$Z = Z_0 \sin(\omega_{SR}X + \phi) \tag{3.4.8}$$

Expanding the sine double angle,

$$Z = Z_0\{\sin(\omega_{SR}X)\cos\phi + \cos(\omega_{SR}X)\sin\phi\} \tag{3.4.9}$$

By comparison with equation 3.4.7,

$$S = Z_0 \cos\phi$$
$$C = Z_0 \sin\phi$$

so

$$Z_0 = \sqrt{S^2 + C^2}$$
$$\tan\phi = \frac{C}{S}$$

as in Figure 3.4.3. The sine and cosine coefficients may be of either sign, and the phase angle must be found in the correct quadrant (use Fortran Atan2(C,S) or Basic Angle(S,C). Note that the order of the arguments differs).

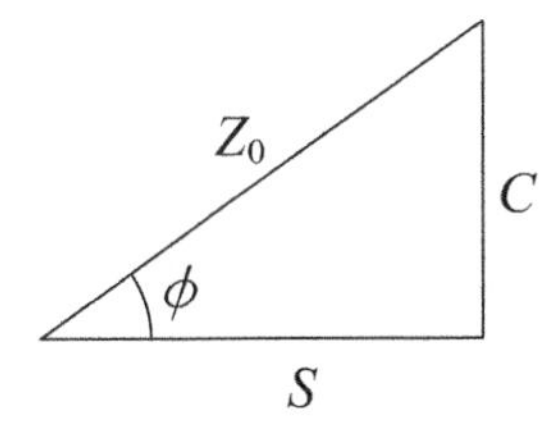

Figure 3.4.3 Sine component amplitude S, cosine component amplitude C, total amplitude Z_0 and phase angle ϕ.

3.5 Sinusoidal Roads

The sinusoidal road is essentially two sinusoidal paths, at a lateral spacing equal to the track (tread) of the vehicle. This introduces several complications. The two sinusoidal paths may be symmetrical, anti-symmetrical or arbitrary, corresponding to relative phase angles of zero, 180° or any value, Figure 3.5.1. Of course, this assumes that the two paths have sinusoids with the same wavelength, which is usual but not actually physically essential.

When the two paths are in phase, then a symmetrical vehicle would respond symmetrically. However, this simple case can be used to investigate the effect of vehicle asymmetries, such as different-length steering track rods.

The purely antisymmetrical case, with 180 degrees of phase difference, stimulates the vehicle strongly in roll. Again, this is a useful test of the effect of bump-steer and roll-steer geometry and of the steering activity required to maintain a straight course.

In general, each path has its own phase angle, ϕ_L and ϕ_R left and right, although the physical effect of the phases on the vehicle depends only on the phase angle difference $\phi_L - \phi_R$. Each path may be expressed in its sine and cosine components:

$$Z_L = S_L \sin(\omega_{SR} X) + C_L \cos(\omega_{SR} X)$$

$$Z_R = S_R \sin(\omega_{SR} X) + C_R \cos(\omega_{SR} X)$$

The road (*viz.* the pair of paths) may be considered as a total symmetrical road part Z_S on the centreline plus an antisymmetrical banking part Z_B, half of the path height difference,

$$Z_S = \tfrac{1}{2}(Z_L + Z_R)$$
$$Z_B = \tfrac{1}{2}(Z_L - Z_R)$$

and in reverse,

$$Z_L = Z_S + Z_B$$
$$Z_R = Z_S - Z_B$$

The centreline road elevation is therefore

$$\begin{aligned} Z_S &= \tfrac{1}{2}(S_L + S_R)\sin(\omega_{SR} X) + \tfrac{1}{2}(C_L + C_R)\cos(\omega_{SR} X) \\ &= S_S \sin(\omega_{SR} X) + C_S \cos(\omega_{SR} X) \end{aligned}$$

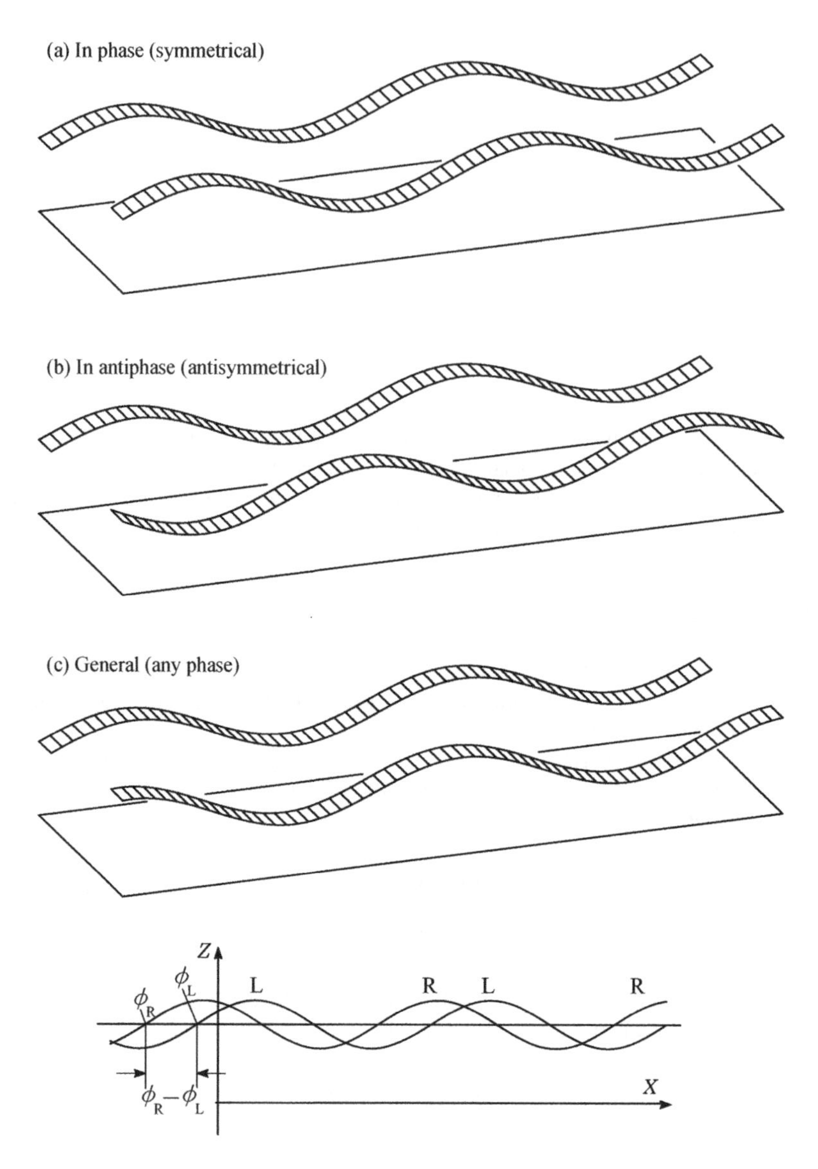

Figure 3.5.1 Sinusoidal two-track roads with various phase relationships: (a) in phase, symmetrical; (b) in antiphase, antisymmetrical; (c) general.

The banking component is

$$
\begin{aligned}
Z_B &= \tfrac{1}{2}(S_L - S_R)\sin(\omega_{SR}X) + \tfrac{1}{2}(C_L - C_R)\cos(\omega_{SR}X) \\
&= S_B \sin(\omega_{SR}X) + C_B \cos(\omega_{SR}X)
\end{aligned}
$$

The amplitudes are therefore related simply by

$$S_S = \tfrac{1}{2}(S_L + S_R)$$
$$C_S = \tfrac{1}{2}(C_L + C_R)$$

$$S_B = \tfrac{1}{2}(S_L - S_R)$$
$$C_B = \tfrac{1}{2}(C_L - C_R)$$

These may in turn be combined into total amplitudes and phases:

$$Z_S = Z_{S0}\sin(\omega_{SR}X + \phi_S)$$
$$Z_B = Z_{B0}\sin(\omega_{SR}X + \phi_B)$$

$$Z_{S0} = \sqrt{S_S^2 + C_S^2}$$
$$Z_{B0} = \sqrt{S_B^2 + C_B^2}$$

The phase angles are

$$\tan\phi_S = C_S/S_S; \quad \phi_S = \text{atan2}(C_S, S_S)$$

$$\tan\phi_B = C_B/S_B; \quad \phi_B = \text{atan2}(C_B, S_B)$$

The antisymmetrical part may be considered to be due to a road bank angle ϕ_{RB}:

$$\tan\phi_{RB} = \frac{2Z_B}{T} = \frac{Z_L - Z_R}{T}; \quad -90^\circ < \phi_{RB} < 90^\circ$$

so

$$\tan\phi_{RB} = \frac{2}{T}\{S_B\sin(\omega_{SR}X) + C_B\cos(\omega_{SR}X)\}$$
$$= \frac{2Z_{B0}}{T}\sin(\omega_{SR}X + \phi_B)$$

The road bank angle is normally quite small, for passenger cars at least, in which case

$$\phi_{RB} \approx \tan\phi_{RB}$$

Then the local road torsion is

$$k_{\phi RS} \equiv \frac{d\phi_{RB}}{dX} = \frac{2Z_{B0}\omega_{SR}}{T}\cos(\omega_{SR}X + \phi_B)$$

The total road torsion angle between the two axles of a vehicle of wheelbase L, with the rear axle at the path position X, is

$$\phi_{RT} = \phi_{RB,X+L} - \phi_{RB,X}$$

which is

$$\phi_{RT} = \frac{2Z_{B0}}{T}\{\sin(\omega_{SR}(X + L_W) + \phi_B) - \sin(\omega_{SR}X + \phi_B)\}$$

3.6 Fixed Waveform

Some other waveforms are sometimes used in testing. A good example is the regularly-repeating rectangular bump, as in Figure 3.6.1. This particular profile is characterised by the wavelength λ, and the rectangular bump length L_B and height H_B. The advantage of this particular path profile is that it is easily constructed by laying down planks of wood on an otherwise level surface. This profile has been used to investigate the effect of profile roughness on the maximum cornering ability of cars, Figure 3.6.2, by driving at near to the limit of cornering ability on a smooth entry section before entering the rough section. The bumps may also be angled to the expected trajectory of the vehicle to introduce an asymmetrical stimulus.

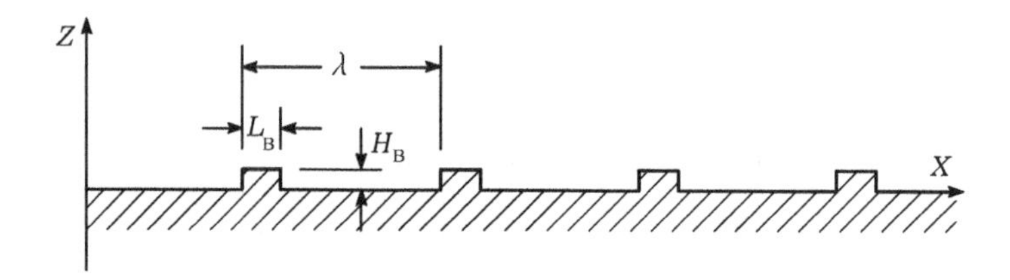

Figure 3.6.1 Path profile with repeating rectangular bump.

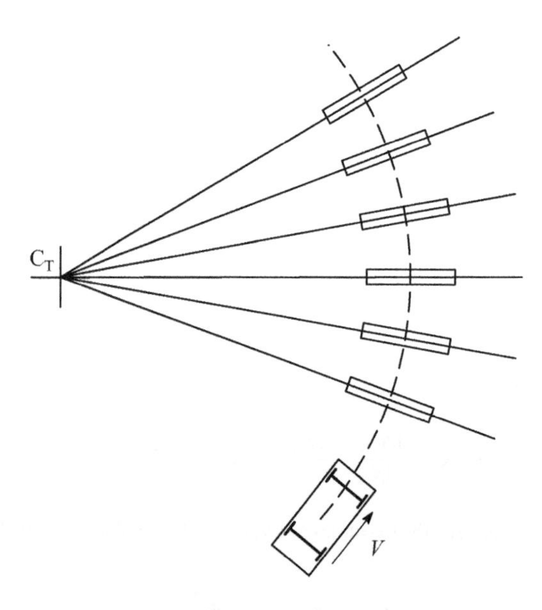

Figure 3.6.2 Rectangular-roughness cornering test.

3.7 Fourier Analysis

Fourier analysis was introduced by French mathematician Jean Baptiste Joseph Fourier (1768–1830) in the context of thermodynamic analysis. (Simultaneously he explicitly introduced the idea that a scientific equation must have consistent units, launching the field of dimensional analysis.) The basic idea of Fourier analysis is that any perfectly repetitive waveform can be represented exactly by the sum of many sinusoidal waves. In Fourier's day this was all done analytically, and in general an infinite number of waves are required for an exact representation, with some lesser number giving an approximation. With a digital representation of a waveform, the resolution is not perfect, and a finite number of sinusoidal waves is sufficient for the perfect representation of the digital wave as sampled. A lesser number than that may still be used if desired, provided that the approximation is acceptable.

Appendix F gives more information on the Fourier transform. For efficient digital calculation the fast Fourier transform (FFT) is used, or specifically the digital Fourier transform (DFT), with a number of points equal to 2^M where M is an integer, for example $1024 = 2^{10}$ points equally spaced at typically 1 m or one quarter of the wheelbase. The DFT then gives 512 frequencies each with one sine coefficient and one cosine coefficient, for a total of 1024 coefficients (approximately, see Appendix F). This requires, in principle, the solution of 1024 simultaneous equations for 1024 coefficients, but fortunately these are decoupled and can be solved independently and therefore efficiently.

Applying the definitive 'slow' Fourier transform, the cosine and sine coefficients for frequency number K are given respectively by

$$C_K = \frac{2}{N}\sum_{J=1}^{N} Y(J)\cos\left(\frac{2\pi JK}{N}\right); \quad K = 0, 1, \ldots, N/2$$

$$S_K = \frac{2}{N}\sum_{J=1}^{N} Y(J)\sin\left(\frac{2\pi JK}{N}\right); \quad K = 1, 2, \ldots, N/2-1$$

where the wavelength of that spatial frequency is L/K, L being the physical length of the sample, not seen in the above equations. The DFT solves for $K = 1$ to $N/2$.

The road profile (e.g. the original 1024 road heights) is recovered from the sine and cosine coefficients for the Jth point by

$$Z(J) = \tfrac{1}{2}A_0 + \sum_{K=1}^{M-1}\left[C_K\cos\left(\frac{2\pi JK}{N}\right) + S_K\sin\left(\frac{2\pi JK}{N}\right)\right] + \tfrac{1}{2}A_M\cos(\pi J)$$

From the above equations, it is apparent that the mean value of the road height can have no effect on the sine or cosine coefficients because they are always complete waves and contribute nothing to the mean value. The mean is therefore accounted for separately:

$$Z_m = \frac{1}{N}\sum_{K=1}^{N} Z_K$$

A more difficult problem is that of the trend of the data. In the specific case of a road, there is generally a gradient from one end of the data set to the other. The important point to appreciate is that a Fourier transform is a transform of a repeated waveform. If path heights at points 1 to 1024 are specified, then the heights of points 1025 onwards are, in the transform, the same as those for point 1 onwards. Therefore, to do a Fourier transform it is implied in the dataset that $Z_{N+K} = Z_K$, although this is not explicitly in the data.

Considering a perfectly smooth road, but with a gradient, so that $Z_N > Z_1$, the Fourier transform assumes that $Z_{N+1} = Z_1$, and the transform will be that of a sawtooth path with a sharp step down from Z_N to Z_{N+1}.

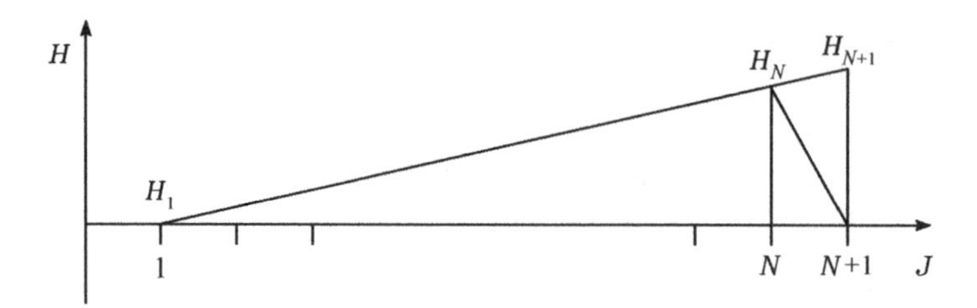

Figure 3.7.1 Degradienting the raw road data.

Therefore the gradient must be removed. To do this, do not set $Z_N = Z_1$. The point Z_{N+1} should be measured to give the gradient explicitly. The degradiented data Z is then derived from the raw data H by

$$Z_K = H_K - \frac{K-1}{N}(H_{N+1} - H_1)$$

as in Figure 3.7.1.

Degradienting is part of a broader process of detrending, ideally removing the effects of all wavelengths greater than the sample. For example, the road actually has significant amplitude content at twice the

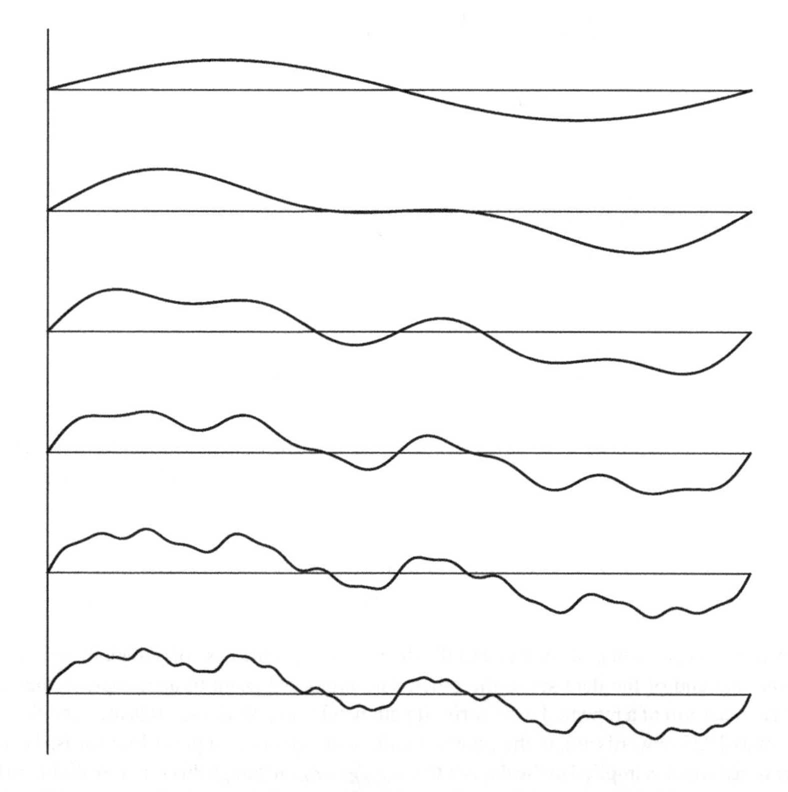

Figure 3.7.2 Illustration of the accumulation of sine components of successive frequencies, showing the first few frequencies only.

sample length, so there will be a consequent curvature in the sample, which will falsely appear as frequency content in the Fourier analysis – so-called aliasing. This frequency content really is present in the assumed repeated waveform, but not desired because it is not present in the real road which is not really an exact repeat of the sample. The Fourier analysis gives the correct frequency content of an infinite repeat of the sample, rather than of the actual road. Failure to detrend the road data will result in considerable spurious frequency content, particularly at high frequencies, with bad results.

When a simulated road is reconstructed from a set of sine and cosine coefficients, then the waveform produced is, naturally, a repeating one outside its basic length.

Anti-aliasing and 'degradienting' is really an area for specialists – vehicle dynamicists can mostly just use the results, such as the ISO standard road spectrum, which encapsulates the characteristics of roads as revealed by Fourier analysis, as discussed in a later section.

Figure 3.7.2 illustrates how a road profile being constructed by an inverse Fourier transform is accumulated from successive frequencies.

3.8 Road Wavelengths

The spectral analysis of roads may be used to separate out the wavelengths into various categories according to their effect, as in Table 3.8.1. At the short wavelengths, affecting tyre friction, the practical division is into macrostructure and microstructure, as seen in Figure 3.8.1. The range of wavelengths from 10 mm to 100 m, covering undulations and roughness, is the main area of interest for Fourier analysis in ride quality studies.

Table 3.8.1 Road spectral analysis

Characteristic	Wavelength	Influence on
Slopes	$100\,\text{m} < \lambda$	Static
Undulations	$1\,\text{m} < \lambda < 100\,\text{m}$	Dynamic ride
Roughness	$10\,\text{mm} < \lambda < 1\,\text{m}$	Dynamic NVH (noise, vibration and harshness)
Macrotexture	$1\,\text{mm} < \lambda < 10\,\text{mm}$	Friction and noise
Microtexture	$10\,\mu\text{m} < \lambda < 1\,\text{mm}$	Friction
Material	molecular	Friction

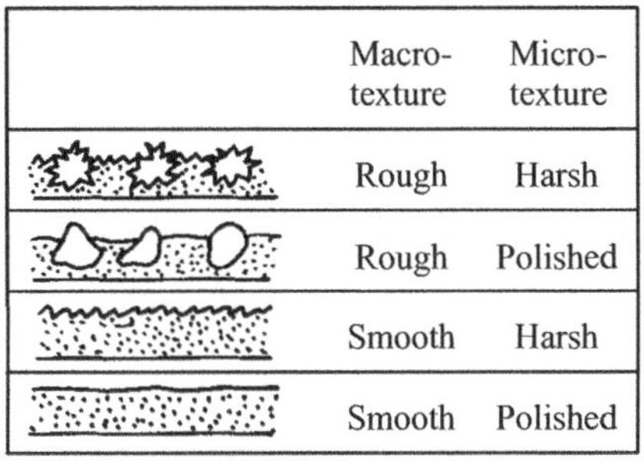

	Macro-texture	Micro-texture
	Rough	Harsh
	Rough	Polished
	Smooth	Harsh
	Smooth	Polished

Figure 3.8.1 Road surface macrostructure and microstructure.

3.9 Stochastic Roads

Real roads, subject to Fourier analysis, are found to have a spectral distribution of roughness declining rapidly with spatial frequency, Figure 3.9.1 giving two examples. A commonly used road model has

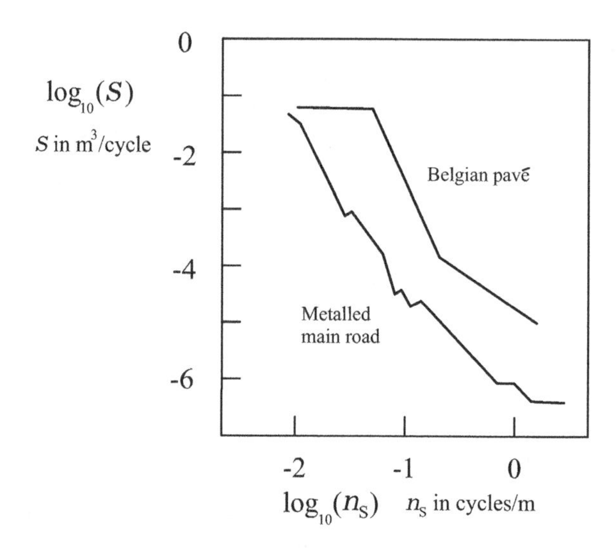

Figure 3.9.1 The spectral analysis of two example roads (spectral density versus spatial frequency).

been that in Figure 3.9.2, with amplitude squared spectral density (so-called power spectral density, PSD)

$$S = S_{\text{ref}}\left(\frac{n_S}{n_{S,\text{ref}}}\right)^{-W}$$

with the negative gradient W usually equal to 2.5 or 2.7. More complex models are also used with two gradients, and various cut-off methods, and sometimes continuous curves.

If the maximum gradient of a sinusoid is constant, then the amplitude is proportional to the wavelength. Then the amplitude squared (the spectral 'power') is proportional to the wavelength squared, and the negative gradient W would take the value 2. Therefore, for representative roads, with $W > 2$, the maximum sinusoidal gradient reduces with frequency.

The ISO road surface model has two linear sections as in Figure 3.9.3, with an average road specified by a break point or reference point at a spatial frequency of 1 rad/m, which is a wavelength of 2π metres. In

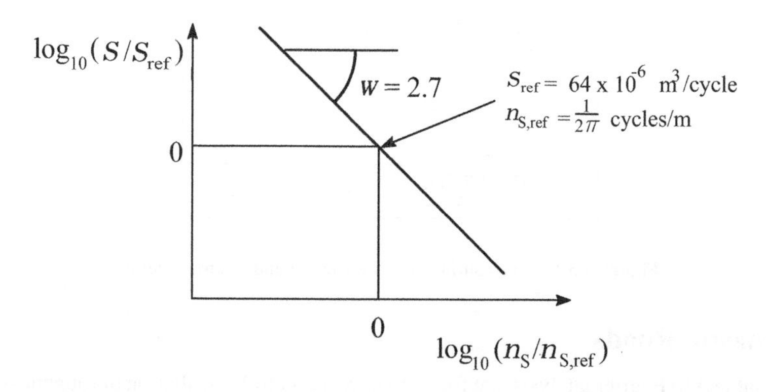

Figure 3.9.2 The basic single gradient spectral model of a representative road surface.

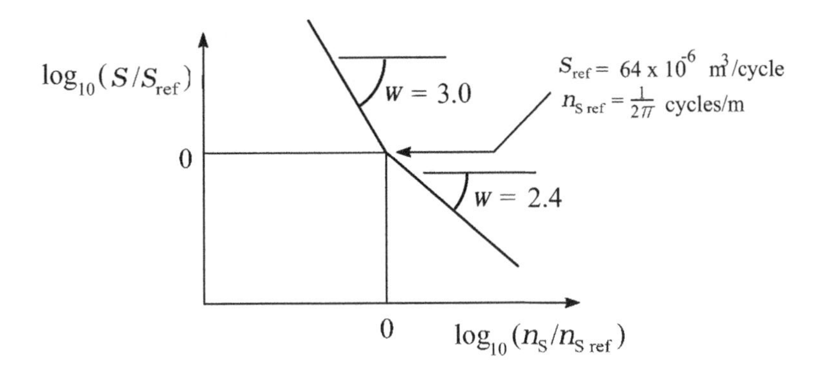

Figure 3.9.3 The ISO spectral road model, having two gradients.

cycles/metre this is

$$n_{S,ref} = \frac{1}{2\pi}\ \mathrm{c/m} = 0.1592\ \mathrm{c/m}$$

At a speed of 20 m/s, this break point at a wavelength of 2π metres corresponds to a frequency of 3.18 Hz, in between the body natural frequency and the wheel hop frequency. The spectral density of the ISO standard road at this frequency is

$$S_{ref} = 64 \times 10^{-6}\ \mathrm{m^3/c} = 64\ \mathrm{cm^3/c}$$

The line negative gradients are

$$W_1 = 3.0; \quad W_2 = 2.4$$

The particular metalled main road in Figure 3.9.1 can be seen to have two gradients, which are about -3.0 and -1.5 with a break point at a wavelength of 10 m. Data published by Verschoore et al. (1996) show a 'perfect asphalt' road to have a single gradient of -2.2 from 100 m to 0.4 m wavelength. Therefore all roads are not ISO roads.

Evidently roads vary considerably in quality. It is useful to consider a grading of roads for simulation purposes. The road model is assumed to have one or two gradients with a reference point at a spatial frequency of $1/2\pi$ c/m. The spectral density at this reference point will then be as in Table 3.9.1. The spectral density is 2^R cm^3/cycle, where R is the road profile roughness rating. The table covers a range from a very good (motorway) quality to a very bad minor road and beyond. The 'average' rating, $R = 6$, is the same as that of the ISO standard road. Road types may actually vary by a factor of up to 8 in the reference spectral density; for example, motorways of typical rating 3, with 8 cm^3/c, may vary from 1 to 64 cm^3/c according to the particular road and maintenance state. However, the table is a useful guide to values.

The spectral 'power' density is proportional to the amplitude squared, so to double the amplitude requires a factor of 4 in the value of S, and an increment of 2 in R. The sequence of ISO letter classes therefore represents a successive doubling of amplitude.

Displacement can be scaled after making the basic profile shape, so one can make the road shape for 1 cm^3/c or 64 cm^3/c and adjust for road quality later. If the vehicle model is linear, the response is proportional to the stimulus, and the transmissibility is unchanged by the roughness grade of the road, which is important only for a non-linear vehicle model or when considering human vibration tolerance.

Table 3.9.1 Road spectral density at spatial frequency $1/2\pi$ c/m

Rating	S mean (cm^3/c)	S range (cm^3/c)	ISO class	ISO description
2	4	<8	A	very good
3	8			
4	16	8–32	B	good
5	32			
6	64	32–128	C	average
7	128			
8	256	128–512	D	poor
9	512			
10	1024	512–2048	E	very poor
11	2048			
12	4096	2048–8192	F	—
14	16384	8192–32768	G	—
16	65536	>32768	H	—

With a negative gradient of 2.5 in a spectral model, doubling the speed of travel increases the roughness amplitude at a given frequency, for example the vehicle heave natural frequency, by a factor of $2^{2.5} = 5.7$. This greatly increases the ride motions, which may become not just uncomfortable but actually hazardous. The provision of high-quality surfaces for high-speed roads is not just a matter of ride quality and comfort, but also one of safety.

Given a spectral distribution for a path, the numerical path profile may be built. This is a process of inverse Fourier transform. It may be done by an inverse FFT, but this is not essential. A 'slow FT' may be used because the road generation need only be done once, the road profile then being stored and used many times. A series of spatial frequencies is chosen, and the spectral distribution condensed down into these frequencies, producing specific amplitudes for each one. In effect, the spectral distribution is converted into a histogram, with one column for each spatial frequency. The area under the spectral distribution within the width of the column, which depends on the adjacent frequencies, gives the column height. The column height is an integral of the spectral density, so having units m^2. The square root of this is the amplitude for that frequency. The wavelengths must all fit exactly into the length of road. The phase angle for each frequency is chosen at random in the range 0 to 2π, this giving amplitudes for the sine and cosine components.

It is possible to study the response of the vehicle to a stochastic road with energy density uniform over the spectrum. This is known more generally as 'white noise'. If the energy density is biased towards low frequencies, then it is known as 'pink noise', by analogy with the visible spectrum of light which has red light at the long-wavelength end. Actual ISO roads are therefore particular cases of pink noise.

In more advanced ride models, using a vehicle model with two wheels on each axle at the track (tread) lateral spacing, the correlation of the road profile at adjacent positions in the two wheel tracks is of interest. This obviously depends on the method of road manufacture and on the types of machine used for finishing, and there could easily be very high coherence at some short wavelengths. However, Figure 3.9.4 shows the coherence found in one study, track value not specified but presumably that of a typical road vehicle. As would be expected, the coherence is unity at long wavelengths, falling to zero at short ones. At a vehicle speed of 20 m/s the resonant wavelength for heave is about 14 m, 0.07 cycles/m, where the coherence may be seen to be quite high, 0.4–1.0 for the wide range of types of the three roads reported. At a wheel hop frequency of about 10 Hz, the wavelength is 2 m (depending on the speed), which is a spatial frequency of 0.5 cycles/m, requiring extrapolation of an undulating curve, with a coherence possibly anywhere in the range 0–0.4, perhaps even more in some cases.

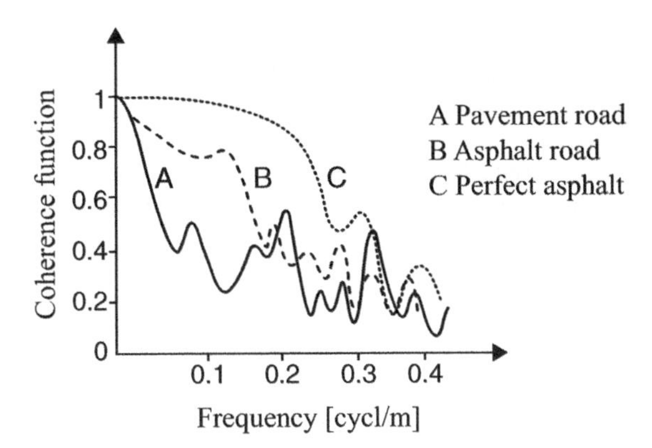

Figure 3.9.4 Coherence between two adjacent tracks (Verschoore et al., (1996). Reproduced with permission from: Verschoore, Duquesne and Kermis (1996) 'Determination of the vibration comfort of vehicles by means of simulation and maeasurement', *EJME* 41(3), pp. 137–143.

The real road is manufactured in a highly directional manner, so there is little reason, *a priori*, to expect the road to be isotropic (properties independent of direction), and in some cases clearly it is not. For example, in soft ground when there are ruts from the passage of the wheels the transverse section has a very strong spectral density at a particular wavelength. Studies of metalled roads do indicate lack of isotropy, but in a highly unsystematic way, so isotropy may still be a reasonable way to anticipate the relationship between adjacent tracks. When there are no actual data, and a road of two or more tracks is to be generated, isotropy is the logical assumption. Figure 3.9.5 shows the coherence between tracks calculated on this basis.

A numerical road of two adjacent tracks is then to be created from a spectral density and a coherence graph. For each spatial frequency in turn, one track is made as for a single track. For a real road, analysis of

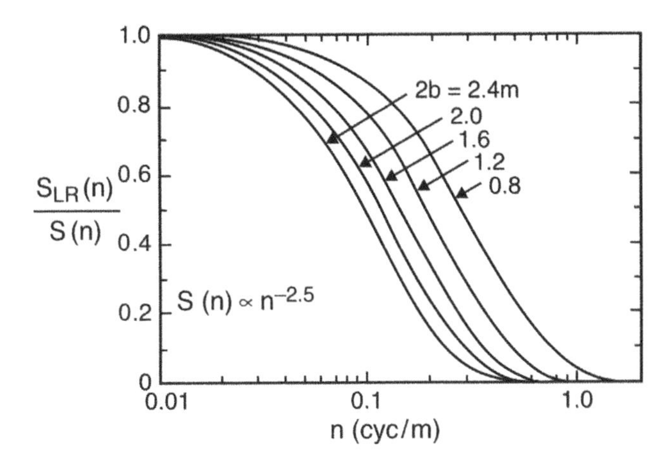

Figure 3.9.5 Coherence between adjacent tracks, on the assumption of a spectral density with W = 2.5 and isotropy. The track spacing is 2b. Reproduced with permission from: J. D. Robson (1979) 'Road surface description and vehicle response', *IJVD* 1(1), pp. 25–35. © Inderscience.

the second path would show a slightly different amplitude, but for the simulated road it is usual, and reasonable, to use the same amplitude for the two tracks. The second track is therefore usually given the same amplitude as the first. The phase difference between the tracks depends on the coherence for that wavelength. However, the coherence is not explicit, and is only a guide to the range of randomness, and actually the distribution is not specified, only the effective mean. For a given wavelength and coherence from the graph, the phase difference ϕ may be taken with reasonable realism as

$$\phi = \pm\,\pi(1 - C_{\mathrm{P}})\mathrm{Rand}$$

with equal probability for either sign, where C_{P} is the coherence between the paths and Rand is a 0–1 random deviate. Alternatively one can use a phase difference of

$$\phi = \pm 2\pi n_{\mathrm{S}} T\ \mathrm{Rand}$$

or

$$\phi = \pi n_{\mathrm{S}} T\{2(\mathrm{Rand} + \mathrm{Rand} + \mathrm{Rand})\}$$

where it should be noted that the sum of three independent Rand calls is not the same as $3 \times$ Rand.

References

Robson J. D. (1979) Road surface description and vehicle response. *International Journal of Vehicle Design*, **1**(1), 25–35.

Verschoore, R., Duquesne, F. and Kermis, L. (1996) Determination of the vibration comfort of vehicles by means of simulation and measurement. *European Journal of Mechnical Engineering*, **41**(3), 137–143.

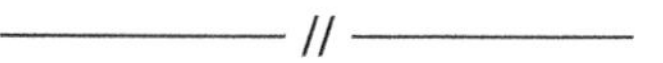

4

Ride Geometry

4.1 Introduction

Ride and handling motions of the body are the results of road roughness and control inputs, and are associated with movement of the suspension. Therefore, it is necessary to specify various aspects of the position and shape of the vehicle body, suspension and tyres. This may be done in a general way applicable to all forms of suspensions, in terms of suspension bump position, body ride heights, etc. The vehicle body ride position is specified by its heave, pitch and roll values. When the car is in combined braking and cornering, the consequent heave, pitch and roll result in a different suspension bump at each wheel.

Bump is upward displacement of a wheel relative to the car body. Droop is negative bump. Heave is symmetrical upward displacement of the body, or of an axle, then often called double bump. These are special cases of the general terminolgy of vehicle positions and velocities, surge (X), sway (Y), heave (Z), yaw (about the vertical axis), pitch (about the lateral horizontal axis) and roll (about the vehicle longitudinal axis).

Bump and body heave influence the wheel camber and steer angles relative to the body and road, and also influence the spring and damper forces and hence the tyre vertical force. These all influence the tyre lateral force. The terms 'bump' and 'droop' may also be applied to velocities to show the direction of motion; in that case it should be borne in mind that a bump velocity may occur in a droop position, and *vice versa*. Bump velocities also affect the slip angle because of the scrub velocity component. Body roll in cornering gives a combination of bump and droop on opposite wheels, relative to the body.

Figure 4.1.1 shows an assortment of tyres with widely varying tread patterns for various applications. For a given vehicle, the best wheel/tyre width and tread form depend on the road material (tarmac, mud, snow, surface water, etc.) and surface shape (roughness, etc.).

4.2 Wheel and Tyre Geometry

Tyre geometry can be an extremely complex subject, particularly for the analysis of carcase reinforcement, which will not be discussed here; this section covers only basic wheel and tyre deflection geometry and terminology. Figure 4.2.1 gives the basic dimensional definitions. Contact patch shapes are generally somewhat more elliptical in shape. The gross contact patch area is the area within the outer non-concave profile. The net area is the area of rubber deemed to be actually in contact with the road, about 75% of the gross area for a normal tyre. The contact patch solidity is the net area over the gross area, and is high for best grip in dry conditions, but low for good water drainage. At a finer scale, of course, even the net area is only partially in actual contact, with the rubber draping over the fine road asperities.

Suspension Geometry and Computation J. C. Dixon

Figure 4.1.1 Some example tyre types and tread forms (Goodyear).

The entire unloaded wheel can be perceived as approximately a section of a sphere, a torus or a cylinder, Figure 4.2.2. For comparison with the highly idealised shapes of Figure 4.2.2, Figure 4.2.3 shows the section of a very early drop-rim tubed tyre, which is almost perfectly toroidal. Figure 4.2.4 shows a wide dry-weather racing tyre section from the days of extreme tyre widths, almost cylindrical in shape to place as much tyre rubber in contact with the ground as possible for maximum grip.

Figure 4.2.5 shows how the sectional shape of the passenger car tyre has changed over the years. This now seems to have largely stabilised, with only sports and high performance cars going below 60% sectional aspect ratio.

The term 'unloaded radius' is self-explanatory, although the value varies a little with inflation pressure and wheel rotational speed. A typical passenger car uses a 12 inch nominal diameter rim (305 mm) carrying a 600 mm outer diameter tyre, i.e. an unloaded radius of 300 mm.

In side view, the tyre is subject to a radial deflection with flattening at the bottom according to the load, Figure 4.2.6. The term 'loaded radius' means the height of the wheel centre above the ground. The loaded radius is therefore the unloaded radius minus the tyre vertical deflection:

$$R_L = R_U - h_T$$

For a passenger car tyre, a typical static deflection is 14 mm.

Unfortunately, the loaded radius is often incorrectly called the rolling radius, which is really the quotient of advance speed of the wheel centre over the angular velocity:

$$R_R = \frac{V_X}{\Omega}$$

The rolling radius is sometimes also known as the effective radius or the effective rolling radius.

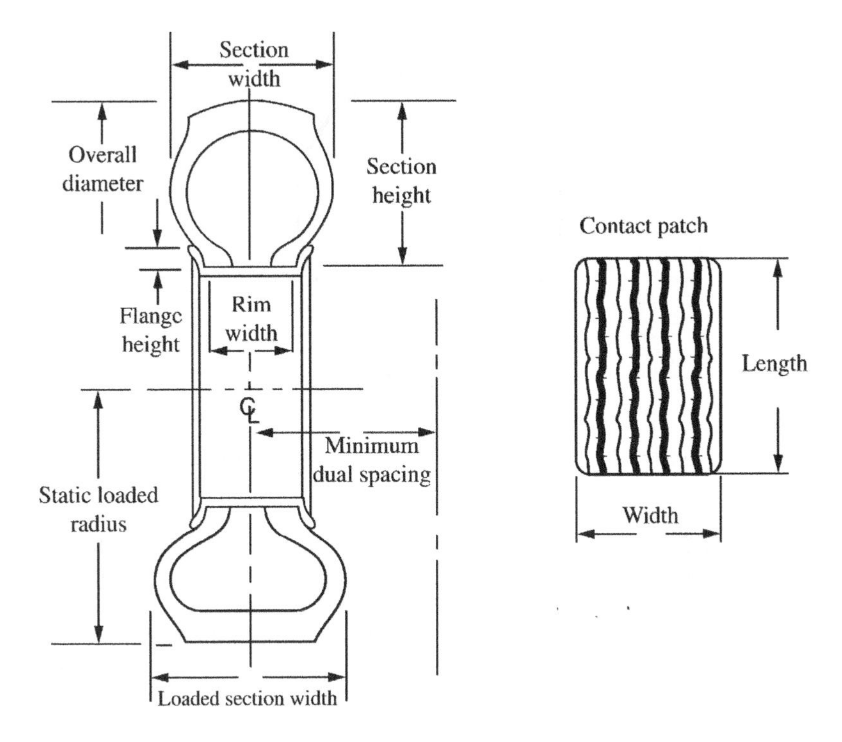

Figure 4.2.1 Tyre and rim leading dimensions.

By simple geometric analysis of Figure 4.2.6, the angle subtended at the wheel centre by the contact patch is

$$\theta_{CP} = 2\,\mathrm{acos}\left(\frac{R_L}{R_U}\right) = 2\,\mathrm{acos}\left(1 - \frac{h_T}{R_U}\right)$$

Typically, this is about 35°. The length of the contact patch is

$$L_{CP} = 2R_U \sin\theta_{CP} \approx \sqrt{8R_U h_T}$$

Typically this gives about 180 mm, in agreement with direct measurement.

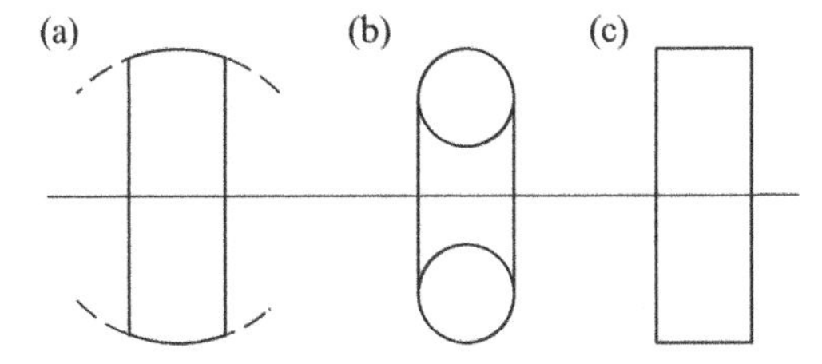

Figure 4.2.2 Basic wheel shapes in front view: (a) spherical; (b) toriodal; (c) cylindrical.

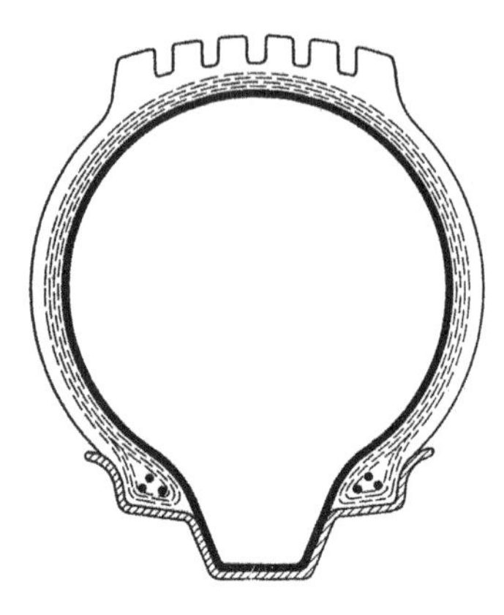

Figure 4.2.3 An early drop-rim wheel with tubed toroidal tyre.

As seen in Figure 4.2.6 and Table 4.2.1, doubling the vertical deflection increases the contact patch length to about 270 mm at a subtended angle of 50°. This covers the usual operating range of varying normal force.

The tyre can be compressed a total of about 70 mm from R_U before it is crushed against the rim, about 56 mm from the static position, although this dimension is sensitive to the tyre aspect ratio. The subtended angle is then 80° with a maximum contact patch length of 385 mm.

Evidently, the contact patch length is not proportional to the deflection. With the tyre normal force mainly (80%) dependent on the inflation pressure, the approximately linear $F_N(h_T)$ characteristic requires that the contact patch width also increases with load, as is reasonable for a spherical tyre form or for a toroidal one, but not for a cylindrical one.

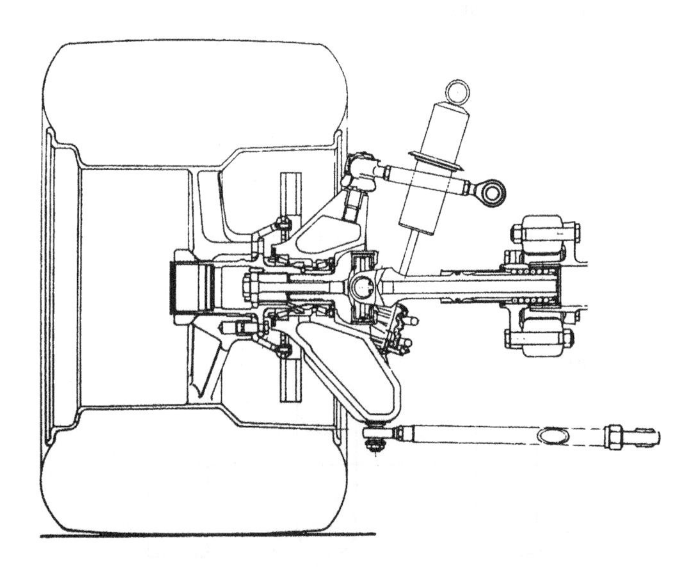

Figure 4.2.4 A sports racing car rear tyre from the ultra-wide era (Porsche 917), exhibiting a near cylindrical form.

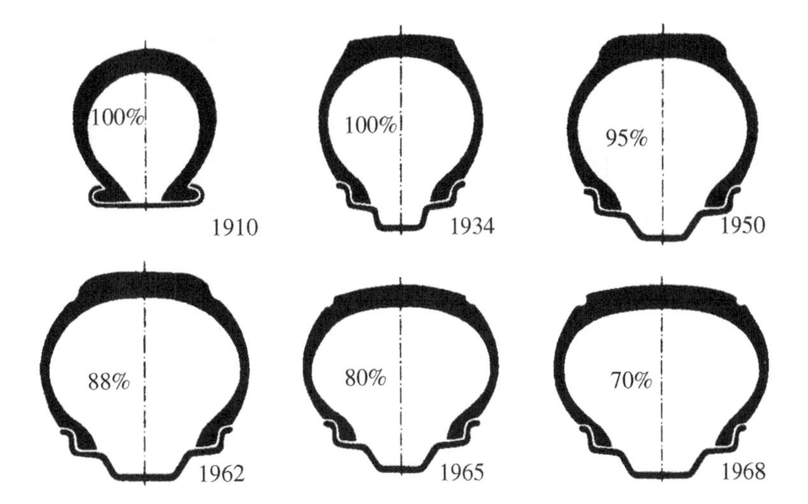

Figure 4.2.5 The progressive development of cross-sectional geometry (profile aspect ratio) of passenger car tyres (Continental).

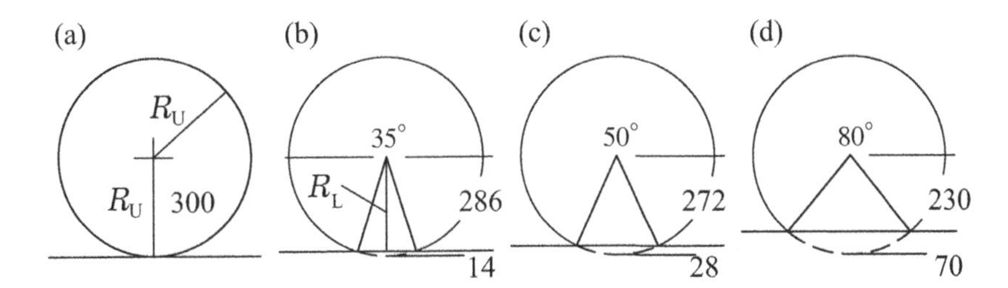

Figure 4.2.6 Idealised tyre deflection geometry.

Where the rolling tyre arrives at and leaves the ground, at the ends of the contact patch, the tread makes a turn through an angle of $\theta/2$. Figure 4.2.7(a) shows the simplified geometry, whilst Figure 4.2.7(b) shows how the idealised tread curvature varies with position, the real graph being somewhat rounded, of course. When the wheel is rolling, at entry to the contact patch the tread is sharply bent into a small radius, and then promptly flattened out to zero curvature. At the rear of the contact patch, the tread is sharply bent and then promptly reverts to the unloaded radius. This severe multiple flexing of the tread is a large factor in the tyre rolling resistance.

Table 4.2.1 Tyre compression geometry (passenger car tyre, $R_U = 300$ mm)

Condition	h_T (mm)	R_L (mm)	θ (deg)	L_{CP} (mm)
Static	14	286	35	181
Double	28	272	50	253
Limit	56	244	80	385

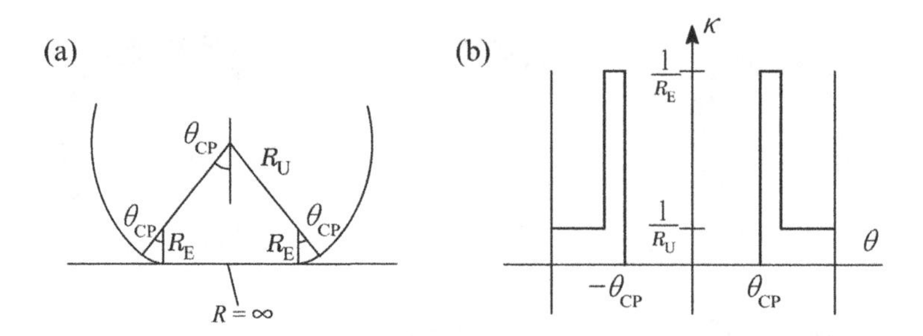

Figure 4.2.7 (a) Tread curvature distortion at the ends of the contact patch. (b) Tread curvature versus position, idealised.

4.3 Suspension Bump

Bump is upward displacement of a wheel relative to the car body, sometimes applied more broadly to mean up or down displacement. It is also known as compression or jounce. The opposite, a lowering of the wheel, is called droop, rebound, extension, or negative bump. Bump is so called because, of course, this is the basic suspension deflection when a single wheel passes over a road bump. When a pair of wheels rises symmetrically this is called double-wheel bump, often referred to simply as double bump. Heave is a vertical upward motion of the body without pitch or roll. A given body heave value then evidently gives equal negative bump all round. Heave is sometimes also known as bounce. Double bump at an axle is equivalent, for suspension deflection, to a negative heave of the body at that axle.

4.4 Ride Positions

The simplest ride model, the heave model (sometimes the so-called quarter-car model), simply has a body mass and wheel mass with associated suspension and tyre stiffnesses and damping. Only heave motion is allowed. For racing cars, the suspension structure compliance may also be significant. The passenger on a seat may be added. Figure 4.4.1 shows such a heave ride model. The terminology ‘quarter-car model’ arises because it appears to have one suspension only. However, the quarter-car model leaves a problem with regard to the mass of the passenger (quarter of a passenger?). It is better to deem it to be a heave model with a mass equal to the whole vehicle body, suspension comprising the total stiffness of four corners, damping comprising the total damping coefficients, and one ‘complete’ passenger on one seat cushion. Even then, however, the effective mass of the passenger resting on the seat is substantially less than the complete human mass because the legs are largely supported by the floor, and the driver’s arms are half supported by the steering wheel.

The parameters of this model, as seen in Figure 4.4.1, are as follows:

(1) passenger (or driver or load) effective mass m_P (usually one);
(2) cushion (seat) vertical stiffness K_C (one);
(3) cushion (seat) vertical damping coefficient C_C (one);
(4) vehicle body (sprung mass) mass m_B (whole car);
(5) suspension vertical stiffness wheel rate K_W (total);
(6) suspension vertical damping coefficient at the wheel C_W (total);
(7) wheel mass (i.e. unsprung mass) m_U (total);
(8) tyre vertical stiffness K_T (total);
(9) tyre vertical damping coefficient C_T (total, usually small).

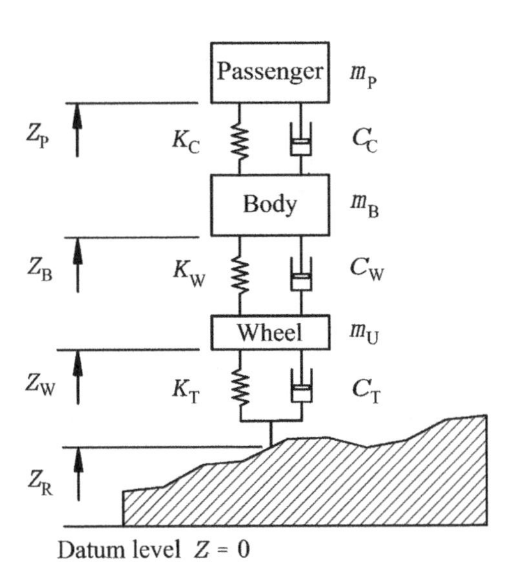

Figure 4.4.1 Heave ride model including passenger, defining inertia, stiffness and damping coefficient notation and ride positions.

Geometrically, measured from the datum level at $Z=0$, there are the following vertical positions of defining points on the masses:

(1) road height (from 'roughness') Z_R;
(2) wheel centre height Z_W;
(3) body ride height Z_B;
(4) passenger ride height Z_P.

The variations of these from the static values are the ride displacements, for example

$$z_B = Z_B - Z_{B0}$$

and are as follows:

(1) passenger ride displacement z_P;
(2) body ride displacement z_B;
(3) wheel ride displacement z_W.

There are also:

(1) tyre ride deflection $z_T = z_R - z_W$;
(2) suspension ride deflection (bump) $z_S = z_W - z_B$;
(3) cushion (seat) ride deflection $z_C = z_B - z_P$.

Expressing the above displacement equations in a summation form,

$$z_R = z_T + z_W$$
$$z_W = z_S + z_B$$
$$z_B = z_C + z_P$$

The ride displacement of the road can therefore be expressed in three simple ways:

$$z_R = z_T + z_W$$

$$z_R = z_T + z_S + z_B$$

$$z_R = z_T + z_S + z_C + z_P$$

4.5 Pitch

Pitch is the angle of the body in rotation about a transverse axis (i.e. nose up and nose down), and is associated with braking and acceleration transients and response to road roughness. Relative to the static position, the body pitch angle is θ_B, often considered positive with the nose up. This is a positive right-hand rotation about a transverse vehicle-fixed axis to the right, as in the Society of Automotive Engineers (SAE) axis system. However, in the ISO system, the transverse (Y) axis is to the left, so arguably positive pitch angle should then be positive nose-down. It is not essential to follow the right-hand rule for rotations, although in dynamic analysis it is safer to do so. However, the signs can easily be adjusted. In the application of equations, care is required with this sign convention, which will be that positive pitch is nose up.

In combination with the road elevations at the axles, the body pitch angle requires ride deflections at the axles, accounted for mainly by the suspension, and partly by tyre deflections.

4.6 Roll

Body roll is rotation about the body longitudinal axis, arising from cornering activity and road roughness. In both SAE and ISO systems, the longitudinal axis is forwards, so roll angle is positive for rotations clockwise seen from the rear, i.e. body down on the right-hand side, as in a left-hand turn. Suspension roll is formally defined (SAE) as rotation of the vehicle sprung mass about a fore–aft axis with respect to a transverse line joining a pair of wheel centres. This is unambiguous provided that the ground is flat and that front and rear wheel centres have parallel transverse lines (e.g. that wheels and tyres are the same size side-to-side). If the ground is not flat then some mean ground plane must be adopted. The roll angle and roll velocity are in practice fairly clear concepts. Asymmetries, such as the driver, mean that the roll angle is not automatically zero under reference conditions, although it is usual practice to work in terms of the roll relative to the static position.

Body roll is accommodated by suspension roll, plus some axle roll from tyre deflection, Figure 4.6.1. On a level road (i.e. zero road bank angle)

$$\phi_B = \phi_S + \phi_A$$

For a passenger car, the axle roll is generally fairly small compared with the suspension roll, but this is certainly not true for many racing cars with stiff suspensions, where inclusion of the axle roll angles is essential.

For a torsionally-rigid vehicle body, in relation to the torsional moments applied, the body torsion angle ϕ_{BT} is negligible. This is applicable to most passenger cars, in which case

$$\phi_{BT} = 0$$

$$\phi_{Bf} = \phi_{Br}$$

$$\phi_{Bf} = \phi_{Sf} + \phi_{Af}$$

$$\phi_{Br} = \phi_{Sr} + \phi_{Ar}$$

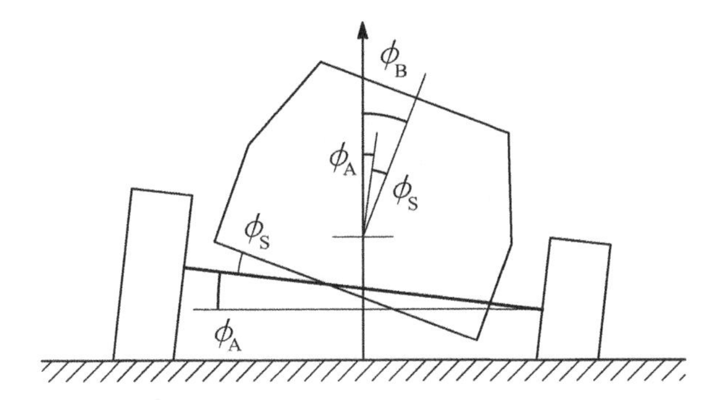

Figure 4.6.1 Body roll, suspension roll and axle roll on a flat, level road, rear view.

In the case of most trucks, and some racing cars, the body torsion angle ϕ_{BT} needs to be considered, in which case

$$\phi_{Bf} = \phi_{Sf} + \phi_{Af}$$
$$\phi_{Br} = \phi_{Sr} + \phi_{Ar}$$
$$\phi_{BT} = \phi_{Bf} - \phi_{Br}$$

The roll gradient is the rate of change of roll angle ϕ with lateral acceleration A:

$$k_\phi = \frac{d\phi}{dA}$$

The body roll gradient is the sum of suspension and axle roll gradients:

$$k_{B\phi} = k_{S\phi} + k_{A\phi}$$

Roll is geometrically equivalent to bump of one wheel and droop of the opposite one, relative to the body. In a rolled position, suspension geometry is generally such that there are changes of wheel steer angles relative to the body. This is roll steer, dealt with in Chapter 5. It is equivalent to bump steer for independent suspension, but not for solid axles. The rolled body position also results in wheel camber relative to the body, and more importantly a wheel camber angle relative to the road, introducing camber forces. In general, in the rolled position the spring stiffnesses on the two sides are different, usually greater on the lower side, so there is some heave of the body in consequence of roll. This is spring jacking. If a droop stop is engaged first, the body may be lowered. There are other jacking effects through the links in cornering and through damper action. Roll speed generally results in a scrub speed of the tyres relative to the ground, causing temporary changes to slip angles and hence to tyre cornering forces.

If the road is not flat and level then additional analysis is required. The first case is a flat but non-level road, as for example on a banked high-speed road. At a road bank angle ϕ_R, Figure 4.6.2, the body roll relative to the datum plane is

$$\phi_B = \phi_R + \phi_A + \phi_S$$

If the road is not flat, then the road bank angles are different at the two axles. Including the possibility of body torsion, then,

$$\phi_{Bf} = \phi_{Rf} + \phi_{Af} + \phi_{Sf}$$
$$\phi_{Br} = \phi_{Rr} + \phi_{Ar} + \phi_{Sr}$$

The road torsion and body torsion angles are

$$\phi_{RT} = \phi_{Rf} - \phi_{Rr}$$
$$\phi_{BT} = \phi_{Bf} - \phi_{Br}$$

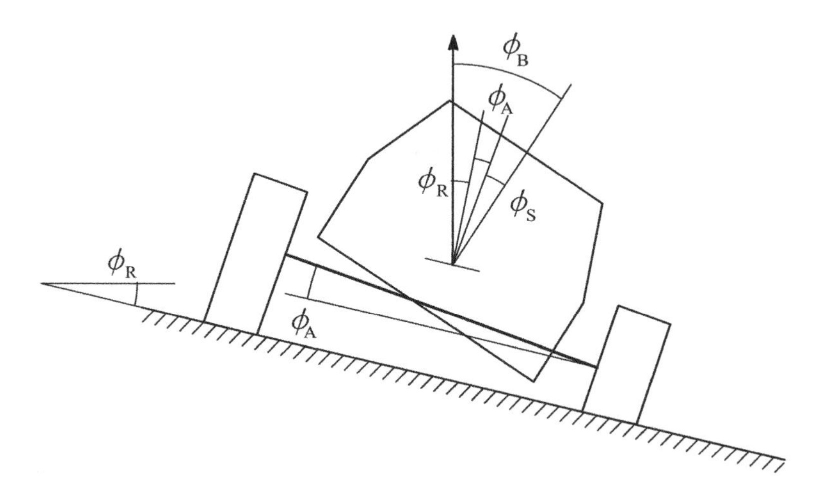

Figure 4.6.2 Body roll, suspension roll and axle roll on a flat, non-level road, rear view.

4.7 Ride Height

Ride height is the position of the body (sprung mass) above the basic ground level datum plane. The dynamic ride height is the ride height relative to the static position. In general, during acceleration, braking or cornering and on rough roads, the ride heights vary continuously and have different values at each wheel. The set-up of the vehicle on an accurate level surface is the static ride height, in practice frequently referred to simply as ride height. Values are specified for the front and rear of the vehicle. Normally it is assumed that the vehicle should have zero body roll in the static position, although some static roll may be specified in special cases, such as racing vehicles on tracks with turns predominantly in one direction.

The front and rear ride heights may be specified at any points that are convenient for measurement. Hence particular points on the front and rear bodywork may be used. However, from the point of view of vehicle dynamics, it is the ride heights at the wheels that are of importance. Hence, for passenger cars it is convenient to measure the ride height at the highest point of the wheel arch opening, the so-called eyebrow level. For ground-effect racing cars, the rules frequently specify a flat, or partially flat, underbody, in which case this flat plane is often used to define the ride heights. Alternatively, measurements may be made to the inner axis of the bottom suspension arm.

Under running conditions, the ride heights or vertical positions of the body (sprung mass), of the wheel and of the local road position are measured from a datum level, typically the mean road plane. Figure 4.7.1 shows this for one suspension unit. Hence there are vertical positions Z_B to some reference point on the body, Z_W to the wheel centre and Z_R to the road height above datum. Each of these has four values, one at each wheel specified by appropriate subscripts, normally either:

(1) f and r for front and rear, with L and R for left and right, or i and o for inner and outer.
(2) 1, . . ., 4 in the order

1: left front
2: right front
3: left rear
4: right rear

as in Figure 4.7.2.

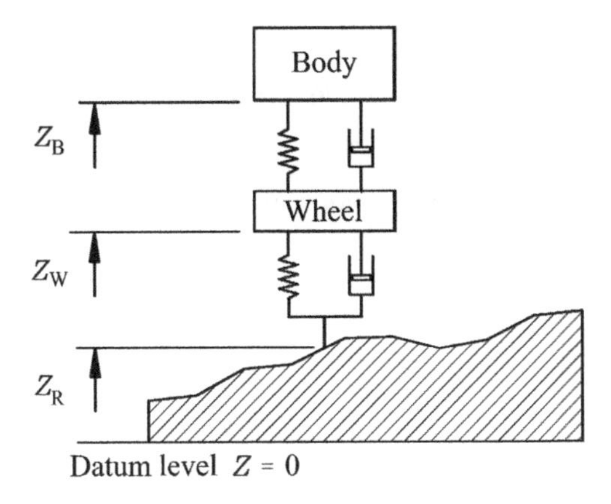

Figure 4.7.1 Body, wheel and road positions above the datum level.

The static (subscript 0) ride-height positions of body, wheel and road are Z_{B0}, Z_{W0}, and Z_{R0}. The increase of these in dynamic, running condition compared with the static condition, represented by lower case z, becomes

$$z_B = Z_B - Z_{B0}$$

$$z_W = Z_W - Z_{W0}$$

In the case of the road, fixed in position, Z_{R0} may most conveniently be taken as zero, so

$$z_R = Z_R - Z_{R0} = Z_R$$

The unloaded tyre radius must be

$$R_U = Z_{W0} + h_{T0}$$

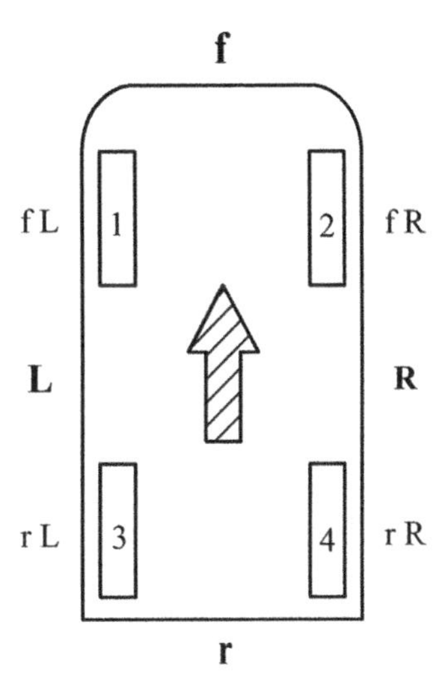

Figure 4.7.2 Notations for the four wheels.

so the static tyre deflection is

$$h_{T0} = R_U - Z_{W0}$$

The static loaded radius is

$$R_{L0} = Z_{W0} = R_U - h_{T0}$$

In running conditions, the loaded radius is

$$R_L = Z_W - Z_R$$

The running tyre deflection is

$$h_T \equiv R_U - R_L$$

which, however, cannot be negative. This becomes

$$h_T = h_{T0} - Z_W + Z_R$$

The increase of tyre deflection z_T is simply

$$z_T = z_R - z_W$$

Frequently in handling analysis, the road is deemed to be smooth and level, giving

$$\begin{aligned} z_R &= 0 \\ h_T &= h_{T0} - z_W \\ z_T &= -z_W \end{aligned}$$

The suspension bump deflection is

$$z_S = z_W - z_B$$

The basic body height is measured at the centre of mass, giving a single value for Z_B. Incorporating the body pitch angle θ_B (relative to static θ_{B1}), the body roll angle ϕ_B (the body not assumed torsionally rigid, the static roll assumed to be zero) then changes the body height at the four wheels, that is, the body ride heights are, with pitch angle sign convention positive nose-up,

$$\begin{aligned} z_{BfL} &= z_B + a \sin\theta_B + \tfrac{1}{2} T_f \sin\phi_{Bf} \\ z_{BfR} &= z_B + a \sin\theta_B - \tfrac{1}{2} T_f \sin\phi_{Bf} \\ z_{BrL} &= z_B - b \sin\theta_B + \tfrac{1}{2} T_r \sin\phi_{Br} \\ z_{BrR} &= z_B - b \sin\theta_B - \tfrac{1}{2} T_r \sin\phi_{Br} \end{aligned}$$

where a and b are the front and rear partial wheelbases. The pitch and roll angles are generally moderately small, with $\sin\theta \approx \theta$, so it is common practice to use the computationally much more efficient

$$\begin{aligned} z_{BfL} &= z_B + a\theta_B + \tfrac{1}{2} T_f \phi_{Bf} \\ z_{BfR} &= z_B + a\theta_B - \tfrac{1}{2} T_f \phi_{Bf} \\ z_{BrL} &= z_B - b\theta_B + \tfrac{1}{2} T_r \phi_{Br} \\ z_{BrR} &= z_B - b\theta_B - \tfrac{1}{2} T_r \phi_{Br} \end{aligned}$$

Normal practice would be to precalculate the front and rear mean ride heights, and the half-track values too of course, with

$$z_{\mathrm{Bf}} = z_{\mathrm{B}} + a\,\theta_{\mathrm{B}}$$
$$z_{\mathrm{Br}} = z_{\mathrm{B}} - b\,\theta_{\mathrm{B}}$$

$$z_{\mathrm{BfL}} = z_{\mathrm{Bf}} + \left(\tfrac{1}{2}T_{\mathrm{f}}\right)\phi_{\mathrm{Bf}}$$
$$z_{\mathrm{BfR}} = z_{\mathrm{Bf}} - \left(\tfrac{1}{2}T_{\mathrm{f}}\right)\phi_{\mathrm{Bf}}$$
$$z_{\mathrm{BrL}} = z_{\mathrm{Br}} + \left(\tfrac{1}{2}T_{\mathrm{r}}\right)\phi_{\mathrm{Br}}$$
$$z_{\mathrm{BrR}} = z_{\mathrm{Br}} - \left(\tfrac{1}{2}T_{\mathrm{r}}\right)\phi_{\mathrm{Br}}$$

There are similar expressions, but with some sign changes, for the small deflections of the unsprung mass considered normally in terms of a pitch angle and front and rear roll angles. These are tyre, unsprung mass or axle, pitch and roll angles and deflections. In practice they are usually called tyre deflections, axle roll angles, and unsprung pitch angle.

The actual body ride positions are given by equations of the form

$$Z_{\mathrm{BfL}} = Z_{\mathrm{BfL0}} + z_{\mathrm{BfL}}$$

With dynamic solution of the tyre deflections, and hence wheel heights, the suspension bump deflections during running are then of the form

$$z_{\mathrm{SfL}} = z_{\mathrm{WfL}} - z_{\mathrm{BfL}}$$

which is positive for bump (suspension compression).

Similar equations to the above apply for velocities. In a time-stepping computer simulation the velocities will usually be obtained by integrating the acceleration over the time step, or possibly in some cases from displacements divided by the time increment.

For any given body position (heave, pitch, roll) the four ride heights and ride velocities can be evaluated as above. In a handling analysis, the frequency range considered will normally be below the wheel hop frequency, so in steady-state handling the wheel will often be assumed to adopt an equilibrium height according to the body ride height, the suspension stiffness and the tyre vertical stiffness. Hence the position and velocities are determined, from which the suspension stiffness and damping forces and tyre forces follow.

4.8 Time-Domain Ride Analysis

The road stimulus considered may be a simple sinusoid to investigate the vehicle behaviour at a given frequency. This may also be done analytically, but with computer time-stepping it is feasible to have non-linear components. In this case the resulting positions and displacements vary sinusoidally (for linear components) or approximately sinusoidally (for non-linear) at the forcing frequency, and the ratio of the response amplitude to the road amplitude gives a transmissibility to each of the wheel, body and passenger for that frequency.

Alternatively, the simulated road may be given a white or pink noise characteristic, at least over a frequency range of interest, say up to 15 Hz; the response of each mass then has a semi-random nature which can be statistically analysed.

With any stochastic road, there will be an overall root-mean-square (rms) displacement of the suspension, which can be divided by the road rms displacement to give an rms transmissibility for these conditions. Spectral analysis will also reveal certain resonances and responses; for example, there will be an increased body response at the body resonant and wheel hop resonant frequencies.

A third option for the model road, frequently favoured, is to give the road a random nature based statistically on the character of real roads, in which case the responses can be regarded as more realistic. The random response of the bodies can then be analysed. The most important results are those for:

(1) the passenger rms response (for passenger comfort, i.e. ride),
(2) the tyre deflection rms variation (for tyre comfort, i.e. handling),
(3) the suspension rms displacement (for suspension comfort, i.e. workspace).

The assessment of passenger ride perception is called the 'passenger discomfort' D_P. This is normally defined as the rms acceleration experienced by the passenger, normalised by dividing by g, standard gravity (ISO 9.80665 m/s^2, Imperial 32.174 ft/s^2 or US 386 in/s^2):

$$D_P = \frac{A_{rms}}{g}$$

Multiplying by the passenger mass, it is seen that this is also equal to the rms variation of vertical force supporting the passenger divided by the mean vertical force. Also, there is a direct geometrical interpretation, For a linear cushion, it is the rms variation of cushion deflection over the mean value.

A ride simulation also gives a figure for the rms variation of tyre vertical force. The tyre discomfort is

$$D_T = \frac{(\Delta_{FVT})_{rms}}{F_{VT,mean}}$$

Here F_{VT} is the vertical force exerted by the road on the tyre. Hence the tyre discomfort is the rms value of the vertical force fluctuations divided by the mean vertical force. Again, for a linear tyre there is a direct geometrical interpretation: it is the rms variation of tyre deflection divided by the mean deflection.

The basic definition of passenger and tyre discomfort corresponds to the statistician's coefficient of variation, or the normalised standard deviation, the standard deviation divided by the mean. The rms value of a Gaussian distribution with zero mean is one standard deviation.

The suspension deflection or workspace can also be expressed in various ways as a suspension discomfort.

4.9 Frequency-Domain Ride Analysis

Frequency-domain analysis considers the behaviour of the vehicle in terms of its response at any given frequency of stimulus. Such analysis therefore produces results such as graphs of the transmissibility against frequency.

For a simple linear system the behaviour at a given frequency may be obtained analytically. For more complex models, or with non-linear components, it may be obtained by time-stepping analysis at each of a series of frequencies of sinusoidal road profile or, more efficiently in general, by a broad spectrum stimulus, using a road with some spectral distribution, with the response being spectrally analysed, the transmissibility then being the amplitude ratio of response to stimulus, for each frequency of interest through the spectrum.

The geometry of the stimulating road, then, may be sinusoidal, with various spatial frequencies taken in turn, or stochastic, being either an actual road profile trace, or, more likely, a profile generated by inverse Fourier transform from a spectral density definition such as the standard ISO road, as discussed in Chapter 3.

4.10 Workspace

The tyre, suspension and seat cushion are often analysed as simple linear components, but even then they have a limited range of action. The tyre will leave the ground at a negative dynamic deflection equal to the static value. The suspension can hit the bump stops or droop stops, and the seat cushion will also have its limits. These define the available workspace for each compliant component. The workspace in use is the result of the dynamic response of the system to the road roughness. Typically the suspension bump deflection will be analysed statistically to give a standard deviation of position for a given road. The workspace requirement will then be considered to be about 3 standard deviations in each direction, which will occasionally be exceeded. Similarly, the workspace requirement of the tyre and seat cushion can be analysed. In practice, the suspension workspace requirement can be reduced by the use of a stiffer spring, so there is a trade-off between the softness and comfort of the central action of the suspension and more frequent limit impacts. A higher speed of travel increases the stimulus, because of the character of the road spectrum, so high-speed vehicles need stiffer suspension for comfort, avoiding limit stop impacts, not just for handling qualities.

With a typical passenger car static tyre deflection of 14 mm, if the 3σ workspace exceeds this value then the tyre will leave the ground intermittently. This is a standard deviation of less than 5 mm.

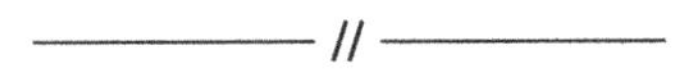

5

Vehicle Steering

5.1 Introduction

Directional control of a road vehicle is normally achieved by steering the front wheels, that is, by rotating them about a roughly vertical axis. The effective steer rotation of the wheels relative to the body centreline is represented by δ (delta). This is mainly the result of steering handwheel movement by the driver, but partly the result of suspension characteristics (bump steer, roll steer and compliance steer). Rear-wheel steering is generally unsafe at high speeds, but because of its convenience in manoeuvring it is sometimes used on specialist low-speed vehicles, such as dumper trucks. This is in contrast to aircraft, which usually have tailplanes and are rear steered by elevators and rudder. Some interest has been shown in variable rear steering for passenger cars, supplementing conventional front steering, and this is has been commercially available.

In the early days of motoring, various hand controls were tried for the driver. Tiller steering, as used on boats, for example as on the 1894 Panhard-Levassor in Figure 5.1.1, was satisfactory at low speed, but often dangerously unstable at high speed. The wheels steered in the same sense as the tiller, so the tiller was moved to the right to make a left turn. The driver then tended to fall centrifugally to the right, exaggerating the turn, sometimes leading to an unstable divergence with disastrous roll-over. Tiller steering was greatly improved by Lanchester who reversed its action to give better stability. It was Benz who introduced the steering wheel, and this was almost universal by 1900. Tests in other control applications show that the hand wheel is the best way to combine rapid large movements with fine precision. For cars, the road wheels steer through a total angle of about 70°, and the steering wheel through three and a half turns, requiring a gear ratio of 18. The average is actually about 17 for power-assisted steering and 21 for unpowered steering. For trucks the steering-wheel movement and gear ratio are about twice those for cars. A high steering gear ratio means a smaller steering force at the hand wheel rim, but a greater angular movement is needed.

The steering of the road wheels is not governed exclusively by the steering hand wheel position. In addition there are:

(1) toe angle – the initial fixed settings;
(2) suspension geometric steer arising from vertical wheel motion coupled with the linkage geometry, expressed through bump steer or roll steer;
(3) suspension compliance steer, resulting from forces or moments and link compliance, mainly in the rubber bushes;
(4) steering system compliance, mainly arising from torsional compliance of the steering column.

Suspension Geometry and Computation J. C. Dixon

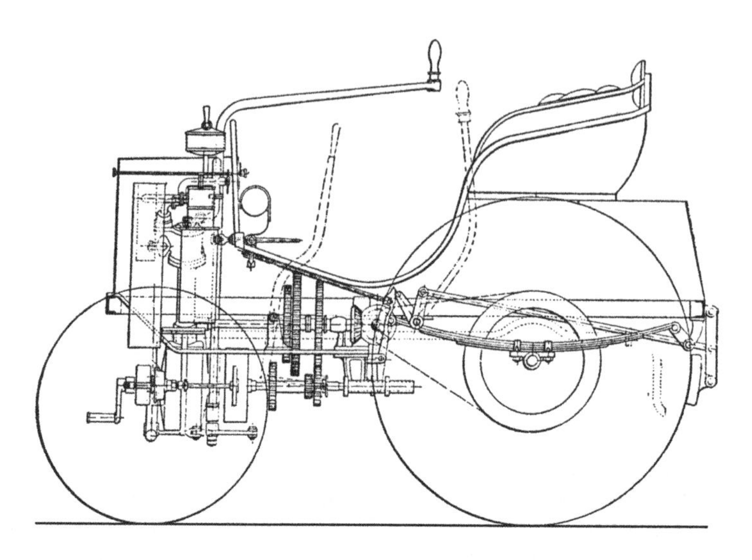

Figure 5.1.1 The front-engined 1894 Panhard-Levassor with tiller steering (uppermost lever) representative of the very early period.

Depending on the design, sometimes other compliances should be included, such as the long tie rod on a rigid axle.

The wheel camber angles are also affected by geometric effects (bump camber and roll camber) and by the flexibility (compliance camber). Bump steer and roll steer are analysed in Chapter 7, and compliance steer in Chapter 9.

Practical truck steering systems are discussed further in Durstine (1973).

5.2 Turning Geometry – Single Track

When a vehicle moves in a curved path at a very low speed the lateral acceleration is very small, so the body roll and the axle lateral forces are negligible. Thus the wheel angles are those arising geometrically, not from the need to produce lateral force. There may, however, be opposing slip angles on the two ends of an axle, giving zero net force. This non-dynamic geometric motion is kinematic turning.

Temporarily neglecting the vehicle width, a single-track model of the vehicle can be used, as in Figure 5.2.1. This is sometimes called a 'bicycle' model, which is correct in the literal sense of bicycle as having two wheels, but possibly misleading in that the intended model vehicle does not have roll banking in cornering as does a common physical bicycle.

The low-speed mean steer angle δ_K required for such a corner is called the kinematic steer angle:

$$\delta_K = \arctan\left(\frac{L}{R_r}\right) \approx \arctan\left(\frac{L}{R}\right) \approx \frac{L}{R} \tag{5.2.1}$$

This is the steer angle of the single front wheel in Figure 5.2.1. It is also the mean steer angle of the two wheels on a real front axle. The SAE standards call this the Ackermann angle, a regrettable choice because it is nothing to do with Ackermann geometry, which relates to the difference of the two front wheel steer angles.

The centres of the two ends of the vehicle corner at different radii. This is one reason why four-wheel-drive vehicles need a centre differential or front over-run clutch. The front cornering radius, as seen in

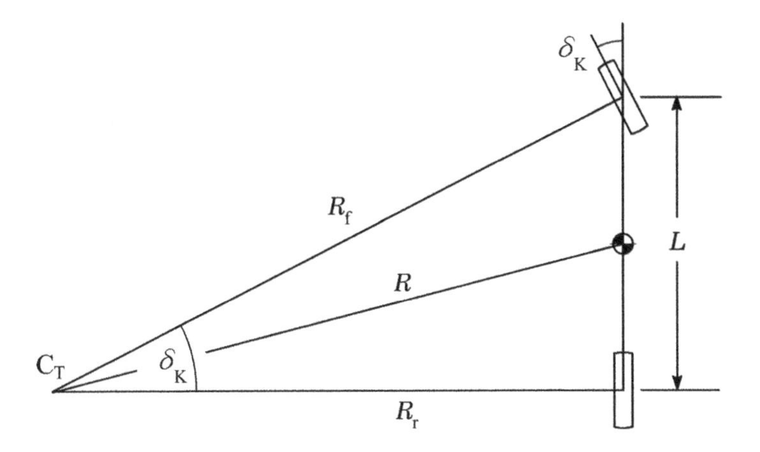

Figure 5.2.1 A single-track (upright bicycle) vehicle model, showing kinematic turning (low speed, no lateral forces, zero tyre slip angles), turning centre C_T.

Figure 5.2.1, is

$$R_f = \frac{L}{\sin\delta_K} \tag{5.2.2}$$

The rear cornering radius is

$$R_r = \frac{L}{\tan\delta_K} \tag{5.2.3}$$

Several other path radius values are also used. The SAE cornering radius is defined to be to the centre of the outer front wheel. The clearance radius is the radius needed to clear the bodywork. In vehicle dynamic analysis, the path radius of the centre of mass is the main one of interest.

The difference of the front and rear central radii is known as the offtracking. The offtracking radius R_{OT} is

$$R_{OT} = R_f - R_r = \frac{L}{\sin\delta} - \frac{L}{\tan\delta} \tag{5.2.4}$$

At low speed, with kinematic turning,

$$R_f^2 = R_r^2 + L^2 \tag{5.2.5}$$

Using

$$L^2 = R_f^2 - R_r^2 = (R_f + R_r)(R_f - R_r) \tag{5.2.6}$$

the offtracking can also be expressed as

$$R_{OT} = R_f - R_r = \frac{L^2}{R_f + R_r} \approx \frac{L^2}{2R} \tag{5.2.7}$$

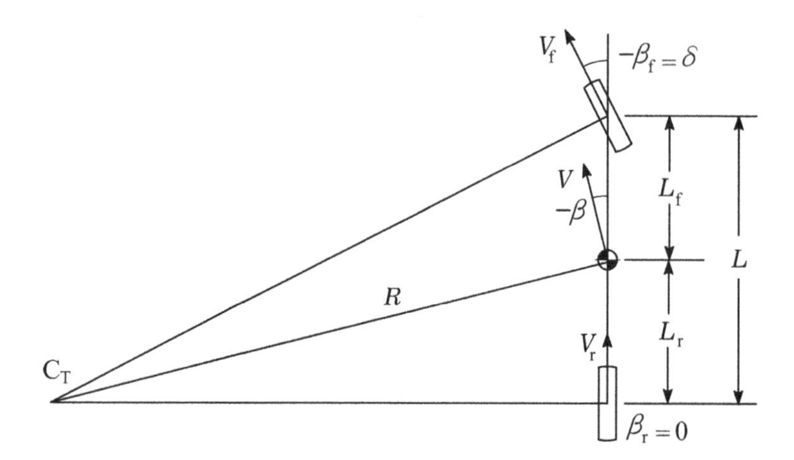

Figure 5.2.2 Attitude angles at a given point of the vehicle depend on the local velocities, as seen here for kinematic turning.

where R is the turn radius of the centre of mass. Example values for a passener car are $L = 2.8$ m, $\delta = 35°$, $R_r = 4$ m, $R_f = 4.88$ m, $R_{OT} = 0.9$ m, so the offtracking can be quite large, leading to occasional kerbing of the rear wheels which 'cut the corner'. It is more problematic for long trucks, and especially for trailers.

The attitude angle β at a given point of the vehicle is the angle between the vehicle centreline and the local velocity vector, which is perpendicular to the radial lines shown in Figure 5.2.2, in which it may be seen that at low speed (negligible lateral acceleration, zero tyre slip angles) the attitude angle β_r at the rear axle is zero. Attitude angle is positive with the vehicle pointing towards the inside of the corner, so the angles are negative in the figure. At the front axle

$$\beta_f = -\arcsin\frac{L}{R} \approx -\frac{L}{R} \tag{5.2.8}$$

At the centre of mass, where the attitude angle is usually measured, the attitude angle is

$$\beta \approx -\frac{L_r}{R} = -\frac{L_r}{L}\delta_K \tag{5.2.9}$$

Hence the attitude angle at the centre of mass takes a negative value under kinematic conditions; however, it is positive-going as lateral acceleration develops, because of the increasing slip angle at the rear axle.

At significant speed, in dynamic cornering, tyre slip angles are required to produce the necessary cornering forces. Then attitude angle develops, and the offtracking also changes, Figure 5.2.3. The front and rear tyre slip angles are denoted by α_f and α_r, respectively. The angles are exaggerated in this figure for clarity of the principle. Tyre slip angles are generally just a few degrees, not often exceeding 3° in everyday motoring. Also, in normal cornering the path radius is much larger in relation to the wheelbase. The turning centre point C_T can be seen to have moved forward from the rear axle line, or the vehicle can be considered to have rotated to a new attitude angle relative to the radius line from the turning centre. The attitude angles are then

$$\begin{aligned} \beta_r &= \alpha_r \\ \beta &= -\frac{L_r}{R} + \alpha_r \\ \beta_f &= -\frac{L}{R} + \alpha_r \end{aligned} \tag{5.2.10}$$

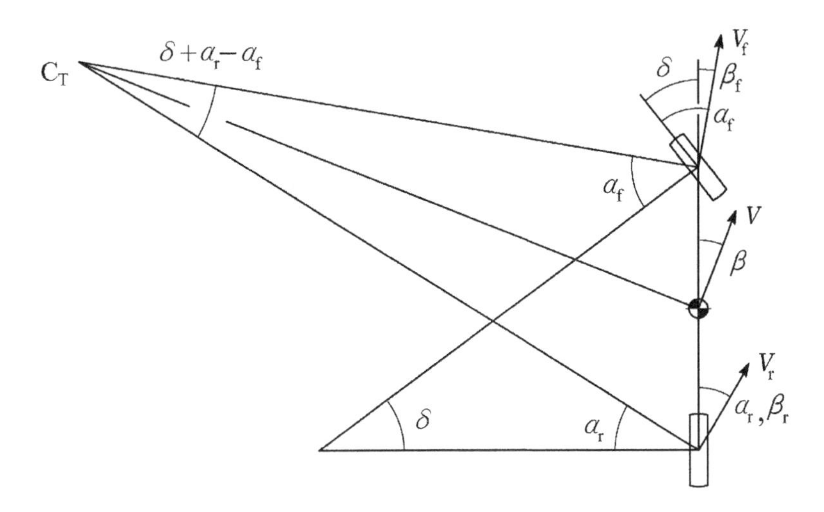

Figure 5.2.3 Single-track model in dynamic turning, with front and rear tyre slip angles α_f and α_r.

The condition of zero offtracking will occur when C_T has moved forward half of the wheelbase. This is possible for some particular speed depending on the cornering radius, according to tyre characteristics etc., and occurs at typically 10 m/s at 30 m radius. At approximately the same speed, the attitude angle at the centre of mass becomes zero. For a greater speed, the offtracking becomes negative, with the rear of the vehicle following a path of greater radius than the front.

5.3 Ackermann Factor

Consider now a vehicle with width, having axle track (tread). For a normal vehicle with front steering, to turn with zero slip angles means that the turning centre C_T must still be in line with the rear axle, Figure 5.3.1. The front wheels must be steered by different amounts, the inner wheel more, in order for both of them to have zero slip angle. Although actually invented by Darwin, Langensperger or others (see Chapter 1), this is widely known as the Ackermann steering concept, and is desirable for low-speed manoeuvring to avoid tyre scrub, minimising motion resistance, tyre wear and surface damage. Various geometries are used in practice, more or less related to the basic Ackermann geometry.

For an axle, the difference between the steer angles of the two wheels to give zero or equal slip angles equals the Langensperger angle λ subtended at the turning centre by the axle. Positive Langensperger angle is with the axle turned out of the corner. In Figure 5.3.1, with kinematic turning, which has zero mean tyre slip angles, evidently the Langensperger angle is positive for the front axle and zero for the rear axle. For the front axle

$$\lambda_{f0} = -\frac{T_f \sin\beta_f}{R_f} \tag{5.3.1}$$

where, kinematically, β_f is negative and λ_{f0} is positive. Substituting for the front attitude angle, kinematically

$$\lambda_{f0} = -T_f\left(-\frac{L}{R_f}\right)\frac{1}{R_f} \approx \frac{T_f L}{R^2} \tag{5.3.2}$$

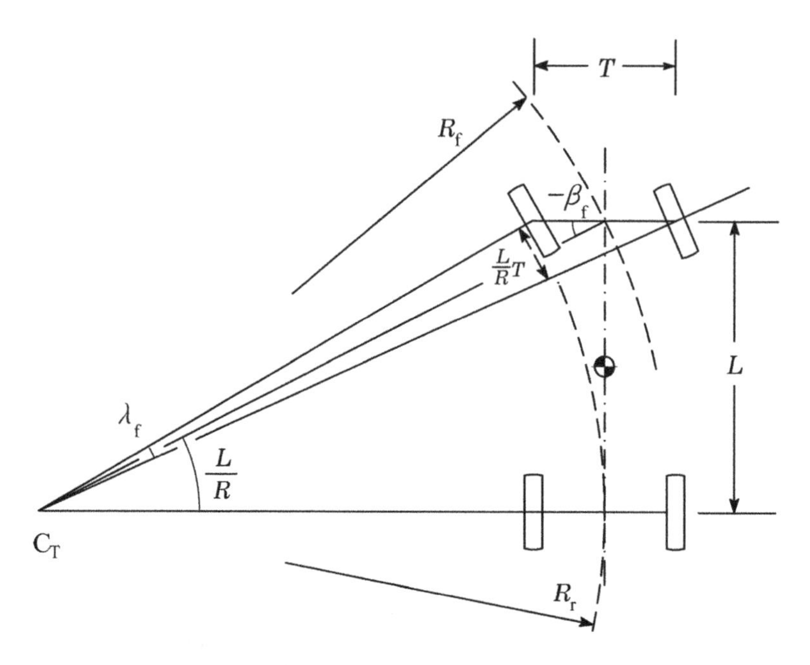

Figure 5.3.1 Turning geometry of a conventional passenger car at very low speed (negligible lateral acceleration).

The basic front Langensperger angle is therefore simply TL/R^2, this giving the angle difference (in radians) required for full Ackermann steering effect. The actual steer angle difference occurring in a given condition, according to the actual steering mechanism geometry, is

$$\delta_{L-R} = \delta_L - \delta_R \tag{5.3.3}$$

For true Ackermann steering, then,

$$\delta_{L-R} = \lambda_{f0} \tag{5.3.4}$$

for all steer angles, where λ_{f0} is the Langensperger angle at zero lateral acceleration. The Ackermann factor is then the ratio of the actual steer angle difference to the ideal difference:

$$f_A = \frac{\delta_{L-R}}{\lambda_{f0}} \tag{5.3.5}$$

which is zero for parallel steering, and 1.0 (100%) for true Ackermann. For a real steering arrangement it is only approximately a constant.

The difference between δ_{L-R} and λ_0 is sometimes called the tie-rod geometry error:

$$\theta_{TRGE} = \delta_{L-R} - \lambda_{f0} = (f_A - 1)\lambda_{f0} \tag{5.3.6}$$

For vehicles that do a great deal of turning or frequently need a very small turning radius, such as purpose-built taxis and urban delivery vehicles, full Ackermann is often used. The Apollo 'Lunar Rover' vehicle had four-wheel steering, giving a minimum turn radius equal to the wheelbase of 2.286 m, with close to full Ackermann at both front and rear, to minimise resistance in tight turns on the soft ground. The traditional London taxi has almost perfect Ackermann over its full 60° of inner road-wheel steer angle.

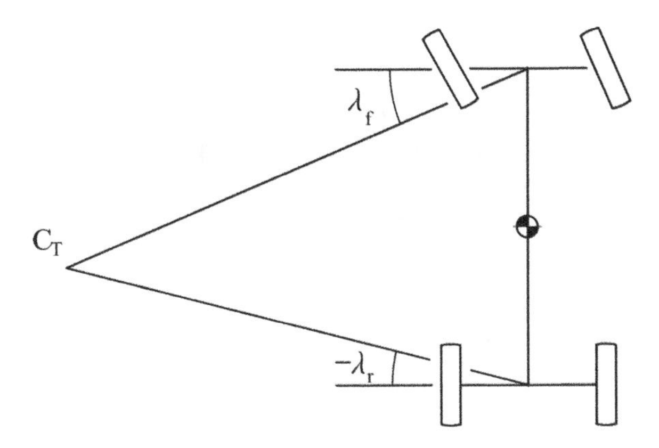

Figure 5.3.2 Under dynamic cornering conditions, the axle Langensperger angles are changed, reducing or going negative as attitude angle is introduced by lateral acceleration.

There is less of a case for Ackermann steering geometry under dynamic cornering conditions. This is because as attitude angle develops the Langensperger angle subtended by the front axle at the turning centre reduces, Figure 5.3.2, and will go negative on fast large-radius corners. At the rear, the Langensperger angle is always zero or negative. Also, because of lateral load transfer, the outer tyre has greater vertical force, hence needing a greater slip angle than the inner tyre to produce its maximum cornering force. Finally, roll steer effects may give a significant steer angle difference. Even anti-Ackermann (negative f_A) has been used on occasion. If the steering is not perfect Ackermann, then at low speed each axle wheel pair must adopt equal and opposite slip angles to give zero net force.

Figure 5.3.3 shows the Ackermann geometry for a left turn. Here, the front 'track' T is measured between the steer pivot axes at ground level – virtually the same as the wheel-centre track. Parallel lines are

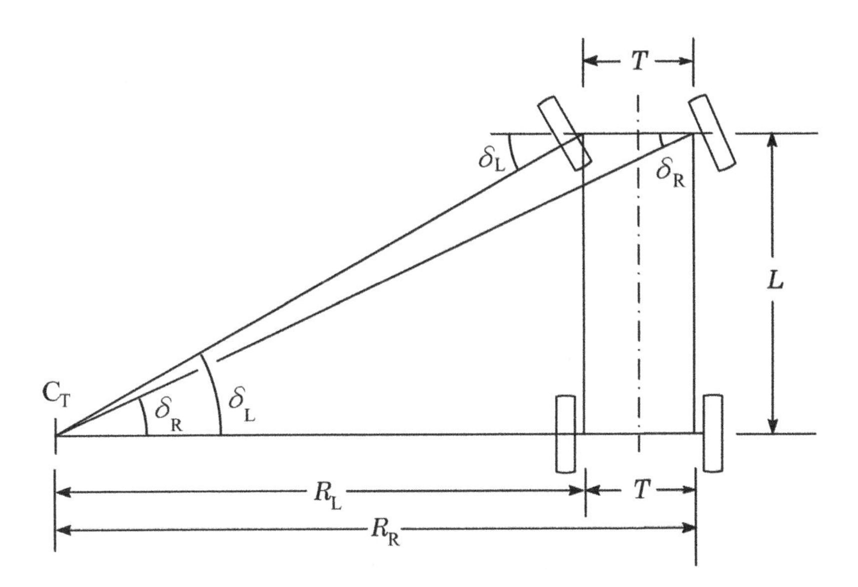

Figure 5.3.3 Geometry for algebraic expression of the Ackermann condition.

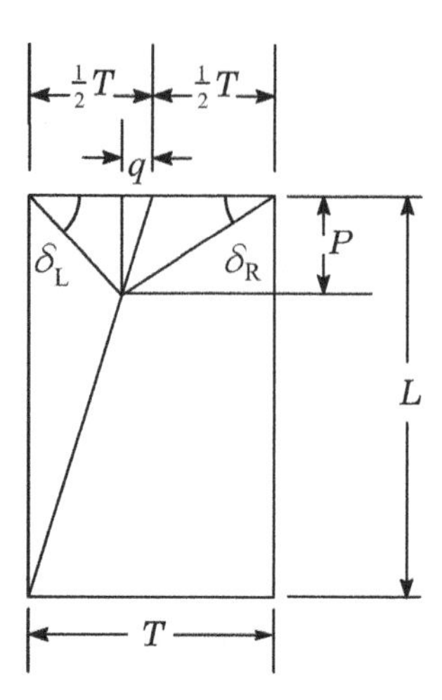

Figure 5.3.4 Alternative geometric construction of the Ackermann condition.

drawn from the pivots to the rear axle, these being R_L and R_R from the turn centre C_T, as shown. The algebraic expression for the Ackermann relationship is

$$\cot\delta_R - \cot\delta_L = \frac{R_R}{L} - \frac{R_L}{L} \tag{5.3.7}$$

so for exact Ackermann effect

$$\cot\delta_R - \cot\delta_L = \frac{T}{L} \tag{5.3.8}$$

This relationship can be expressed in an interesting geometric way, as in Figure 5.3.4 (see also Heldt, 1948). A line is drawn from the centre of the front axle to the left rear axle point (for a left turn). The steer angles, laid out as shown, should meet on this line. This can be proved as follows. In the diagram,

$$p\cot\delta_R - p\cot\delta_L = \left(\tfrac{1}{2}T + q\right) - \left(\tfrac{1}{2}T - q\right) = 2q = 2\frac{p}{L}\frac{T}{2}$$

so

$$\cot\delta_R - \cot\delta_L = \frac{T}{L}$$

as required.

One convenient way to obtain complete or partial Ackermann steer angles is to angle the steering arms inwards (for a rack or tie rod behind the kingpins), as in Figure 5.3.5, so that as steer is applied there is a progressive difference in the effective moment arms of the steering arms. This slanting of the arms also helps with wheel and brake clearance. It is widely believed that aligning the steering arms so that their lines intersect at the rear axle will give true Ackermann steering (the Jeantaud diagram). However, this is far from true, the actual Ackermann factor varying in a complex way with the arm angle, rack length, rack offset forward or rearward of the arm ends, whether the rack is forward or rearward of the kingpins, and with the actual mean steer angle. The Jeantaud arrangement with a single track rod and the steering arms projected meeting at the rear axle actually gives about 60% of full Ackermann effect. To obtain an Ackermann factor close to 1.0 may require the projected steering arm intersection point to be at about 60% of the distance to the rear axle.

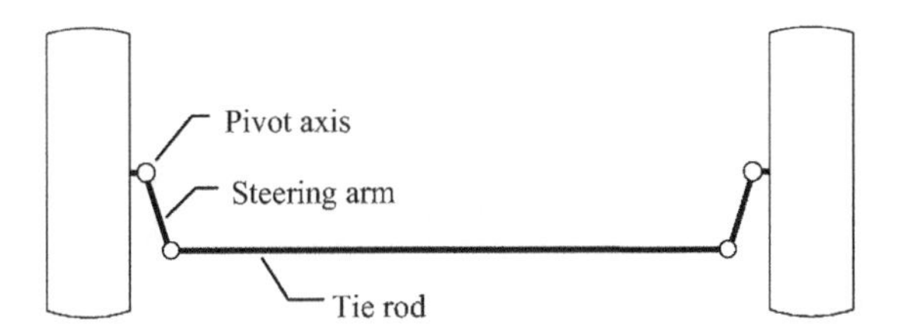

Figure 5.3.5 Basic Ackermann steering mechanism, using a single tie rod with steering arms inclined inwards, in plan view. The section is at the height of the steering system, with the corresponding pivot axis point.

Moving a steering rack forwards or backwards to change the track-rod angle relative to the steering arm can be a useful way to obtain the desired Ackermann factor, the most important single variable being the angle between the track rod and the steering arm in plan view, the Ackermann factor being proportional to the deviation of this angle from 90°. Figure 5.3.6(a) shows the steering arm angle usually considered important, as indeed it is with the aligned rack and track rods. Figure 5.3.6(b) shows the angle that is actually important, the deviation from 90° at the steering arm to track rod joint. Figure 5.3.6(c) shows this angle opened up by moving the rack to the rear, increasing the Ackermann effect. Evidently, inclination of the steering arms is not required at all, it can all be done by positioning the rack to give the necessary included angle. With this method, forward steering arms need not be angled out, and can even be angled inwards, as in Figure 5.3.6(d), although this increases the loads and wear on the rack.

As an interesting comparison, Figure 5.3.7 shows the La Mancelle/Benz type of mechanism, in which the desired steering geometry is instead obtained by the triangular centre plate.

Considering the included angle between the track rod and the steering arm, with a conventional system, as seen in Figure 5.3.6, denoting the deviation from 90° ($\pi/2$ rad) as the steering linkage Ackermann effect angle θ_{StA}, ideal (100%) Ackermann requires

$$\theta_{StA} = \theta_{StA,I} \approx 1.6\frac{T}{2L} = 0.8\frac{T}{L} \qquad (5.3.9)$$

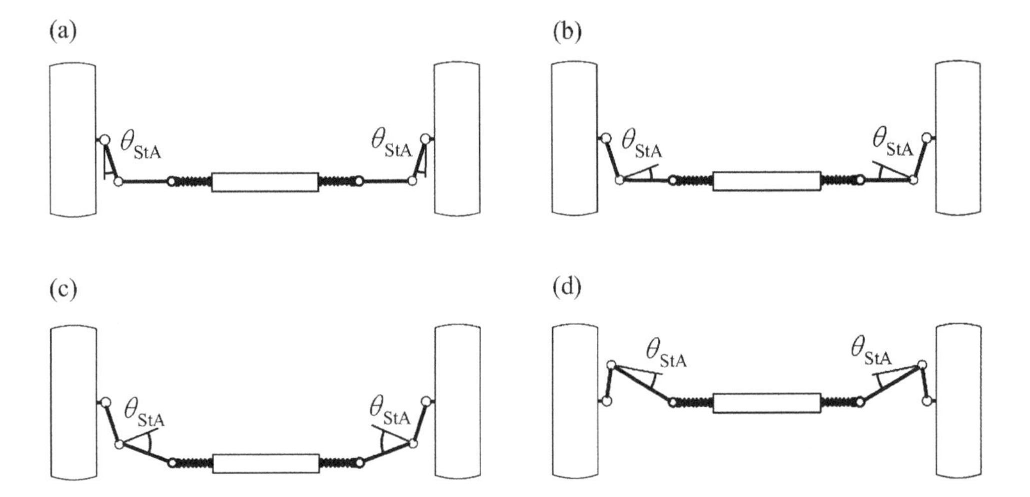

Figure 5.3.6 Ackermann effect achieved or altered by rack position.

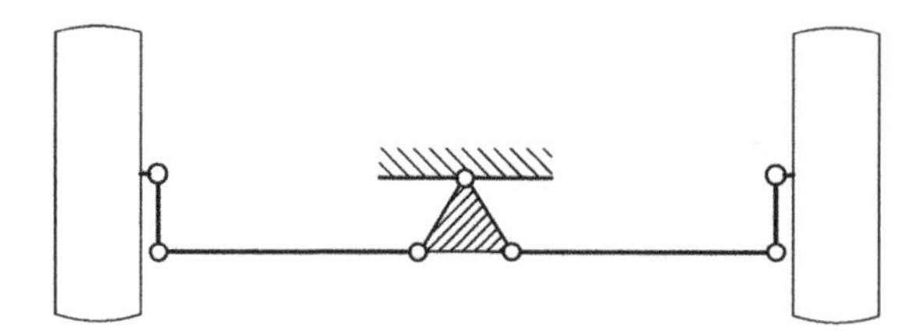

Figure 5.3.7 The La Mancelle/Benz type steering mechanism, using a triangular centre member to give the desired Ackermann effect.

For a passenger car this is about 0.4 rad or 23°. The Ackermann factor is then

$$f_A \approx \frac{\theta_{StA}}{\theta_{StA,I}} \approx 2.5\,\theta_{StA} \qquad (5.3.10)$$

with the angle in radians.

5.4 Turning Geometry – Large Vehicles

Large vehicles basically fall into two categories: articulated and non-articulated. The latter are typified by the (omni)bus. Here the vehicle length creates a manoeuvrability problem. A long wheelbase with a given steer angle gives a large turning radius and large offtracking. Therefore the wheelbase is typically made shorter, with large body overhangs, Figure 5.4.1. Also, the steer angle is made larger than usual, possibly 50° as in Figure 5.4.2.

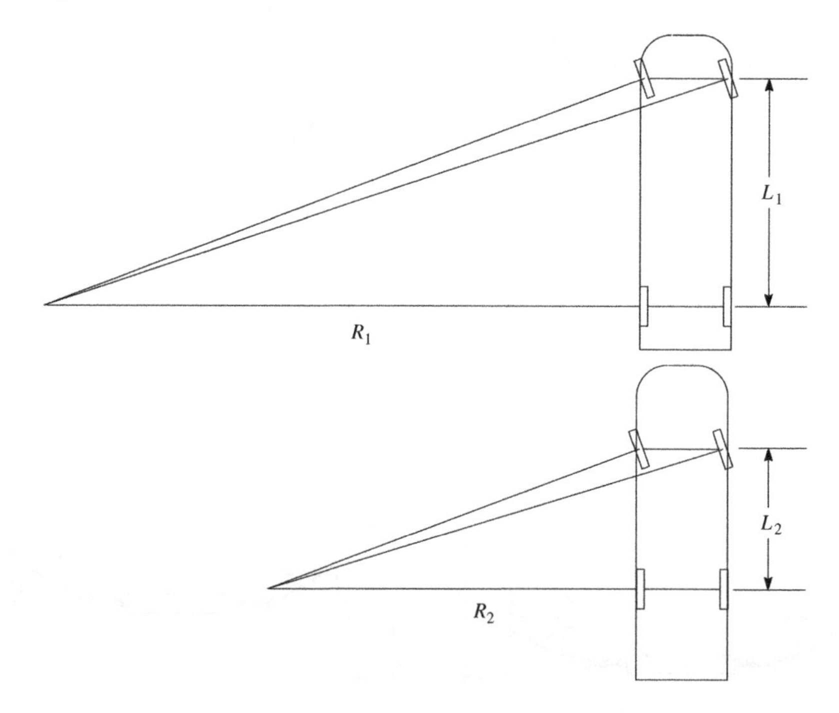

Figure 5.4.1 Turning radius of a bus improved at constant steer angle by the use of a reduced wheelbase with body overhangs.

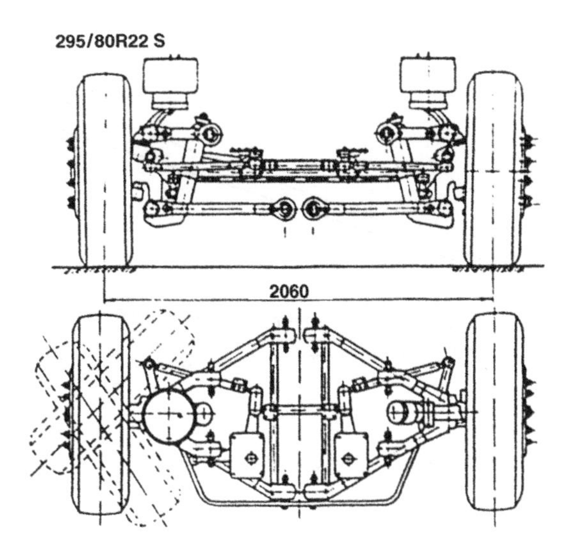

Figure 5.4.2 Bus steering system for increased steering angle. Reproduced with permission from: J. Nemeth (1989) 'Optimization on steering mechanisms for high speed coaches' in *Second International Conference on New Developments in Powertrain and Chassis Engineering*, p325–330. Mechanical Engineering Publications.

Some large front-heavy trucks have two steered front axles, as in Figure 5.4.3. Here, the steer angles required on the second front axle are somewhat less than those of the first front axle.

More commonly, on a non-articulated truck, there are two rear axles, as in Figure 5.4.4. Here the kinematic turning centre is aligned with the mid-point of the rear axle lines, although the precision of this depends on the axle loads and tyre characteristics. The result of the geometry is slip angles at the rear axles, with all the rear tyres fighting each other but giving zero net cornering force. To minimise this, the axles should be as close together as possible. This design aspect is not always dealt with correctly.

When two wheels are used at each end of a single axle, as in Figure 5.4.5, even with a differential to allow different wheel speeds at the two ends of the axle, in cornering the pair of wheels at each single end of the axle fight each other by producing longitudinal forces because of the difference of longitudinal

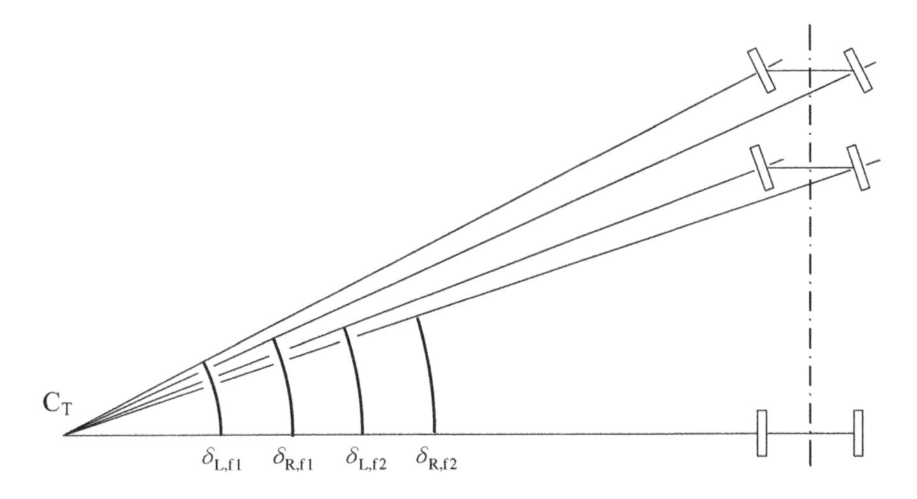

Figure 5.4.3 Kinematic turning of a truck with two front axles, both steered.

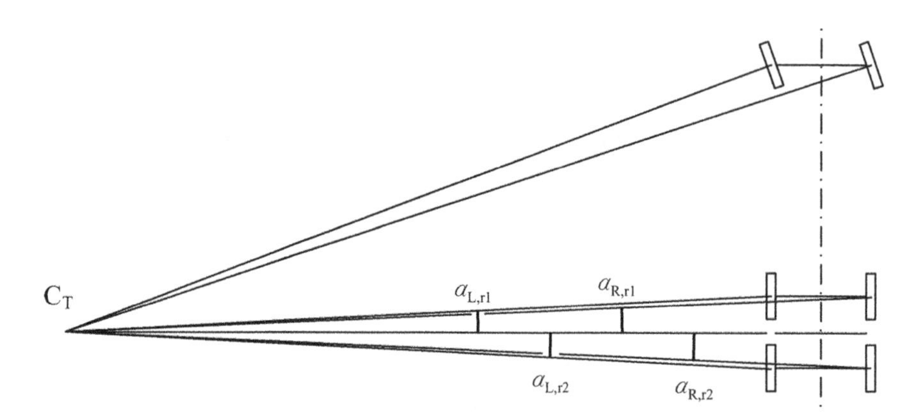

Figure 5.4.4 Kinematic turning of a truck with two rear axles, with slip angles at the rear despite no net cornering force.

velocity, this being proportional to the path radius experienced by the individual wheel. To minimise this problem, the wheels must have minimum axial spacing, just sufficient for the tyres to remain clear of each other under load distortion at the contact patch. This longitudinal slip problem is less severe than the slip angle problem of multiple axles. Double wheels have now largely been superseded by high-load-capacity wide-single tyres.

Heavy articulated trucks now have typically three rear axles with wide-single tyres, as in Figure 5.4.6, obviously creating large slip angles and conflicting lateral forces even when cornering slowly, because the first and last of the rear axles have, of necessity, a long spacing, necessarily greater than the wheel diameter if the wheels are to have the same track. The pivot point on the tractor unit, the so-called 'fifth wheel', is a little in front of the tractor unit rear axle. In some cases, the tractor unit itself has three axles, one steered at the front and two load-bearing at the rear, in which case the fifth wheel is typically about 30% back from the first rear axle to the second one. In either case, the trailer is in effect steered by the cart-steering effect of the rotated tractor unit. The tractor unit itself is steered conventionally on a short wheelbase, requiring quite a small kinematic steer angle

$$\delta_{K,T} = \frac{L_1}{R} \tag{5.4.1}$$

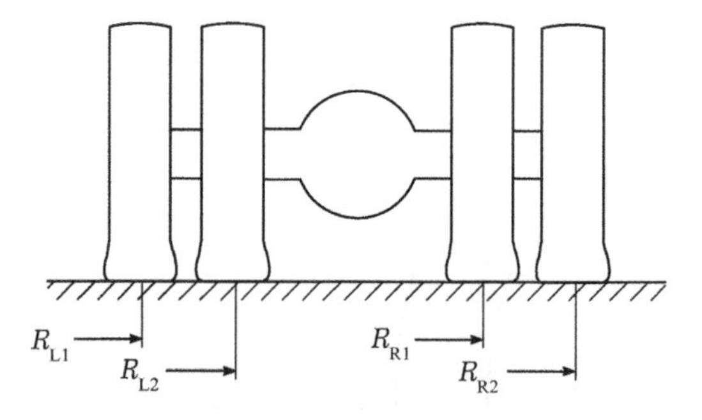

Figure 5.4.5 A single axle with wheel pairs has longitudinal slip and longitudinal forces in low-speed cornering, according to the path radii.

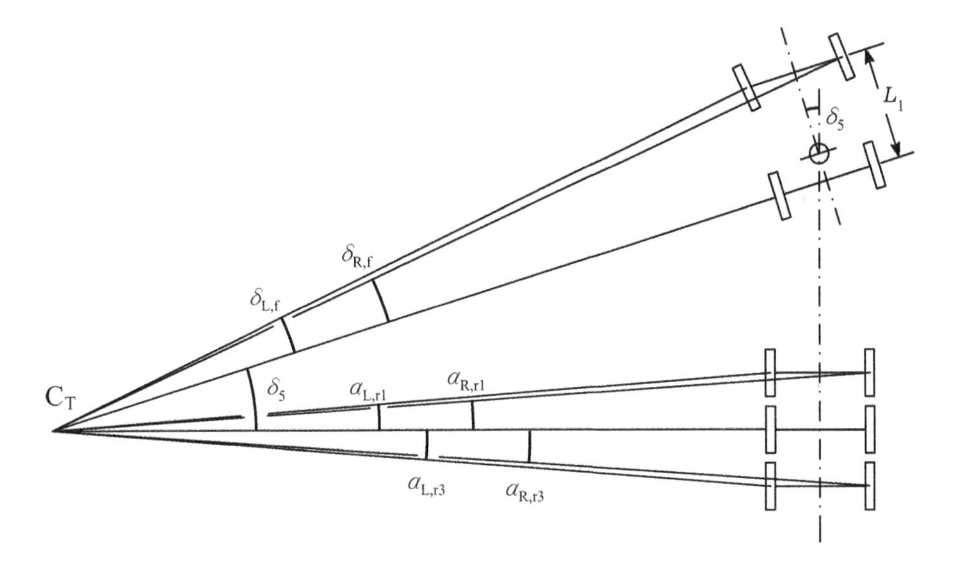

Figure 5.4.6 Kinematic turning of an articulated vehicle with three rear axles.

The tractor unit can hinge to a wide angle on the trailer, so manoeuvrability is good despite the long trailer wheelbase.

A car with a single-axle trailer or caravan, Figure 5.4.7, again can achieve good manoeuvrability by hinging.

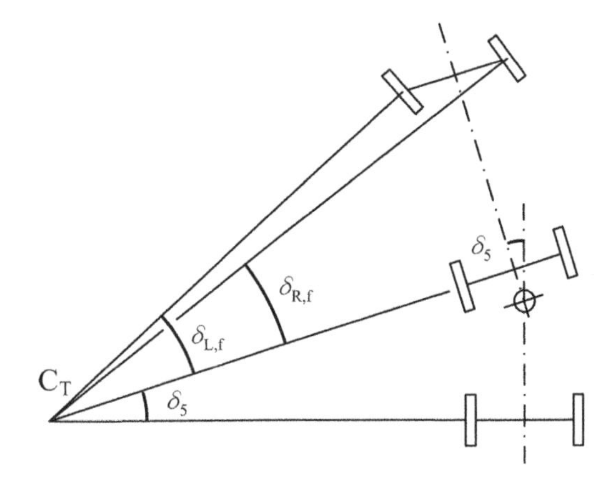

Figure 5.4.7 Kinematic turning of a car with a single-axle trailer.

5.5 Steering Ratio

The steering hand wheel angle δ_S is the angular displacement of the handwheel from the straight-ahead position. The overall steering ratio G is the rate of change of steering-wheel angle with respect to the average steer angle of the steered wheels, with negligible forces in the steering system, or assuming a perfectly rigid system, and with zero suspension roll:

$$G = \frac{d\,\delta_S}{d\,\delta} \tag{5.5.1}$$

The mean overall steering ratio is

$$G_m = \frac{\delta_S}{\delta} \tag{5.5.2}$$

For a linear system G and G_m are equal and constant. In this case it is sometimes convenient to introduce the reference steer angle

$$\delta_{ref} = \frac{\delta_S}{G} \tag{5.5.3}$$

This is not the same as δ because it incorporates the effect of steering compliance. However, it has the advantage over δ_S that it is based on the road wheel angles and therefore it is not very sensitive to G, unlike δ_S.

5.6 Steering Systems

The steering system must connect the hand steering wheel to the road wheels with the appropriate ratio, and also meet other geometric constraints, such as limits on bump steer. It is desirable for the forward efficiency of the system to be high, in order to keep the steering forces low. On the other hand, a low reverse efficiency helps to reduce the transmission of road roughness disturbances back to the driver, at the cost of some loss of the important feel that helps a driver to sense the frictional state of the road. Finally, an appropriate degree of Ackermann effect must be incorporated. Hence, there is a conflict in steering design, which must be resolved according to the particular application.

Precision is of great importance in the steering system, and the rack system has a superior reputation to the steering box, although it is quite difficult to observe any substantial difference between a rack and a good box system in comparative driving tests. To prevent play in the various inter-link ball joints, they are spring-loaded. Where the suspension is mounted on a subframe which has some compliance relative to the body, in the interests of steering precision it is desirable that the steering rack or box also be mounted on the subframe.

On trucks it is still common to use a rigid axle at the front, mounted on two longitudinal leaf springs. Usually the wheel steering arms are connected together by a single tie rod, Figure 5.6.1. This was the system used on passenger cars before the introduction of front independent suspension. Steering is effected by operating a steering control arm A on one of the wheels, by the horizontal drag link B from a vertical Pitman arm C. This acts from the side of the steering box, which is mounted on the sprung mass. The steering column is inclined to the vertical at a convenient angle, about 20° for commercial vehicles and 50° for cars.

For independent suspension, two principal steering systems have been used, one based on a steering box, the other on a rack and pinion. In the typical steering box system, Figure 5.6.2, known as the parallelogram linkage, the steering wheel operates the Pitman arm A via the steering box. The box itself is usually a cam and roller or a recirculating ball worm-and-nut system. The gear ratio of the box alone is usually somewhat less than that of the overall ratio, because of the effect of the links. Symmetrical with the Pitman arm is an idler arm B, connected by the relay rod C, so that the whole linkage is geometrically symmetrical. From appropriate points on the relay rod, the track rods D connect to the steering arms E. The length and alignment of the track rods are critical in controlling bump steer effects. The relay rod layout provides a convenient basis for allowing any required length of the track rods. The steering box system has the advantage of a suitable reverse efficiency, but this has become less important than in the earlier days of motoring because of the improved quality of roads.

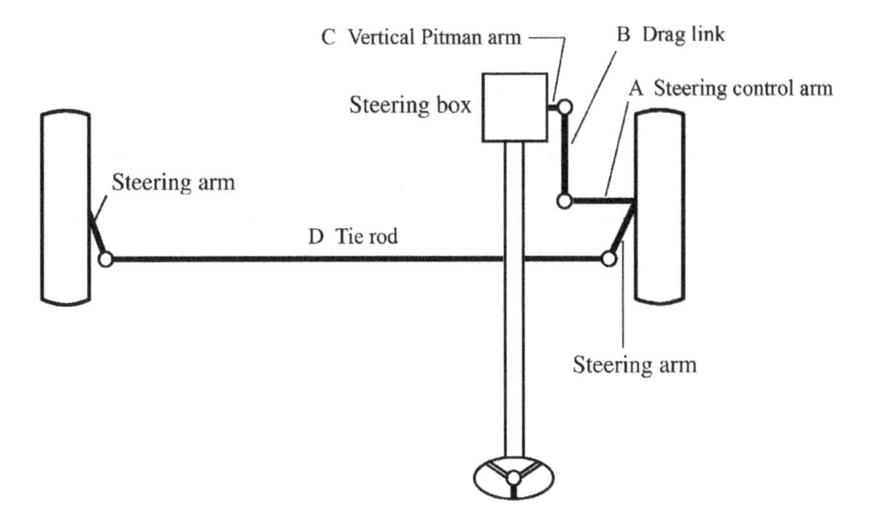

Figure 5.6.1 Truck steering for leaf-spring-supported rigid front axle, plan view, right-hand drive.

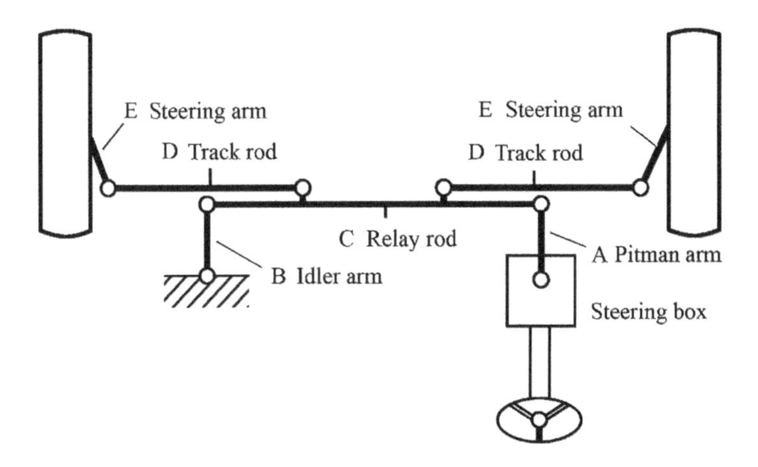

Figure 5.6.2 Independent suspension steering box with parallelogram linkage, plan view, right-hand drive.

In the steering rack system, the steering column is connected directly to a pinion acting on a laterally moving rack. The track rods may be connected to the ends of the rack, or they are sometimes attached close to the centre where the geometry favours this, for example when the rack is high up with strut suspension. Earlier figures in this section, and in Chapter 1, illustrate such systems. Road shock feedback can be controlled to some extent by choosing a suitable gearing helix angle, minimising wheel offset, or increasing handwheel inertia. Flexible mounting of the rack has sometimes been used, but this causes loss of steering precision. A steering damper may be helpful.

5.7 Wheel Spin Axis

To investigate the variation of steer and camber angles, it is first necessary to be clear about the definition of these angles and their relationship to the axis direction cosines. This is complicated, a little, by variations of sign because of the different ISO and SAE axis system definitions, by the left and right sides of the vehicle, and by the choice of two directions for the unit vector along any line, and adjustments must be made to allow for these. Here, generally, the ISO axes will be used (X forwards, Y to the left, Z upwards), with a left-hand wheel, shown from the rear, Figure 5.7.1.

It is easiest to think of the wheel beginning with zero steer and camber. In that case the wheel rotation axis is parallel to the vehicle Y axis. The direction cosines of the axis in this case are $(0, -1, 0)$ or $(0, 1, 0)$ depending on the direction chosen for the axis unit vector, inwards or outwards respectively, for the left wheel. Either direction could be used, but here the inward direction will be preferred; that is, the unit vector points from the wheel generally towards the vehicle centreline. At zero steer and camber, the wheel axis direction cosines for a left wheel are therefore $l=0$, $m=-1$ and $n=0$.

Consider the wheel now to be rotated in steer, and then in camber. The order of these rotations is important, affecting the equations. The steer rotation is first. Toe-out is taken as positive. The wheel is a

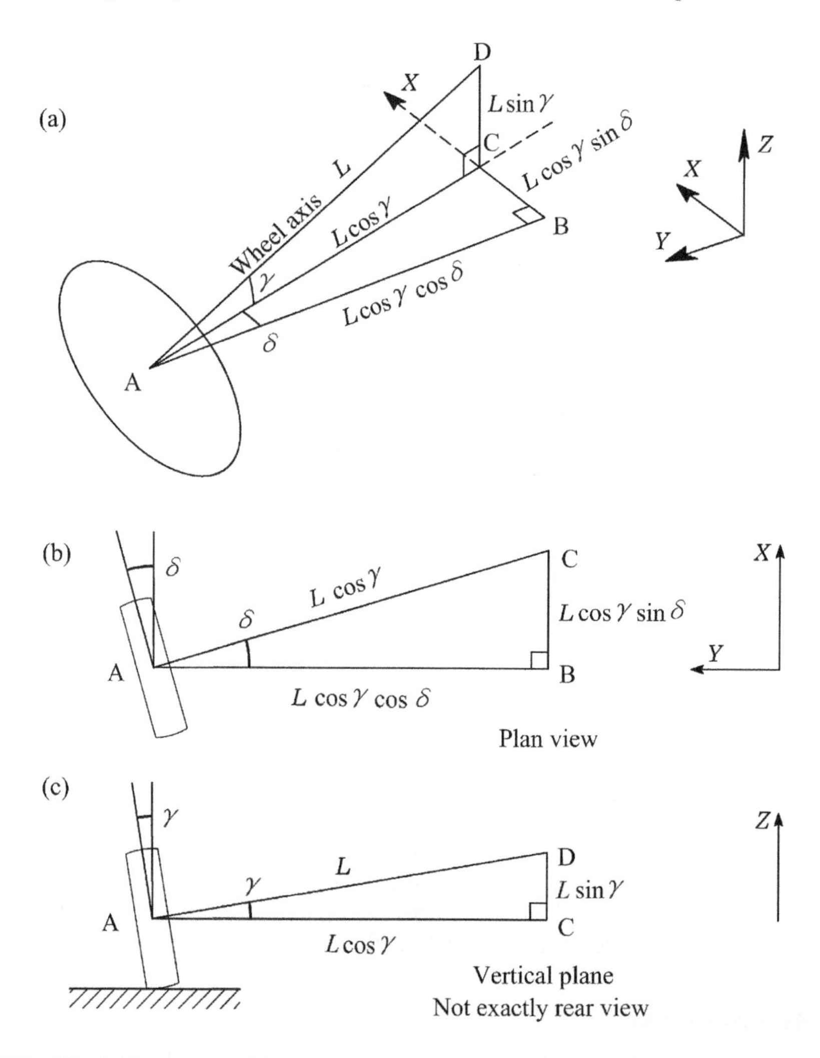

Figure 5.7.1 Wheel axis geometry, with steer and camber angles: (a) rear three-quarter view; (b) plan view showing steer angle; (c) view on ACD with camber angle.

left-hand one, and positive steer is anticlockwise in plan view, and positive camber (negative inclination for a left wheel) is anticlockwise in rear view. The wheel is steered by angle δ (delta) and then cambered by angle γ (gamma); see Figure 5.7.1(b) and (c).

In Figure 5.7.1(a), consider a segment AD of the wheel axis, of length L. The triangle ABC is in the horizontal plane, with AB parallel to the Y axis and BC parallel to the X axis, so ABC is a right angle. The five lengths in the figure are:

$$\begin{aligned} \mathrm{AD} &= L \\ \mathrm{CD} &= L \sin \gamma \\ \mathrm{AC} &= L \cos \gamma \\ \mathrm{BC} &= L \cos \gamma \sin \delta \\ \mathrm{AB} &= L \cos \gamma \cos \delta \end{aligned}$$

Now it follows directly that the direction cosines (l, m, n) of the wheel axis, which are the (X, Y, Z) components of the unit vector along the line AD, are

$$\begin{aligned} l &= \cos \gamma \sin \delta \\ m &= -\cos \gamma \cos \delta \\ n &= \sin \gamma \end{aligned} \tag{5.7.1}$$

It is easily confirmed that, as is necessary,

$$l^2 + m^2 + n^2 = \cos^2 \gamma (\sin^2 \delta + \cos^2 \delta) + \sin^2 \gamma = 1$$

Given particular values for the steer and camber angles, the direction cosines of the wheel axis are thereby easily calculated. If it is desired to obtain the angles from the direction cosines, equations (5.7.1) provide three simultaneous equations for only two variables, which are therefore overspecified, but the equations have the necessary consistency. The solution is

$$\begin{aligned} \gamma &= \operatorname{asin} n \\ \delta &= \operatorname{atan}\left(-\frac{l}{m}\right) \end{aligned} \tag{5.7.2}$$

We also have

$$\delta = \operatorname{asin}\left(\frac{l}{\cos \gamma}\right)$$

$$\delta = \operatorname{acos}\left(\frac{-m}{\cos \gamma}\right)$$

but for small steer angles the latter is poorly conditioned. These conversion equations are summarised in Table 5.7.1.

Equations (5.7.1) and (5.7.2) are sufficient for nearly all purposes, but it is possible to write the lengths and angles generally in terms of the direction cosines, as shown in Table 5.7.2.

Table 5.7.1 Steer and camber angle direction cosine conversion equations

$$l = \cos\gamma \sin\delta$$

$$m = -\cos\gamma \cos\delta$$

$$n = \sin\gamma$$

$$\gamma = \operatorname{asin} n$$

$$\delta = \operatorname{atan}\left(-\frac{l}{m}\right)$$

Table 5.7.2 Axis segment lengths and angles in terms of the direction cosines

Lengths

$$\mathrm{AD} = L$$

$$\mathrm{CD} = Ln$$

$$\mathrm{AC} = L\sqrt{l^2+m^2} = L\sqrt{1-n^2}$$

$$\mathrm{BC} = Ll$$

$$\mathrm{AB} = L(-m)$$

Angles

$$\sin\delta = \frac{l}{\sqrt{l^2+m^2}} = \frac{l}{\sqrt{1-n^2}}$$

$$\cos\delta = \frac{-m}{\sqrt{l^2+m^2}} = \frac{-m}{\sqrt{1-n^2}}$$

$$\tan\delta = -\frac{l}{m}$$

$$\sin\gamma = n$$

$$\cos\gamma = \sqrt{l^2+m^2} = \sqrt{1-n^2}$$

$$\tan\gamma = \frac{n}{\sqrt{l^2+m^2}} = \frac{n}{\sqrt{1-n^2}}$$

As far as the signs are concerned, using instead the outward unit vector inverts the sign of m, the Y component. For a right-hand wheel, the inward axis unit vector is towards positive Y, so this has inverted sign of m. A right-hand wheel with outward vector retains the signs.

The SAE axis system has Z downwards, so the sign of n is inverted. The lateral axis Y is still to the left as in the ISO system.

Wheel inclination angles (Γ) are the same sign as the camber for a right-hand wheel, but of opposite sign for the left-hand wheel considered as the standard here.

All direction cosine magnitudes are the same in the various systems and for the two sides of the vehicle (given that the axis systems are parallel and perpendicular to each other).

5.8 Wheel Bottom Point

It may be required to calculate the bottom point of a wheel disc, i.e. the ground contact point, given the wheel centre coordinates (x_C, y_C, z_C), the axis direction cosines (l, m, n) and the wheel disc radius R. This is

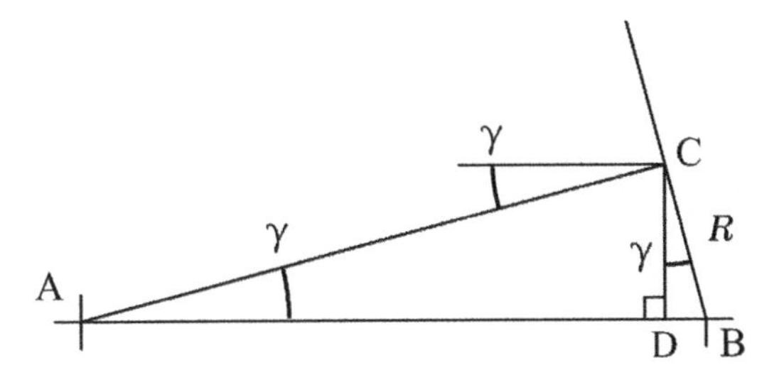

Figure 5.8.1 The bottom point of a wheel (left wheel, rear view, positive camber).

simple in two dimensions, but a little trickier in three dimensions. The bottom point is defined as the point on the disc edge circle with least Z value (not necessarily zero). In two dimensions, Figure 5.8.1, with camber angle γ, consider the point A at the intersection of the wheel axis with the horizontal plane at the level of the bottom point B, and D directly below the centre C, in the same plane.

The lengths are:

$$\begin{aligned}
L_{AC} &= \frac{R}{\tan\gamma} \\
L_{AB} &= \frac{R}{\sin\gamma} \\
L_{AD} &= L_{AC}\cos\gamma = \frac{R\cos^2\gamma}{\sin\gamma} \\
L_{CD} &= R\cos\gamma \\
L_{BD} &= R\sin\gamma
\end{aligned}$$

In a three-dimensional computation, with the axis direction cosines to the inner side of the wheel, use the factor

$$f = \frac{1}{l^2 + m^2}$$

With the direction cosines from C towards A, the camber angle is given by

$$\tan\gamma = \frac{-n}{\sqrt{f}}$$

and

$$L_{AC} = \frac{R}{\tan\gamma} = -\frac{R\sqrt{f}}{n}$$

The coordinates of point A then follow:

$$x_A = x_C + l\,L_{AC}; \quad y_A = y_C + m\,L_{AC}; \quad z_A = z_C + n\,L_{AC}$$

Point D has the same x and y coordinates as C, and $z_D = z_A$. Point B is then given vectorially by

$$\boldsymbol{P}_B = \boldsymbol{P}_A + f(\boldsymbol{P}_D - \boldsymbol{P}_A)$$

expressed in the three coordinates. In implementing this, the possibility of negative camber angle must be considered, and the possibility that the direction cosines are away from A.

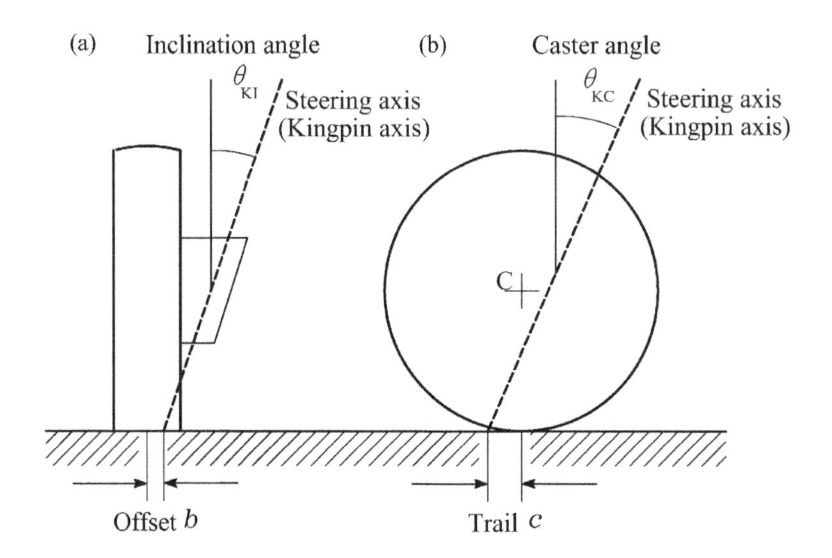

Figure 5.9.1 Geometry of the steering axis of a wheel.

5.9 Wheel Steering Axis

For all systems, the wheels and hubs are pivoted about a steering axis, commonly called the 'kingpin' axis, Figure 5.9.1. Nowadays, on cars at least, physical kingpins are no longer used, the steering axis now usually being defined by upper and lower ball joints on a double-wishbone suspension, or by one ball joint and the upper pivot point of the strut on a strut system. In front view, the axis is at the kingpin inclination angle θ_{KI}, usually from zero to 20°, giving a reduced kingpin offset b at the ground, measured from the undistorted wheel centre plane. The inclination angle helps to give space for the brakes. Where the steering arms are forward (rack in front of the wheel centres) it also gives room to angle the steering arms for Ackermann geometry. Sometimes a negative offset is used, this giving straighter braking when surface friction varies between tracks. Zero offset is called centrepoint steering. If the kingpin inclination angle is also zero, it is called centreline steering. Centrepoint steering gives no steering moment from longitudinal F_X forces at the contact patch; centreline steering gives, in addition, zero moment for longitudinal forces at the wheel axle height. This latter effect is important because of the variation of rolling resistance on rough roads, and traction effects.

5.10 Caster Angle

In the side view of Figure 5.9.1(b), the kingpin axis is slanted at the kingpin caster angle θ_{KC}, with a value generally in the range of 0–5°. This introduces a mechanical trail, called the caster trail, which acts in concert with the tyre pneumatic trail. On cars and trucks it is usual for the kingpin axis to pass through the wheel spin axis C in side view, but this is not essential, and some offsetting of the axis from the centre enables the caster angle and trail to be independently varied. The caster angle and the associated caster trail are important in the feel of the steering. In general, the caster angle varies relative to the body, in suspension bump z_S, with

$$\theta_{KC} = \theta_{KC0} + \varepsilon_{BCas1} z_S + \varepsilon_{BCas2} z_S^2 \tag{5.10.1}$$

where ε_{BCas1} is the linear bump caster coefficient (SI units rad/m in equations, practical units deg/dm and deg/inch; to be distinguished from ε_{BC1}, the linear bump camber coefficient), and ε_{BCas2} is the quadratic bump caster coefficient (rad/m^2). Relative to the road, the castor angle also varies with the body pitch angle θ_B, due to acceleration or braking:

$$\theta_{KC,R} = -\theta_B + \theta_{KC0} + \varepsilon_{BCas1} z_S + \varepsilon_{BCas2} z_S^2 \tag{5.10.2}$$

Caster variation is generally undesirable, but is used deliberately in some cases, for example to offset the effect of body pitch in braking. The caster angle and axis offset (the kingpin axis to wheel centre separation in side view) give the caster trail, which, in conjunction with the tyre pneumatic trail, is very important in giving the steering a suitable feel, and also has a significant effect on directional stability because of steering compliance. The steering feel is adjusted to obtain a desired relationship between lateral force and total aligning torque, so that the experienced driver can tell from the steering hand wheel torque when the tyres are approaching their lateral force limit, as expressed in the Gough plot, a graph of F_Y versus M_Z. Adding caster trail moves the point of maximum steering torque closer to the point of maximum lateral force, or even beyond it – that is, the steering goes light later.

During cornering, the steering must also support the centrifugal compensation forces on the steering mechanism, for example the rack and the track rods. This is called centrifugal caster, and reaches a typical moment about the kingpin of 10 Nm.

On a bumpy road, as the wheel rolls the loaded radius constantly varies. The associated change of effective rolling radius causes changes of wheel angular speed, with associated longitudinal forces on the tyre, required to provide the angular acceleration. These forces, of magnitude about 1 kN, act on the hub at wheel axis height, and hence disturb the steering. This can be eliminated only by centreline steering.

Another steering disturbance, for front drive, is the side-to-side difference of the component of the driveshaft torque along the kingpin axis; this is a problem where the driveshafts have different inclinations, because they have different lengths, or where they are momentarily differently inclined because of rough roads. This is worst for large torques, and hence during acceleration. This also applies to inboard brakes.

5.11 Camber Angle

When the wheel is steered, the kingpin caster angle and kingpin inclination angle affect the camber angle, and this can therefore influence turn-in, and especially can influence handling in small-radius corners. For realistic steer angles, a positive kingpin inclination angle causes a positive camber on the outer wheel, growing roughly with the steer angle squared, and being typically 0.15° of camber per degree of inclination at 30° of steer. Positive caster angle causes a negative camber on the outer wheel, approximately proportional to steer angle, and is typically −0.50° of camber per degree of caster at 30° of steer. The actual camber angle is

$$\gamma = \gamma_0 + \arccos(\sin\theta_{KI}\cos\delta) + \theta_{KI} + \arccos(\sin\theta_{KC}\sin\delta) - 180° \tag{5.11.1}$$

The effect of steer angle on camber angle may be represented approximately by

$$\gamma_S = \varepsilon_{SC1}\,\delta + \varepsilon_{SC2}\delta^2 \tag{5.11.2}$$

where ε_{SC1}, the linear steer camber coefficient, depends upon the caster angle ($-\theta_{KC}$ in radians), and ε_{SC2}, the quadratic steer camber coefficient, depends on the kingpin angle ($\frac{1}{2}\,\theta_{KC}$ in radians), but changes sign with the sign of δ, that is,

$$\varepsilon_{SC1} = -\theta_{KC} \tag{5.11.3}$$

$$\varepsilon_{SC2} = \frac{1}{2}\,\theta_{KI}\mathrm{sign}(\delta) \tag{5.11.4}$$

5.12 Kingpin Angle Analysis

The steering axis (kingpin axis), illustrated in Figure 5.12.1, is defined by two points A and B, or L and U, lower and upper respectively, possibly but not essentially the actual ball joints, and has four associated angles. The two normally used are the kingpin inclination angle θ_{KI} and the kingpin caster angle θ_{KC}. An alternative pair, sometimes useful, is the total kingpin angle θ_{KP} with the kingpin sweep angle ψ_{KP}.

In Figure 5.12.1, the kingpin axis AB is of length

$$L_{KP} = \sqrt{(x_U - x_L)^2 + (y_U - y_L)^2 + (z_U - z_L)^2}$$

The kingpin height is $Z_{KP} = DB$. The various lengths in the diagram include the kingpin height, the kingpin caster spacing and the kingpin inclination spacing, which, for a left wheel, with direction cosines from the lower point to the upper point, are

$$\begin{aligned} Z_{KP} &= z_U - z_L = nL_{KP} \\ X_{KC} &= -(x_U - x_L) = -l\,L_{KP} \\ Y_{KI} &= -(y_U - y_L) = -m\,L_{KP} \end{aligned}$$

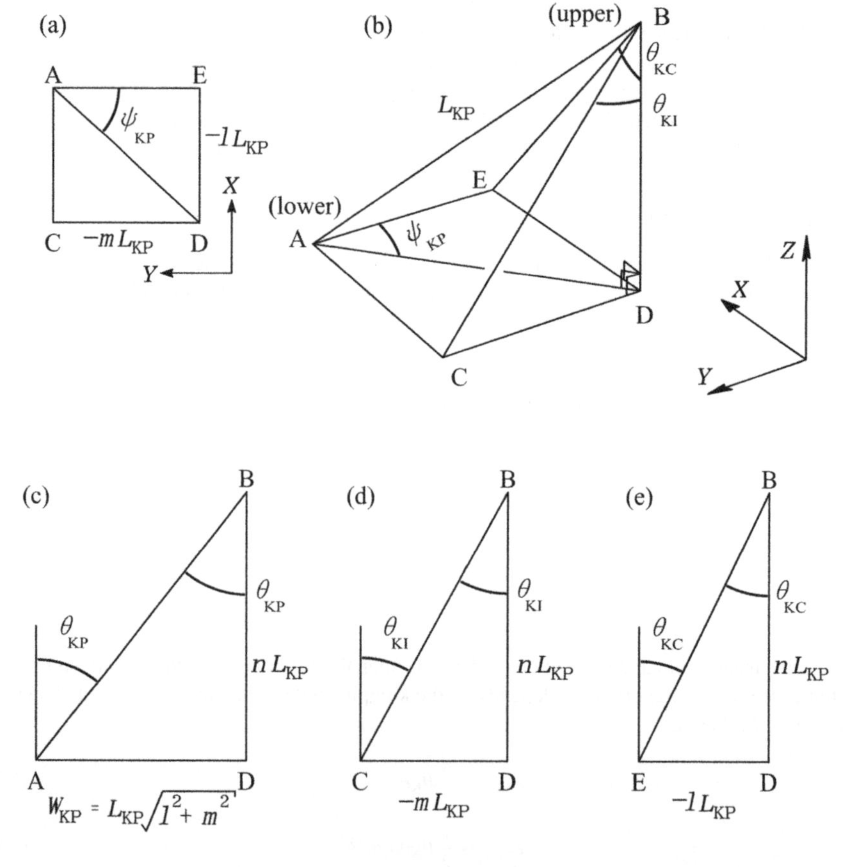

Figure 5.12.1 The kingpin axis and its associated angles, with the direction cosines (l, m, n).

The signs are chosen to make the variables normally positive. The diagonal AD is

$$W_{KP} = \sqrt{X_{KC}^2 + Y_{KI}^2} = \sqrt{l^2 + m^2} L_{KP} = \sqrt{1 - n^2} L_{KP}$$

The angles may be written as

$$\tan\theta_{KP} = \frac{W_{KP}}{Z_{KP}} = \frac{\sqrt{1-n^2}}{n}$$

$$\tan\psi_{KP} = \frac{X_{KC}}{Y_{KI}} = \frac{l}{m}$$

$$\tan\theta_{KC} = \frac{X_{KI}}{Z_{KP}} = -\frac{l}{n}$$

$$\tan\theta_{KI} = \frac{Y_{KI}}{Z_{KP}} = -\frac{m}{n}$$

The relationships between the angles may be deduced from Figure 5.12.1(b):

$$Z_{KP} \tan\theta_{KP} \cos\psi_{KP} = \text{AE} = \text{CD} = Z_{KP} \tan\theta_{KI}$$

so

$$\tan\theta_{KI} = \tan\theta_{KP} \cos\psi_{KP}$$

Similarly,

$$\tan\theta_{KC} = \tan\theta_{KP} \sin\psi_{KP}$$

Given θ_{KI} and θ_{KC} initially, the above two equations are easily solved for θ_{KP} and ψ_{KP}. The conversion equations are summarised in Table 5.12.1. Based on these, various sets of equations may be put together for the dimensions according to the starting information, for example as in Tables 5.11.2–5.11.5.

Table 5.12.1 Kingpin angle conversion equations

$\tan\theta_{KI} = \tan\theta_{KP} \cos\psi_{KP}$
$\tan\theta_{KC} = \tan\theta_{KP} \sin\psi_{KP}$
$\tan\psi_{KP} = \dfrac{\tan\theta_{KC}}{\tan\theta_{KI}}$
$\tan\theta_{KP} = \dfrac{\tan\theta_{KI}}{\cos\psi_{KP}} = \dfrac{\tan\theta_{KC}}{\sin\psi_{KP}}$

Table 5.12.2 Static kingpin analysis given lower point L (x_L, y_L, z_L) and upper point U (x_U, y_U, z_U)

$$Z_{KP} = z_U - z_L$$

$$\theta_{KI} = \mathrm{atan}\left(\frac{-m}{n}\right) = \mathrm{atan}\left(\frac{y_L - y_U}{Z_{KP}}\right)$$

$$\theta_{KC} = \mathrm{atan}\left(\frac{-l}{n}\right) = \mathrm{atan}\left(\frac{x_L - x_U}{Z_{KP}}\right)$$

$$\psi_{KP} = \mathrm{atan}\left(\frac{l}{m}\right) = \mathrm{atan}\left(\frac{x_L - x_U}{y_L - y_U}\right)$$

$$\theta_{KP} = \mathrm{atan}\left(\frac{\sqrt{(x_L - x_U)^2 + (y_L - y_U)^2}}{Z_{KP}}\right)$$

$$n = \frac{1}{\sqrt{1 + \tan^2\theta_{KI} + \tan^2\theta_{KC}}}$$

$$l = -n \tan\theta_{KC}$$

$$m = -n \tan\theta_{KI}$$

Table 5.12.3 Static kingpin analysis given lower point L (x_L,y_L,z_L), H_{KP}, θ_{KI} and θ_{KC}

$$x_U = x_L - Z_{KP} \tan\ \theta_{KC}$$

$$y_U = y_L - Z_{KP} \tan\ \theta_{KI}$$

$$z_U = z_L + Z_{KP}$$

$$W_{KP} = \sqrt{X_{KC}^2 + Y_{KI}^2}$$

$$\psi_{KP} = \mathrm{atan}\left(\frac{\tan\theta_{KC}}{\tan\theta_{KI}}\right) = \mathrm{atan}\left(\frac{X_{KC}}{Y_{KI}}\right)$$

$$\theta_{KP} = \mathrm{atan}\left(\frac{W_{KP}}{Z_{KP}}\right)$$

Table 5.12.4 Static kingpin analysis given lower points L (x_L, y_L, z_L), L_{KP}, θ_{KI} and θ_{KC}

$$\psi_{KP} = \mathrm{atan}\left(\frac{\tan\theta_{KC}}{\tan\theta_{KI}}\right)$$

$$\theta_{KP} = \mathrm{atan}\left(\frac{\tan\theta_{KC}}{\sin\psi_{KP}}\right) = \mathrm{atan}\left(\frac{\tan\theta_{KI}}{\cos\psi_{KP}}\right)$$

$$Z_{KP} = L_{KP}\cos\theta_{KP}$$

$$x_U = x_L - Z_{KP}\tan\theta_{KC}$$

$$y_U = y_L - Z_{KP}\tan\theta_{KI}$$

$$z_U = z_L + Z_{KP}$$

Table 5.12.5 Static kingpin analysis given lower point L (x_L,y_L,z_L), L_{KP}, θ_{KP} and ψ_{KP}

$$\theta_{KI} = \text{atan}(\tan\theta_{KP}\cos\psi_{KP})$$
$$\theta_{KC} = \text{atan}(\tan\theta_{KP}\sin\psi_{KP})$$
$$H_{KP} = L_{KP}\cos\theta_{KP}$$
$$x_U = x_L - Z_{KP}\tan\theta_{KC}$$
$$y_U = y_L - Z_{KP}\tan\theta_{KI}$$
$$z_U = z_L + Z_{KP}$$

5.13 Kingpin Axis Steered

When a wheel is steered, in a coordinate system rotating with wheel steer angle the caster and inclination angles, as seen by the wheel, change. With a left wheel and left steer angle δ, the wheel sees a kingpin sweep angle

$$\psi_{KPw} = \psi_{KP} + \delta$$

so

$$X_{KCw} = X_{KC}\cos\delta + Y_{KI}\sin\delta$$

$$Y_{KIw} = -X_{KC}\sin\delta + Y_{KI}\cos\delta$$

The variables θ_{KP}, L_{KP}, Z_{KP} and W_{KP} are invariant in this transformation. The kingpin angles seen by the wheel are therefore, for example,

$$\tan\theta_{KCw} = \frac{X_{KCw}}{Z_{KP}} = \frac{X_{KC}\cos\delta + Y_{KI}\sin\delta}{Z_{KP}}$$

giving

$$\tan\theta_{KCw} = \tan\theta_{KC}\cos\delta + \tan\theta_{KI}\sin\delta$$

and

$$\tan\theta_{KIw} = -\tan\theta_{KC}\sin\delta + \tan\theta_{KI}\cos\delta$$

Approximating the above equations by a quadratic steer angle gives

$$\theta_{KCw} = \theta_{KC}\left(1 - \tfrac{1}{2}\delta^2\right) + \theta_{KI}\,\delta = \theta_{KC} + \theta_{KI}\,\delta - \tfrac{1}{2}\theta_{KC}\,\delta^2$$

$$\theta_{KIw} = -\theta_{KC}\,\delta + \theta_{KI}\left(1 - \tfrac{1}{2}\delta^2\right) = \theta_{KI} - \theta_{KC}\,\delta - \tfrac{1}{2}\theta_{KI}\,\delta^2$$

Representing the kingpin angles seen by the wheel as quadratic expressions

$$\theta_{\mathrm{KCw}} = \theta_{\mathrm{KC}} + \varepsilon_{\mathrm{SKC1}}\ \delta + \varepsilon_{\mathrm{SKC2}}\,\delta^2$$

$$\theta_{\mathrm{KIw}} = \theta_{\mathrm{KI}} + \varepsilon_{\mathrm{SKI1}}\ \delta + \varepsilon_{\mathrm{SKI2}}\,\delta^2$$

the linear and quadratic steer caster coefficients of a left wheel are

$$\varepsilon_{\mathrm{SKC1}} = \theta_{\mathrm{KI}}$$

$$\varepsilon_{\mathrm{SKC2}} = -\tfrac{1}{2}\theta_{\mathrm{KC}}$$

and the steer inclination coefficients are

$$\varepsilon_{\mathrm{SKI1}} = -\,\theta_{\mathrm{KC}}$$

$$\varepsilon_{\mathrm{SKI2}} = -\tfrac{1}{2}\theta_{\mathrm{KI}}$$

These apply to individual wheels, with contrary steer angles on the opposite sides of the vehicle.

5.14 Steer Jacking

Steering of the wheels, in conjunction with the kingpin inclination angle θ_{KI}, the kingpin caster angle θ_{KC}, the caster offset y_{CO} and the caster trail x_{CT}, has the effect of lifting or lowering the steered corner of the vehicle slightly, affecting the steering feel. The steer jacking effect is positive for a raising of the body relative to the road, or as a lowering of the wheel relative to the body. Because of the total kingpin inclination angle θ_{KP}, the rotation required about the kingpin axis to give a wheel steer angle δ is $\delta\cos\theta_{\mathrm{KP}}$.

Considering first a case of zero inclination angle, we have

$$z_{\mathrm{SJ}} = -x_{\mathrm{CT}}\cos\theta_{\mathrm{KC}}\left(1-\cos\left(\frac{\delta}{\cos\theta_{\mathrm{KP}}}\right)\right)\sin\theta_{\mathrm{KC}} - y_{\mathrm{CO}}\sin\delta\,\sin\theta_{\mathrm{KC}}$$

Now considering small kingpin angles,

$$z_{\mathrm{SJ}} = -x_{\mathrm{CT}}\,\theta_{\mathrm{KC}}\,\tfrac{1}{2}\delta^2 - y_{\mathrm{CO}}\,\theta_{\mathrm{C}}\,\delta$$

Considering now zero caster angle, the effect of the inclination angle is

$$z_{\mathrm{SJ}} = y_{\mathrm{CO}}\cos\theta_{\mathrm{KI}}\left(1-\cos\left(\frac{\delta}{\cos\theta_{\mathrm{KP}}}\right)\right)\sin\theta_{\mathrm{KI}} + x_{\mathrm{CT}}\sin\delta\,\sin\theta_{\mathrm{KI}}$$

With the small-angle approximation, this becomes

$$z_{\mathrm{SJ}} = y_{\mathrm{CO}}\,\theta_{\mathrm{KI}}\,\tfrac{1}{2}\delta^2 + x_{\mathrm{CT}}\,\theta_{\mathrm{KI}}\ \delta$$

Considering a quadratic model of the jacking effect,

$$z_{\mathrm{SJ}} = \varepsilon_{\mathrm{SJ1}}\ \delta + \varepsilon_{\mathrm{SJ2}}\,\delta^2$$

then the single wheel steer jacking coefficients are

$$\varepsilon_{\text{SJ1}} = x_{\text{CT}}\,\theta_{\text{KI}} - y_{\text{CO}}\,\theta_{\text{KC}}$$
$$\varepsilon_{\text{SJ2}} = \tfrac{1}{2}y_{\text{CO}}\,\theta_{\text{KI}} - \tfrac{1}{2}x_{\text{CT}}\,\theta_{\text{KC}}$$

For a complete axle it is necessary to consider the effects of the two wheels with corresponding steer angles, negated on one side. The steer jacking effect of the axle is

$$z_{\text{SJA}} = 2\varepsilon_{\text{SJ2}}\,\delta^2$$

with no linear effect. The roll effect is

$$\phi_{\text{SJA}} = 2\frac{\varepsilon_{\text{SJ1}}}{T}\,\delta$$

with no quadratic effect. The terms tend to compensate each other. The linear term gives asymmetrical jacking, i.e. no front-end lift but a small roll effect. The second-order term gives a symmetrical effect, so giving a small front-end lift. Substituting practical values, the effects are small other than at large steering angles, but can nevertheless be felt by the driver.

References

Durstine, J.W. (1973) *The Truck Steering System from Hand Wheel to Road Wheel*, SAE 730039, The 19th Ray Buckendale Lecture, January, SAE SP-374. New York: Society of Automotive Engineers.

Heldt, P.M. (1948) *The Automotive Chassis*, 2nd edition, Published by P. M. Heldt, 600pp.

Nemeth, J. (1989) Optimization on steering mechanisms for high speed coaches. In *Second International Conference on New Developments in Powertrain and Chassis Engineering*, pp. 325–330. Bury St Edmunds: Mechanical Engineering Publications. ISBN 0852986890.

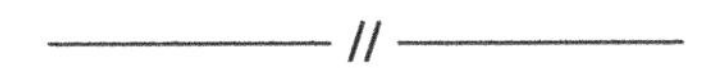

6

Bump and Roll Steer

6.1 Introduction

A car is usually steered by changes to the angles of the front wheels, controlled by the driver's handwheel. However, the geometry of the suspension linkages also results in changes of the steer angle of the wheels when they move up or down, effects called 'bump steer' and 'roll steer'. All types of steer angle are represented by δ with an appropriate subscript. In dynamic cornering, the tyre lateral forces are associated with the tyre slip angles, denoted by α. Careful distinction must be drawn between steer angles δ, which are the angle of the wheel relative to the vehicle centreline, and slip angles α, which are the angle of the wheel compared with the local direction of motion.

The term 'bump steer' means changes of steer angle for a single wheel of an independent suspension when the wheel is moved up or down relative to the body, in suspension bump and droop. The term 'roll steer' refers to changes of steer of the pair of wheels (i.e. of the axle) when the body rolls. These are obviously related, but bump steer is the basic form of data for independent suspension, and roll steer is the basic form for rigid axles. Some forms of bump and roll steer can cause poor straight-line stability, and very unpredictable and unpleasant vehicle behaviour, the handling being sensitive to small changes of wheel steer angles. To the driver, a vehicle with significant bump or roll steer, particularly oversteer, may seem to wander and to be unpredictable.

An axle may have steer effects because of suspension roll or symmetrical suspension double bump. The whole vehicle may have steer effects because of heave, pitch or roll of the suspension. All of these effects can be expressed as steer coefficients relating the wheel angle changes to the stimulus. In summary, the main geometric steer and camber coefficients are those in Table 6.1.1. The angle units are specified in radians. These are the correct units for use in equations, although the actual values are often expressed in degrees. The bump steer coefficient is conveniently expressed as deg/dm (deg/inch US). Angle responses to angle stimulus have no units, being expressible as rad/rad, which is the same as deg/deg.

6.2 Wheel Bump Steer

The steer angle of a single wheel may be represented as

$$\delta = \delta_{T0} + \delta_S + \varepsilon_{BS1}\, z_S + \varepsilon_{BS2}\, z_S^2 + \cdots \qquad (6.2.1)$$

Here δ is the complete road wheel steer angle, δ_{T0} is the initial (static) toe-out angle at zero bump, δ_S is the wheel steer angle contributed by movement of the driver's handwheel, ε_{BS1} is the linear bump steer coefficient, and ε_{BS2} is the quadratic bump steer coefficient. More terms could be added to the polynomial if desired, although this is only necessary for large suspension motions and high accuracy.

Table 6.1.1 The main wheel, axle and vehicle linear steer variables

(1) δ_{BS}	rad	single-wheel bump steer angle
(2) δ_{RS}	rad	axle roll steer angle
(3) δ_{DBS}	rad	axle double-bump steer angle
(4) δ_{RT}	rad	axle roll toe angle
(5) δ_{DBT}	rad	axle double-bump toe angle
(6) ε_{BS1}	rad/m	single-wheel bump steer coefficient
(7) ε_{RS1}	rad/rad	axle roll steer coefficient
(8) ε_{RU1}	rad/m	axle roll understeer coefficient
(9) ε_{DBS1}	rad/m	axle double-bump steer coefficient
(10) ε_{RT1}	rad/m	axle roll toe coefficient
(11) ε_{DBT1}	rad/m	axle double-bump toe coefficient
(12) ε_{VHU1}	rad/m	vehicle heave understeer coefficient
(13) ε_{VPU1}	rad/rad	vehicle pitch understeer coefficient
(14) ε_{VRU1}	rad/rad	vehicle roll understeer coefficient

For understanding bump steer, the linear and quadratic terms are sufficient. The linear bump steer coefficient ε_{BS1} has units rad/m but it is usually expressed in deg/dm (deg/inch US). The quadratic coefficient ε_{BS2} has fundamental units rad/m^2, but is often expressed as deg/dm^2. It is normal practice to express the values in degrees, but radian units should be used in the equations (1 radian $= 180/\pi =$ 57.29578 deg).

In general, the linear bump steer coefficient is defined as the rate of change of wheel steer angle with vertical wheel position (suspension bump deflection), with the steering hand wheel fixed, of course:

$$\varepsilon_{BS} = \frac{d\delta}{dz_S} \tag{6.2.2}$$

This will be taken as positive for toe-out with a rising wheel. The first linear bump steer coefficient ε_{BS1} is the value of this at the static position. A positive quadratic coefficient ε_{BS2} gives toe-out in both bump and droop, proportional to the square of the bump value.

For any independent front suspension, e.g. the double wishbone of Figure 6.2.1, consider the vehicle body fixed and the wheel to move in bump with the steering disconnected at B, in a controlled ideal way with no steer angle change of the wheel. The steering-arm ball joint B will then move in an arc with centre at some point A, called the ideal point or ideal centre, where the position of A depends on the particular suspension geometry. For most real suspensions, the movement of B is not a perfectly circular arc, but an arc is a good approximation. With the steering now connected, if the track rod to rack ball joint, or track rod to relay rod joint, C, is actually at A then there will be no bump steer. This possibility of achieving an accurate steering motion is an important advantage of independent suspension over a steered rigid axle. Predictable and precise handling is particularly important for competition and high-performance vehicles, and in such cases it is considered to be of paramount importance that the rack is mounted accurately at the correct height, with a tolerance of only 1 or 2 mm, and that the rack and track rods should be of the appropriate length. In practice, for passenger cars there are often significant production tolerance error discrepancies of the height, length or alignment of the rack. For ordinary road vehicles complete accuracy is often not even attempted, deliberate discrepancies being introduced. These are sometimes claimed to give less wheel response to rough roads or to have handling advantages, although this is a controversial, and interesting, issue.

The above is a two-dimensional analysis, of course, and in reality the steering and suspension mechanism is a three-dimensional one. The 'ideal point' is really extended in space, usually roughly

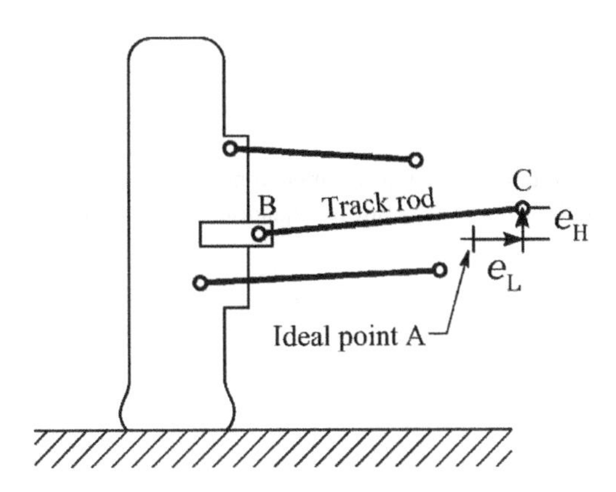

Figure 6.2.1 Rear view of left wheel, showing the track rod BC, and the ideal centre A, with 'errors' e_L and e_H, possibly deliberate.

along the vehicle, forming an approximately straight line, the 'ideal axis'. With parallel-pivot-axis double transverse arms (double wishbones), the ideal axis is roughly parallel to the pivot axes, which may be inclined somewhat to the horizontal plane, and, even more so, possibly swept outwards to the rear. If the rack is moved forwards or backwards, then, the ideal rack and track-rod lengths will change somewhat. The angles of the ideal axis are such that the axis is perpendicular to the plane in which the track-rod end ball joint centre moves with steering detached. The ideal centre then is the point at which the ideal axis penetrates the transverse vertical plane through the rack.

Returning to a two-dimensional analysis, as seen in Figure 6.2.1, the two kinds of position error of the rack end point are the height error e_H of the actual joint C, and the length error e_L of the track rod BC (given correct adjustment of the static toe angle value). With the wheel static toe angle correctly adjusted, too long a track rod corresponds to too short a steering rack. The dimensions of importance are then as in Table 6.2.1. The moment arm length of the steering arm, L_{AX}, is the effective length of the steering arm, perpendicular to the track rod.

A positive height error is defined to be with the rack high relative to the ideal wheel steering-arm point (ball joint centre). This may arise from a high rack, a low steering arm, or from a tilted rack with opposite height errors on the two sides. The usual arrangement is with the track rod behind the steering axis, although this is not invariable, in which case the consequence of a height error e_H is a first-order bump steer coefficient

$$\varepsilon_{BS1} = -\frac{e_H}{L_{TR}L_{AX}} \tag{6.2.3}$$

The negative sign follows from Figure 6.2.1, where in bump point B will be forced outwards, toeing the wheel in. If the track rod is in front of the steer axis then the sign in the above equation is positive.

Table 6.2.1 Bump steer dimensions

e_H	height error, rack high positive
e_L	length error, track rod long positive
L_{AX}	moment arm length of steering arm
L_{ITR}	length of ideal track rod
L_{TR}	length of actual track rod

A positive length error is defined to be with the track rod too long, when adjusted for correct static toe:

$$e_L = L_{TR} - L_{ITR} \tag{6.2.4}$$

This could be due to a rack too short, a steering arm misaligned, or a rack laterally misplaced on the body, giving antisymmetrical length errors. The result is a quadratic bump steer coefficient

$$\varepsilon_{BS2} = -\frac{e_L}{2L_{AX}L_{ITR}^2} \tag{6.2.5}$$

For an independent suspension, calculation of the ideal pivot centre A for the track rod, Figure 6.2.1, to give zero bump steer is a purely geometrical problem. However, it really requires a three-dimensional solution.

In the case of strut suspension, the ideal track-rod length is highly sensitive to the vertical position of the steering arm, tending to infinity when it is at the strut top, Figure 6.2.2. A common solution is to use a rack with centre-mounted track rods, and then to choose a rack height for which these are the correct length, which roughly matches up with the main spring seat. There are various other independent front suspensions, not often seen nowadays, for which there exist suitable ideal steering layouts free of bump steer.

If the measured wheel steer angle is plotted against bump, a result such as Figure 6.2.3 is typically obtained. Actually there is normally also a toe angle at zero bump; this is the static toe, and does not usually appear in bump steer plots because it is readily adjustable and is measured separately; bump steer is usually measured only as the change of angle from static.

The measured bump steer can then be characterised in two main ways – the linear and quadratic components. At zero bump there is a gradient of steer change with bump, $d\delta/dz_S$, the tangent to the curve at zero bump, which in this case can be measured from the graph as about 0.3° toe-out in 75 mm bump, that is, a bump steer coefficient of $+4°/m$ or 0.4°/dm. This linear effect results from the rack height, say about 5 mm too low in this case with the rack behind the wheels, or high for the rack in front. Secondly, there is a curvature of the graph. In this case the curvature is towards toe-in, which would result from a track rod

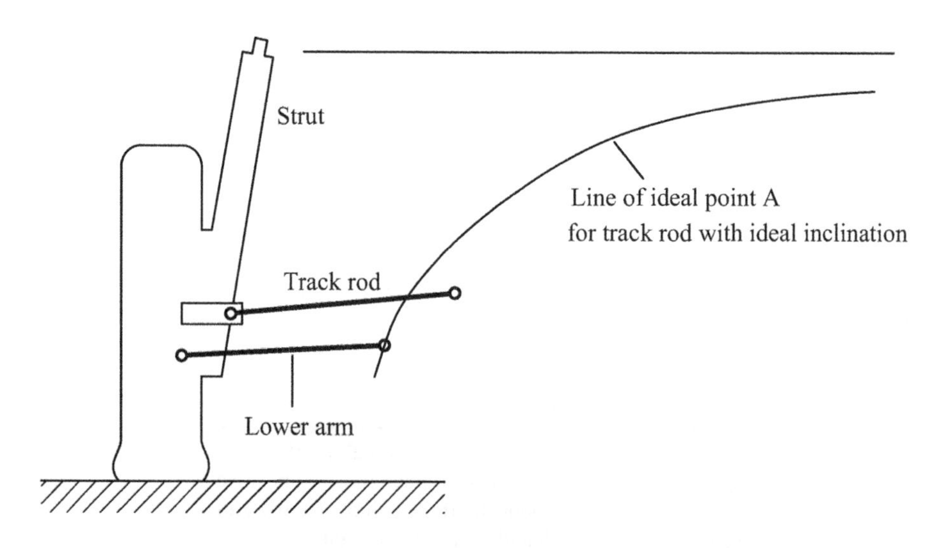

Figure 6.2.2 For strut suspension, the correct length of the steering arm varies considerably with the vertical position of the arm.

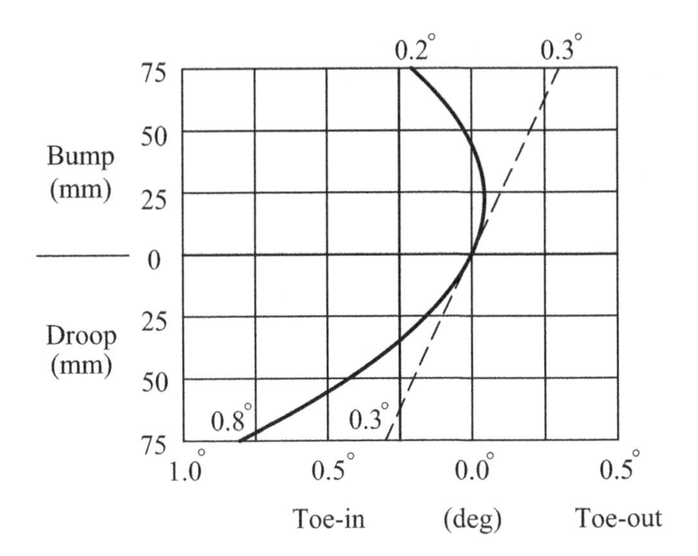

Figure 6.2.3 Example bump steer graph.

longer than ideal for a rack behind the wheels, and shorter than ideal for a rack in front of the wheels, as can be seen by imagining the ideal and actual arcs of the track-rod end. Relative to the tangent at zero bump, the quadratic bump steer is about 0.5° toe-in in 75 mm bump or droop, so the quadratic bump steer coefficient here is about $-90°/m^2$ or $-0.90°/dm^2$. This would be caused by a rack length error of about 20 mm (at each end).

The initial toe-out tendency in bump means that the wheel itself tends to recede from the bump, and so possibly reduces the steering fight on rough roads. The quadratic-term curvature toes in the inner wheel in cornering, reducing its slip angle. This can help to prevent undue wear because of excessive slip on the lightly loaded tyre, an example of the abandonment of ideal Ackermann steering under dynamic conditions.

Because of the nominal symmetry of the whole steering and suspension system, even if the above deliberate steer deviations are adopted, symmetrical axle heave (double bump) should not cause net steering effects, which is an advantage over a steered rigid axle. Nevertheless, the toe-in varies with load which can affect steering feel and tyre wear, for which reason if bump steer is deliberately introduced it is best to design it around a light-load position. Where present, bump steer is often a palliative for some other fault such as bad weight distribution, and is likely to give poor straight-line stability and tyre wear. At best such effects compromise the basic handling in order to gain some rough road or other small advantages, and so should be used with caution.

Many cars are designed with zero theoretical bump steer, but positioning of the steering rack is critical and sufficiently close tolerances are rarely held in production, particularly if this aspect was not considered adequately during the design process. Even for a given design, bump steer varies from car to car and even from side to side of one car, sometimes to the extent that one wheel toes in with bump and the other toes out. Unequal bump steer left and right, which is heave steer or double-bump steer, is especially bad for straight-line stability.

6.3 Axle Steer Angles

Given the individual steer angles of the two independent wheels on an axle, the axle can be considered to have combined steer angles, these being the symmetrical and antisymmetrical components of the individual wheel steer angles. The symmetrical part is the mean steer angle of the two wheels. The antisymmetrical part is the toe angle of the axle. Hence, the axle steer angle is

$$\delta_A = \tfrac{1}{2}(\delta_L + \delta_R) \tag{6.3.1}$$

and the axle toe angle is

$$\delta_T = \tfrac{1}{2}(\delta_L - \delta_R) \tag{6.3.2}$$

Given these axle angles, the individual wheel angles may then be deduced as

$$\delta_L = \delta_A + \delta_T \tag{6.3.3}$$

$$\delta_R = \delta_A - \delta_T \tag{6.3.4}$$

It is apparent from this that bump steer angles of individual wheels can alternatively be expressed as angle properties of the complete axle.

If the individual wheel static toe angles of an axle are unequal, normally attributable to improper adjustment rather than production tolerance error, then the axle will have a static steer angle. On an unsteered axle, at the rear, this will cause crabbing of the vehicle.

6.4 Roll Steer and Understeer

For a rigid axle, in cornering the suspension roll angle ϕ_S can cause a steer angle of the whole axle as a unit. This can also occur for an axle with independent wheels, with the additional complication of the effect of suspension roll on the axle toe angle. Suspension heave (double bump) may also cause steer effects. The axle steer angle may then be expressed as

$$\delta_A = \delta_{A0} + \varepsilon_{RS}\phi_S + \varepsilon_{DBS}z_S + \cdots \tag{6.4.1}$$

with other possible terms, where ε_{RS} is the linear roll-steer coefficient and ε_{DBS} is the linear double-bump steer coefficient.

The axle roll steer coefficient ε_{RS} is the rate of change of axle mean wheel steer angle with respect to suspension roll angle, and is usually expressed as deg/deg (i.e. degrees of steer per degree of roll), so it is dimensionless. Formally, the units are rad/rad. For small roll angles, the roll steer angle δ_{RS} is given in terms of the roll steer coefficient and suspension roll angle by

$$\delta_{RS} = \varepsilon_{RS}\phi_S \tag{6.4.2}$$

For independent suspension, the roll steer coefficient is closely related to the bump steer coefficient, but is also influenced by vehicle track width. For a rigid axle it relates to the steer angle of the complete axle, whereas for independent suspension it relates to the mean value for the pair of wheels.

The axle roll understeer angle δ_{RU} must be clearly distinguished from the axle roll steer angle δ_{RS}. They are numerically equal but the sign convention is different. The roll understeer angle is positive when it requires increased handwheel angle to compensate, that is, when it requires increased driver-applied steering.

Hence the relationship between roll understeer angle and roll steer angle is that they are the same at the front, but have change of sign at the rear:

$$\begin{aligned} \delta_{RUf} &= +\,\delta_{RSf} \\ \delta_{RUr} &= -\,\delta_{RSr} \end{aligned} \tag{6.4.3}$$

This also applies to the relationship between roll understeer and roll steer coefficients:

$$\begin{aligned} \varepsilon_{\text{RUf}} &= +\varepsilon_{\text{RSf}} \\ \varepsilon_{\text{RUr}} &= -\varepsilon_{\text{RSr}} \end{aligned} \tag{6.4.4}$$

Corresponding roll oversteer coefficients could be defined, but these would simply be negatives of the roll understeer values.

Axle roll steer is a property of a single axle, whereas axle roll understeer is a property of an axle in the context of a vehicle, the context defining the attributable sign. Roll understeer may also be a property of the vehicle as a whole, the vehicle roll understeer resulting from the characteristics of both axles.

The advantage of referring to roll understeer coefficients rather than roll steer coefficients is that in evaluating the total understeer coefficient or understeer angle the effects are simply added without constant consideration of a front/rear sign convention. For understeer coefficients,

$$\varepsilon_{\text{RU}} = \varepsilon_{\text{RUf}} + \varepsilon_{\text{RUr}} \tag{6.4.5}$$

whereas for roll steer coefficients the total understeer coefficient is

$$\varepsilon_{\text{RU}} = \varepsilon_{\text{RSf}} - \varepsilon_{\text{RSr}} \tag{6.4.6}$$

The sum of the front and rear roll steer coefficients has no simple physical meaning in terms of the steering. If the front and rear axles exhibit equal positive roll steer, the roll understeer is zero, and the steady-state steering is largely unaffected, but the dynamic response may be altered because suspension roll will cause transient slip angles.

The smooth-road-condition ideal of zero roll steer may be abandoned in some cases, but there are definite limitations to the amount of roll steer that is acceptable. Roll oversteer of independent front suspension (i.e. wheel toe-in in bump) gives severe wheel-fight on rough roads. Roll oversteer of the rear gives increased body attitude angles, a very unpleasant uncertain feeling for the driver, and bad directional response to rough roads and to side winds.

It does not seem to be effective to balance an oversteering rear with an understeering front, and really there is little reason to try to do so. In short, there is good reason to avoid roll oversteer at either end. Some would argue for rear roll understeer for its reduced attitude angle and hence possibly faster response, especially for large cars which tend to have a larger dynamic index, but this can lead to problems with rough roads or wind, and may result in engine torque steering because of body roll in strong acceleration, unless the power is transmitted through an independent or de Dion axle, or an offset lift bar is used.

6.5 Axle Linear Bump Steer and Roll Steer

Although the geometry of steering variation is, for independent suspension, expressed by the bump steer coefficients, from a practical point of view it is the combined action of left and right sides that matters when the body moves in roll. Hence, it is necessary to deduce the axle roll steer from the single-wheel bump steer. Of course this may be done by calculating the individual suspension bump values on each side from the suspension roll angle ϕ_S, the bump steer angles from the suspension bump values, and the roll steer angles from the bump steer angle values. However, it is also of interest to obtain the roll steer coefficients directly in terms of the bump steer properties.

Given a linear bump steer angle, with constant bump steer coefficient,

$$\delta_{\text{BS}} = \varepsilon_{\text{BS1}} z_{\text{S}} \tag{6.5.1}$$

and the relationship between roll angle and individual suspension bump

$$\begin{aligned} z_{S,R} &= +\tfrac{1}{2}T\phi_S \\ z_{S,L} &= -\tfrac{1}{2}T\phi_S \end{aligned} \tag{6.5.2}$$

then the individual wheel bump steer angles, toe-out positive, are

$$\begin{aligned} \delta_{BS,R} &= +\varepsilon_{BS1}\tfrac{1}{2}T\phi_S \\ \delta_{BS,L} &= -\varepsilon_{BS1}\tfrac{1}{2}T\phi_S \end{aligned} \tag{6.5.3}$$

To convert these to roll steer angles, there is a sign change on the left, because bump steer is positive for toe-out on both sides, but roll steer must have the same sign for the same sense of rotation on the two sides of the car:

$$\begin{aligned} \delta_{RS,R} &= +\ \delta_{BS,R} \\ \delta_{RS,L} &= -\ \delta_{BS,L} \end{aligned} \tag{6.5.4}$$

The wheel roll steer angles are therefore

$$\begin{aligned} \delta_{RS,R} &= +\varepsilon_{BS1}\tfrac{1}{2}T\phi_S \\ \delta_{RS,L} &= +\varepsilon_{BS1}\tfrac{1}{2}T\phi_S \end{aligned} \tag{6.5.5}$$

The axle mean roll steer angle is therefore

$$\begin{aligned} \delta_{RS} &= \tfrac{1}{2}(\delta_{RS,R}+\delta_{RS,L}) \\ &= \tfrac{1}{2}2\varepsilon_{BS1}\tfrac{1}{2}T\phi_S \\ &= \left(\tfrac{1}{2}T\varepsilon_{BS1}\right)\phi_S \end{aligned} \tag{6.5.6}$$

Comparing this with equation (6.5.1), the linear roll steer coefficient is

$$\varepsilon_{RS1} = \tfrac{1}{2}T\varepsilon_{BS1} \tag{6.5.7}$$

This is the fundamental relationship between roll steer and bump steer for an independent-suspension axle.

Because the bump steer angles are equal and opposite in this simple symmetrical case, there is no axle toe angle resulting from the roll angle.

6.6 Axle Non-Linear Bump Steer and Roll Steer

Consider now the more complex case, with quadratic bump steer included, and also vehicle asymmetries. The individual suspension bumps are still

$$\begin{aligned} z_{S,R} &= +\tfrac{1}{2}\phi_S T \\ z_{S,L} &= -\tfrac{1}{2}\phi_S T \end{aligned} \tag{6.6.1}$$

The wheel bump steer angles are

$$\begin{aligned} \delta_{BS,R} &= \varepsilon_{BS1,R}\, z_{S,R} + \varepsilon_{BS2,R}\, z_{S,R}^2 \\ \delta_{BS,L} &= \varepsilon_{BS1,L}\, z_{S,L} + \varepsilon_{BS2,L}\, z_{S,L}^2 \end{aligned} \tag{6.6.2}$$

which are therefore

$$\begin{aligned}\delta_{\text{BS,R}} &= \varepsilon_{\text{BS1,R}}\left(+\tfrac{1}{2}\phi_{\text{S}}T\right)+\varepsilon_{\text{BS2,R}}\left(+\tfrac{1}{2}\phi_{\text{S}}T\right)^2\\ \delta_{\text{BS,L}} &= \varepsilon_{\text{BS1,L}}\left(-\tfrac{1}{2}\phi_{\text{S}}T\right)+\varepsilon_{\text{BS2,L}}\left(-\tfrac{1}{2}\phi_{\text{S}}T\right)^2\end{aligned} \tag{6.6.3}$$

These bump steer angles are positive for toe-out on each side, so the equivalent mean roll steer angle for the complete axle is half of the difference of the bump steer angles of the two wheels:

$$\delta_{\text{RS}} = \tfrac{1}{2}(\delta_{\text{BS,R}} - \delta_{\text{BS,L}})$$

Since individual wheel toe-out is positive, this gives positive roll steer corresponding to the axle effectively turning out of the curve. Hence, this is an understeer effect at the front and oversteer at the rear. Substituting the above expressions, and noting that a negative valve squared is positive, the axle roll steer angle becomes

$$\delta_{\text{RS}} = \tfrac{1}{4}\phi_{\text{S}}T(\varepsilon_{\text{BS1,R}}+\varepsilon_{\text{BS1,L}})+\tfrac{1}{2}\left(\tfrac{1}{2}\phi_{\text{S}}T\right)^2(\varepsilon_{\text{BS2,R}}-\varepsilon_{\text{BS2,L}}) \tag{6.6.4}$$

Hence the first-order and second-order roll-steer coefficients are

$$\begin{aligned}\varepsilon_{\text{RS1}} &= \tfrac{1}{4}T(\varepsilon_{\text{BS1,R}}+\varepsilon_{\text{BS1,L}})\\ \varepsilon_{\text{RS2}} &= \tfrac{1}{8}T^2(\varepsilon_{\text{BS2,R}}-\varepsilon_{\text{BS2,L}})\end{aligned} \tag{6.6.5}$$

Therefore, any linear bump steer gives linear roll steer. However, symmetrical second-order bump steer does not give second-order roll steer, which only arises from asymmetrical ε_{BS2}. Hence, this will tend to be fairly small, although possibly not insignificant, because it will usually arise from production tolerance errors rather than from design intent. One feature that can cause it is unequal track-rod lengths in conjunction with certain types of suspension, a feature which has been used on some passenger cars.

The roll toe angle of the axle is

$$\begin{aligned}\delta_{\text{RT}} &= \tfrac{1}{2}(\delta_{\text{BS,R}}+\delta_{\text{BS,L}})\\ &= \tfrac{1}{4}T\phi_{\text{S}}(\varepsilon_{\text{BS1,R}}-\varepsilon_{\text{BS1,L}})+\tfrac{1}{2}\left(\tfrac{1}{2}T\phi_{\text{S}}\right)^2(\varepsilon_{\text{BS2,R}}+\varepsilon_{\text{BS2,L}})\end{aligned} \tag{6.6.6}$$

Hence, the first and second roll toe coefficients are

$$\begin{aligned}\varepsilon_{\text{RT1}} &= \tfrac{1}{4}T(\varepsilon_{\text{BS1,R}}-\varepsilon_{\text{BS1,L}})\\ \varepsilon_{\text{RT2}} &= \tfrac{1}{8}T^2(\varepsilon_{\text{BS2,R}}+\varepsilon_{\text{BS2,L}})\end{aligned} \tag{6.6.7}$$

Therefore, linear roll toe is due to asymmetry of the linear bump steer, due to production tolerance errors (e.g. a tilted rack). The quadratic roll toe coefficient is due to quadratic bump steer.

Equations (6.6.5) and (6.6.7) also make it possible to deduce the bump steer coefficients from roll steer and toe steer values. The equations are

$$\begin{aligned}\varepsilon_{\text{BS1,R}} &= \frac{2}{T}(\varepsilon_{\text{RS1}}+\varepsilon_{\text{RT1}})\\ \varepsilon_{\text{BS1,L}} &= \frac{2}{T}(\varepsilon_{\text{RS1}}-\varepsilon_{\text{RT1}})\\ \varepsilon_{\text{BS2,R}} &= \frac{4}{T^2}(\varepsilon_{\text{RS2}}+\varepsilon_{\text{RT2}})\\ \varepsilon_{\text{BS2,L}} &= \frac{4}{T^2}(\varepsilon_{\text{RS2}}-\varepsilon_{\text{RT2}})\end{aligned} \tag{6.6.8}$$

6.7 Axle Double-Bump Steer

Consider now an axle bump deflection, which is a suspension double bump, z_A, which might occur on landing after a symmetrical crest, as frequently occurs with some severity on rally cars. The right and left suspension bumps on the axle are

$$z_{S,R} = z_{S,L} = z_A \tag{6.7.1}$$

The individual bump steer angles are

$$\begin{aligned} \delta_{BS,R} &= \varepsilon_{BS1,R}\, z_A + \varepsilon_{BS2,R}\, z_A^2 \\ \delta_{BS,L} &= \varepsilon_{BS1,L}\, z_A + \varepsilon_{BS2,L}\, z_A^2 \end{aligned} \tag{6.7.2}$$

These are both positive for toe-out, so the axle (mean double bump) steer angle is given by

$$\delta_{DBS} = \tfrac{1}{2}(\delta_{BS,R} - \delta_{BS,L}) \tag{6.7.3}$$

which is

$$\delta_{DBS} = \tfrac{1}{2} z_A(\varepsilon_{BS1,R} - \varepsilon_{BS1,L}) + \tfrac{1}{2} z_A^2(\varepsilon_{BS2,R} - \varepsilon_{BS2,L}) \tag{6.7.4}$$

The linear and quadratic double-bump steer coefficients are therefore

$$\varepsilon_{DBS1} = \tfrac{1}{2}(\varepsilon_{BS1,R} - \varepsilon_{BS1,L}) \tag{6.7.5}$$

$$\varepsilon_{DBS2} = \tfrac{1}{2}(\varepsilon_{BS2,R} - \varepsilon_{BS2,L}) \tag{6.7.6}$$

Hence, both of the double-bump steer coefficients depend upon the side-to-side difference between the individual bump steer coefficients, and hence upon vehicle asymmetries, normally arising from production tolerance errors, for example a tilted rack, although possibly from design asymmetrical track-rod lengths, etc. The first-order double-bump steer coefficient ε_{DBS1} is particularly critical for straight-line behaviour on dips and crests, and suitable design and control of production tolerances must be applied to ensure only a small resulting value.

6.8 Vehicle Roll Steer

Roll of the entire sprung mass gives suspension roll of the front and rear axles, with total steer effects. The vehicle understeer coefficients follow easily from the axle roll steer coefficients, allowing for the understeer sign convention for axle position at the front or rear:

$$\begin{aligned} \varepsilon_{VRU1} &= \varepsilon_{RUf1} + \varepsilon_{RUr1} \\ \varepsilon_{VRU2} &= \varepsilon_{RUf2} + \varepsilon_{RUr2} \end{aligned} \tag{6.8.1}$$

Assuming equal tracks front and rear, by substitution the vehicle roll understeer coefficients are

$$\begin{aligned} \varepsilon_{VRU1} &= \tfrac{1}{4}\, T(\varepsilon_{BS1,fR} + \varepsilon_{BS1,fL} - \varepsilon_{BS1,rR} - \varepsilon_{BS1,rL}) \\ \varepsilon_{VRU2} &= \tfrac{1}{8}\, T^2(\varepsilon_{BS2,fR} - \varepsilon_{BS2,fL} - \varepsilon_{BS2,rR} + \varepsilon_{BS2,rL}) \end{aligned} \tag{6.8.2}$$

For the linear coefficient, the units are none, rad/rad or deg/deg as preferred, these having the same meaning. The quadratic coefficient has units rad^{-1} or rad/rad^2.

6.9 Vehicle Heave Steer

Heave of the entire sprung mass gives double bump of the front and rear axles, with possible steer effects. For the front axle,

$$\begin{aligned}\varepsilon_{\mathrm{DBS1,f}} &= \tfrac{1}{2}\left(\varepsilon_{\mathrm{BS1,fR}} - \varepsilon_{\mathrm{BS1,fL}}\right)\\ \varepsilon_{\mathrm{DBS2,f}} &= \tfrac{1}{2}\left(\varepsilon_{\mathrm{BS2,fR}} - \varepsilon_{\mathrm{BS2,fL}}\right)\end{aligned} \tag{6.9.1}$$

For the rear axle,

$$\begin{aligned}\varepsilon_{\mathrm{DBS1,r}} &= \tfrac{1}{2}\left(\varepsilon_{\mathrm{BS1,rR}} - \varepsilon_{\mathrm{BS1,rL}}\right)\\ \varepsilon_{\mathrm{DBS2,r}} &= \tfrac{1}{2}\left(\varepsilon_{\mathrm{BS2,rR}} - \varepsilon_{\mathrm{BS2,rL}}\right)\end{aligned} \tag{6.9.2}$$

Summing the effects, allowing for the signs, the vehicle heave understeer coefficients are

$$\begin{aligned}\varepsilon_{\mathrm{VHU1}} &= \varepsilon_{\mathrm{DBS1,f}} - \varepsilon_{\mathrm{DBS1,r}}\\ \varepsilon_{\mathrm{VHU2}} &= \varepsilon_{\mathrm{DBS2,f}} - \varepsilon_{\mathrm{DBS2,r}}\end{aligned} \tag{6.9.3}$$

giving the vehicle heave understeer coefficients finally as

$$\begin{aligned}\varepsilon_{\mathrm{VHU1}} &= \tfrac{1}{2}\left(\varepsilon_{\mathrm{BS1,fR}} - \varepsilon_{\mathrm{BS1,fL}} - \varepsilon_{\mathrm{BS1,rR}} + \varepsilon_{\mathrm{BS1,rL}}\right)\\ \varepsilon_{\mathrm{VHU2}} &= \tfrac{1}{2}\left(\varepsilon_{\mathrm{BS2,fR}} - \varepsilon_{\mathrm{BS2,fL}} - \varepsilon_{\mathrm{BS2,rR}} + \varepsilon_{\mathrm{BS2,rL}}\right)\end{aligned} \tag{6.9.4}$$

For the linear coefficient, the units are rad/m for use in equations, possibly expressed as deg/m. The quadratic coefficient has units rad/m^2, possibly expressed as deg/m^2.

Body heave will also cause axle toe angle effects. The required expressions for these are easily derived as above.

6.10 Vehicle Pitch Steer

For body pitching about the centre of mass, with suspension pitch angle θ_{S}, the axle double bumps are in the ratio of the partial wheelbases:

$$\begin{aligned}z_{\mathrm{Af}} &= +\,a\,\theta_{\mathrm{S}}\\ z_{\mathrm{Ar}} &= -\,b\,\theta_{\mathrm{S}}\end{aligned} \tag{6.10.1}$$

The vehicle pitch understeer coefficients are therefore

$$\varepsilon_{\mathrm{VPU1}} = \tfrac{1}{2}a\left(\varepsilon_{\mathrm{BS1,fR}} - \varepsilon_{\mathrm{BS1,fL}}\right) - \tfrac{1}{2}b\left(\varepsilon_{\mathrm{BS1,rR}} - \varepsilon_{\mathrm{BS1,rL}}\right) \tag{6.10.2}$$

$$\varepsilon_{\mathrm{VPU2}} = \tfrac{1}{2}a\left(\varepsilon_{\mathrm{BS2,fR}} - \varepsilon_{\mathrm{BS2,fL}}\right) - \tfrac{1}{2}b\left(\varepsilon_{\mathrm{BS2,rR}} - \varepsilon_{\mathrm{BS2,rL}}\right) \tag{6.10.3}$$

For the linear coefficient, the units are none, rad/rad or deg/deg as preferred, these having the same meaning. The quadratic coefficient has units rad^{-1} or rad/rad^2.

By an adverse accumulation of production tolerance errors, these coefficients may be significant, even though they are zero in principle for a symmetrical design of vehicle. The main result is a steer effect in braking, which, combined with reduced stability under braking, can be problematic in extreme cases. The solution is to hold sufficiently close tolerances on critical dimensions.

Body pitch will also cause toe effects on each axle. The coefficients may be derived as above.

Two other pitch understeer coefficients can be derived. These are for the combined-heave-and-pitch ride oscillation modes, which have nodes at particular points, depending on the body inertia and suspension stiffnesses. The coefficient for a mode follows by methods similar to the above, once the nodal position for that mode has been determined (e.g. Dixon, 2007).

6.11 Static Toe-In and Toe-Out

A small static toe-in at the rear often has a surprisingly large effect on handling, increasing understeer; front-wheel-drive vehicles, being lightly loaded at the rear, and therefore having a large rear tyre cornering stiffness coefficient, are especially sensitive to this. Front toe-out might be expected to have the same effect. However, practical experience shows the reverse: front toe-out gives a vague steering feel, whereas front toe-in gives a favourable feel and leads to increased understeer that can be measured on the skid pad. This seems to be caused by the combination of load transfer, aligning torque, lateral force on caster trail, and steering compliance. Too much front toe-in affects corner turn-in, giving an unprogressive and imprecise steering feel.

Static toe settings are governed within quite narrow bands by tyre wear. For the least wear, toe settings should be arranged to give minimal steer angles when running, regardless of small static camber angles. This means a small static toe-in for undriven wheels, and a small toe-out for driven ones, so that when the vehicle is running normally the tractive forces and compliances act to bring the toe angles close to zero. Within the allowable band for low wear (a total range of about 1°), there is only limited scope to use toe angles to tune the handling characteristics.

6.12 Rigid Axles with Link Location

Rigid rear axles can be considered in two groups: those with link location and those with longitudinal leaf springs. In the case of link location, there are lateral location points A and B according to the particular linkages, defining an axis of rotation of the axle relative to the body. Figure 6.12.1 shows a general four-link axle. The method is based on studying the support links to find two points A and B where forces are exerted by the axle on the body. The line through the two points is the axis of rotation of the axle relative to the body. The roll centre lies on this line, at the point where it penetrates the transverse vertical

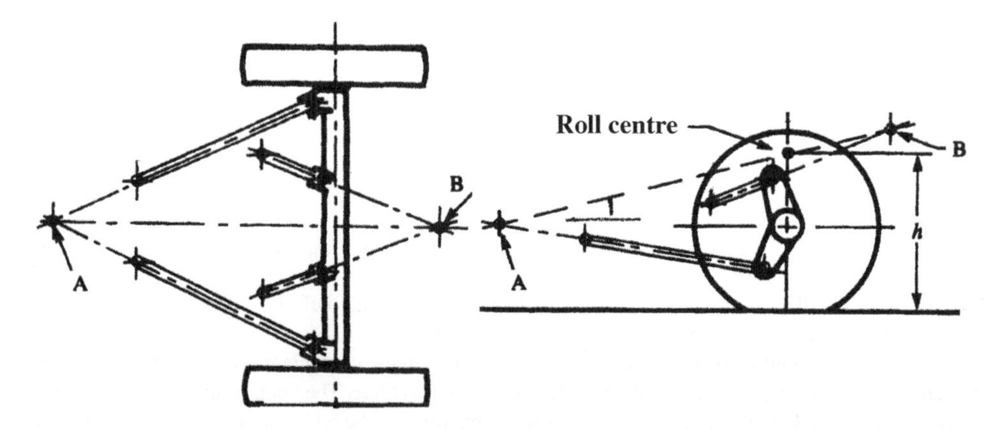

Figure 6.12.1 A rigid axle with link location to the body has two location points which define the relative axis of rotation. This position and inclination angle of this axis gives the roll centre and the roll steer coefficient.

plane of the wheel centres. One link pair has an intersection point at A, so the combined force exerted by these links on the body must act through A (neglecting bush torques and link weight). Similarly, the other link pair exerts a force through B. The resultant of the two forces at A and B acts through a point somewhere on the line AB. Suitable points A and B can be found for other axle link layouts. For example, if the lower links are parallel then the point B is at infinity, so AB is parallel to the bottom links. If the bottom link pair is replaced by a torque tube or similar system, then point B is the front ball joint. If transverse location is by a Panhard rod, then point A is the point at which the rod intersects the vertical central plane.

Because these points define the lateral location of the axle relative to the body, they also define the axis about which the axle will roll relative to the body if the road is considered to be rolled.

Considering the vehicle body to be rolled about a longitudinal axis parallel to the road, then if the front location point A is lower than the rear point B the different sideways movements of A and B, A more inward, will result in a steer rotation of the axle, such that the axle tries to increase its slip angle. This is at the rear, so it is an understeer effect. If the axle axis is inclined at ρ_A radians, positive angle being down at the front, a suspension roll angle of ϕ_S results in a rear axle roll steer angle, out of the curve, of

$$\delta_{RS} = -\rho_A \phi_S \tag{6.12.1}$$

The axle is actually turned into the curve, for positive ρ_A, reducing the required attitude angle. This is an understeer effect for a rear axle, so

$$\delta_{RU} = \rho_A \phi_S \tag{6.12.2}$$

The roll understeer coefficient (the rate of change of axle roll understeer angle with suspension roll angle) is therefore equal to the value of ρ_A in radians, that is, in this case value of the first roll understeer coefficient is

$$\varepsilon_{RU1} = \rho_A \tag{6.12.3}$$

This is often expressed as a percentage roll understeer, $\rho_A \times 100\%$. The axis angle may be around 0.05–0.10 rad, giving the roll understeer coefficient a value of 0.05–0.10 (rad/rad or deg/deg).

The variation of roll steer coefficient with axle load is important. The variation of roll steer with axle suspension bump is

$$k_{RUZA} = \frac{d\varepsilon_{RU}}{dz_A} \tag{6.12.4}$$

This gives the relationship to load (in newtons) according to the total axle suspension stiffness K_A:

$$k_{RUFV} = \frac{d\varepsilon_{RU}}{dF_{VA}} = \frac{1}{K_A}\frac{d\varepsilon_{RU}}{dz_A} \tag{6.12.5}$$

It can be examined easily by considering the change of the axis angle ρ_A from the motion of points A and B with increasing load and axle bump deflection. If the points move equally in the same direction then there is a change of roll centre height, but no change of roll steer. In some cases, for example the convergent four-link suspension of Figure 6.12.1, when the body moves down then A moves down and B moves up, giving a small change of roll centre height but a large increase of roll steer coefficient. Some positive sensitivity (i.e. increasing the axis angle ρ_A), may be desirable to help to compensate for the otherwise general trend towards oversteer with increasing load that occurs because of the tyre characteristics. This can help with primary understeer but does not help with final understeer or oversteer, for which the roll centre height is the dominant factor.

It might appear in the above discussion that the axle axis angle should be measured relative to the vehicle roll axis rather than to the horizontal. However, the vehicle does not in a real sense roll about its roll axis. Rolling the body about the inclined roll axis implies a roll about a horizontal axis plus a yaw movement which will affect front and rear suspensions equally, and so will have no net result on the steering angle required.

6.13 Rigid Axles with Leaf Springs

In the case of longitudinal leaf springs, the roll steer coefficient depends on the inclination of the equivalent link AB that describes the motion arc of the wheel centre, Figure 6.13.1. This equivalent link is directed towards the unshackled end, is about three-quarters of the length of that end of the spring, and is roughly parallel to it. When the body rolls, point A rises on the inner side and falls on the outer, thus tending to steer the axle. Horizontal equivalent links give no steer, because both sides move forward equally. Having point A higher than B gives roll oversteer for a rear axle (i.e. reduced slip angle tendency), B higher gives roll understeer. For a front axle the effects are opposite. The roll steer coefficient is equal to the AB line inclination angle expressed in radians, independent of the spring length or separation. On the other hand, the spring length affects the influence of load variation on roll steer coefficient. Too high a coefficient, apart from being bad for handling, also leads to harshness on rough roads because of the wheel path in bump, and other problems. Early Hotchkiss axles, before 1930, were given a negative roll-steer coefficient because this resulted in less sensitivity to road roughness. In the early 1930s a positive coefficient was first used (i.e. roll oversteer was replaced by roll understeer) and a dramatic increase in directional stability was found.

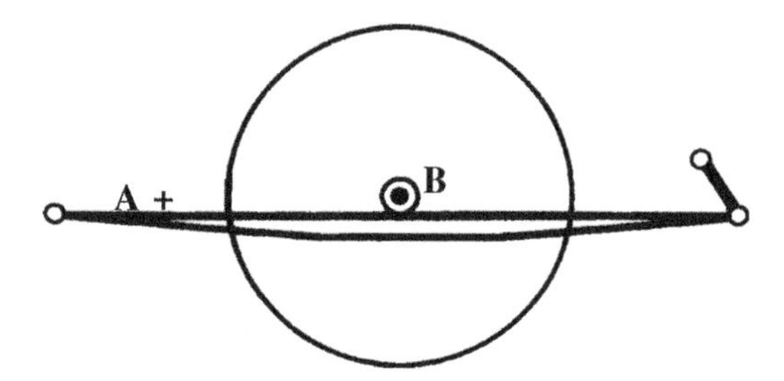

Figure 6.13.1 Rigid axle with location by leaf springs, without steering.

6.14 Rigid Axles with Steering

Before the introduction of independent front suspension, all cars had a rigid front axle with steering. They suffered from a variety of rather severe steering problems.

When the rigid axle with longitudinal leaf springs is used at the front, as on some trucks and off-road vehicles even today, it is subject to all the effects described earlier, plus additional effects because of the steering linkage. The critical steering link is always the one that connects the sprung and unsprung parts of the steering. Figure 6.14.1 shows a typical arrangement with front steering and a rear spring shackle. Here CD is the drag link, D is the connection to the wheel hub, and C is the connection to the Pitman arm on the steering box, fixed to the sprung mass. When the axle moves, D has an ideal no-steer arc centred on the ideal centre E. If E and C do not coincide, there will be steering errors. As early as the 1920s, it was attempted to match the E and C positions for roll motions, but with disappointing results. The reason is that the arc of D is different for each of roll, single-wheel bump, and heave, and different again with braking

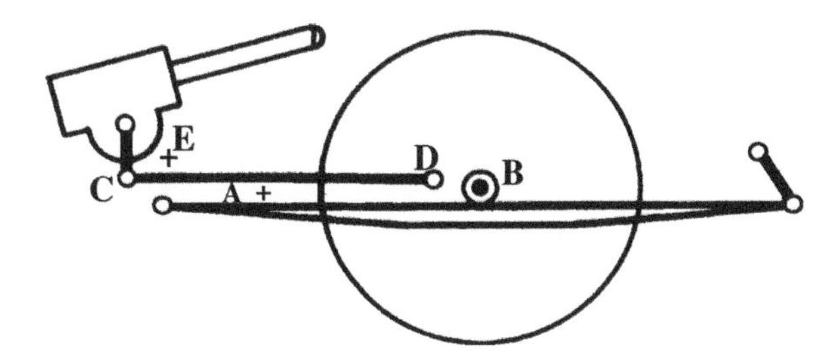

Figure 6.14.1 Rigid axle with leaf-spring location and steering.

because of axle wind-up. Also there are differences because of production variability of springs, and variation between spring options. One improvement that is sometimes adopted is to use an unsprung steering box, the sprung to unsprung connecting linkage being through a splined steering column.

In the early days, before the introduction of front independent suspension, the steering errors of rigid front axles were a real problem. The usual way to deal with the problem was to use stiff springs to minimise the axle movement. When independent suspension was introduced, as discussed in Chapter 1, it became possible to use much softer front suspension springing having greater deflections without unacceptable bump steer and roll steer problems, allowing implementation of the now better-understood relationship between front and rear stiffnesses that give good ride qualities – with the front springs softer than the rear to give the desired pitch and heave oscillation mode qualities and mode nodal positions.

References

Dixon, J.C. (2007) *The Shock Absorber Handbook*. Chichester: John Wiley & Sons, Ltd. ISBN 9780470510209.

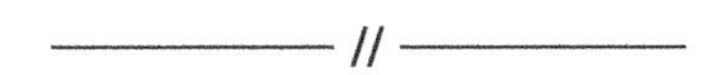

7

Camber and Scrub

7.1 Introduction

Camber is angling of the wheel from the vertical, in front or rear view. Scrub is lateral movement of the bottom point of the wheel. These are controlled by similar factors in the suspension geometry, and so are dealt with together.

Wheel camber relative to the road surface affects the tyre lateral forces. The camber angle variation with suspension bump depends on the position of the swing centre, introduced here. The swing centre also affects the scrub rate and the roll centre position. For an independent suspension, the wheel camber variation is described by the bump camber. For a complete independent axle, the coefficients can be expressed as symmetrical and antisymmetrical properties of the axle. The main angles and coefficients are as in Table 7.1.1. The units shown are the fundamental SI ones, as normally required for use in equations. Actual values are, however, often quoted in other units, particularly degrees instead of radians, and inches instead of metres in the USA. There are also several important bump scrub coefficients, leading to heave, double bump and roll scrub coefficients.

7.2 Wheel Inclination and Camber

The symbol for tyre slip angle is α, but a different symbol, δ, is used for steer angles, including bump steer. There is no accepted symbol for a deliberate alteration of camber angle, since steering is not controlled in that way. Hence, γ is used here and elsewhere for all types of wheel camber. The distinction should be borne in mind, however, between the different effects represented by γ, especially those corresponding to the distinction between δ and α. There can be camber angles relative to the vehicle body, corresponding to δ, and camber angles relative to the road surface, corresponding to α. The body roll angle ϕ then corresponds to the attitude angle β.

A distinction must also be made between the inclination angle and the camber angle. For a given wheel, these have the same magnitude but with a different sign convention. Inclination is positive for right-hand rotation about the forward longitudinal axis, which is clockwise in rear view. Camber is positive with the top of the wheel outward from the vehicle centreline, Figure 7.2.1. Here, the wheel inclination angle will be represented by Γ (Greek capital gamma), whereas the wheel camber angle will be denoted by γ (Greek lower case gamma).

The basic relationship between the inclination and camber angles for the two sides of the vehicle is therefore simply that on the right-hand side they are the same, but on the left-hand side there is a change

Table 7.1.1 Camber and inclination angles and coefficients

(1) γ_0	the static wheel camber angle (rad)
(2) γ_{BC}	the wheel bump camber angle (rad)
(3) Γ_A	the axle inclination angle (rad)
(4) γ_A	the axle camber angle (rad)
(5) ε_{BC}	the wheel bump camber coefficient (rad/m)
(6) ε_{ARI}	the axle roll inclination coefficient (—, rad/rad, deg/deg),
(7) ε_{ARC}	the axle roll camber coefficient (rad/rad)
(8) Γ_P	the path section inclination angle (rad).

of sign:

$$\begin{aligned} \Gamma_R &= +\gamma_R \\ \Gamma_L &= -\gamma_L \end{aligned} \tag{7.2.1}$$

There is a small static camber setting on a wheel, denoted γ_0. On a passenger car this is chosen to minimise wear. On a racing car it is almost invariably negative. The wheel camber relative to the body changes as the suspension moves in bump, Figure 7.2.2. The bump camber angle of a single wheel, γ_{BC}, is the increase of camber angle, compared with the static value, due to suspension bump deflection z_S. For a linear model, the total camber angle at a given suspension bump is

$$\gamma = \gamma_0 + \gamma_{BC} \tag{7.2.2}$$

Linearly, the bump camber angle is

$$\gamma_{BC} = \varepsilon_{BC1} z_S \tag{7.2.3}$$

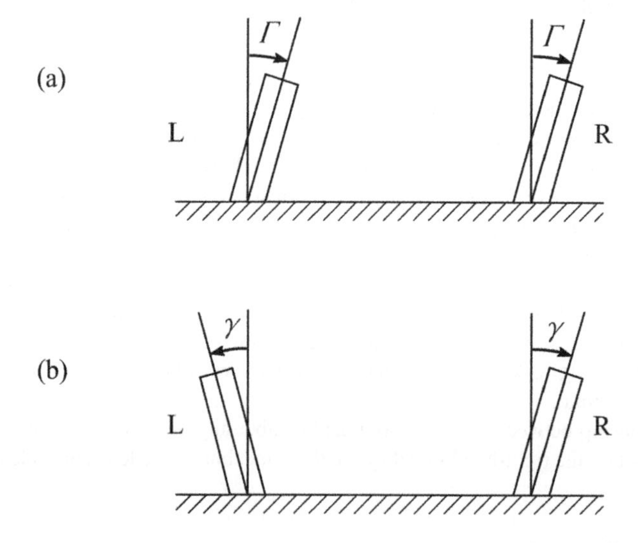

Figure 7.2.1 Wheel inclination angle and camber angle sign convention: (a) positive inclination, rear view only; (b) positive camber, front or rear view.

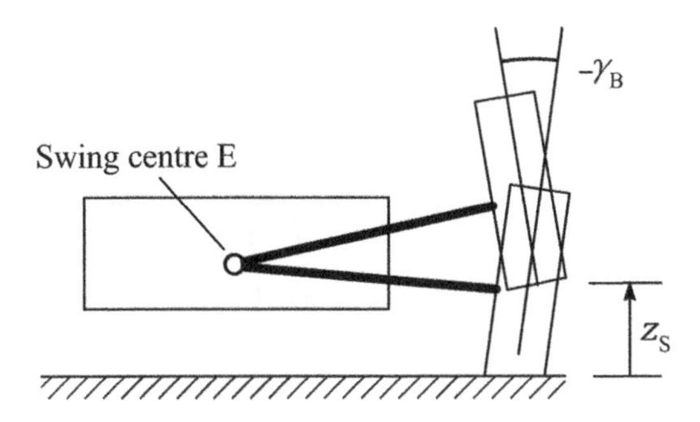

Figure 7.2.2 Suspension bump causes wheel camber angle change (rear view).

where the linear bump camber coefficient ε_{BC1} depends on the suspension geometry, and would be positive for positive-going camber on a rising wheel. The basic units of the bump camber coefficient are rad/m, although commonly expressed as deg/dm (deg/inch in the USA). The bump camber coefficient is usually negative for practical designs. More elaborately, including a second-order term,

$$\gamma = \gamma_0 + \varepsilon_{BC1} z_S + \varepsilon_{BC2} z_S^2 + \cdots \tag{7.2.4}$$

where ε_{BC2} is the quadratic bump camber coefficient.

7.3 Axle Inclination and Camber

The axle line is the line joining the centres of the two wheels at the opposite ends of the axle. Here, the term 'axle' is applied broadly to include the two wheels of an independent suspension. The wheel inclination angle and camber angle are measured from the perpendicular to the axle line. Axle roll angle (the roll angle of the axle line) and road surface transverse-section angles also affect the camber of the wheel relative to the road, but are accounted for separately, Figure 7.3.1. When there is suspension roll, the individual wheels are in suspension bump and have camber changes relative to the body. For an independent suspension, these wheel cambers can be accounted for separately, but it is also possible to work in terms of the symmetrical and antisymmetrical axle values, the axle mean inclination and the axle mean camber, respectively. In Figure 7.3.1(d) note that, with clockwise rotation positive, a negative path inclination gives a positive wheel inclination relative to the road.

The axle mean inclination angle (nothing to do with the axle roll angle) is the mean of the inclination angles of the two wheels of the axle:

$$\Gamma_A = \tfrac{1}{2}(\Gamma_R + \Gamma_L) \tag{7.3.1}$$

For an ideally symmetrical vehicle, this is zero. By substitution, it is equal to the semi-difference of the wheel camber angles, arising from constructional imperfections or compliance camber:

$$\Gamma_A = \tfrac{1}{2}(\gamma_R - \gamma_L) = \gamma_d \tag{7.3.2}$$

The antisymmetrical component, called the axle camber angle, is the semi-difference of the inclination angles,

$$\gamma_A = \tfrac{1}{2}(\Gamma_R - \Gamma_L) \tag{7.3.3}$$

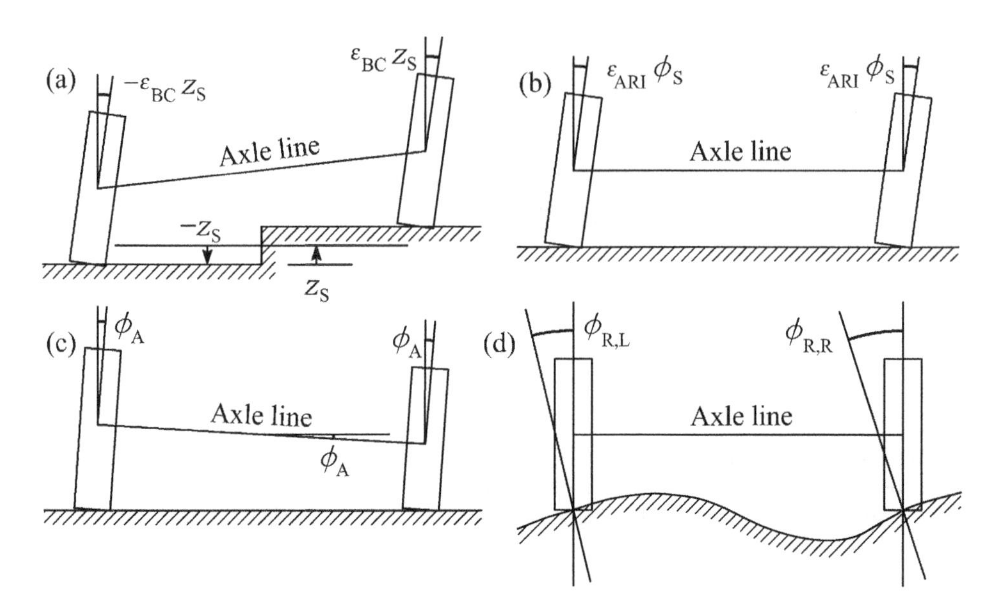

Figure 7.3.1 Wheel/road camber caused by: (a) antisymmetrical suspension bump; (b) suspension roll camber; (c) rigid axle roll on tyre vertical stiffness; (d) road section shape.

This is also equal to the mean of the camber angles:

$$\gamma_A = \tfrac{1}{2}(\gamma_R + \gamma_L) = \gamma_m \tag{7.3.4}$$

Then the individual right and left wheel inclination angles are

$$\begin{aligned} \Gamma_R &= \Gamma_A + \gamma_A \\ \Gamma_L &= \Gamma_A - \gamma_A \end{aligned} \tag{7.3.5}$$

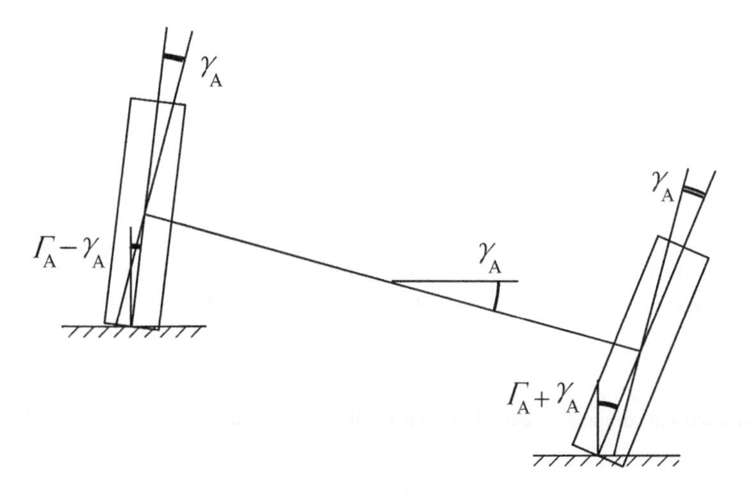

Figure 7.3.2 Axle mean inclination Γ_A and axle camber γ_A together give the individual wheel inclinations.

The static camber angles are:

$$\begin{aligned}\gamma_{RO} &= +\,\Gamma_{AO} + \gamma_{AO}\\ \gamma_{LO} &= -\,\Gamma_{AO} + \gamma_{AO}\end{aligned} \tag{7.3.6}$$

The axle camber angle (Figure 7.3.2, the mean camber angle) is the intended design value, once being normally positive but nowadays often negative, and usually negative on racing cars to give the best cornering grip and to equalise the temperature distribution across the tread. The antisymmetrical component, the difference of (static) camber angle values, arises on a passenger car only because of production tolerances, although it is used deliberately on some racing cars, essentially those that operate on circuits with turns in one direction only, for example US left-turn 'oval' tracks.

7.4 Linear Bump and Roll

In suspension roll, there are suspension bump deflections, $+\frac{1}{2}T\phi_S$ on the right and $-\frac{1}{2}T\phi_S$ on the left, with wheel camber changes, altering the axle inclination and camber values, Figure 7.4.1.

The axle roll inclination angle is the mean inclination angle for the two wheels, relative to the road, resulting from suspension roll. The axle roll inclination coefficient ε_{ARI} is the rate of change of the mean inclination of the two wheels with respect to suspension roll. It is dimensionless, being the ratio of an angle over an angle, possibly expressed in units rad/rad or deg/deg. It is related to the bump camber coefficient but also depends on the vehicle track. For small roll angles, linearly, the axle roll inclination angle (mean of the two wheels) is

$$\Gamma_{ARI} = \varepsilon_{ARI}\phi_S \tag{7.4.1}$$

Considering a suspension roll angle ϕ_S, with corresponding suspension bumps $\pm\frac{1}{2}T\phi_S$ relative to the body, the single wheel camber and inclination angles due to suspension roll (axle roll accounted separately), relative to the axle line, are

$$\gamma_R = +\,\phi_S + \tfrac{1}{2}T\phi_S\varepsilon_{BC} \tag{7.4.2}$$

$$\gamma_L = -\,\phi_S - \tfrac{1}{2}T\phi_S\varepsilon_{BC} \tag{7.4.3}$$

$$\Gamma_R = +\,\phi_S + \tfrac{1}{2}T\phi_S\varepsilon_{BC} \tag{7.4.4}$$

$$\Gamma_L = +\,\phi_S + \tfrac{1}{2}T\phi_S\varepsilon_{BC} \tag{7.4.5}$$

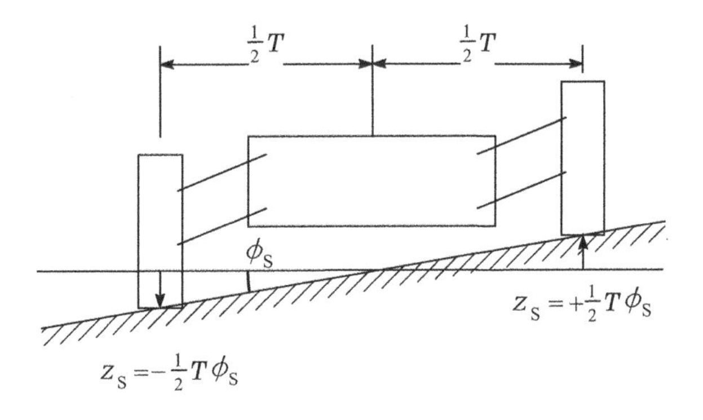

Figure 7.4.1 Individual wheel bump resulting from suspension roll.

The axle mean inclination angle is

$$\begin{aligned}\Gamma_A &= \tfrac{1}{2}(\Gamma_R + \Gamma_L) \\ &= \phi_S + \tfrac{1}{2}T\phi_S\varepsilon_{BC} \\ &= (1 + \tfrac{1}{2}T\varepsilon_{BC})\phi_S\end{aligned} \tag{7.4.6}$$

Hence the fundamental relationship between the linear axle roll inclination coefficient and the linear wheel bump camber coefficient is

$$\varepsilon_{ARI} = 1 + \tfrac{1}{2}T\varepsilon_{BC} \tag{7.4.7}$$

Bump camber coefficients are usually made negative in order to reduce ε_{ARI}, to offset the effect of body roll on wheel inclination, Figure 7.4.2. To be explicit, equation (7.4.7) may be inverted into the design equation for the necessary bump camber for a desired roll inclination coefficient:

$$\varepsilon_{BC} = \frac{2}{T}(\varepsilon_{ARI} - 1) \tag{7.4.8}$$

For example, the axle roll inclination coefficient would typically be 0.5 (rad/rad), requiring a bump camber coefficient value of $-1/T$ (rad/m), about -0.7 rad/m.

A rigid axle is not subject to suspension roll inclination or camber, but it does roll because suspension roll leads to axle roll as a result of load transfer on the tyre vertical stiffness. Very loosely, then, a solid axle might be said to have an axle roll camber coefficient, typically of about 0.12 (deg/deg) for a passenger car, although this is not a real geometric effect. This axle roll on the tyres is also applicable to independent suspensions. However, this is not normally accounted for as an axle inclination angle change. It is a roll of the wheel-centres axle line:

$$\phi_A = k_{\phi A \phi S}\phi_S \tag{7.4.9}$$

The complete inclination angles of the wheels relative to the path surface are therefore, Figure 7.4.3, including path inclination angles, given by the path angle contact point to contact point, minus the local path inclination angle, plus the axle roll angle plus the wheel inclination relative to the axle:

$$\Gamma_{W/P} = (\phi_P + \phi_A + \Gamma_W) - (\phi_P + \Gamma_P) \tag{7.4.10}$$

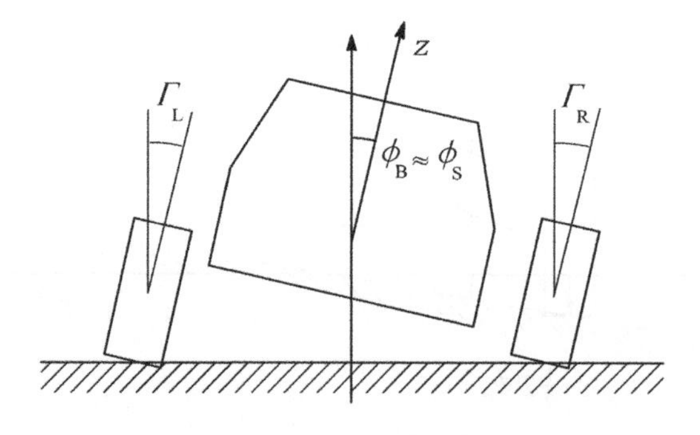

Figure 7.4.2 Suspension roll causes wheel inclination.

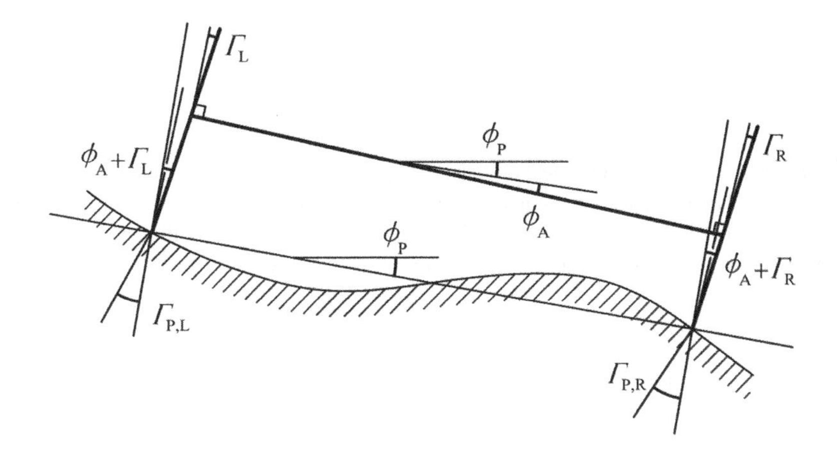

Figure 7.4.3 Wheel inclination angles relative to the local path surface.

where the path inclination angle Γ_P is measured relative to the path bank angle ϕ_P. To be specific, noting that the path bank angle is eliminated, the left and right inclination angles relative to the path (road) are:

$$\Gamma_{W,R} = \phi_A + \Gamma_R - \Gamma_{P,R} = \phi_A + (\gamma_{R0} + \phi_S + \tfrac{1}{2}T\varepsilon_{BC}\phi_S) - \Gamma_{P,R} \quad (7.4.11)$$

$$\Gamma_{W,L} = \phi_A + \Gamma_L - \Gamma_{P,L} = \phi_A + (-\gamma_{L0} + \phi_S + \tfrac{1}{2}T\varepsilon_{BC}\phi_S) - \Gamma_{P,L} \quad (7.4.12)$$

7.5 Non-Linear Bump and Roll

Considering non-linear bump and roll suspension characteristics, and also asymmetrical characteristics, the basic wheel camber angles are

$$\begin{aligned}\gamma_R &= \gamma_{R0} + \phi_S + \varepsilon_{BC1,R}z_S + \varepsilon_{BC2,R}z_S^2 \\ \gamma_L &= \gamma_{L0} - \phi_S + \varepsilon_{BC1,L}z_S + \varepsilon_{BC2,L}z_S^2\end{aligned} \quad (7.5.1)$$

Substituting $z_S = \pm\tfrac{1}{2}\phi T_S$, the inclination angles are

$$\begin{aligned}\Gamma_R &= +\gamma_{R0} + \phi_S + \tfrac{1}{2}T\varepsilon_{BC1,R}\phi_S + \tfrac{1}{4}T^2\varepsilon_{BC2,R}\phi_S^2 \\ \Gamma_L &= -\gamma_{L0} + \phi_S + \tfrac{1}{2}T\varepsilon_{BC1,L}\phi_S - \tfrac{1}{4}T^2\varepsilon_{BC2,R}\phi_S^2\end{aligned} \quad (7.5.2)$$

The axle mean inclination angle is

$$\begin{aligned}\Gamma_A &= \tfrac{1}{2}(\Gamma_R + \Gamma_L) \\ &= \tfrac{1}{2}(\gamma_{R0} - \gamma_{L0}) + \phi_S + \tfrac{1}{4}T(\varepsilon_{BC1,R} + \varepsilon_{BC1,L})\phi_S + \tfrac{1}{8}T^2(\varepsilon_{BC2,R} - \varepsilon_{BC2,L})\phi_S^2\end{aligned} \quad (7.5.3)$$

This may be expressed in terms of the axle roll inclination coefficients as

$$\Gamma_A = \Gamma_{A0} + \varepsilon_{ARI1}\phi_S + \varepsilon_{ARI2}\phi_S^2 \quad (7.5.4)$$

with the three terms

$$\Gamma_{A0} = \tfrac{1}{2}(\gamma_{R0} - \gamma_{L0}) \quad (7.5.5)$$

$$\varepsilon_{\mathrm{ARI1}} = 1 + \tfrac{1}{4}T(\varepsilon_{\mathrm{BC1,R}} + \varepsilon_{\mathrm{BC1,L}}) \tag{7.5.6}$$

$$\varepsilon_{\mathrm{ARI2}} = \tfrac{1}{8}T^2(\varepsilon_{\mathrm{BC2,R}} - \varepsilon_{\mathrm{BC2,L}}) \tag{7.5.7}$$

The first of these, Γ_{A0}, is the asymmetrical production tolerance error on static camber. The second, $\varepsilon_{\mathrm{ARI1}}$, is the linear axle roll inclination coefficient, and depends on the sum of the wheel first-order bump camber coefficients. The final term, $\varepsilon_{\mathrm{ARI2}}$, is the quadratic axle roll inclination coefficient, which therefore depends on design asymmetry and production tolerance asymmetry giving different second-order bump camber coefficients.

The axle camber is

$$\begin{aligned}\gamma_{\mathrm{A}} &= \tfrac{1}{2}(\Gamma_{\mathrm{R}} - \Gamma_{\mathrm{L}})\\ &= \tfrac{1}{2}(\gamma_{\mathrm{R0}} + \gamma_{\mathrm{L0}}) + \tfrac{1}{4}T(\varepsilon_{\mathrm{BC1,R}} - \varepsilon_{\mathrm{BC1,L}})\phi_{\mathrm{S}} + \tfrac{1}{8}T^2(\varepsilon_{\mathrm{BC2,R}} + \varepsilon_{\mathrm{BC2,L}})\phi_{\mathrm{S}}^2\end{aligned} \tag{7.5.8}$$

This may be expressed as

$$\gamma_{\mathrm{A}} = \gamma_{\mathrm{A0}} + \varepsilon_{\mathrm{ARC1}}\phi_{\mathrm{S}} + \varepsilon_{\mathrm{ARC2}}\phi_{\mathrm{S}}^2 \tag{7.5.9}$$

with the three terms

$$\gamma_{\mathrm{A0}} = \tfrac{1}{2}(\gamma_{\mathrm{R0}} + \gamma_{\mathrm{L0}}) \tag{7.5.10}$$

$$\varepsilon_{\mathrm{ARC1}} = \tfrac{1}{4}T(\varepsilon_{\mathrm{BC1,R}} - \varepsilon_{\mathrm{BC1,L}}) \tag{7.5.11}$$

$$\varepsilon_{\mathrm{ARC2}} = \tfrac{1}{8}T^2(\varepsilon_{\mathrm{BC2,R}} + \varepsilon_{\mathrm{BC2,L}}) \tag{7.5.12}$$

The first of these, γ_{A0}, is the mean static camber of the two wheels, arising from production tolerances. The second, $\varepsilon_{\mathrm{ARC1}}$, is the linear axle roll camber coefficient, and depends on the production tolerance difference of the wheel first-order bump camber coefficients. The final term, $\varepsilon_{\mathrm{ARC2}}$, is the quadratic axle roll camber coefficient which depends on the sum of the wheel second-order bump camber coefficients.

An ideal symmetrical vehicle can therefore be expected to exhibit zero second-order axle roll inclination and zero first-order axle roll camber.

7.6 The Swing Arm

At any given bump position, for a further infinitesimal increment of the suspension in bump, the wheel has an instantaneous centre of rotation relative to the body, point E in Figure 7.6.1. In general kinematics, an instantaneous centre of rotation is known as a 'centro'. In vehicle dynamics, in this particular case it is called the swing arm centre or just swing centre. For some suspensions (e.g. a swing axle) this centro is a fixed point on the body, while for others (e.g. wishbones, struts) the point depends on the suspension geometry in a more complex way, and moves relative to the body, so the coordinates of E depend on the suspension bump position z_{S}.

Point F is the notional wheel contact point with the ground. This is in the vertical transverse section of the suspension, that is, in the section shown in Figure 7.6.1, in the longitudinal plane of the notionally undistorted wheel, and at the level of the bottom point of the wheel. In cornering, the tyre has substantial lateral distortion, but point F excludes this, being in the wheel centre plane. Point H is the foot of the perpendicular from the swing centre E to the level of the bottom of the wheel. Length FE is called the swing arm length, whereas the horizontal length FH is called the swing arm radius (SAE definitions).

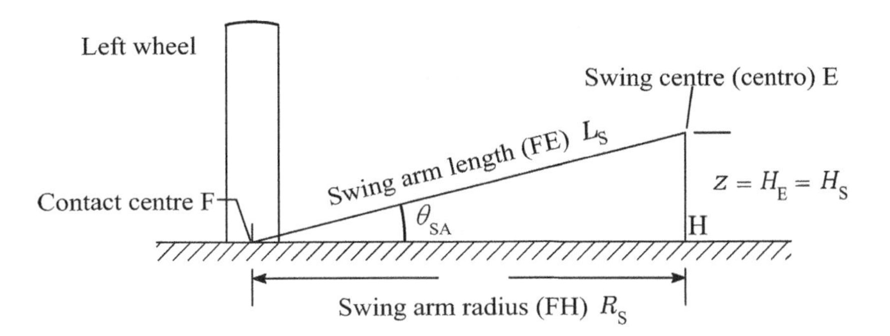

Figure 7.6.1 The swing arm FE and swing centre E. The swing centre is the instantaneous centre of rotation of the wheel relative to the body, in suspension bump movement. The diagram shows a rear elevation of the left wheel (a front elevation of the right wheel is the same).

For simplicity of notation here,

$$y = \text{FH}; \quad z = \text{HE}$$

The swing arm angle is

$$\theta_{SA} = \text{atan}\frac{z}{y}$$

and in some cases is negative, that is, the swing centre is sometimes below ground level. The tangent of θ_{SA} is called the bump scrub rate coefficient, discussed in detail later. The swing arm angle plays an important part in vehicle handling characteristics, controlling, *inter alia*, the roll centre position. Its value is typically about 8°, but it varies widely for different types of suspension, from −8° to 30°.

Considering the vehicle body to be fixed, when the wheel moves in suspension bump, the bottom point of the wheel moves perpendicularly to the swing arm line EF. Therefore the bottom of the wheel 'scrubs' across the ground, unless point E is at ground level. With the vehicle moving longitudinally, this scrub motion momentarily alters the tyre slip angle and cornering force. The ratio of the horizontal velocity component at the contact patch, V_{CH}, to the vertical bump velocity for the bottom of the wheel (the contact point) equals the bump scrub rate coefficient:

$$\varepsilon_{BScd} = \frac{V_{CH}}{V_S} = \tan\theta_{SA} = \frac{\text{EH}}{\text{FH}} = \frac{z}{y} \tag{7.6.1}$$

Approximating the tangent, the bump scrub rate coefficient equals the swing arm angle when the latter is expressed in radians, therefore being typically about 0.1 and varying from −0.1 to 0.4. A positive value is a track increase in suspension bump.

Variation of the position of E in bump – and in particular the variation of θ_{SA} or ε_{BScd} – is important. The variation of the swing arm may be analysed as in the following sections. By way of introduction, in Figure 7.6.1 consider the swing arm in a coordinate system (y, z) fixed to the body but with origin at the original wheel contact point. The initial swing arm centre is at (y_0, z_0). The linear variation of the swing centre position relative to the body is

$$\begin{aligned} y &= y_0 + c_y z_S \\ z &= z_0 + c_z z_S \end{aligned} \tag{7.6.2}$$

7.7 Bump Camber Coefficients

Referring again to Figure 7.6.1, the incremental change of single-wheel camber angle for an increment of suspension bump is

$$\mathrm{d}\gamma = -\frac{1}{y}\mathrm{d}z_{\mathrm{S}}$$

so the rate of change of camber with suspension bump is

$$\frac{\mathrm{d}\gamma}{\mathrm{d}z_{\mathrm{S}}} = \frac{-1}{y_0 + c_y z_{\mathrm{S}}} \tag{7.7.1}$$

In the special case of $c_y = 0$, with the swing centre moving only vertically,

$$\gamma = \gamma_0 - \frac{z_{\mathrm{S}}}{y_0} \tag{7.7.2}$$

Otherwise, by integration of equation (7.7.1) (e.g. Spiegel, 1968, 14.59), from $z_{\mathrm{S}} = 0$,

$$\gamma = \gamma_0 - \frac{1}{c_y}\log_e\left(1 + \frac{c_y z_{\mathrm{S}}}{y_0}\right) \tag{7.7.3}$$

The vertical variation of the swing centre position, coefficient c_z, plays no part in this. Expanding the logarithm as a series,

$$\log_e(1+x) = x - \tfrac{1}{2}x^2 + \tfrac{1}{3}x^3 - \tfrac{1}{4}x^4 + \cdots$$

gives

$$\gamma = \gamma_0 - \frac{z_{\mathrm{S}}}{y_0} + \tfrac{1}{2}c_y\left(\frac{z_{\mathrm{S}}}{y_0}\right)^2 - \cdots \tag{7.7.4}$$

which for $c_y = 0$ reduces to the earlier simplified case solution. Matching the polynomial expression for camber, from equation (7.2.4), which was

$$\gamma = \gamma_0 + \varepsilon_{\mathrm{BC1}} z_{\mathrm{S}} + \varepsilon_{\mathrm{BC2}} z_{\mathrm{S}}^2$$

gives the bump camber coefficients in terms of the swing-arm centre initial position and its movement coefficient c_y:

$$\varepsilon_{\mathrm{BC1}} = -\frac{1}{y_0} \tag{7.7.5}$$

$$\varepsilon_{\mathrm{BC2}} = \frac{c_y}{2y_0^2} \tag{7.7.6}$$

7.8 Roll Camber Coefficients

Substituting equations (7.7.5) and (7.7.6) into (7.5.6) onwards, the linear and quadratic roll inclination and roll camber coefficients may be expressed in terms of the swing centre movement coefficients. The results are:

$$\varepsilon_{\mathrm{ARI1}} = 1 - \tfrac{1}{4}T\left(\frac{1}{y_{0,\mathrm{R}}} + \frac{1}{y_{0,\mathrm{L}}}\right) \tag{7.8.1}$$

$$\varepsilon_{\text{ARI2}} = \tfrac{1}{8}T^2\left(\frac{c_{y,\text{R}}}{2y_{0,\text{R}}^2} - \frac{c_{y,\text{L}}}{2y_{0,\text{L}}^2}\right) \tag{7.8.2}$$

$$\varepsilon_{\text{ARC1}} = -\tfrac{1}{4}T\left(\frac{1}{y_{0,\text{R}}} - \frac{1}{y_{0,\text{L}}}\right) \tag{7.8.3}$$

$$\varepsilon_{\text{ARC2}} = \tfrac{1}{8}T^2\left(\frac{c_{y,\text{R}}}{2y_{0,\text{R}}^2} + \frac{c_{y,\text{L}}}{2y_{0,\text{L}}^2}\right) \tag{7.8.4}$$

7.9 Bump Scrub

The basic bump scrub polynomial, for the actual wheel lateral scrub value s, using the basic bump scrub coefficients ε_{BSc}, is

$$s = y_{\text{C}} - y_{\text{C0}} = \varepsilon_{\text{BSc1}} z_{\text{S}} + \varepsilon_{\text{BSc2}} z_{\text{S}}^2 + \varepsilon_{\text{BSc3}} z_{\text{S}}^3 + \cdots \tag{7.9.1}$$

with the initial scrub being zero at static ride height. The local bump scrub rate is the derivative of the above:

$$\varepsilon_{\text{BScd}} = \frac{\text{d}s}{\text{d}z_{\text{S}}} = \varepsilon_{\text{BSc1}} + 2\,\varepsilon_{\text{BSc2}} z_{\text{S}} + 3\,\varepsilon_{\text{BSc3}} z_{\text{S}}^2 + \cdots \tag{7.9.2}$$

However, this is usually expressed in terms of the bump scrub rate coefficients $\varepsilon_{\text{BScd0}}$, $\varepsilon_{\text{BScd1}}$, $\varepsilon_{\text{BScd2}}$, etc.:

$$\varepsilon_{\text{BScd}} = \varepsilon_{\text{BScd0}} + \varepsilon_{\text{BScd1}} z_{\text{S}} + \varepsilon_{\text{BScd2}} z_{\text{S}}^2 + \cdots \tag{7.9.3}$$

The bump scrub rate coefficients are more convenient than the actual scrub polynomial coefficients when the roll centre is being analysed, so it is easier to use the rate coefficients throughout, with, when required,

$$\begin{aligned} \varepsilon_{\text{BSc1}} &= \varepsilon_{\text{BScd0}} \\ \varepsilon_{\text{BSc2}} &= \tfrac{1}{2}\varepsilon_{\text{BScd1}} \\ \varepsilon_{\text{BSc3}} &= \tfrac{1}{3}\,\varepsilon_{\text{BScd2}} \\ &\ldots \end{aligned} \tag{7.9.4}$$

so the actual scrub then becomes

$$s = \varepsilon_{\text{BScd0}} z_{\text{S}} + \tfrac{1}{2}\varepsilon_{\text{BScd1}} z_{\text{S}}^2 + \tfrac{1}{3}\,\varepsilon_{\text{BScd2}} z_{\text{S}}^3 + \cdots \tag{7.9.4}$$

The initial bump scrub rate coefficient, Figure 7.9.1(a), at static ride height, is

$$\varepsilon_{\text{BScd0}} = \frac{z_0}{y_0} \tag{7.9.5}$$

Allowing for movement of the suspension and of the wheel itself, in terms of the body-fixed swing-centre coordinates (y,z) Figure 7.9.1(b), the instantaneous bump scrub rate is

$$\varepsilon_{\text{BScd}} = \frac{z - z_{\text{S}}}{y}$$

Using

$$\begin{aligned} z &= z_0 + c_z z_{\text{S}} \\ y &= y_0 + c_y z_{\text{S}} \end{aligned}$$

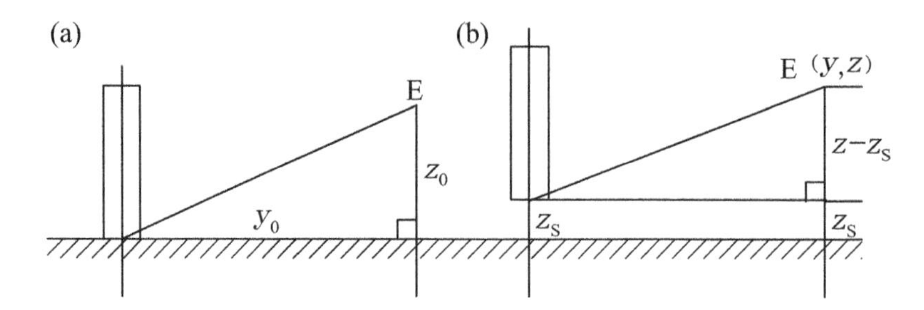

Figure 7.9.1 Variation of the bump scrub coefficient with bump: (a) initial position; (b) in bump the point E rises by $z = c_z z_S$ whilst the wheel rises by z_S.

then

$$\varepsilon_{\mathrm{BScd}} = \frac{z_0 + (c_z - 1)z_S}{y_0 + c_y z_S} \tag{7.9.6}$$

Consider initially the simplified case with swing centre movement being vertical only ($c_y = 0$), giving

$$\varepsilon_{\mathrm{BScd}} = \frac{z_0 + (c_z - 1)z_S}{y_0} \tag{7.9.7}$$

The variation of the bump scrub variation coefficient is then

$$\frac{\mathrm{d}\varepsilon_{\mathrm{BScd}}}{\mathrm{d}z_S} = \frac{c_z - 1}{y_0} \tag{7.9.8}$$

Therefore, when the suspension geometry is such that $c_z = 1$ it follows that the derivative of the bump scrub variation is zero, and the bump scrub variation coefficient will remain constant in bump. Some designers hold this to be a desirable condition, as it fixes the roll centre height.

More generally, including c_y, by the standard derivative of the quotient expression in equation (7.9.7),

$$\begin{aligned}\frac{\mathrm{d}\varepsilon_{\mathrm{BScd}}}{\mathrm{d}z_S} &= \frac{(y_0 + c_y z_S)(c_z - 1) - (z_0 + (c_z - 1)z_S)c_y}{(y_0 + c_y z_S)^2} \\ &= \frac{(c_z - 1)y_0 - c_y z_0}{(y_0 + c_y z_S)^2}\end{aligned} \tag{7.9.9}$$

Setting $c_y = 0$, this agrees with the simpler earlier expression.

Consider the general bump scrub rate variation, from equation (7.9.3),

$$\varepsilon_{\mathrm{BScd}} = \varepsilon_{\mathrm{BScd0}} + \varepsilon_{\mathrm{BScd1}} z_S + \varepsilon_{\mathrm{BScd2}} z_S^2 + \cdots$$

The implications of a linear variation, at least, will be explained further later. The coefficient values in the polynomial may be found in terms of the swing centre variation, *viz.*:

$$\varepsilon_{\mathrm{BScd0}} = \frac{z_0}{y_0}$$

$$\varepsilon_{\mathrm{BScd1}} \approx \frac{d\varepsilon_{\mathrm{BScd}}}{dz_S} = \frac{(c_z - 1)y_0 - c_y z_0}{(y_0 + c_y z_S)^2}$$

At $z_S = 0$, then,

$$\varepsilon_{\mathrm{BScd1}} = \frac{(c_z - 1)y_0 - c_y z_0}{y_0^2} \tag{7.9.10}$$

Alternatively, use a curve fit to computer-solved point values.

The bump scrub variation, using equation (7.9.7), may be expressed as

$$\begin{aligned}\frac{\mathrm{d}s}{\mathrm{d}z_S} &= \varepsilon_{\mathrm{BScd}} = \frac{z - z_S}{y} = \frac{z_0 + (c_z - 1)z_S}{y_0 + c_y z_S} \\ &= (z_0 + (c_z - 1)z_S)(y_0 + c_y z_S)^{-1} \\ &= (z_0 + (c_z - 1)z_S)\left(\frac{1}{y_0}\right)\left(1 + \frac{c_y}{y_0}z_S\right)^{-1}\end{aligned}$$

The series expansion of $(1+x)^{-1}$ (e.g. Spiegel, 1968, equation 20.8) is

$$\frac{1}{1+x} = 1 - x + x^2 - x^3 + \cdots$$

so

$$\begin{aligned}\frac{\mathrm{d}s}{\mathrm{d}z_S} &= \left(\frac{1}{y_0}\right)(z_0 + (c_z - 1)z_S)\left(1 - \frac{c_y}{y_0}z_S + \left(\frac{c_y}{y_0}\right)^2 z_S^2 - \cdots\right) \\ &= \left(\frac{1}{y_0}\right)\left\{z_0 + \left((c_z - 1) - \frac{c_y z_0}{y_0}\right)z_S + \cdots\right\}\end{aligned}$$

giving

$$\frac{\mathrm{d}s}{\mathrm{d}z_S} = \frac{z_0}{y_0} + \left(\frac{(c_z - 1)y_0 - c_y z_0}{y_0^2}\right)z_S + \cdots$$

Integrating, with zero scrub at zero suspension bump,

$$s = \left(\frac{z_0}{y_0}\right)z_S + \left(\frac{(c_z - 1)y_0 - c_y z_0}{2y_0^2}\right)z_S^2 + \cdots$$

Comparing this with the actual scrub polynomial (7.9.5) in terms of the scrub rate coefficients, which was

$$s = \varepsilon_{\mathrm{BScd0}} z_S + \tfrac{1}{2}\varepsilon_{\mathrm{BScd1}} z_S^2 + \tfrac{1}{3}\varepsilon_{\mathrm{BScd2}} z_S^3 + \cdots$$

then

$$\begin{aligned}\varepsilon_{\mathrm{BScd0}} &= \frac{z_0}{y_0} \\ \varepsilon_{\mathrm{BScd1}} &= \frac{(c_z - 1)y_0 - c_y z_0}{y_0^2}\end{aligned} \tag{7.9.11}$$

agreeing with equation (7.9.11). Further terms are easily obtained.

7.10 Double-Bump Scrub

In symmetrical double-wheel bump, equivalent to a negative suspension heave, for independent suspension there is track change due to scrub and camber change. For a symmetrical axle the axle inclination is unchanged, with the introduction of symmetrical negative camber (on an independent axle). Positive scrub causes a track increase. There may be asymmetries due to production tolerance errors, but these are generally small compared with the bump steer effects.

7.11 Roll Scrub

Geometrically, roll of the suspension will change the axle track due to bump scrub. The total track change, to second order, is

$$\begin{aligned}\Delta T &= T(\cos\phi_{\mathrm{S}}-1)+\left(\varepsilon_{\mathrm{BSc0,R}}\left(\tfrac{1}{2}T\phi_{\mathrm{S}}\right)+\tfrac{1}{2}\varepsilon_{\mathrm{BSc1,R}}\left(\tfrac{1}{2}T\phi_{\mathrm{S}}\right)^{2}\right)+\left(\varepsilon_{\mathrm{BSc0,L}}\left(-\tfrac{1}{2}T\phi_{\mathrm{S}}\right)+\tfrac{1}{2}\varepsilon_{\mathrm{BSc1,L}}\left(-\tfrac{1}{2}T\phi_{S}\right)^{2}\right)\\ &= T\left(-\tfrac{1}{2}\phi_{\mathrm{S}}^{2}\right)+(\varepsilon_{\mathrm{BSc0,R}}-\varepsilon_{\mathrm{BSc0,L}})\left(\tfrac{1}{2}T\phi_{\mathrm{S}}\right)+\tfrac{1}{2}(\varepsilon_{\mathrm{BSc1,R}}+\varepsilon_{\mathrm{BSc1,L}})\left(\tfrac{1}{2}T\phi_{\mathrm{S}}\right)^{2}\\ &= \left\{\tfrac{1}{2}T\left(\varepsilon_{\mathrm{BSc0,R}}-\varepsilon_{\mathrm{BSc0,L}}\right)\right\}\phi_{\mathrm{S}}+\left\{-\tfrac{1}{2}T+\left(\tfrac{1}{8}T^{2}\right)\left(\varepsilon_{\mathrm{BSc1,R}}+\varepsilon_{\mathrm{BSc1,L}}\right)\right\}\phi_{\mathrm{S}}^{2}\end{aligned} \tag{7.11.1}$$

For an ideal symmetrical vehicle the first-order roll scrub is therefore zero, the linear wheel camber angles compensating each other. There is a second-order term, partially due to the cosine effect on wheel spacing and partially due to the second-order wheel bump scrub. For an ideal symmetrical vehicle,

$$\Delta T = -\tfrac{1}{2}T\left\{1-\tfrac{1}{2}T\varepsilon_{\mathrm{BSc1}}\right\}\phi_{\mathrm{S}}^{2} \tag{7.11.2}$$

For an asymmetrical vehicle,

$$\begin{aligned}\varepsilon_{\mathrm{RSc1}} &= \tfrac{1}{2}T(\varepsilon_{\mathrm{BSc0,R}}-\varepsilon_{\mathrm{BSc0,L}})\\ \varepsilon_{\mathrm{RSc2}} &= -\tfrac{1}{2}T+\tfrac{1}{8}T^{2}(\varepsilon_{\mathrm{BSc1,R}}+\varepsilon_{\mathrm{BSc1,L}})\end{aligned} \tag{7.11.3}$$

7.12 Rigid Axles

The nature of non-independent live axles effectively precludes substantial misalignment of the wheels, which remain effectively parallel, with zero geometric axle inclination and camber. The one exception to this is in some racing classes, now mainly historical, where the axles were 'tweaked' in a hydraulic press to give a small amount of static negative camber, and possibly some toe angle.

De Dion axles, where the wheels are rigidly connected to each other but the drive differential is attached to the sprung mass, may have static negative camber and toe angle, but this is constant, not varying as for independent suspension.

References

Spiegel, M.R. (1968) *Mathematical Handbook of Formulas and Tables.* New York: McGraw-Hill.

8

Roll Centres

8.1 Introduction

This chapter deals with the roll centre for independent suspension. Rigid axles are covered in Chapter 13. Actually, the idea of the roll centre was introduced early in the days of vehicle dynamics, when all cars had a rigid axle at both front and rear. The roll axis is a straight line through the front and rear roll centres. With the introduction of independent suspension, these concepts were carried over despite some confusing complications, such as track change in bump and roll, and continued to be the basis of dynamic cornering analysis. Over the years since then, the roll centre concept has been subject to some criticism, but clarified and improved.

Really, for a thorough understanding, it is necessary to appreciate that the idea of a single roll centre for a suspension is only an approximation. There are actually several different types of roll centre, and these definitions are not exactly equivalent. Also, different authors have used different definitions for the roll centre of any particular type, none of which are necessarily wrong – each may be good if used correctly. Third, the roll centre, in general, however defined, is not necessarily a single point, it moves around when there is suspension bump or roll. Finally, the roll centre as defined may not even be a point, but may be a range of positions.

Originally, the roll centre idea was, in practice, essential for manual computation of vehicle cornering dynamics. With the introduction of computers, it became entirely practical to make the necessary calculations without reference to a roll centre at all. However, the idea of a roll centre remains a valuable tool for human understanding of the suspension characteristics, in the design process, and as a basis for simple calculations or as a summary of the suspension characteristics as found by computer-produced results.

In the simple model of suspension, there is a single roll centre, fulfilling several functions. In reality, there are several distinct functions, with distinct roll centres:

(1) the geometric roll centre (GRC);
(2) the kinematic roll centre (KRC);
(3) the force roll centre (FRC);
(4) the moment roll centre (MRC).

Historically, for the most part, as a useful practical approximation, these have all been treated as one fixed point of the suspension. For a thorough understanding, it is necessary to appreciate the differences.

The GRC is a property of the suspension geometry, a point found by specified geometric methods.

The KRC is the point about which the body rolls relative to the axle. This then controls the consequent lateral position of the sprung centre of mass, affecting the total load transfer.

Suspension Geometry and Computation J. C. Dixon

The FRC is a point at which the suspension links exert a lateral force on the sprung mass. This is represented, in essence, by the FRC height, which affects the roll angle of the body and the means of lateral load transfer through the suspension links.

The MRC is the point about which moments are taken in dynamic analysis of the body. For a true dynamic analysis, the front and rear MRCs should generally be on the longitudinal principal axis of inertia.

The traditional method of working, for steady-state analysis at least, was that the various roll centres on an axle were all deemed to be at one point. This point was found by solving the GRC position, which was then used, with some justification, as the FRC and the KRC, and even the MRC, allowing practical solution of vehicle dynamics problems.

The assumptions behind the various roll centres are somewhat different. There is an analogy here with the lever, for which the mechanical advantage equals the velocity ratio only for zero friction. The usual analysis of the FRC neglects joint friction and stiffness.

In this book, the emphasis is on geometric properties of roads and suspensions, so this is not the place for a thorough analysis of all roll centre issues. Here, only the GRC will be investigated in detail. The FRC and KRC will be reviewed, however, to show their relationship to the GRC. For a more detailed explanation of these, see, for example, Dixon (1996).

8.2 The Swing Arm

Figure 8.2.1 shows the basic swing arm constructions, in the two dimensions of the transverse vertical plane of the suspension, for the main types of independent suspension. Point F is the wheel-to-road contact point, considering the wheel and tyre to be undistorted (i.e. in the centre plane of the wheel). Tyre distortion effects such as overturning moment are dealt with separately. Point E is the swing centre. For an unrolled symmetrical vehicle, the swing arms for the two sides intersect at the centre, this intersection being the GRC for the symmetrical unrolled position (i.e. at small lateral accelerations only). In practice, for a simple analysis, the roll centre is assumed not to move relative to the body. The GRC found by this means (geometric analysis as in Figure 8.2.1) is then considered to indicate the KRC and FRC positions. In fact, these are not really the same, and the GRC moves relative to the body, as discussed later in this chapter.

Briefly, the swing arm constructions are as follows. Figure 8.2.1(a) shows the double-wishbone suspension (double A-arm), where the motion of one end of a link relative to the other end must be perpendicular to the link line between centres. This neglects rubber bush compliance. Therefore, the motion of the wheel relative to the body, in bump, at each link connection, must be perpendicular to the link. Therefore the link lines are radii from the instantaneous centre of rotation, which is found by intersecting the link lines, giving the swing centre E. For the symmetrical case, the roll centre is always in the centre plane, so it is at the intersection of the swing arm with the centreline.

For the strut of Figure 8.2.1(b), the top slider is like an infinitely long wishbone perpendicular to the slider. For the pure straight slider suspension of Figure 8.2.1(c), the bottom point of the wheel obviously moves parallel to the slider, so the swing centre is at infinity, with the swing arm perpendicular to the slider direction.

Figure 8.2.1(d) shows a driven swing axle, where the wheel is located by the driveshaft, with a universal joint at the side of the final drive unit. This joint acts as the swing centre, giving point E. This is unusual in that the swing centre is on the same side of the vehicle as the wheel. The initial roll centre follows at R, which is notable for being very high. Figure 8.2.1(e) shows a lateral swinging arm with longitudinal axis. This is known as low-pivot swing arm, producing a desired lower roll centre.

Figures 8.2.1(f,g,h) show various forms of trailing arm. The simplest case is (f) in which the pivot axis is perpendicular to the vehicle centre plane, and therefore parallel to the ground. The motion of the wheel relative to the body is a linear parallel action, so the swing centre is at infinity, and the initial roll centre is at ground level. In case (g), the trailing arm axis is inclined in rear view only. The body-fixed wheel motion is parallel but inclined, so the swing centre is again at infinity, but now with an inclined swing arm. Finally, for the general-case trailing arm of case (h), the arm axis is not perpendicular to the vehicle

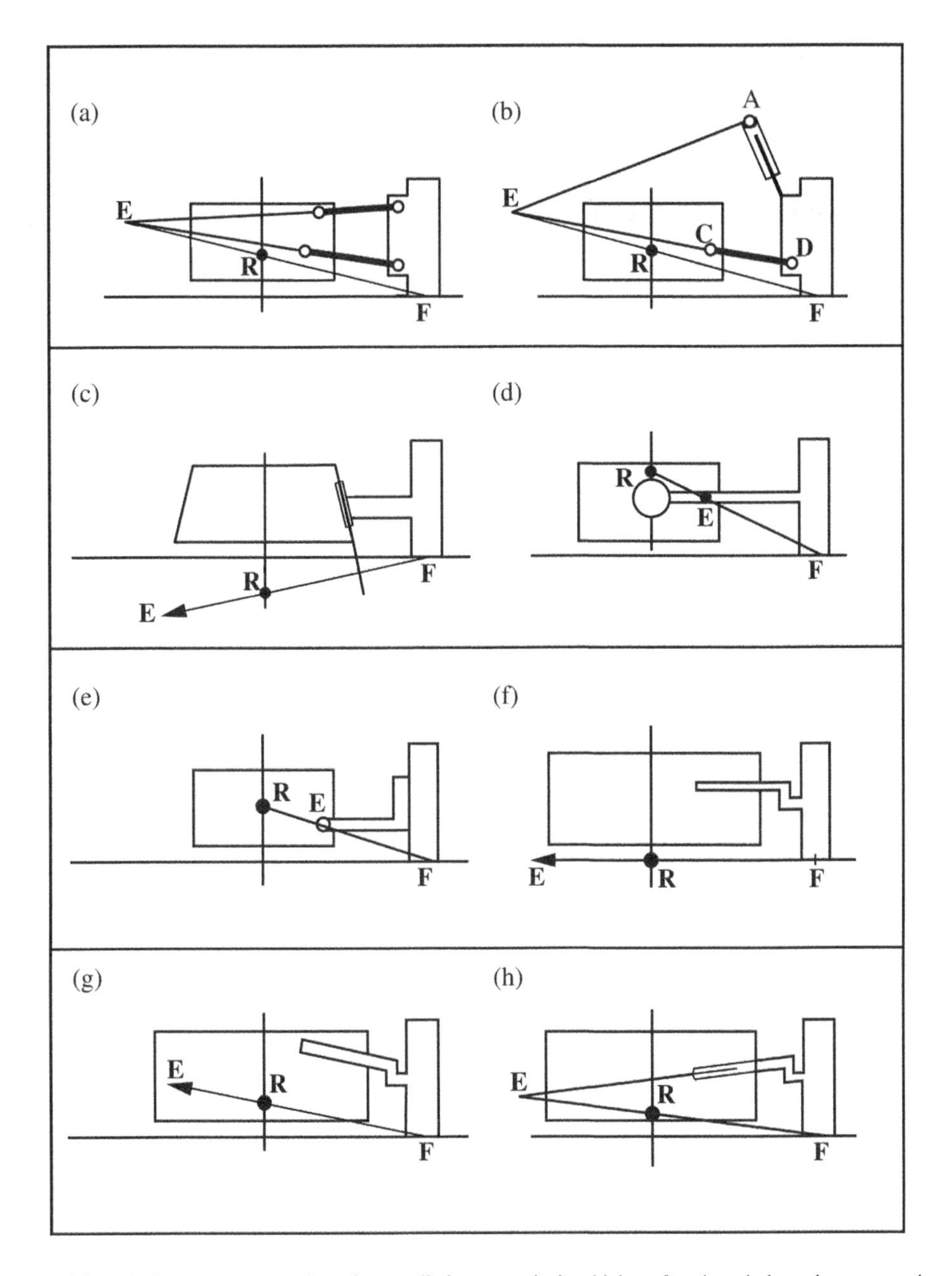

Figure 8.2.1 Swing arm constructions for unrolled symmetrical vehicles of various independent suspensions: (a) double wishbones; (b) strut; (c) slider; (d) swing axle; (e) swing arm; (f) trailing arm; (g) trailing arm with inclination; (h) trailing arm with inclination and plan-view axis rotation. F is the (undistorted) contact patch centre point, E is the swing centre, R is the roll centre (Dixon, 1991).

centreline in plan view. The projected axis therefore intersects the vertical transverse plane of the suspension at a non-infinite swing arm length, at swing centre E.

Over the last century of vehicle suspension analysis, many versions of the above swing arm and roll centre diagrams have been published. The fact that they have often differed shows that even basic suspension analysis is by no means trivial. Figure 8.2.1, which first appeared in Dixon (1991), seems to have stood the test of time. Admittedly, however, this is not exactly correct. It is a two-dimensional approximation to a problem which should really be solved in three dimensions. That said, the approximation is quite a good one.

8.3 The Kinematic Roll Centre

The very name 'roll centre' suggests a kinematic concept, and many authors do introduce the roll centre as a point about which the vehicle body rolls. Unfortunately, this is not really true, and the more one thinks about it the more confusing it can become, considering lateral velocity of the vehicle in cornering and track changes of the axle.

The essential point of the KRC is that it could be defined as follows (temporary definition):

> The kinematic roll centre is the point, in the transverse vertical plane of the suspension, about which the body rolls in cornering, relative to the axle.

Here, the term 'axle' is used in a broad sense to mean the pair of wheels, even if independently sprung. For independent suspension, the axle track (tread) can change, so the axle lateral position is deemed to be the mean lateral position of the bottom of the two wheels, considering their undistorted wheel planes. For a rigid axle, track change is not an issue.

The simplest case is a rigid axle using a sliding block lateral location. This is not used on passenger cars, but has been used in racing, particularly as a modification to lower the roll centre of some designs of rigid rear-axle linkage. There is a vertical channel on the axle, carrying a sliding block. The centre of the block carries a pin about which the body may rotate in roll, Figure 8.3.1. Alternatively, the channel may be on the body, and the block and pin on the axle. There is a difference when the body is rolled, because heave of the body perpendicular to the road may or may not cause axle sway (lateral motion) according to the system chosen. The channel on the axle is usually preferred. Independent suspensions have a similar problem, according to the design.

Here, it is clear that the body can pivot (i.e. roll) about the pin. Equally apparent is that the block and pin can, and generally will, move up or down in cornering. Typically the suspension springs are rising rate, so in roll the body tends to rise, an effect called spring jacking. If the body rolls and rises, then the complete motion relative to the axle is a rotation about a point offset to one side, Figure 8.3.2.

For simplicity, neglect the tyre deflections. Coordinate Y_P is positive to the right. For a suspension roll angle ϕ_S and a body heave z_B, the mean of the two wheel suspension droops, the lateral offset of the effective pivot point is

$$Y_P = \frac{z_B}{\tan \phi_S} \tag{8.3.1}$$

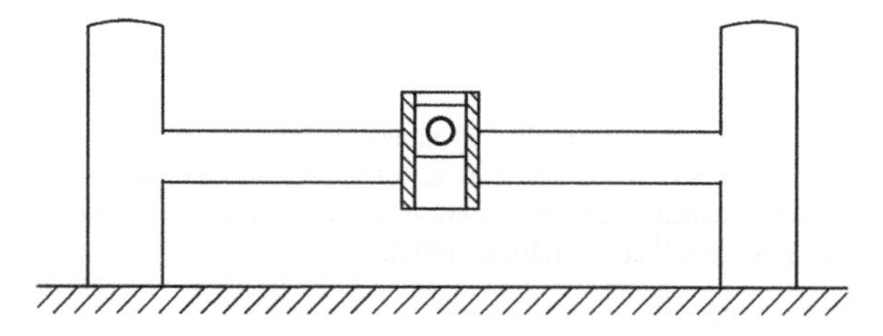

Figure 8.3.1 Sliding block lateral location of the body on a rigid axle. The body pivots on a pin in the block, which slides in the channel attached to the axle.

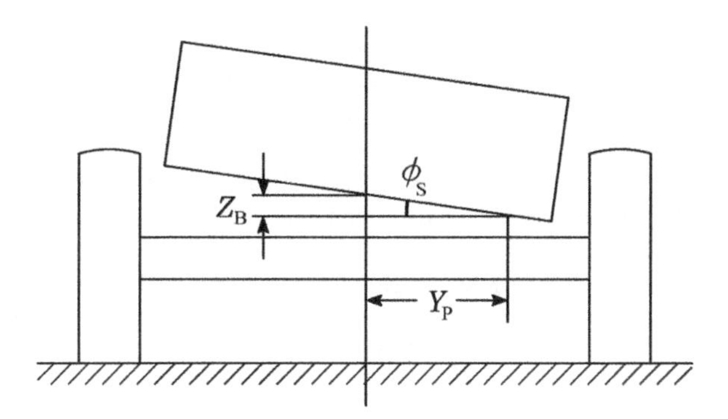

Figure 8.3.2 Relative rotation point of the body relative to the axle when the body rises as it rolls, front view.

This is evidently, in a sense, the roll centre of the body relative to the axle, but the lateral offset is governed by the properties of the springs, not the suspension links.

In practical application, the lateral offset of the pivot point, as seen above, is not considered as part of the roll centre action. The application of the KRC is in calculating the lateral offset of the centre of mass relative to the axle, a dimension which governs the additional load transfer. In practice therefore, a different definition is applied:

At a suspension roll angle ϕ_S, with resulting centre of mass offset Y_{CM} relative to the axle mean wheel position, the KRC is in the vehicle centre plane and the axle vertical transverse plane, at the KRC height H_{KRC}, where

$$H_{CM} - H_{KRC} = \frac{Y_{CM}}{\tan \phi_S} \tag{8.3.2}$$

in which H_{CM} is the initial height of the body centre of mass.

As actually used, the KRC position is derived from the GRC position, and the centre of mass offset, positive to the right, is calculated as

$$Y_{CM} = (H_{CM} - H_{KRC})\tan \phi_S \tag{8.3.3}$$

as in Figure 8.3.3.

This offset of the centre of mass offsets the body weight force $m_B g$, causing a supplementary roll moment

$$M_{YCM} = m_B g Y_{CM} \tag{8.3.4}$$

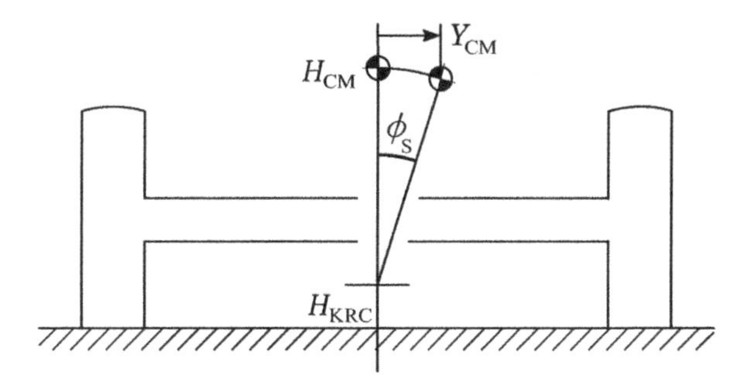

Figure 8.3.3 Lateral displacement of the centre of mass due to roll angle about the KRC at height H_{KRC}.

This is usually much smaller than the load transfer moment directly due to lateral forces but may still be significant, depending on the suspension roll stiffness and consequent roll angle. A high roll centre at both ends of the vehicle therefore reduces the roll angle and the total load transfer. However, in contrast, a property of the force roll centre is that a higher roll centre at one axle increases the load transfer on that axle and reduces it on the other axle, this being an important factor in limit handling adjustment. These conflicting effects of the KRC and FRC have caused some confusion, and even controversy.

The common practice is to take the KRC height to be the same as the GRC height, in the vehicle coordinates, neglecting any GRC lateral offset for the reasons stated above, with the GRC found by geometric analysis of the suspension linkage. In the case of a simple rigid axle, as in Figure 8.3.1, combined with the defining principle of Figure 8.3.3, this is reasonable. This is also fairly clear for other rigid axle location systems, such as the Panhard rod.

In the case of independent suspension, the justification is less clear-cut, and reference must be made to other texts (e.g. Dixon, 1996) for a fuller discussion. The problem is the complication introduced by the track change of an independent suspension in roll. The swing arm defines the motion of each wheel relative to the body, so with some approximations it is reasonable to take the intersection point of the swing arms (i.e. the GRC), as the putative KRC, and then neglect any a lateral offset (i.e. again to take the height of the KRC as being the same as the height of the GRC, in vehicle coordinates).

8.4 The Force Roll Centre

The height of the FRC controls two important factors:

(1) the suspension link lateral load transfer;
(2) the suspension link jacking force.

Figure 8.4.1(a) shows the tyre normal (Z) and lateral (Y) force components on the two wheels of an independent-type axle. This is after removal of forces that support and accelerate the unsprung mass, so it is for the forces supporting and accelerating the sprung mass (body) only. The coordinate system used here is the standard ISO one, with X forward, Y to the left, and Z upwards. The tyre forces are shown in vehicle coordinates. The tyre forces are usually expressed in road normal and lateral coordinates. The road surface is in general rotated by the roll angle, and also by the two road camber angles. The vehicle-axis components shown are easily derived. Figure 8.4.1(b) shows these forces resolved instead along (F_2) and perpendicular (F_1) to the swing arms. The F_1 parts, perpendicular to the swing arms, must be supported by the springs (and dampers in unsteady state). The F_2 parts act through the links, and, as may be seen in the figure, will exert no moment about the intersection point of the swing arm lines. This conclusion may be drawn, for a double-wishbone suspension for example, on the basis that the swing arm is the intersection of the lines of the links, and the links carry link-aligned forces only, that is, that the joints at the ends of the links are free of stiffness moments (e.g. rubber bushes) and free of friction moments (e.g. rotating metal sleeve bushes). This is a standard principle of the analysis of pin-jointed structures. These are evidently engineering approximations, which could be improved upon in a more detailed analysis, including bush friction or stiffness moments. The lateral position of the GRC is B, positive to the left.

Figure 8.4.1(c) shows the link force alone resolved back into normal and lateral components. The link jacking force, which is the total vertical force in the links, relieves the springs of load, causing jacking of the body. This link jacking force is

$$F_{\rm LJ} = F_{2,\rm R} \sin\theta_{\rm SA,R} - F_{2,\rm L} \sin\theta_{\rm SA,L} \tag{8.4.1}$$

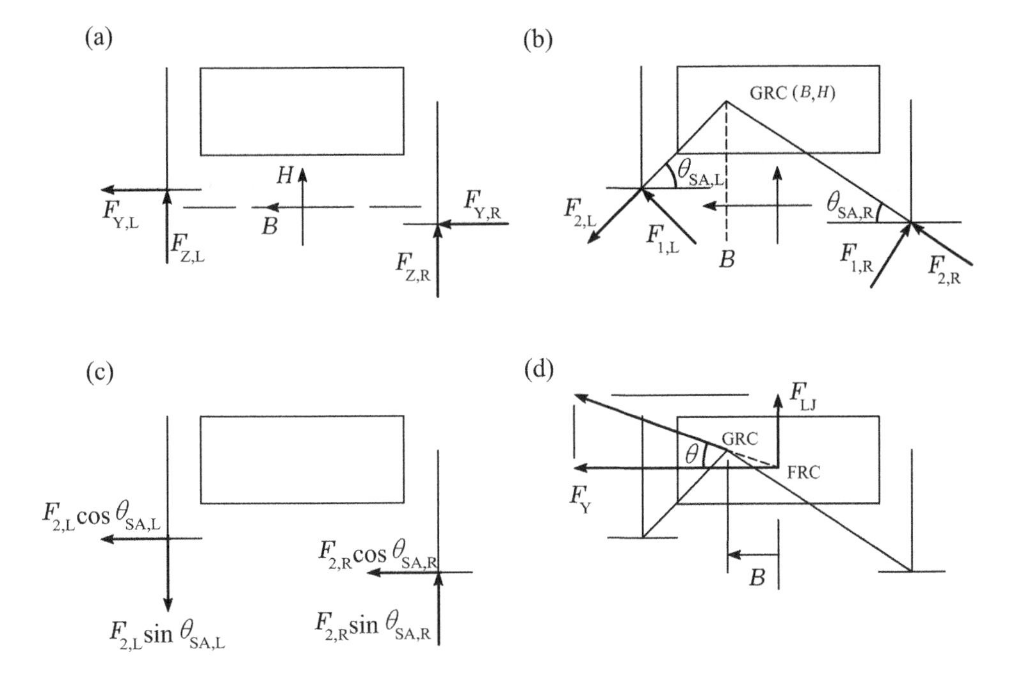

Figure 8.4.1 Various representations of the body-applied forces in independent suspension links, rear view, in vehicle coordinates, in left-hand cornering: (a) normal and lateral components of tyre forces at the contact patch; (b) resolved in swing arm coordinates; (c) the link forces alone resolved back into normal and lateral components; (d) the single resultant at the GRC, and components F_{LJ} (link jacking force) and F_Y (lateral force) at the FRC.

The lateral transfer of normal force through the links (colloquially, the link load transfer) is half of the difference between the upwardly-directed link normal forces:

$$F_{LT} = \tfrac{1}{2}(F_{2,R}\sin\theta_{SA,R} + F_{2,L}\sin\theta_{SA,L}) \tag{8.4.2}$$

The link-only upward normal forces for each side are then

$$\begin{aligned} F_{Z2,R} &= \tfrac{1}{2}F_{LJ} + F_{LT} \\ F_{Z2,L} &= \tfrac{1}{2}F_{LJ} - F_{LT} \end{aligned} \tag{8.4.3}$$

Figure 8.4.1(d) shows the resultant of the link forces, combined vectorially at their intersection point (nominally the GRC). Considering the line of action of the resultant, where this crosses the vehicle centreline the resultant can be resolved into vertical (jacking) and lateral components, F_{LJ} and F_Y, respectively. This point on the centreline is the one usually accepted as the FRC. The angle of the total resultant link force to the horizontal is

$$\theta = \operatorname{atan}\left(\frac{F_{LJ}}{F_Y}\right) \tag{8.4.4}$$

so the height of the FRC is

$$H_{\mathrm{FRC}} = H_{\mathrm{GRC}} - B\left(\frac{F_{\mathrm{LJ}}}{F_{\mathrm{Y}}}\right) \tag{8.4.5}$$

In Figure 8.4.1(d), B is positive, so H_{FRC} is less than H_{GRC}.

The roll moment exerted by the lateral link force about a central axis at ground level is

$$M_{\mathrm{FYL}} = F_{\mathrm{Y}} H_{\mathrm{FRC}} = F_{\mathrm{LT}} T \tag{8.4.6}$$

so the FRC height is

$$H_{\mathrm{FRC}} = \frac{M_{\mathrm{FYL}}}{F_{\mathrm{Y}}} = \frac{F_{\mathrm{LT}} T}{F_{\mathrm{Y}}} \tag{8.4.7}$$

The load transfer factor f_{LT} is defined as

$$f_{\mathrm{LT}} = \frac{F_{\mathrm{LT}}}{F_{\mathrm{Y}}} = \frac{H_{\mathrm{FRC}}}{T} \tag{8.4.8}$$

so

$$\begin{aligned} H_{\mathrm{FRC}} &= f_{\mathrm{LT}} T \\ F_{\mathrm{LT}} &= f_{\mathrm{LT}} F_{\mathrm{Y}} \end{aligned} \tag{8.4.9}$$

This is sufficient not only to illustrate the close relationship between the FRC and the GRC, but also to illustrate that they are not identical because of the necessary assumptions and approximations, and the lateral offset.

The actual distribution of forces left and right depends on the tyre characteristics, so only the approximate position of the FRC can be calculated on the basis of the links alone, the height difference from the GRC depending on the inclination of the link force resultant.

The anti-roll coefficient J_{AR} is defined as

$$J_{\mathrm{AR}} = \frac{H_{\mathrm{FRC}}}{H_{\mathrm{B}}} \tag{8.4.10}$$

and is usually expressed as a percentage, $J_{\mathrm{ARpc}} = J_{\mathrm{AR}} \times 100\%$.

8.5 The Geometric Roll Centre

The GRC has already been mentioned several times. Its basic definition is simply as follows:

> The geometric roll centre is a point in the transverse vertical plane of the suspension, being the intersection point of the two swing arm lines, one from each side.

Therefore, once the geometric constructions for the swing arms are understood, the GRC follows easily, as in Figure 8.4.1(b).

Difficulties arise with the above simple definition when the swing arms are parallel. This divides into two cases. The first case is as for simple unrolled trailing arms – the swing arms may be overlaid in the ground surface, with the intersection point anywhere along the swing arm lines. The second case arises, for example with simple trailing arms in the rolled position, when the swing arms are parallel and do not intersect at any point. This may occur with or without infinitely long swing arms, which is not important in this context.

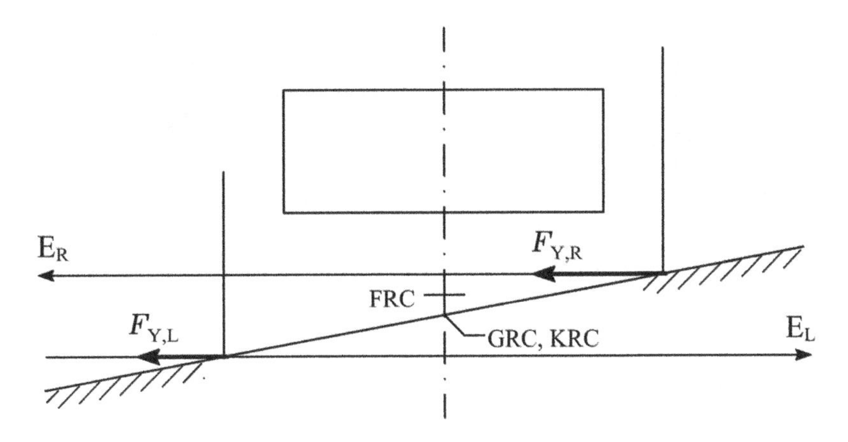

Figure 8.5.1 Parallel swing arms on a simple rolled trailing-arm suspension.

In the first case, the problem is solved simply by extending the definition to say that when the swing arms overlap, in the road surface, the intersection point is to be taken as on the vehicle centreline, in the road surface.

When the swing arms are parallel and non-overlapping, as occurs for simple rolled trailing arms or for rolled equal parallel wishbones, regardless of swing arm length, the GRC is at infinity, either to the left or right. An infinitesimal change of one swing arm angle, say due to a slight asymmetry, can shift it from one side to the other. If they are even slightly away from parallel, then the GRC exists in the usual way. Projecting an infinite intersection point back into the vehicle centreline does not solve the problem of the KRC, because the vertical position of the infinite GRC, in vehicle coordinates, may be anywhere within the range of the swing arm heights. In this specific case, then, the GRC is really of no help.

The case of parallel swing arms can occur for most suspension types, at one double bump position for any given roll angle. It is exemplified by the simple parallel trailing arm suspension in the rolled position, as in Figure 8.5.1. The KRC is on the centreline at ground level. The FRC is found as the point in the centre plane about which the link forces exert zero moment. The link forces are simply parallel, and the outer wheel will generate more cornering force, in general, so the FRC tends to be up towards the swing arm of the outer wheel, according to the relative magnitude of the forces.

8.6 Symmetrical Double Bump

The simplest case of GRC movement is for a symmetrical vehicle in symmetrical double bump, Figure 8.6.1. With a symmetrical axle bump z_A each suspension bump is z_A. The local bump scrub rate coefficients are equal, by symmetry, for the linear case

$$\varepsilon_{BScd} = \varepsilon_{BScd0} + \varepsilon_{BScd1} z_A$$

The GRC height, in the body-fixed coordinates, is

$$H = z_A + \tfrac{1}{2} T \varepsilon_{BScd} \tag{8.6.1}$$

Substituting for the bump scrub coefficient, linear in this case, then

$$H = z_A + \tfrac{1}{2} T(\varepsilon_{BScd0} + \varepsilon_{BScd1} z_A) \tag{8.6.2}$$

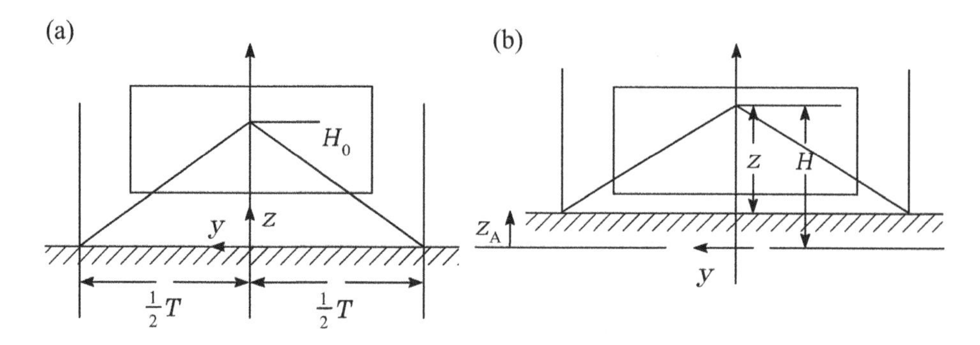

Figure 8.6.1 A symmetrical vehicle shown in vehicle-fixed coordinates: (a) static; (b) with symmetrical double bump.

with

$$H_0 = \tfrac{1}{2}T\varepsilon_{\mathrm{BScd0}} \tag{8.6.3}$$

Typically, the linear bump scrub variation coefficient $\varepsilon_{\mathrm{BScd1}}$ is negative, but the GRC rises relative to the body, remaining on the centreline. The vertical motion of the GRC, relative to the body, in relation to the suspension double bump, by differentiating equation (8.6.2), is

$$\varepsilon_{\mathrm{H1,DB}} = \frac{\mathrm{d}H}{\mathrm{d}z_{\mathrm{A}}} = 1 + \tfrac{1}{2}T\varepsilon_{\mathrm{BScd1}} \tag{8.6.4}$$

This value occurs frequently in GRC equations. Relative to the road, the GRC height is

$$Z = H - z_{\mathrm{A}} \tag{8.6.5}$$

so

$$Z = \tfrac{1}{2}T(\varepsilon_{\mathrm{BScd0}} + \varepsilon_{\mathrm{BScd1}} z_{\mathrm{A}}) \tag{8.6.6}$$

With a negative linear bump scrub rate coefficient $\varepsilon_{\mathrm{BScd1}}$, the GRC falls relative to the road. The vertical motion of the GRC, relative to the road, in response to suspension double bump, is

$$\varepsilon_{\mathrm{Z1,DB}} = \frac{\mathrm{d}Z}{\mathrm{d}z_{\mathrm{A}}} = \tfrac{1}{2}T\varepsilon_{\mathrm{BScd1}} \tag{8.6.7}$$

Evidently, the relationship between the variation relative to the body and that relative to the road is

$$\varepsilon_{\mathrm{H1,DB}} = 1 + \varepsilon_{\mathrm{Z1,DB}} \tag{8.6.8}$$

For example, with $T = 1.40\,\mathrm{m}$, $\varepsilon_{\mathrm{BScd1}} = -0.6\,\mathrm{m}^{-1}$, $z_{\mathrm{A}} = 50\,\mathrm{mm}$, relative to the body the GRC goes up by

$$\begin{aligned} H - H_0 &= z_{\mathrm{A}}(1 + \tfrac{1}{2}T\varepsilon_{\mathrm{BScd1}}) \\ &= 50(1 + 0.5 \times 1.4 \times (-0.6)) = 50(1 - 0.42) = 29\,\mathrm{mm} \end{aligned}$$

On the other hand, relative to the ground the GRC goes up by

$$\begin{aligned} Z - Z_0 &= z_{\mathrm{A}}\left(\tfrac{1}{2}T\varepsilon_{\mathrm{BScd1}}\right) \\ &= 50(0.5 \times 1.4 \times (-0.6)) = 50(-0.42) = -21\,\mathrm{mm} \end{aligned}$$

that is, it goes down by 21 mm relative to the road.

If non-linear bump scrub variation is included, then

$$H = z_A + \tfrac{1}{2}T(\varepsilon_{BScd0} + \varepsilon_{BScd1}z_A + \varepsilon_{BScd2}z_A^2) \tag{8.6.9}$$

$$Z = \tfrac{1}{2}T(\varepsilon_{BScd0} + \varepsilon_{BScd1}z_A + \varepsilon_{BScd2}z_A^2) \tag{8.6.10}$$

and the relationship between the two variation rates is preserved. The height H can also be expressed as

$$\begin{aligned} H &= \tfrac{1}{2}T\varepsilon_{BScd0} + \left(1 + \tfrac{1}{2}T\varepsilon_{BScd1}\right)z_A + \tfrac{1}{2}T\varepsilon_{BScd2}z_A^2 \\ &= H_0 + \varepsilon_{Z1,DB}z_A + \left(\tfrac{1}{2}T\varepsilon_{BScd2}\right)z_A^2 \end{aligned} \tag{8.6.11}$$

8.7 Linear Single Bump

Figure 8.7.1 shows the geometry for movement of the GRC relative to the body when there is single-wheel bump, here taken to be on the left-hand side. Here, for simplicity, the small track variation is neglected. Vehicle bump scrub asymmetry is included. The GRC moves upwards or downwards, and to one side, generally towards the rising wheel. In body coordinates, the GRC is at (B, H), with B positive to the left. It is easy to write two simultaneous equations for H:

$$\begin{aligned} H &= \left(\tfrac{1}{2}T + B\right)\varepsilon_{BScd,R} \\ H &= \left(\tfrac{1}{2}T - B\right)\varepsilon_{BScd,L} + z_L \end{aligned} \tag{8.7.1}$$

These must be solved for B and H, in the usual way. Eliminating H by subtraction,

$$0 = \tfrac{1}{2}T(\varepsilon_{BScd,R} - \varepsilon_{BScd,L}) + B(\varepsilon_{BScd,R} + \varepsilon_{BScd,L}) - z_L \tag{8.7.2}$$

Using the notation subscript L + R to denote the addition of terms, L − R for subtraction (and later L × R for multiplication), i.e.

$$\begin{aligned} \varepsilon_{BScd,L+R} &= +\varepsilon_{BScd,R+L} = \varepsilon_{BScd,L} + \varepsilon_{BScd,R} \\ \varepsilon_{BScd,L-R} &= -\varepsilon_{BScd,R-L} = \varepsilon_{BScd,L} - \varepsilon_{BScd,R} \\ \varepsilon_{BScd,L\times R} &= +\varepsilon_{BScd,R\times L} = \varepsilon_{BScd,L} \times \varepsilon_{BScd,R} \end{aligned} \tag{8.7.3}$$

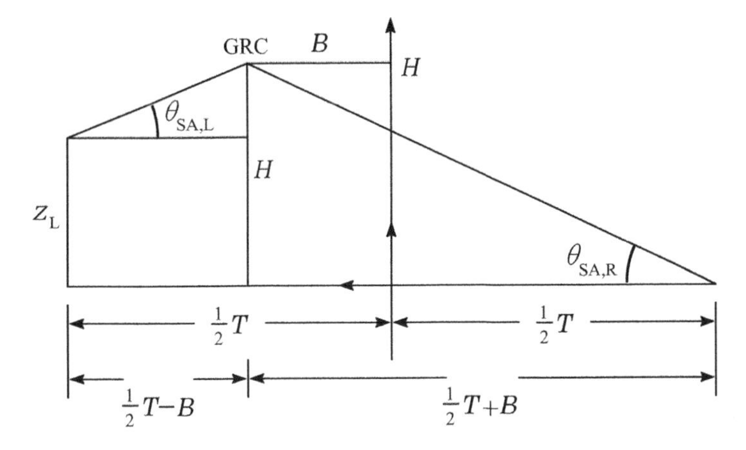

Figure 8.7.1 An asymmetrical vehicle with single (left) bump, rear view, in vehicle coordinates.

equation (8.7.2) is more compactly expressed as

$$0 = \tfrac{1}{2}T\varepsilon_{\mathrm{BScd,R-L}} + B\varepsilon_{\mathrm{BScd,R+L}} - z_{\mathrm{L}}$$

so the lateral GRC position becomes (with L − R, not R − L, and a corresponding sign change)

$$B = \frac{z_{\mathrm{L}} + \tfrac{1}{2}T\varepsilon_{\mathrm{BScd,L-R}}}{\varepsilon_{\mathrm{BScd,R+L}}} \tag{8.7.4}$$

Eliminating B from equations (8.7.1), by appropriate multiplication and addition of equations,

$$H\varepsilon_{\mathrm{BScd,L+R}} = \tfrac{1}{2}T2\varepsilon_{\mathrm{BScd,R\times L}} + z_{\mathrm{L}}\varepsilon_{\mathrm{BScd,R}}$$

so

$$H = \frac{T\varepsilon_{\mathrm{BScd,R\times L}} + z_{\mathrm{L}}\varepsilon_{\mathrm{BScd,R}}}{\varepsilon_{\mathrm{BScd,R+L}}} \tag{8.7.5}$$

From equation (8.7.4) with $z_{\mathrm{L}} = 0$, the initial GRC lateral offset due to asymmetry at zero bump is

$$B_0 = \tfrac{1}{2}T\left(\frac{\varepsilon_{\mathrm{BScd0,L-R}}}{\varepsilon_{\mathrm{BScd0,R+L}}}\right) \tag{8.7.6}$$

In the symmetrical-vehicle case, this reduces to zero. From equation (8.7.5), the initial height of the GRC, in body-fixed coordinates, is

$$H_0 = T\left(\frac{\varepsilon_{\mathrm{BScd0,R\,\times\,L}}}{\varepsilon_{\mathrm{BScd0,R\,+\,L}}}\right) \tag{8.7.7}$$

In the symmetrical case, this reduces to

$$H_0 = \tfrac{1}{2}T\varepsilon_{\mathrm{BScd0}} \quad \text{(symmetrical vehicle)} \tag{8.7.8}$$

To go into further detail, the general form of bump scrub rate $\varepsilon_{\mathrm{BScd}}$ must be replaced by actual values, or by more detailed equations. Considering the case of linear variation of bump scrub rate, with bump on the left wheel only, we may write

$$\begin{aligned} \varepsilon_{\mathrm{BScd,L}} &= \varepsilon_{\mathrm{BScd0,L}} + \varepsilon_{\mathrm{BScd1,L}}z_{\mathrm{L}} \\ \varepsilon_{\mathrm{BScd,R}} &= \varepsilon_{\mathrm{BScd0,R}} \end{aligned} \tag{8.7.9}$$

Using equation 8.7.4, the GRC lateral position is then

$$B = \frac{z_{\mathrm{L}} + \tfrac{1}{2}T(\varepsilon_{\mathrm{BScd0,L-R}} + \varepsilon_{\mathrm{BScd1,L}}z_{\mathrm{L}})}{(\varepsilon_{\mathrm{BScd0,R+L}} + \varepsilon_{\mathrm{BScd1,L}}z_{\mathrm{L}})} \tag{8.7.10}$$

The variation of this with left bump z_{L} may be found by the standard derivative of a quotient:

$$\frac{\mathrm{d}}{\mathrm{d}x}\left(\frac{u(x)}{v(x)}\right) = \frac{v\dfrac{\mathrm{d}u}{\mathrm{d}x} - u\dfrac{\mathrm{d}v}{\mathrm{d}x}}{v^2}$$

This can be applied to equation 8.7.10, then setting $z_L = 0$ to obtain the value at static ride height, which is the GRC lateral position variation in single bump:

$$\varepsilon_{B1,SB} = \left(\frac{dB}{dz_L}\right)_{z_L=0} \tag{8.7.11}$$

The result is

$$\varepsilon_{B1,SB} = \frac{1 + T\varepsilon_{BScd0,L}\varepsilon_{BScd1,L}}{\varepsilon_{BScd0,R+L}} \tag{8.7.12}$$

In the symmetrical-vehicle case this reduces to

$$\varepsilon_{B1,SB} = \frac{1}{2\varepsilon_{BScd0}} + \tfrac{1}{2} T\varepsilon_{BScd1} \quad \text{(symmetrical vehicle)} \tag{8.7.13}$$

Applying the same form of analysis to the GRC vertical position variation in left bump, substituting equation (8.7.9) into (8.7.5) gives

$$H = \frac{T\varepsilon_{BScd0,R}(\varepsilon_{BScd0,L} + \varepsilon_{BScd1,L}z_L) + z_L\varepsilon_{BScd0,R}}{\varepsilon_{BScd0,R+L} + \varepsilon_{BScd1,L}z_L} \tag{8.7.14}$$

Taking the derivative of a quotient and then setting $z_L = 0$ gives the GRC vertical position variation in single bump:

$$\varepsilon_{H1,SB} = \left(\frac{dH}{dz_L}\right)_{z_L=0} \tag{8.7.15}$$

with the value

$$\varepsilon_{H1,SB} = \frac{\varepsilon_{BScd0,R}(\varepsilon_{BScd0,R+L} + T\varepsilon_{BScd0,R}\varepsilon_{BScd1,L})}{\varepsilon^2_{BScd0,R+L}} \tag{8.7.16}$$

In the symmetrical-vehicle case, this reduces to

$$\varepsilon_{H1,SB} = \tfrac{1}{2}(1 + \tfrac{1}{2}T\varepsilon_{BScd1}) \quad \text{(symmetrical vehicle)} \tag{8.7.17}$$

which is just half of the symmetrical double-bump effect.

8.8 Asymmetrical Double Bump

Figure 8.8.1 shows the geometry for movement of the GRC relative to the body when there is asymmetrical double-wheel bump. The small track variation is neglected. Vehicle bump scrub asymmetry is included. In body coordinates, the GRC is at (B,H). Again, it is easy to write two simultaneous equations for H, the GRC height in vehicle-fixed coordinates:

$$\begin{aligned} H &= z_R + \left(\tfrac{1}{2}T + B\right)\varepsilon_{BScd,R} \\ H &= z_L + \left(\tfrac{1}{2}T - B\right)\varepsilon_{BScd,L} \end{aligned} \tag{8.8.1}$$

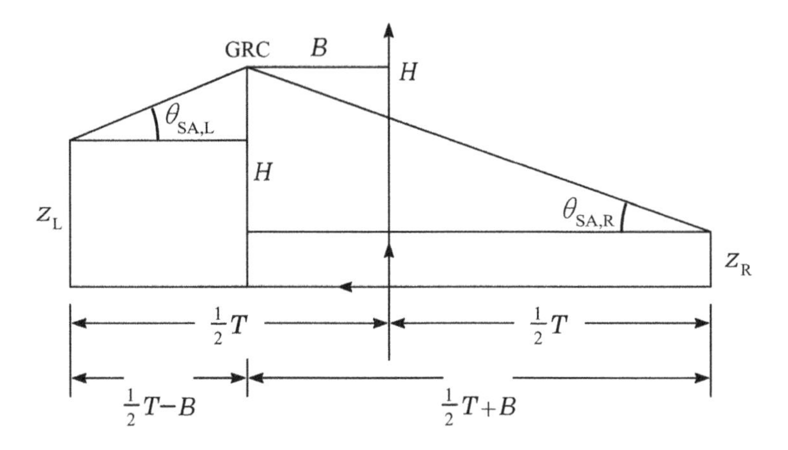

Figure 8.8.1 A vehicle with asymmetrical double bump, rear view, in vehicle coordinates.

Eliminating H by subtraction,

$$0 = z_R - z_L + \tfrac{1}{2}T\varepsilon_{BScd,R-L} + B\varepsilon_{BScd,R+L}$$

so

$$B = \frac{(z_L - z_R) + \frac{1}{2}T\varepsilon_{BScd,L-R}}{\varepsilon_{BScd,R+L}} \tag{8.8.2}$$

To obtain H, eliminate B by multiplication and addition of equations (8.8.1),

$$H\varepsilon_{BScd,L+R} = z_R\varepsilon_{BScd,L} + z_L\varepsilon_{BScd,R} + \tfrac{1}{2}T2\varepsilon_{BScd,R\times L}$$

so

$$H = \frac{(z_R\varepsilon_{BScd,L} + z_L\varepsilon_{BScd,R}) + T\varepsilon_{BScd,R\times L}}{\varepsilon_{BScd,R+L}} \tag{8.8.3}$$

To use equation (8.8.2) or (8.8.3) in specific cases, it is necessary to substitute specific values, or equations, for the bump scrub rate coefficients (e.g. constant, linear variation, or other polynomials). The variation coefficients for B and H with the left and right bumps may be derived, but are not particularly useful. However, the above results are very general, and can be used to obtain results for specific cases of combined heave and roll, expressed as equivalent bumps.

In the case of a symmetrical vehicle, equation (8.8.2) for the GRC lateral position becomes

$$B = \frac{(z_L - z_R)\left(1 + \frac{1}{2}T\varepsilon_{BScd1}\right)}{2\varepsilon_{BScd0} + (z_L + z_R)\varepsilon_{BScd1}} \quad \text{(symmetrical vehicle)} \tag{8.8.4}$$

Equation (8.8.3) for the GRC height does not simplify usefully, so it is better to evaluate the bump scrub rates at the two bump positions and use equation (8.8.3) as it is.

8.9 Roll of a Symmetrical Vehicle

Figure 8.9.1 shows the geometry for movement of the GRC relative to the body when the vehicle is rolled with no heave, just a positive suspension roll angle. Positive suspension roll is clockwise in the figure, looking along the positive longitudinal axis, so the corresponding suspension bumps are

$$z_R = +\tfrac{1}{2}T\phi_S; \quad z_L = -\tfrac{1}{2}T\phi_S = -z_R \tag{8.9.1}$$

For convenience, it is easier to work initially in terms of z_R and $-z_R$ rather than the actual roll angle expressions, in other words to analyse an antisymmetrical double bump.

The small track variation is neglected. In body coordinates, the GRC is at (B,H). As usual, it is easy to write two simultaneous equations for H, the GRC height in vehicle-fixed coordinates:

$$\begin{aligned} H &= +z_R + \left(\tfrac{1}{2}T + B\right)\varepsilon_{\text{BScd,R}} \\ H &= -z_R + \left(\tfrac{1}{2}T - B\right)\varepsilon_{\text{BScd,L}} \end{aligned} \tag{8.9.2}$$

Eliminating H by subtraction,

$$0 = 2z_R + \tfrac{1}{2}T\varepsilon_{\text{BScd,R}-\text{L}} + B\varepsilon_{\text{BScd,R}+\text{L}}$$

so

$$B = \frac{-2z_R + \tfrac{1}{2}T\varepsilon_{\text{BScd,L}-\text{R}}}{\varepsilon_{\text{BScd,R}+\text{L}}} \tag{8.9.3}$$

To obtain H, eliminate B by multiplication and addition of equations (8.9.2),

$$H\varepsilon_{\text{BScd,L}+\text{R}} = z_R\varepsilon_{\text{BScd,L}} - z_R\varepsilon_{\text{BScd,R}} + \tfrac{1}{2}T2\varepsilon_{\text{BScd,L}\times\text{R}}$$

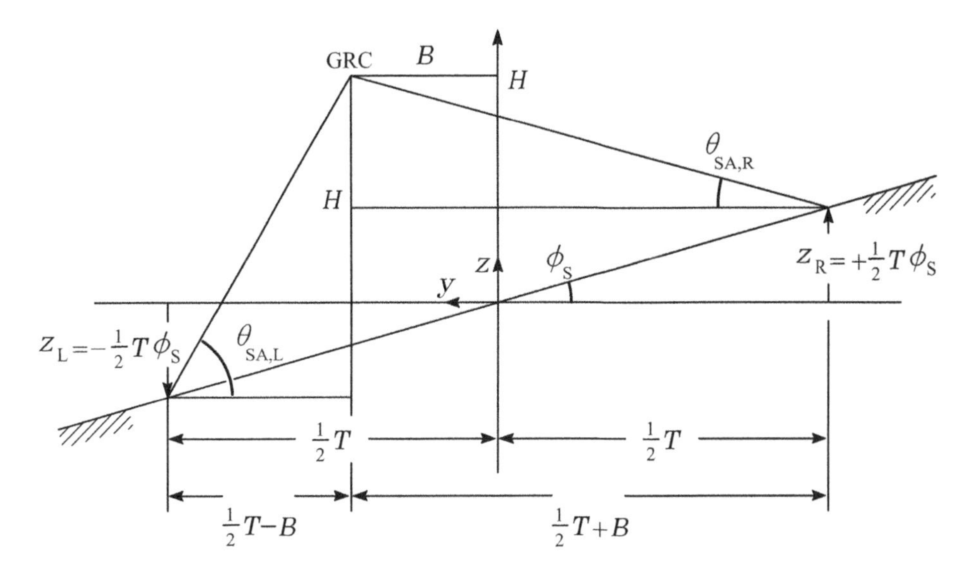

Figure 8.9.1 A vehicle with rolled suspension, rear view, in vehicle coordinates, shown with positive B.

so

$$H = \frac{z_R \varepsilon_{BScd,L-R} + T\varepsilon_{BScd,R\times L}}{\varepsilon_{BScd,R+L}} \tag{8.9.4}$$

This completes the basic solution of B and H. To progress the analysis, consider linear bump scrub variation for a symmetrical vehicle:

$$\begin{aligned} \varepsilon_{BScd} &= \varepsilon_{BScd0} + \varepsilon_{BScd1} z_S \\ \varepsilon_{BScd,R} &= \varepsilon_{BScd0} + \varepsilon_{BScd1} z_R \\ \varepsilon_{BScd,L} &= \varepsilon_{BScd0} - \varepsilon_{BScd1} z_R \end{aligned} \tag{8.9.5}$$

with further relationships

$$\begin{aligned} \varepsilon_{BScd,R+L} &= 2\varepsilon_{BScd0} \\ \varepsilon_{BScd,R-L} &= 2\varepsilon_{BScd1} z_R \\ \varepsilon_{BScd,R\times L} &= \varepsilon^2_{BScd0} - \varepsilon^2_{BScd1} z^2_R \end{aligned} \tag{8.9.6}$$

Using these in equation (8.9.3) gives

$$B = \frac{-2z_R + \frac{1}{2}T(-2\varepsilon_{BScd1} z_R)}{2\varepsilon_{BScd0}}$$

so

$$B = -\frac{\left(1 + \frac{1}{2}T\varepsilon_{BScd1}\right)}{\varepsilon_{BScd0}} z_R \tag{8.9.7}$$

and, in terms of the suspension roll angle ϕ_S,

$$B = -\tfrac{1}{2}T\frac{\left(1 + \frac{1}{2}T\varepsilon_{BScd1}\right)}{\varepsilon_{BScd0}} \phi_S \tag{8.9.8}$$

By differentiation, the lateral GRC movement coefficient in roll is

$$\varepsilon_{B1,Roll} = -\tfrac{1}{2}T\frac{1 + \frac{1}{2}T\varepsilon_{BScd1}}{\varepsilon_{BScd0}} \tag{8.9.9}$$

If desired, this can be made zero by designing the suspension such that

$$\varepsilon_{BScd1} = -\frac{2}{T} \tag{8.9.10}$$

With positive initial roll centre height, for $\varepsilon_{BScd1} > -2/T$ (e.g. zero), the GRC moves towards the outside of the curve. For $\varepsilon_{BScd1} < -2/T$, the GRC moves towards the inside of the curve, left in Figure 8.9.1.

The GRC height, from equation (8.9.4), by substitution of the linear symmetrical bump scrub rates, becomes

$$\begin{aligned} H &= \frac{-z_R 2\varepsilon_{BScd1} z_R + T(\varepsilon^2_{BScd0} - \varepsilon^2_{BScd1} z^2_R)}{2\varepsilon_{BScd0}} \\ &= \tfrac{1}{2}T\varepsilon_{BScd0} - \frac{\varepsilon_{BScd1}}{\varepsilon_{BScd0}}\left(1 + \tfrac{1}{2}T\varepsilon_{BScd1}\right) z^2_R \end{aligned} \tag{8.9.11}$$

so, in terms of the suspension roll angle ϕ_S,

$$H = \tfrac{1}{2} T\varepsilon_{BScd0} - \tfrac{1}{4} T^2 \frac{\varepsilon_{BScd1}}{\varepsilon_{BScd0}} \left(1 + \tfrac{1}{2} T\varepsilon_{BScd1}\right)\phi_S^2 \qquad (8.9.12)$$

The linear bump scrub variation, therefore, does not produce a linear factor in the GRC height, but a quadratic effect:

$$\begin{aligned} H &= H_0 + \varepsilon_{H1,Roll}\phi_S + \varepsilon_{H2,Roll}\phi_S^2 \\ H_0 &= \tfrac{1}{2} T\varepsilon_{BScd0} \\ \varepsilon_{H1,Roll} &= 0 \\ \varepsilon_{H2,Roll} &= -\tfrac{1}{4} T^2 \frac{\varepsilon_{BScd1}}{\varepsilon_{BScd0}} \left(1 + \tfrac{1}{2} T\varepsilon_{BScd1}\right) \end{aligned} \qquad (8.9.13)$$

with the quadratic coefficient having units m/rad^2. For ε_{BScd0} positive, and ε_{BScd1} positive or large negative, less than $-2/T$, then $\varepsilon_{H2,Roll}$ is negative, but for ε_{BScd1} in the range $-2/T$ to zero, then $\varepsilon_{H2,Roll}$ is positive. The magnitude is sensitive to the initial roll centre height, or ε_{BScd0}, the value of which may be very small and of either sign, according to details of the suspension design and operation.

In conjunction with the linear lateral movement, the quadratic height change gives a parabolic trajectory to the GRC in roll. Both movements can be eliminated by setting $\varepsilon_{BScd1} = -2/T$. From equations (8.9.9) and (8.9.13),

$$\varepsilon_{H2,Roll} = \tfrac{1}{2} T\varepsilon_{BScd1}\varepsilon_{B1,Roll} \qquad (8.9.14)$$

The equation of the parabola is

$$H = H_0 + c_P B^2 \qquad (8.9.15)$$

with

$$c_P = \frac{\varepsilon_{H2,Roll}}{\varepsilon_{B1,Roll}^2} = -\frac{\varepsilon_{BScd0}\varepsilon_{BScd1}}{1 + \tfrac{1}{2} T\varepsilon_{BScd1}} \qquad (8.9.16)$$

For example values, $T = 1.4$ m, $\varepsilon_{BScd0} = 0.14$, $\varepsilon_{BScd1} = -0.70\,\text{m}^{-1}$, $\varepsilon_{B1,Roll} = -2.55$ m/rad, $\varepsilon_{H2,Roll} = 1.495$ m/rad^2, $c_P = 0.1922\,\text{m}^{-1}$.

8.10 Linear Symmetrical Vehicle Summary

For a symmetrical vehicle, the geometric suspension equations for linear bump scrub variation may be summarised as in Table 8.10.1.

Example values are: $T = 1.46$ m, $\varepsilon_{BScd0} = 0.06$, $\varepsilon_{BScd1} = -0.70\,\text{m}^{-1}$, $H_0 = 0.044$ m, $\varepsilon_{B1,SB} = 7.82$, $\varepsilon_{H1,SB} = 0.2445$, $\varepsilon_{Z1,DB} = -0.511$, $\varepsilon_{H1,DB} = 0.489$, $\varepsilon_{B1,Roll} = -5.95$ m/rad, $\varepsilon_{H2,Roll} = 3.04$ m/rad^2.

Table 8.10.1 Summary of equations for symmetrical vehicle GRC

$\varepsilon_{\mathrm{BScd}} = \varepsilon_{\mathrm{BScd0}} + \varepsilon_{\mathrm{BScd1}} z_{\mathrm{SB}}$	(8.10.1)
$H_0 = \frac{1}{2} T \varepsilon_{\mathrm{BScd0}}$	(8.10.2)
$B_0 = 0$	(8.10.3)
$\varepsilon_{\mathrm{B1,SB}} = \dfrac{1}{2\varepsilon_{\mathrm{BScd0}}} + \frac{1}{2} T \varepsilon_{\mathrm{BScd1}}$	(8.10.4)
$\varepsilon_{\mathrm{H1,SB}} = \frac{1}{2}\left(1 + \frac{1}{2} T \varepsilon_{\mathrm{BScd1}}\right)$	(8.10.5)
$\varepsilon_{\mathrm{B1,DB}} = 0$	(8.10.6)
$\varepsilon_{\mathrm{Z1,DB}} = \frac{1}{2} T \varepsilon_{\mathrm{BScd1}}$	(8.10.7)
$\varepsilon_{\mathrm{H1,DB}} = 1 + \frac{1}{2} T \varepsilon_{\mathrm{BScd1}}$	(8.10.8)
$\varepsilon_{\mathrm{B1,Roll}} = -\frac{1}{2} T \dfrac{\left(1 + \frac{1}{2} T \varepsilon_{\mathrm{BScd1}}\right)}{\varepsilon_{\mathrm{BScd0}}}$	(8.10.9)
$\varepsilon_{\mathrm{H1,Roll}} = 0$	(8.10.10)
$\varepsilon_{\mathrm{H2,Roll}} = -\frac{1}{4} T^2 \left(\dfrac{\varepsilon_{\mathrm{BScd1}}}{\varepsilon_{\mathrm{BScd0}}}\right)\left(1 + \frac{1}{2} T \varepsilon_{\mathrm{BScd1}}\right)$	(8.10.11)

8.11 Roll of an Asymmetrical Vehicle

Figure 8.11.1 shows the geometry for movement of the GRC relative to the body when the asymmetrical-suspension-geometry vehicle is rolled with no heave. The small track variation is neglected. Vehicle bump scrub asymmetry is now included, although this does not show in the diagram. The simultaneous equations for B and H in general form are as in the previous section, and the solution is the same:

$$B = \frac{-2z_{\mathrm{R}} + \frac{1}{2} T \varepsilon_{\mathrm{BScd,L-R}}}{\varepsilon_{\mathrm{BScd,R+L}}} \tag{8.11.1}$$

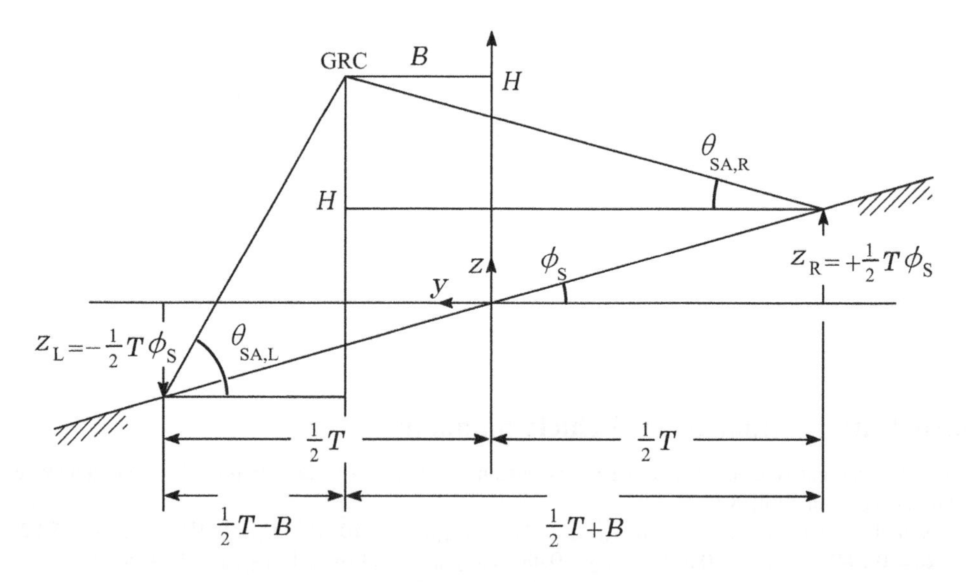

Figure 8.11.1 An asymmetrical suspension geometry vehicle with rolled suspension, rear view, in vehicle coordinates, but still with zero heave, the same diagram as Figure 8.9.1.

$$H = \frac{z_R \varepsilon_{BScd,L-R} + T\varepsilon_{BScd,R\times L}}{\varepsilon_{BScd,R+L}} \tag{8.11.2}$$

Now consider asymmetrical linear bump scrub variation (still with equal and opposite bumps, $z_L = -z_R$):

$$\begin{aligned} \varepsilon_{BScd,R} &= \varepsilon_{BScd0,R} + \varepsilon_{BScd1,R} z_R \\ \varepsilon_{BScd,L} &= \varepsilon_{BScd0,L} - \varepsilon_{BScd1,L} z_R \end{aligned} \tag{8.11.3}$$

with further relationships

$$\begin{aligned} \varepsilon_{BScd,R+L} &= \varepsilon_{BScd0,R+L} + \varepsilon_{BScd1,R-L} z_R \\ \varepsilon_{BScd,R-L} &= \varepsilon_{BScd0,R-L} + \varepsilon_{BScd1,R+L} z_R \end{aligned} \tag{8.11.4}$$

Inserting the linear variations into equation (8.10.1),

$$B = \frac{-2z_R - \frac{1}{2}T\varepsilon_{BScd0,R-L} - \frac{1}{2}T\varepsilon_{Bscd1,R+L} z_R}{\varepsilon_{BScd0,R+L} + \varepsilon_{BScd1,R-L} z_R} \tag{8.11.5}$$

Taking the derivative of the quotient, and substituting $z_R = 0$ gives the GRC lateral position variation in roll as a function of the associated suspension bump deflection:

$$\varepsilon_{B1,Roll,ZB} = \frac{-2\varepsilon_{BScd0,R+L} + \frac{1}{2}T\left(\varepsilon_{BScd0,R+L}\varepsilon_{BScd1,R+L} - \varepsilon_{BScd0,R-L}\varepsilon_{BScd1,R-L}\right)}{\varepsilon^2_{BScd0,R+L}} \tag{8.11.6}$$

This coefficient has units m/m, that is, no dimensions. To obtain this in terms of the actual roll angle, in radians,

$$\varepsilon_{B1,Roll} = \frac{1}{2}T\varepsilon_{B1,Roll,ZB} \tag{8.11.7}$$

This has units m/rad, lateral motion of the GRC per radian of suspension roll angle.

8.12 Road Coordinates

The preceding analysis has been performed in the vehicle-body-fixed coordinates (B, H) because this is much clearer than a suspension analysis in Earth-fixed road coordinates (Y, Z). However, for application, it is desired to know the values in the Earth-fixed system, so a transformation of coordinates must be applied, as seen in Figure 8.12.1. The vehicle body position is defined by a heave Z_B (a negative suspension bump) followed by a suspension roll ϕ_S. The order of this sequence is significant; if the movement is defined in the other order then the transformation equations are altered. Previous to the displacement, the two axis systems (B, H) and (Y, Z) are coincident.

A general point P is at vehicle-fixed (B, H) and at Earth-fixed (Y, Z). By simple geometry,

$$\begin{aligned} Y &= B\cos\phi_S - H\sin\phi_S \\ Z &= B\sin\phi_S + H\cos\phi_S + Z_B \end{aligned} \tag{8.12.1}$$

This may be arranged as two simultaneous equations in B and H, and solved by the usual methods, giving

$$\begin{aligned} B &= Y\cos\phi_S + (Z - Z_B)\sin\phi_S \\ H &= (Z - Z_B)\cos\phi_S - Y\sin\phi_S \end{aligned} \tag{8.12.2}$$

The vehicle also heaves on the tyres because of changing total normal force and because of variation of the effective tyre normal stiffness with lateral force, and the axle rolls on the tyres because of the lateral transfer of normal force in cornering combined with the tyre vertical compliance. An additional transformation, similar to the above, may be applied for this.

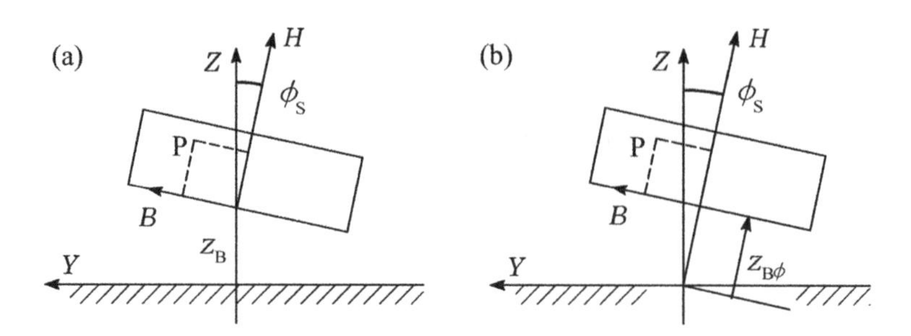

Figure 8.12.1 Vehicle-fixed (B, H) and Earth-fixed (Y, Z) coordinates: (a) with heave before roll; (b) with roll before heave.

When roll is applied before heave, the equations are

$$\begin{aligned} Y &= -(H + Z_{B\phi})\sin\phi_S + B\cos\phi_S \\ Z &= (H + Z_{B\phi})\cos\phi_S + B\sin\phi_S \end{aligned} \tag{8.12.3}$$

with

$$\begin{aligned} B &= Y\cos\phi_S + Z\sin\phi_S \\ H &= -Y\sin\phi_S + Z\cos\phi_S - Z_{B\phi} \end{aligned} \tag{8.12.4}$$

Figure 8.12.2 shows the rolled vehicle, with swing arm angles in vehicle coordinates $\theta_{SA,L}$ and $\theta_{SA,R}$. Relative to the road, the swing arm angles are

$$\begin{aligned} \theta_{SAR,L} &= \theta_{SA,L} - \phi_S \\ \theta_{SAR,R} &= \theta_{SA,R} + \phi_S \end{aligned} \tag{8.12.5}$$

By this means, the GRC may be solved directly in road coordinates, if desired.

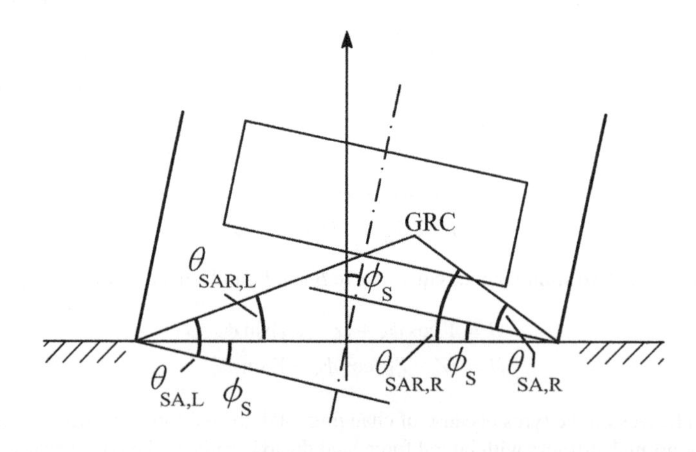

Figure 8.12.2 Rolled vehicle showing swing arm angles θ_{SA} in vehicle coordinates and θ_{SAR} in road coordinates.

The bump scrub rate coefficients may be expressed easily in road coordinates for small angles:

$$\begin{aligned} \varepsilon_{\mathrm{BScdR,R}} &= \varepsilon_{\mathrm{BScd,R}} + \phi_{\mathrm{S}} \\ \varepsilon_{\mathrm{BScdR,L}} &= \varepsilon_{\mathrm{BScd,L}} - \phi_{\mathrm{S}} \end{aligned} \tag{8.12.6}$$

Using this method, all vehicles become asymmetrical.

8.13 GRC and Latac

For steady-state handling analysis, the roll centre height may be represented approximately by

$$\begin{aligned} H &= H_0 + k_{\mathrm{H1}}A_{\mathrm{Y}} + k_{\mathrm{H2}}A_{\mathrm{Y}}^2 \\ B &= B_0 + k_{\mathrm{B1}}A_{\mathrm{Y}} + k_{\mathrm{B2}}A_{\mathrm{Y}}^2 \end{aligned} \tag{8.13.1}$$

In a linear model, the suspension double bump (axle bump z_{A}) and suspension roll angle ϕ_{S} are related to the lateral acceleration (latac) by

$$\begin{aligned} z_{\mathrm{A}} &= k_{\mathrm{ZA}}A_{\mathrm{Y}} \\ \phi_{\mathrm{S}} &= k_{\phi\mathrm{S}}A_{\mathrm{Y}} \end{aligned} \tag{8.13.2}$$

In simple roll, then, the GRC movements become

$$\begin{aligned} H &= H_0 + \varepsilon_{\mathrm{H1,Roll}}k_{\phi\mathrm{S}}A_{\mathrm{Y}} + \varepsilon_{\mathrm{H2,Roll}}k_{\phi\mathrm{S}}^2A_{\mathrm{Y}}^2 \\ B &= \varepsilon_{\mathrm{B1,Roll}}k_{\phi\mathrm{S}}A_{\mathrm{Y}} + \varepsilon_{\mathrm{B2,Roll}}k_{\phi\mathrm{S}}^2A_{\mathrm{Y}}^2 \end{aligned} \tag{8.13.3}$$

The linear effect on GRC height is usually quite small, being the result of vehicle asymmetries, so k_{H1} is small. The coefficient k_{H2} is the quadratic roll-centre height variation coefficient, and again tends to be small with a value possibly around plus or minus 0.001 m/(m s^{-2})^2 (0.1 m/g^2). In roll, k_{B1} may be around -0.10 s^2 [m/(m s^{-1})].

8.14 Experimental Roll Centres

Various methods have been suggested at times for the experimental measurement of roll-centre position, none of which are really very satisfactory. One method is the observation of body displacement with application of a lateral force at the centre of mass, for example using a laterally slanted ramp, but this is definitely not suitable. The large lateral displacement at the tyres is a problem which may be eliminated by using solid wheels, but there will still be inappropriate lateral displacements within the suspension compliances. A correct method of finding the SAE-defined roll centre would be to apply lateral forces at various heights. With the lateral force applied at the height of the roll axis, there will be a lateral displacement of the body because of compliances, and a vertical displacement because of the jacking force, but no rotation. In this sense, the FRC is directly analogous to a shear centre, which would probably be a better name for it, although the term 'roll centre' is no doubt too entrenched by use to change now.

As in the case of a shear centre, there is a more direct method of measurement. If the displacement is observed when a moment is applied, the shear centre will be the centre of rotation, i.e. the point of zero displacement. The most convenient way to apply a moment is by a couple, for example by a joist through the doors, with a load first on one end then on the other. This does involve a vertical load which is acceptable if the vehicle is to be tested in a loaded condition. Otherwise it is necessary to arrange for a vertical upward force to be applied as half of the couple, say by a hydraulic jack or a weight over a pulley. The actual deflection has been analysed by double-exposure photographs, but in the interests of accuracy it is much better to take measurements of the displacement of specific points by dial gauges. If these points are not in the vertical transverse plane of the wheel centres, then the data must be used to find the roll axis, and this then used to find the roll-axis position at each suspension.

This method is still subject to serious errors and problems. Under real cornering conditions there are actually large forces in the suspension links, causing geometric changes through bush distortions. These are largely absent from the test. Any small inappropriate lateral displacement of the body will give a large error in the measured roll-centre height. The applied couple will change the load on the tyres, giving an additional rotation of the complete chassis, including the wheels, about a point at ground level roughly midway between the tyres. The result is a false lateral motion of the body at roll centre height. This must be guarded against by using solid disc wheels, although this exacerbates the problem of allowing scrub, because when finding the roll centre for large angular displacements, there will be a consequent track change. This must be permitted, perhaps with air bearing pads, but without allowing any inappropriate lateral motion of the body, which is critical. It is difficult to say how the body should be located.

In summary, it is fairly easy to observe the displacement of a vehicle when it is subject to loads or couples, but it is very much more difficult to obtain results that have a worthwhile roll centre interpretation, especially for large latacs. Probably the only satisfactory experimental method is to use an instrumented vehicle in real cornering to find the displacements and actual load transfer for each axle, for example with strain gauges on the links, and to deduce the KRC from the lateral displacements, and the net link load transfer and FRC height from the forces.

References

Dixon, J.C. (1991) *Tyres, Suspension and Handling*, 1st edition. Cambridge: Cambridge University Press.

Dixon, J.C. (1996) *Tires, Suspension and Handling*, 2nd edition. London: Edward Arnold.

Analysis of the GRC using a variable bump scrub rate does not seem to have been published previously in any books or research literature.

9

Compliance Steer

9.1 Introduction

This chapter discusses compliance steer effects in general terms. Specific designs of suspension are covered in later chapters.

The wheel position, for a given suspension bump, is determined initially by the basic geometry of the suspension, but this is modified by the forces and moments acting on the wheel, which cause compliance deflections. These are generally small, but never zero. They may be a problem, with inadequate design, or may be kept small or put to good effect by good design. For example, sometimes rear compliance oversteer is deliberately introduced. Under tractive or braking forces, there may also be compliant changes of front caster angle. These will augment the caster change relative to the ground caused by vehicle pitch due to longitudinal load transfer.

In general, the compliance may be in the suspension or in the steering linkage. The main effects are changes of wheel angles, known as compliance steer angles (δ_C) and compliance camber angles (γ_C). The fundamental units are radians, and these are the units normally used in equations and computer programs, but the usual form for human consumption is degrees.

For the design of ordinary passenger vehicles, handling must be seen in the context of the ride–handling compromise. To provide comfort and to avoid noise, vibration and harshness, it is necessary to have significant compliance of wheel motion, not just vertically but also in longitudinal and lateral directions, achieved by tyre compliance, by the springs and by the extensive use of rubber bushes. Unfortunately, suspension link compliance has often led to considerable angular compliance of the wheels, resulting in unfavourable or unpredictable handling because of compliance camber deflection and especially because of compliance steer deflection. Through better understanding, this conflict is now largely resolvable, by allowing the wheel relatively generous movement in translation, but little angular movement in steer or camber for the forces and moments that it actually experiences. The method is to bring the shear centre of an independent suspension, in plan view, close to the centre of tyre contact, or for a rigid axle near to the mid-point between the tyres. Tyre forces then have little moment about the shear centre, and so, although the wheel still has angular compliance, there is little angular response, so that steer angle changes are controlled. Careful location of the shear centre can even be used to introduce favourable small deflections. A significant characteristic of compliance steer is that, unlike roll steer, it occurs almost immediately.

On low-friction surfaces such as ice or snow, the tyre characteristics are different, and there is a low limit to lateral acceleration, so load transfer distribution becomes less significant, and roll steer and side-force steer become more critical in determining limit handling, especially at the rear.

Suspension Geometry and Computation J. C. Dixon

9.2 Wheel Forces and Moments

The basic coordinate system for the wheel is shown in Figures 9.2.1 and 9.2.2. At the base of the wheel, at the notional centre of the undistorted contact patch, there are three forces and three moments acting on the wheel, Figures 9.2.3 and 9.2.4. More accurately, one should say that there is one resultant wrench, which is a force with a coaxial moment, which may usefully be considered in terms of these six components. The origin of the wheel coordinate system is the intersection of the ground plane, the undistorted wheel

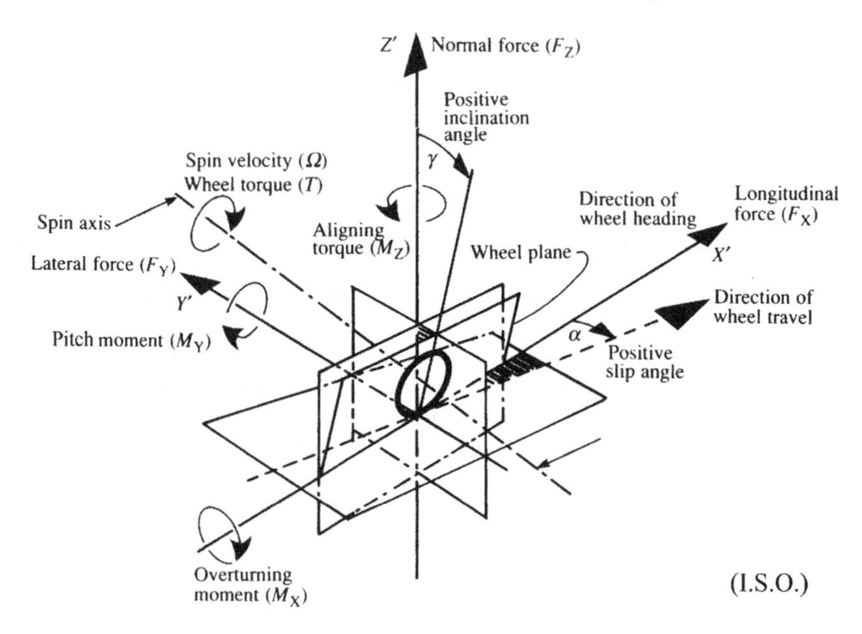

Figure 9.2.1 The ISO coordinate system for a wheel.

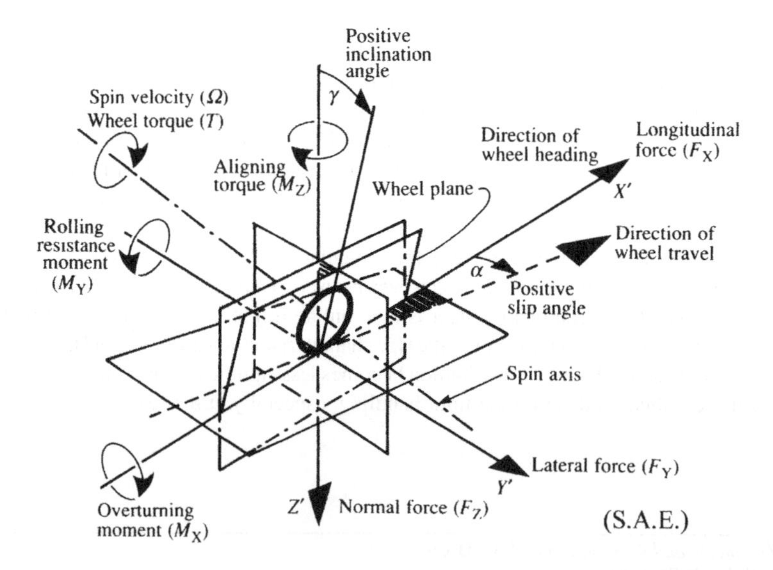

Figure 9.2.2 The SAE coordinate system for a wheel (SAE, 1976).

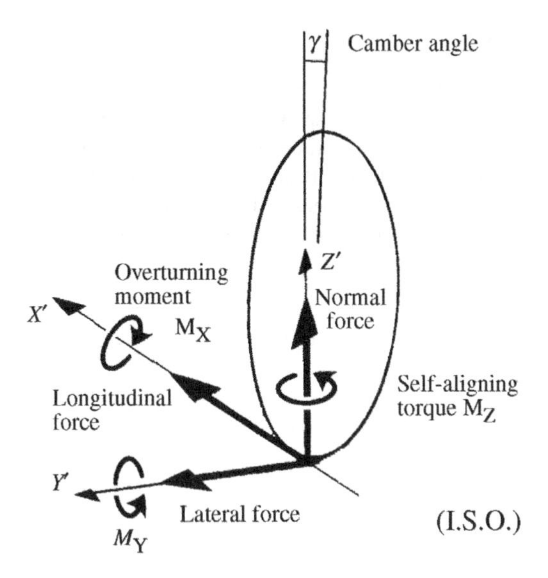

Figure 9.2.3 The total wrench at the base of a wheel, exerted by the ground on the wheel, analysed as six force–moment components, ISO system.

plane and the transverse vertical plane through the wheel centre. The axis directions are:

(1) the X axis, forward in the ground and wheel planes;
(2) the Y axis, in the ground and transverse vertical planes;
(3) the Z axis, perpendicular to the ground in the transverse plane.

Note that the Z axis is not in the wheel plane, unless the wheel has zero inclination.

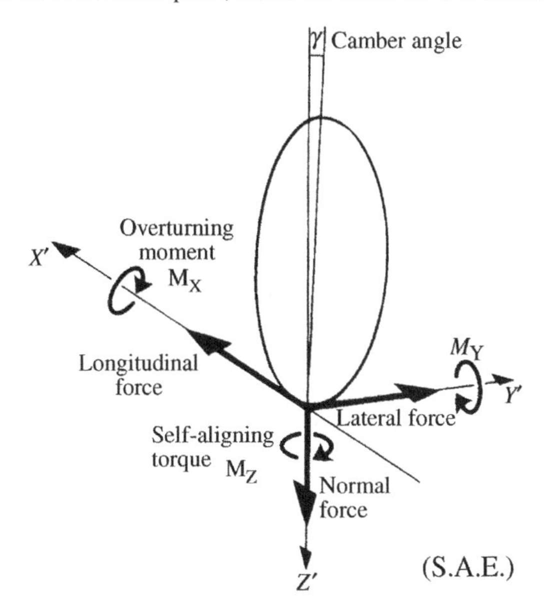

Figure 9.2.4 The total wrench at the base of a wheel, exerted by the ground on the wheel, analysed as six force–moment components, SAE system.

9.3 Compliance Angles

Apart from the ride-improving properties of the suspension compliance, the most important compliance effects are those directly affecting the handling qualities, specifically the compliance steer angle and the compliance camber angle or compliance inclination angle, Figure 9.3.1. Considering these angles, for a given design of suspension location system with its rubber bushes, reveals the critical aspects of the suspension in this regard. The steer angles are measured from a line parallel to the vehicle centreline, with a total steer angle which is simply the sum of the geometric and compliance angles:

$$\delta = \delta_G + \delta_C \tag{9.3.1}$$

Similarly, the inclination or camber angles may be measured from the perpendicular to the road surface, with

$$\gamma = \gamma_G + \gamma_C \tag{9.3.2}$$

Also, for pitch angle (caster angle) of the wheel hub,

$$\theta = \theta_G + \theta_C \tag{9.3.3}$$

The pitch angle variation changes the caster angle of the wheel upright. With positive rotation as right-hand about the Y-axis, the static caster angle would be negative, so this sign convention is rarely applied. Figure 9.3.1(c) shows positive values of θ_G and θ_C, caster being simply taken as positive when the steering axis leans backwards.. The wheel hub pitch angle is not necessarily identical to the caster angle, the former possibly being defined as zero in the static position, although the difference will be constant at the static caster minus static hub pitch angle.

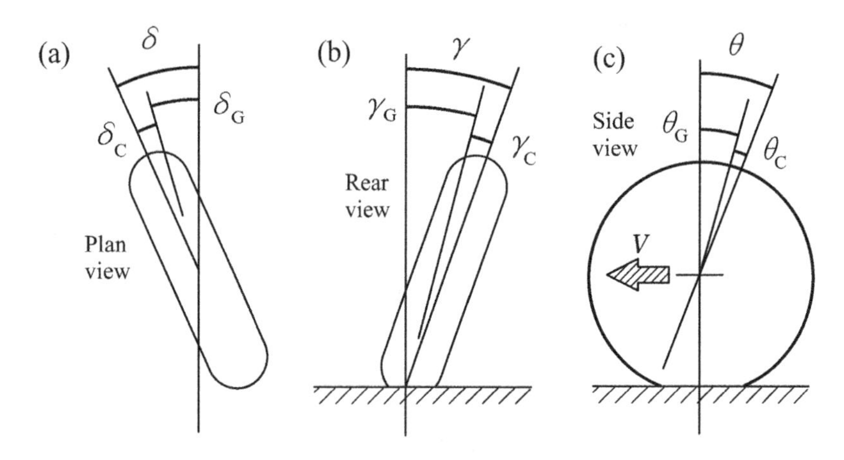

Figure 9.3.1 Compliance steer and compliance inclination angles for a single wheel: (a) steer angle in plan view; (b) inclination angle in rear view; (c) hub pitch angle in side view.

9.4 Independent Suspension Compliance

The total force and moment wrench at each independent wheel contact patch is considered as three force and three moment components: F_X, F_Y, F_Z, M_X, M_Y, M_Z. Also there are six possible displacement responses of the wheel: X, Y, Z, γ, θ, δ. Each displacement response may be affected by each of the six stimuli, so, considering even the simplest linear model, there will be 36 coefficients in total. This can be combined into a matrix equation

$$\boldsymbol{X} = \boldsymbol{CF} \tag{9.4.1}$$

where $\boldsymbol{X}$ is the six-component linear and angular compliance displacement column vector

$$\boldsymbol{X} = (X_C, Y_C, Z_C, \gamma_C, \theta_C, \delta_C)^T \tag{9.4.2}$$

in which the suffix T means transpose (in this case, just change the row to a column); $\boldsymbol{F}$ is the six-component force–moment column vector,

$$\boldsymbol{F} = (F_X, F_Y, F_Z, M_X, M_Y, M_Z)^T \tag{9.4.3}$$

and $\boldsymbol{C}$ is the compliance matrix. This matrix equation has no special properties, it is just a simple and concise way of writing the combined linear equations. In more detail

$$\begin{pmatrix} X_C \\ Y_C \\ Z_C \\ \gamma_C \\ \theta_C \\ \delta_C \end{pmatrix} = \begin{pmatrix} C_{X,FX} & C_{X,FY} & C_{X,FZ} & C_{X,MX} & C_{X,MY} & C_{X,MZ} \\ C_{Y,FX} & C_{Y,FY} & C_{Y,FZ} & C_{Y,MX} & C_{Y,MY} & C_{Y,MZ} \\ C_{Z,FX} & C_{Z,FY} & C_{Z,FZ} & C_{Z,MX} & C_{Z,MY} & C_{Z,MZ} \\ C_{\gamma,FX} & C_{\gamma,FY} & C_{\gamma,FZ} & C_{\gamma,MX} & C_{\gamma,MY} & C_{\gamma,MZ} \\ C_{\theta,FX} & C_{\theta,FY} & C_{\theta,FZ} & C_{\theta,MX} & C_{\theta,MY} & C_{\theta,MZ} \\ C_{\delta,FX} & C_{\delta,FY} & C_{\delta,FZ} & C_{\delta,MX} & C_{\delta,MY} & C_{\delta,MZ} \end{pmatrix} \times \begin{pmatrix} F_X \\ F_Y \\ F_Z \\ M_X \\ M_Y \\ M_Z \end{pmatrix} \tag{9.4.4}$$

This is simply, then, a concise way of representing six equations of the form

$$X_C = C_{X,FX}F_X + C_{X,FY}F_Y + C_{X,FZ}F_Z + C_{X,MX}M_X + C_{X,MY}M_Y + C_{X,MZ}M_Z \tag{9.4.5}$$

The units of the compliance matrix are in four equal 3×3 sections:

$$\text{units}\,(\boldsymbol{C}) = \begin{pmatrix} \text{m/N} & \text{m/Nm} = \text{N}^{-1} \\ \text{rad/N} & \text{rad/Nm} \end{pmatrix}$$

Fortunately, in view of the large number of coefficients at each wheel, many of the effects are insignificant in practice. This occurs if the stimulus is small, if the compliance is small, or if the resulting deflection is unimportant. Expressed in partial derivatives, the more interesting and important compliances of the matrix are as follows:

$$\boldsymbol{C} = \begin{bmatrix} \dfrac{\partial X}{\partial F_X} & - & - & - & \dfrac{\partial X}{\partial M_Y} & - \\ - & \dfrac{\partial Y}{\partial F_Y} & - & - & - & - \\ - & \dfrac{\partial Z}{\partial F_Y} & \dfrac{\partial Z}{\partial F_Z} & - & \dfrac{\partial Z}{\partial M_Y} & - \\ - & \dfrac{\partial \gamma}{\partial F_Y} & \dfrac{\partial \gamma}{\partial F_Z} & \dfrac{\partial \gamma}{\partial M_X} & - & - \\ \dfrac{\partial \theta}{\partial F_X} & - & - & - & \dfrac{\partial \theta}{\partial M_Y} & - \\ \dfrac{\partial \delta}{\partial F_X} & \dfrac{\partial \delta}{\partial F_Y} & \dfrac{\partial \delta}{\partial F_Z} & \dfrac{\partial \delta}{\partial M_X} & \dfrac{\partial \delta}{\partial M_Y} & \dfrac{\partial \delta}{\partial M_Z} \end{bmatrix} \tag{9.4.6}$$

Here it is apparent that the rows are the various coefficients for the effects on one displacement (e.g. the sixth row is for changes to the wheel steer angle δ), and columns are the coefficients for the effects of a given force or moment (e.g. the second column is for the effects of F_Y, the lateral force). The values in this matrix are normally positive, but this is contingent upon the sign conventions adopted. A matrix of derivatives such as this is known mathematically as a Jacobean matrix (with emphasis on the 'oh').

9.5 Discussion of Matrix

To understand the terms of the Jacobean sensitivity matrix, it is necessary to define clearly the conditions of the partial derivatives. Basically, the body is held fixed, and the wheel is initially held at its normal position by a vertical force, which would normally support the vehicle. The wheel is locked to the hub in rotation. Additional forces and moments are then applied and the displacements are measured. The meaning and significance of the results are not always completely obvious, and the results require interpretation. For example, as tested with a wheel and tyre, the lateral displacement response to a lateral force shows a large value due to the tyre lateral compliance. Normally, this would be treated separately, so the tyre compliance would be excluded from the compliance matrix, leaving only the structural compliance. Similarly, the vertical displacement response to vertical force is dominated by the suspension spring compliance and tyre vertical compliance, which would be deducted to leave the structural compliance only. In the case of formula racing cars with long wishbones, this may be important.

In some applications of the compliance matrix, the compliances of the spring, etc., may be included. This is just a matter of methodology, but it is important to be clear what the matrix compliances do include in any particular case.

In the first row, X displacements, sensitivity to F_X is deliberately introduced by appropriate rubber bushes. This is to prevent a significant harshness on hitting bumps. Bias-ply tyres (diagonal-ply) are more compliant, but with the widespread use of radial-ply tyres which are much less compliant in the longitudinal direction it is really essential to make appropriate use of this suspension compliance in modern road vehicles. Exceptions are, naturally, off-road vehicles which generally use diagonal-ply tyres because these have sidewalls of greater strength. Sensitivity of X to M_Y is of interest because it relates sensitivity to wheel contact-patch height forces to the sensitivity to axle-height forces. These moments are produced by variations in rolling resistance and by traction or braking forces produced through the action of a driveshaft.

The second row has an entry for $\partial Y/\partial F_Y$, the lateral compliance. This is generally not important, not least because the tyre itself exhibits a larger lateral compliance, which would normally be eliminated from an experimental measure, but it is easy to see physically how the residual lateral structural compliance at the contact patch arises from linear lateral compliance of two wishbones.

In the third row, $\partial Z/\partial F_Y$ is the change of vertical position with lateral force. This may appear to be the suspension link jacking effect, but usually this would be separated out and dealt with as an independent item, not classified as compliance. Link jacking can occur even with ideally-rigid links. The residual term is small and due to link compliance only, allowing a small vertical motion of the wheel with the body fixed. The next entry in the third row is $\partial Z/\partial F_Z$, the change of vertical position with vertical force. This is the main suspension action, and again this would normally be separated out and dealt with by other means. The basic compliance here is the large ride compliance, the sum of the spring (wheel rate) and tyre normal compliances:

$$C_R = \frac{1}{K_S} + \frac{1}{K_{ZT}} \tag{9.5.1}$$

The matrix compliance coefficient here, $C_{Z,FZ}$, would normally apply to any additional deflection, above that due to the spring and tyre action, due to distortion of the links and deflection of rubber bushes.

The third term shown in the third row is $\partial Z/\partial M_Y$, which is the effect of pitching moment on vertical position. This is mentioned because such an effect can arise in the case of trailing arms, but is then analogous to link jacking, and, similarly, would be dealt with separately.

In the fourth row, the effects on inclination/camber angle, significant effects are shown for F_Y, F_Z and M_X. The way in which these can alter a camber angle through compliance of the links is easy to imagine.

The fifth row, for changes of pitch angle, has two entries. The pitch angle of the wheel is not a handling or ride issue, because the wheel rotates anyway, but the pitch angle should be considered to be the angle of the wheel hub. This can deflect in pitch compliance due to pitch moments or longitudinal forces from traction or braking, changing the steering caster angle.

The sixth row, that for steer angle effects, which is most critical, is full, although in practice some of these terms are small. The first term, $\partial\delta/\partial F_X$, is the important steer effect of tractive and braking forces. The second term, $\partial\delta/\partial F_Z$, is the important steer effect of lateral forces. The third term, $\partial\delta/\partial F_Y$, is the steer effect of vertical forces. The fourth term, $\partial\delta/\partial M_X$, is the steer effect of the tyre overturning moment, usually small. The fifth term, $\partial\delta/\partial M_Y$, is the steer effect of pitch moment. The final term, $\partial\delta/\partial M_Z$, is the steer effect of wheel yaw moments (e.g. due to mechanical or pneumatic trail combined with lateral forces), and may be significant.

9.6 Independent-Suspension Summary

The main structural compliance effect in ride is:

(1) $C_{X,FX}$ (m/N), the longitudinal compliance to longitudinal force.

For formula racing cars, the structural vertical compliance to vertical force may be as great as the basic spring compliance, because of the use of stiff suspensions but long thin links and body compliance at the connection points.

The main structural compliance effects in handling are:

(1) $C_{\delta,FY}$ (deg/N), the lateral force compliance steer coefficient;
(2) $C_{\delta,MZ}$ (deg/Nm), the aligning moment compliance steer coefficient;
(3) $C_{\gamma,FY}$ (deg/N), the lateral force compliance camber coefficient;
(4) $C_{\gamma,MX}$ (deg/Nm), the overturning moment compliance camber coefficient.

The last of these is usually neglected, and often the last two or three. The actual compliance steer angle is typically expressed as

$$\delta_C = \delta_{FY} + \delta_{MZ} = C_{\delta,FY}F_Y + C_{\delta,MZ}M_Z \qquad (9.6.1)$$

and the compliance inclination angle is

$$\gamma_C = \gamma_{FY} + \gamma_{MX} = C_{\gamma,FY}F_Y + C_{\gamma,MX}M_X \qquad (9.6.2)$$

In the case of compliance camber, some symbol other than γ might be desirable, but there is no commonly accepted symbol. The use of γ is acceptable provided that the corresponding preference for δ rather than α is clearly understood in the case of compliance steer.

The total effect of compliance in the linear cornering regime may be summarised by the compliance steer gradient $k_{\delta C}$, which has SI units rad/m.s^{-2}, but is usually expressed in deg/g. Because the lateral force effective in creating distortions is the sprung mass force rather than the total lateral force, the compliance gradients are as follows:

$$\begin{aligned} k_{\delta,\mathrm{C}} &= k_{\delta,\mathrm{FY}} + k_{\delta,\mathrm{MZ}} \\ k_{\delta,\mathrm{FY}} &= m_{\mathrm{Bf}} C_{\delta,\mathrm{FY,f}} + m_{\mathrm{Br}} C_{\delta,\mathrm{FY,r}} \\ k_{\delta,\mathrm{MZ}} &= m_{\mathrm{Bf}} t_{\mathrm{f}} C_{\delta,\mathrm{MZ,f}} + m_{\mathrm{Br}} t_{\mathrm{r}} C_{\delta,\mathrm{MZ,r}} \end{aligned} \tag{9.6.3}$$

where m_{B} is the body mass (sprung mass) and t is the tyre pneumatic trail. Similar equations can be written for the camber. Typical values of compliance understeer gradient are 1.0 deg/g at the front and 0.2 deg/g at the rear.

9.7 Hub Centre Forces

For a driven wheel, or with inboard brakes, the driveshaft transmits a moment M_{D} which, for constant wheel angular speed, generates a longitudinal force

$$F_{\mathrm{X}} = \frac{M_{\mathrm{D}}}{R_{\mathrm{L}}} \tag{9.7.1}$$

where R_{L} is the loaded radius of the wheel. As shown in Figure 9.7.1, this combination of force and moment is equivalent to a force only, at wheel centre height. It is also equivalent to a force F_{X} at the contact patch plus a pitching moment M_{Y} there, where $M_{\mathrm{Y}} = M_{\mathrm{D}}$. Comparing Figure 9.7.1(a) and (c), the moment has been moved. This is permissible, as the application point, or axis, of a moment has no effect on the influence of the moment on a rigid body. The configuration of Figure 9.7.1(c) can then be used with the compliance matrix to obtain the compliance deflections for the hub-height force. This illustrates the principle, although, of course, in a practical case there may be complications; for example, with an inclined driveshaft there will be an additional moment M_{Z}.

It is also of interest to obtain the coefficients for the force F_{X} applied directly at the wheel centre. The compliance steer angle is

$$\delta_{\mathrm{C}} = C_{\delta,\mathrm{FX}} F_{\mathrm{X}} + C_{\delta,\mathrm{MY}} F_{\mathrm{X}} R_{\mathrm{L}} \tag{9.7.2}$$

so the compliance steer coefficient for the wheel-centre-height force is

$$C_{\delta,\mathrm{FX,WCH}} = C_{\delta,\mathrm{FX}} + C_{\delta,\mathrm{MY}} R_{\mathrm{L}} \tag{9.7.3}$$

This may be elaborated to include other effects such as an inclined driveshaft. Similar methods may be applied to other coefficients.

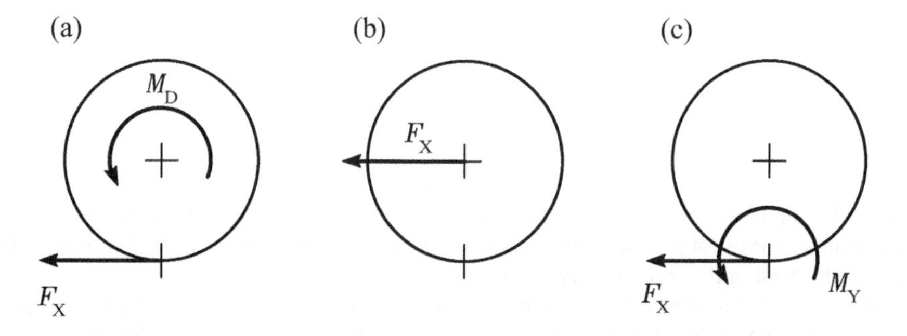

Figure 9.7.1 Equivalent force systems for tractive force generated by driveshaft torque.

9.8 Steering

The situation at the front of the vehicle is complicated by the steering. The outside wheel in cornering is subject to an inward lateral force, putting the lower wishbone in compression and the upper one in a smaller tension. Because of the bushes this results in a wheel camber. There will be no resultant lateral displacement of the wheel at some height between the wishbones, depending on the bush stiffnesses. If the steering arm is at this height then there will be no steering effect from this cause, whether the steering tie rod is in front or behind the kingpin axis.

In plan view the lateral force acts behind the kingpin because of pneumatic and caster trail. For front or rear tie rods, the steering compliance then results in a side-force understeer. This is the most significant of all side-force steer effects because of the considerable torsional compliance of the steering column. The consequence of this relatively high compliance is that for side force the shear centre (the point where a force will not cause a steer deflection) is a rather small distance behind the kingpin axis. For tractive forces the shear centre is again rather closely aligned with the kingpin axis, so in practice steering not far removed from centre-point steering is used. For front drive, careful tuning of the system is required to prevent power steer, although the situation is complicated by the dominant influence of the driveshaft torque component.

9.9 Rigid Axles

A complete rigid axle may be analysed for compliance steer effects in the same way as an independent suspension. The axis origin is taken at the centre of the axle at ground level, Figure 9.9.1. The characteristics naturally depend very much on the particular type of link location system adopted, or the use of leaf springs. Just as for independent suspensions, there has been considerable development of link systems and carefully controlled compliance to achieve desired handling characteristics.

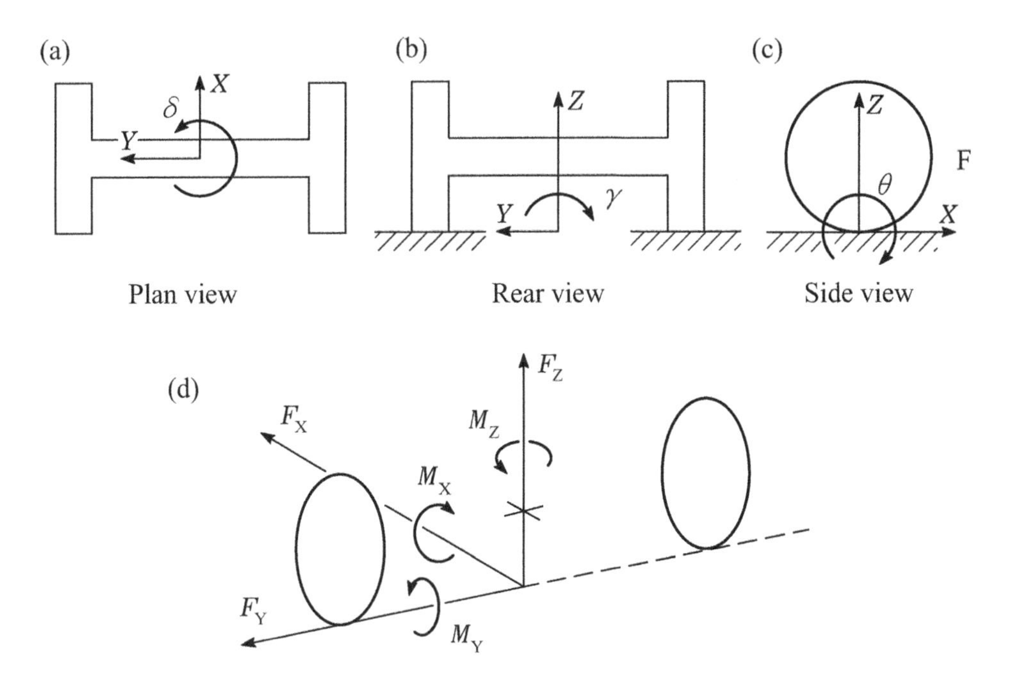

Figure 9.9.1 Rigid axle coordinates, with forces and moments, for compliance analysis.

9.10 Experimental Measurements

Various rigs have been built for the measurement of suspension compliance, typically combined with measurement of geometric bump steer and bump camber. Basically, such a rig comprises a rigid frame replacement of the vehicle body, to which the suspension arms are connected in the usual way. Hydraulic actuators apply forces or moments in the desired way, with deflection measurements made by dial gauges or some automated system. By this means the linear coefficients can be determined, along with non-linear effects and the limits of linearity. Usually, the suspension spring is replaced by a rigid link, and the wheel is replaced by a rigid disc or structure, to isolate the structural compliances. The wheel compliance, excluding the tyre, can be tested separately, and is relatively small.

References

SAE (1976) *Vehicle Dynamics Technology*, SAE J670e. Warrendale, PA: Society of Automotive Engineers (reprinted as an appendix to J.C. Dixon (1996) *Tires, Suspension and Handling*, 2nd edition. London: Edward Arnold).

———— // ————

10

Pitch Geometry

10.1 Introduction

Vehicle longitudinal dynamics dictate certain aspects of suspension geometry. As a result of aerodynamic pitching moment, hill climbing, braking or accelerating, or steady-state cornering with an attitude angle, there is a longitudinal load transfer, that is, longitudinal transfer of tyre vertical force. It is considered positive when the rear reactions are increased. The change of vertical force in the suspension results in ride-height changes at the front and rear suspensions. The suspension bump deflections have geometric effects on the wheels, such as bump camber and bump steer angle changes, and also affect the caster angles significantly. As in the case of lateral load transfer, the longitudinal load transfer may be achieved partly through the springs and partly through the links. Transfer of some of the load through the links rather than through the springs is called 'anti-dive' at the front and 'anti-rise' at the rear in the case of braking, and 'anti-lift' at the front and 'anti-squat' at the rear in the case of traction. Having a roll centre above ground could correspondingly be described as 'anti-roll'. These all depend on the geometry of the suspension links.

10.2 Acceleration and Braking

The total longitudinal load transfer moment reacted by the tyres at a longitudinal acceleration A_x, for a total vehicle mass m and centre-of-mass height H, is

$$M_{TXT} = mHA_x$$

Similar additional effects occur for a longitudinally sloping road. The basic load transfer moment on the suspension is due to the sprung mass:

$$M_{TXS} = m_S H_S A_x$$

However, the unsprung mass contributes to the longitudinal load transfer, and part of this acts through the sprung mass, having some effect on the pitch angle, again depending on details of the suspension geometry. Therefore the second equation above is not strictly correct. Also, of course, the pitch angle resulting from tyre deflections depends on the entire load transfer.

The pitch stiffness (as Nm/rad) of the tyres along the wheelbase, with the vehicle pitching about its centre of mass, is

$$k_{TP} = 2a^2 K_{tf} + 2b^2 K_{tr}$$

Suspension Geometry and Computation J. C. Dixon

where K_t is the tyre vertical stiffness (one wheel) and a and b are the partial wheelbases, and two wheels per axle are assumed. This gives an unsprung-mass pitch angle (i.e. the pitch angle of the line through the wheel centres in side view) of

$$\theta_U = \frac{m_S H_S A_x}{k_{TP}}$$

The suspension pitch stiffness (Nm/rad) is

$$k_{SP} = 2a^2 K_{Sf} + 2b^2 K_{Sr}$$

where K_S is the effective suspension vertical stiffness at one wheel, giving a suspension pitch angle, with no anti-dive or anti-rise effect, of

$$\theta_S = \frac{m_S H_S A_x}{k_{SP}}$$

The complete body pitch angle, for steady acceleration, is the total pitch angle on the tyres and suspension:

$$\theta_B = \theta_U + \theta_S$$

with an associated pitch angle gradient $d\theta/dA_x$ around 4 deg/g (longitudinal).

10.3 Anti-Dive

Really the whole suspension is a geometric problem to be solved in three dimensions, as in later chapters, but for human understanding it is useful to consider approximate two-dimensional simplifications. The principle of anti-dive for a twin-wishbone front suspension is shown in two-dimensional side view in Figure 10.3.1. Each wishbone has a geometric plane through three points (e.g. two on the body plus one at the centre of the ball joint on the wheel upright). There is a line where the plane of each wishbone intersects the longitudinal plane of the wheel, these lines being shown in the figure. These two lines are arranged to converge to a point E, possibly at infinity. The line AE at angle θ_{gf} (gf for ground front) joins this point to the base of the wheel. The suspension geometry is characterised primarily by the angle θ_{gf} and by the pitch arm length L_{AE} or the horizontal pitch arm radius L_{AD}. With the usual approximation of zero stiffness and friction moment about the pivot axis, the suspension links can exert no moment about the point E. When, resulting from brake action, a horizontal force is applied by the road to the wheel at the bottom, this can be resolved into components along and perpendicular to the line AE. The component F_1 along AE is reacted by the links. The component F_2 perpendicular to AE must be reacted by the springs: at the front a

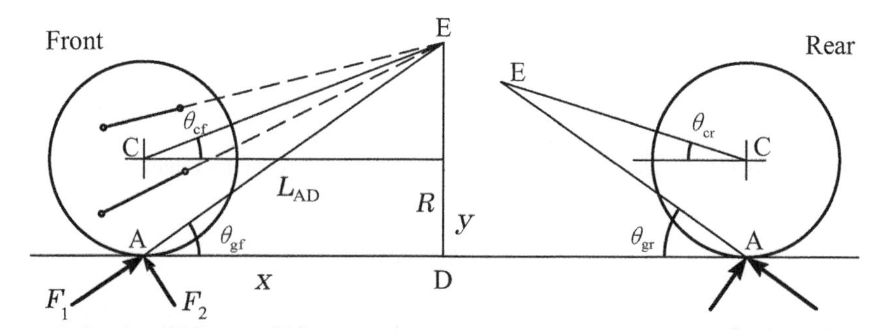

Figure 10.3.1 Anti-dive and anti-squat geometry (only braking forces shown).

component of the braking force acts downwards on the wheel, extending the spring and so opposing the usual compression of the front spring in braking. This is anti-dive.

Figure 10.3.1 assumes that the brakes are outboard. If they are inboard then the torque transferred by the driveshafts must be added to the brake force. The result is that the brake force is effectively applied at the wheel centre height, so it is necessary to arrange for appropriate inclination of line CE instead. Similar principles can be applied to the rear suspension, but in this case the force centre must be in front of the wheel, to discourage the usual extension of the rear springs in braking or compression in acceleration.

The pitch properties of the suspension therefore depend on the position of the point E, and so on the suspension geometry.

The notation for angles in Figure 10.3.1 is that subscripts c and g mean from the centre and ground respectively, and subscripts f and r mean front and rear as usual. The anti coefficients will be denoted by J (in practice often expressed as $J \times 100\%$). For example, J_{AL} is the anti-lift coefficient. Table 10.3.1 summarises the parameters, showing which angle is relevant to which coefficient. As an *aide-mémoire* to distinguish between rise and lift, it is convenient to recall that *r*ise is at the *r*ear.

Inboard brakes are relatively rare, being used on a few high-performance vehicles and some racing cars. Outboard drive is very unusual, achieved on some specialist vehicles by hydraulic or electrical drive, in which case there is no driveshaft torque, analogous to conventional brakes.

In braking, the total deceleration depends on the total longitudinal tyre force from the two ends of the vehicle, but the horizontal longitudinal suspension force on each end depends on the braking proportion at that end. Accurate calculation of the anti-dive coefficient requires consideration of the proportion of brake force at the front, the sprung and unsprung masses, and the sprung centre-of-mass height. The analysis is usually simplified by neglecting the inertial effect of the unsprung mass, as in the following. As an example, consider the braking force

$$F = F_f + F_r = pmA_B + (1-p)mA_B$$

where p is the proportion of braking at the front, which may be acceleration-dependent, for example if there is a rear pressure limiter, and A_B is a braking deceleration, positive to the rear. The resulting vehicle deceleration is

$$A_B = \frac{F_f + F_r}{m}$$

The total longitudinal transfer of vertical tyre force to the rear is

$$F_{TX} = -\frac{mA_B H}{L}$$

Table 10.3.1 Anti-dive/rise/lift/squat coefficients and angles

Action	Symbol	Direction	End	Relevant angle	
				Inboard brake/drive	Outboard brake/drive
Anti-dive	J_{AD}	braking	front	θ_{cf}	θ_{gf}
Anti-rise	J_{AR}	braking	rear	θ_{cr}	θ_{gr}
Anti-lift	J_{AL}	driving	front	θ_{cf}	θ_{gf}
Anti-squat	J_{AS}	driving	rear	θ_{cr}	θ_{gr}

The vertical force exerted by the ground on the front axle is therefore

$$F_{\mathrm{Vf}} = W_{\mathrm{f}} - F_{\mathrm{TX}} = W_{\mathrm{f}} + \frac{mA_{\mathrm{B}}H}{L}$$

The forces on the suspension from the body have no moment about E, so taking moments about E for the front wheel shows the effective front spring force (sum of two wheels) to be approximately

$$F_{\mathrm{S}} = W_{\mathrm{f}} - W_{\mathrm{Uf}} + \frac{mA_{\mathrm{B}}H}{L} - \frac{pmA_{\mathrm{B}}y}{x}$$

Hence the front anti-dive against forces exerted on the wheel at ground level (e.g. for outboard brakes) is

$$J_{\mathrm{gf}} = \frac{p(y/x)}{H/L} = \frac{\tan\theta_{\mathrm{gf}}}{\tan\theta_{\mathrm{gfi}}}$$

where θ_{gfi} is the 'ideal' angle for full anti-dive, given by

$$\tan\theta_{\mathrm{gf}i} = \frac{H}{pL}$$

10.4 Anti-Rise

In braking, with outboard brakes, at the rear the anti-rise is correspondingly given by

$$J_{\mathrm{gr}} = \frac{\tan\theta_{\mathrm{gr}}}{\tan\theta_{\mathrm{gri}}}$$

where

$$\tan\theta_{\mathrm{gri}} = \frac{H}{(1-p)L}$$

is the full anti-rise angle.

10.5 Anti-Lift

In the case of traction, with tractive force

$$F = F_{\mathrm{f}} + F_{\mathrm{r}} = tmA_{\mathrm{x}} + (1-t)mA_{\mathrm{X}}$$

where t is the tractive force fraction on the front wheels, the tractive force is produced because of torque in the driveshafts, totalling for both wheels $tmAr$ at the front and $(1-t)mAr$ at the rear, neglecting rotational inertia of the wheels. This means that the effective line of action of the tractive force is transferred to the centre of the wheel. The front anti-lift is

$$J_{\mathrm{cf}} = \frac{t[(y-r)/x]}{H/L} = \frac{\tan\theta_{\mathrm{cf}}}{\tan\theta_{\mathrm{cfi}}}$$

where θ_{cf} is the angle of the line CE at the front:

$$\tan\theta_{\mathrm{cf}} = \frac{y-r}{x}$$

and θ_{cfi} is the angle for full anti-lift, given by

$$\tan\theta_{\text{cfi}} = \frac{H}{tL}$$

10.6 Anti-Squat

The traction anti-squat at the rear is

$$J_{\text{cr}} = \frac{(1-t)[(y-r)/x]}{H/L} = \frac{\tan\theta_{\text{cr}}}{\tan\theta_{\text{cri}}}$$

where the full anti-squat angle is

$$\tan\theta_{\text{cri}} = \frac{H}{(1-t)L}$$

In the case of inboard brakes, the braking force is associated with a driveshaft torque, so the inclination of line CE gives the relevant angle. All the above equations can be extended to include the effect of the translational and rotational inertia of the unsprung masses. For example, during linear vehicle deceleration, the wheels also have angular deceleration, with angular momentum change that must be provided by the moment of a longitudinal load transfer.

10.7 Design Implications

In practice there are objections to anti-dive geometry. It tends to cause harshness of the front suspension on rough roads because the wheel moves forward as it rises, attacking the bump; it may also cause steering kick-back and wander under braking. It also becomes more difficult to achieve good-quality steering geometry, so instead of full anti-dive a proportion is often used, usually expressed as a percentage. Up to 50% anti-dive has been used on passenger cars. Amounts in excess of 50% have been used on ground-effect racing cars because of their extreme sensitivity to pitch angle. However, a large amount may be problematic on small-radius turns, where the large steer angle plus cornering force results in significant jacking forces. The most successful applications of anti-dive seem to be those in which the geometry is arranged to minimise changes of caster angle, and the anti-dive quantity is more moderate at 20% to 25%.

Considering again Figure 10.3.1, characterised by the angle θ_{gf} and horizontal anti-dive arm radius L_{AD}, it is apparent that vertical motion of the wheel (i.e. suspension bump) will give an effective rotation about some point. In the case of a rigid leading arm, this point is simply the body-fixed point E. Any such rotation changes the effective angle of motion θ_{gf} so the anti-dive coefficient will vary significantly with the suspension bump position. The pitch angle of the body, perhaps 4–5° in strong braking, will also influence θ_{gf}. Any effective rotation also changes the caster angle relative to the body; this can be eliminated, if desired, by using parallel arms, although this still leaves the body pitch angle affecting the caster angle, that is, the anti-dive and the geometric caster change can be chosen independently.

From Section 5.10, the caster angle relative to the ground, expressed as a linear model equation, is

$$\theta_{\text{C}} = \theta_{\text{C0}} - \theta_{\text{B}} + \varepsilon_{\text{BCas}} z_{\text{Sf}}$$

where θ_{C0} is the static caster angle, θ_{B} is the body pitch angle (positive nose down), and $\varepsilon_{\text{BCas}}$ is the bump caster coefficient. The possibility exists of maintaining θ_{C} constant by using the bump caster coefficient to counteract the effect of the body pitch angle.

The bump caster coefficient is

$$\varepsilon_{\text{BCas}} = \frac{1}{L_{\text{AD}}}$$

which is the reciprocal of the pitch arm radius. The front and rear suspensions are of approximately equal stiffness, so the front suspension bump is

$$z_{\text{Sf}} = \tfrac{1}{2} L_{\text{W}} \, \theta_{\text{B}}$$

where L_{W} is the wheelbase. The bump caster angle is then

$$\theta_{\text{BCas}} = \varepsilon_{\text{BCas}} z_{\text{Sf}} = \varepsilon_{\text{BCas}} \tfrac{1}{2} L_{\text{W}} \, \theta_{\text{B}}$$

To fully offset the body pitch angle effect,

$$\theta_{\text{BCas}} - \theta_{\text{B}} = 0$$

so requiring

$$\varepsilon_{\text{BCas}} = \frac{2}{L_{\text{W}}}$$

Considered even more simply, the vehicle pitches roughly about its mid-wheelbase position, so if the point E is at mid-wheelbase then the caster angle relative to the road will not be affected. Therefore we desire the pitch arm radius L_{AD} to be $L_{\text{W}}/2$. This simple analysis can easily be made more accurate for particular cases by calculation of the actual pitch angle and suspension bump values.

In the case of the pitch-up caused by traction forces, with rear-wheel drive it is quite common to have some anti-squat at the rear. It is not possible to provide anti-rise at the front of a rear-wheel-drive vehicle because there is no associated horizontal force applied to the wheel. Anti-squat may have detrimental effects on traction on rough surfaces. Anti-squat effects may also be achieved by other arrangements of the rear suspension, such as a lift bar, which is rigidly attached to the rear axle, protruding forward and acting upward on a rubber block on the body; the axle reaction torque therefore provides an upward force on the rear body, depending on the length of the lift bar, relieving the rear springs.

Having chosen desired 'anti' coefficients, at least a partial geometric specification for the suspension side view can be made.

For the front suspension, the anti-dive gives θ_{gf}. If the vehicle is front-wheel drive, then the anti-lift gives θ_{cf} and the point E is thereby determined for the front. For the front suspension with rear-wheel drive, θ_{cf} is unspecified, and the length of the anti-squat arm could be chosen freely, possibly infinite, or on other criteria, such as the caster angle effect as described above.

For the rear suspension with rear-wheel drive, the required anti-rise and rear-wheel-drive anti-squat together specify E. For the rear suspension of front-wheel drive, only the anti-rise is relevant, and then the arm length may be freely chosen.

The preceding simplified description and analysis of longitudinal behaviour is based on approximate two-dimensional considerations to convey the principle. A more complete quantitative analysis requires consideration of the suspension in three dimensions, as in later chapters.

—————— // ——————

11

Single-Arm Suspensions

11.1 Introduction

This chapter deals with suspensions in which the wheel hub is located by a single rigid arm (various type of trailing arm, swing axles, etc.). The alternatives, in later chapters, are to use location by two arms, as for double wishbones, or by one wishbone and a trunnion, as on struts, and some other types. Pictures of many examples of these various types of suspension appear in Chapter 1.

Considering rigid single-arm suspensions, then, there are two rotational axes to consider, the wheel rotation axis and the arm pivot axis. The wheel rotation axis is fixed relative to a single arm, which is pivoted on the vehicle body. For a single-arm suspension, this pivot axis is fixed relative to the body. The fixity of this axis is a distinctive feature from other suspensions in which the pivot axis is virtual and moves relative to the members and the body. A fixed position of the axis generally simplifies the analysis, and certainly changes the details of the methods of solution.

These are all independent suspensions, so the geometric steer effects are expressed in basic form by the bump coefficients (bump camber, bump steer, etc.). In general, a slope angle of the pivot axis, out of the horizontal plane, introduces linear bump steer. A sweep angle, in plan view, affects the quadratic bump steer coefficient. Camber coefficients are also affected. The small static wheel toe and camber angles also interact with the axis angles to give some bump steer and camber effects, not always negligible. The pivot axis height also introduces small steer effects in some cases. In general, then, it is necessary to investigate the effect of:

(1) the pivot axis plan-view sweep angle ψ_{Ax};
(2) the pivot axis slope angle (out-of-horizontal) ϕ_{Ax};
(3) the pivot axis height H_{Ax};
(4) the wheel static toe angle δ_0;
(5) the wheel static camber angle γ_0.

These may influence, in the present context:

(1) the bump steer coefficients ε_{BS1} and ε_{BS2};
(2) the bump camber coefficients ε_{BC1} and ε_{BC2};
(3) the bump scrub rate coefficient ε_{BScd0} (roll centre height);
(4) the bump scrub rate variation coefficient ε_{BScd1} (roll centre movement).

For the occasional cases of a rigid arm with steering, additional factors are:

(5) the bump kingpin inclination angle coefficients ε_{BKI1} and ε_{BKI2};
(6) the bump kingpin caster angle coefficients ε_{BKC1} and ε_{BKC2}.

For the bump steer, for example, various mathematical models may be adopted, depending on the purpose of the analysis. The main models are:

(1) exact analytical geometric model;
(2) approximate geometric model;
(3) linear bump steer coefficient;
(4) quadratic model (linear and quadratic bump steer coefficients);
(5) polynomial bump steer expression;
(6) accurate numerical (computer) solution.

Exact geometric models are sometimes useful, but require some skill to achieve, and are, except for the very simple cases, rather unwieldy for most purposes. They have the advantage of being accurate if correctly solved, but do not necessarily provide good insight because of the complexity of the ultimate algebraic expression.

Approximate geometric models are simple to use, and are usefully close to correct over the practical range of application.

Linear bump steer coefficients, usually derived from simplified geometric models, have been widely used for many years. These, naturally, give considerable insight to linear bump steer effects, but omit the second-order effects.

The quadratic bump steer expressions, including linear and second-order terms (the first and second bump steer coefficients), give excellent insight, and normally include all important effects, that is, they are accurate enough. They are likely to be used as a convenient summary, for human consumption, of the results of an accurate numerical computer analysis, or as the design intent, as a suspension specification.

The polynomial bump steer equation adds cubic, quartic, etc., terms to the quadratic form above. These are found by fitting a polynomial expression, of whatever order is desired, to numerical results. The additional terms may occasionally be of interest, but are not usually very important, so not very helpful to the suspension designer.

The accurate numerical solution of a suspension is achieved by computer, using sequential geometric analyses, each exact algebraically and relatively simple, without ever needing a collected algebraic expression for the result. This method is very effective, particularly for a general case where the direct geometric solution may be tricky and unwieldy. A computer can easily produce graphs of bump steer, etc, against suspension bump, but for human consumption it is usually best to also produce the first and second bump steer coefficients by a quadratic curve fit to the computer generated points.

Frequently used here will be the series expansions for $\sin\theta$ and $\cos\theta$, truncated to second-order terms only, to give the first- and second-order bump coefficients. The series are

$$\sin\theta = \theta - \frac{\theta^3}{6} + \frac{\theta^5}{120} - \cdots$$

$$\cos\theta = 1 - \frac{\theta^2}{2} + \frac{\theta^4}{24} - \frac{\theta^6}{720} + \cdots$$

11.2 Pivot Axis Geometry

Single-arm suspensions are characterised mainly by the angle of the pivot axis to the vehicle body, ψ_{Ax} and θ_{Ax} in Figure 11.2.1. Note here that the arm length, pitch arm length and swing arm length are from the

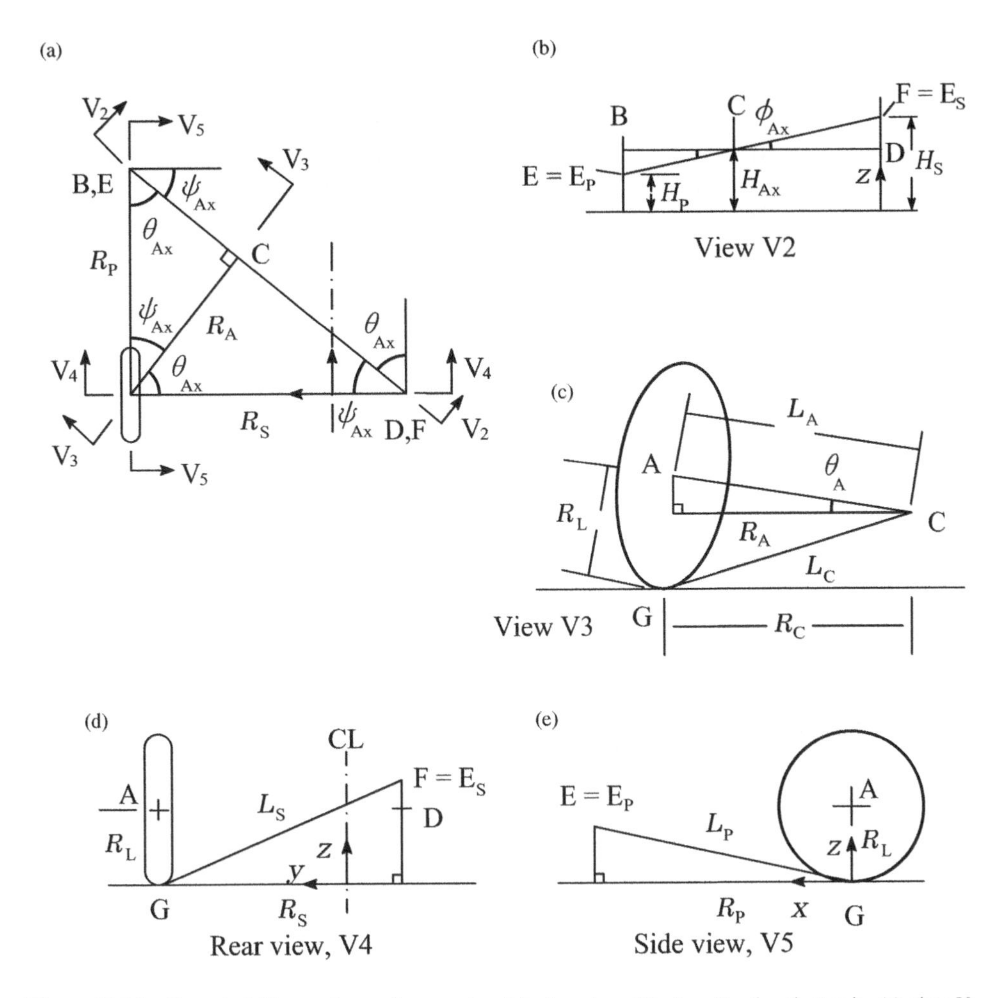

Figure 11.2.1 General rigid-arm pivot axis geometry: (a) plan view; (b) view V_2, the pivot axis; (c) view V_3, transverse to the pivot axis; (d) view V_4, the rear view; (e) view V_5, the side view.

axis to a point, and the horizontal projection of these lengths is called the radius, following the terminology of SAE J670e (e.g. the swing arm radius). For a pure trailing arm the axis is transverse, perpendicular to the vehicle longitudinal centre plane. If the axis is perpendicular to the centreline in plan view but inclined in rear view, the result is a sloped-axis trailing arm. If the axis is swept in plan view, up to about 30° from the transverse Y axis (i.e. at least 60° from the longitudinal X axis), the result is called a semi-trailing arm. The semi-trailing arm may also have its pivot axis out of the horizontal plane, giving the sloped-axis semi-trailing arm. Also, the height of the pivot axis may vary.

If the arm pivot axis is strictly longitudinal, parallel to the X axis, and at the normal height of the wheel centre, the result is a simple transverse-arm suspension. If the pivot axis is slightly to the same side of the centre plane as the wheel, and the lateral location is by the length of the driveshaft, it is a swing-axle suspension. There is no basic geometric difference between the transverse arm and a swing axle. If the pivot axis is parallel to the X axis but lower than the static wheel centre, the result is a low-pivot transverse arm, or a low-pivot swing axle, depending on exactly how the lateral location is achieved. If the pivot axis is angled to the horizontal in side view, the type is a sloped-axis transverse arm or a sloped-axis swing axle. If the axis is angled in plan view, not exactly parallel to the X axis, but

Table 11.2.1 Basic rigid-arm variables (see Figure 11.0.1)

(1)	ψ_{Ax}	the pivot axis sweep angle
(2)	θ_{Ax}	the pivot axis yaw angle, $= \pi/2 - \psi_{Ax}$
(3)	ϕ_{Ax}	the pivot axis slope angle
(4)	H_P	height of the pitch arm centre
(5)	H_S	height of the swing arm centre
(6)	H_{Ax}	height of the pivot axis (at C)
(7)	L_A	the arm length to the wheel centre
(8)	R_A	the arm radius (in plan view)
(9)	L_C	length to the ground contact point
(10)	R_C	arm radius to the ground contact point
(11)	L_P	the pitch arm length
(12)	R_P	the pitch arm radius (in plan view)
(13)	L_S	the swing arm length
(14)	R_S	the swing arm radius (in plan view)
(15)	θ_A	the arm angle, wheel end up
(16)	R_L	the wheel static loaded radius

up to about 30° from it (i.e. at least 60° from the transverse *Y* axis), the type is a semi-swing-axle or a semi-transverse arm.

In the most general case, simply a general rigid arm, the axis is at some angle to the *X* axis and at some angle to the horizontal plane, and also at no particular height relative to the wheel centre. Naturally this is the most complex case to analyse, and subsumes all the other special cases above.

Figure 11.2.1 shows the pivot axis geometry for the general case. The pivot axis is EF. Point $E = E_P$ is the pitch arm centre, point $F = E_S$ is the swing arm centre. The main variables are listed in Table 11.2.1.

For numerical analysis the pivot axis is best expressed in parametric form, by one point and direction cosines. In Figure 11.2.2, the pivot axis is PQ. Consider this axis also to be specified by the coordinates of point P, and the direction cosines of PQ, that is, from P to Q, tending inwards and towards the rear of the vehicle. The dimensions are

$$\begin{aligned}
\text{PQ} &= L \\
\text{QR} &= L \sin\phi_{Ax} \\
\text{PR} &= L \cos\phi_{Ax} \\
\text{PS} &= L \cos\phi_{Ax} \cos\theta_{Ax} \\
\text{SR} &= L \cos\phi_{Ax} \sin\theta_{Ax}
\end{aligned}$$

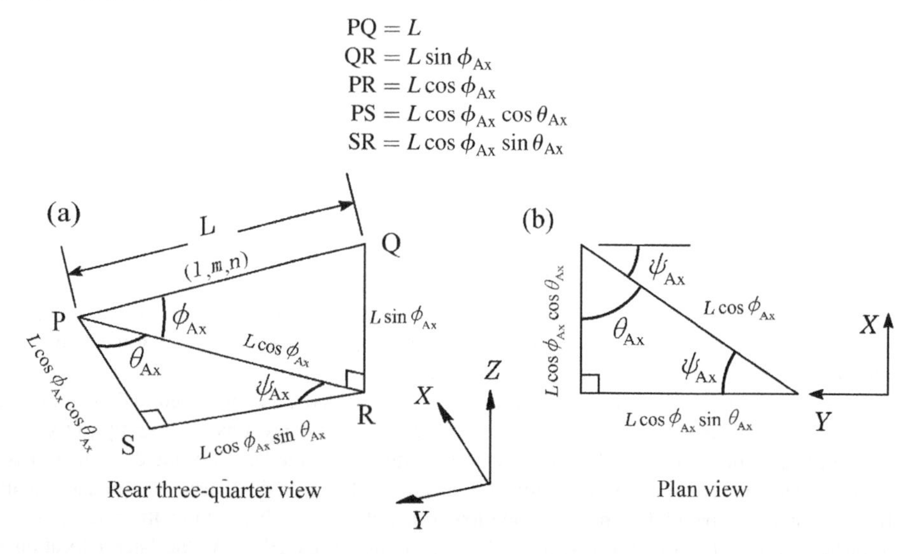

Figure 11.2.2 Pivot axis geometry for direction cosines. QR is parallel to *Z*. Triangle PRS is in the horizontal plane with PS parallel to *X* and RS parallel to *Y*.

The direction cosines (l, m, n) in positive (X, Y, Z) directions are therefore

$$l = -\frac{\text{PS}}{\text{PQ}} = -\cos\phi_{\text{Ax}}\cos\theta_{\text{Ax}}$$

$$m = -\frac{\text{SR}}{\text{PQ}} = -\cos\phi_{\text{Ax}}\sin\theta_{\text{Ax}}$$

$$n = \frac{\text{QR}}{\text{PQ}} = \sin\phi_{\text{Ax}}$$

It is easily confirmed that, as is necessary,

$$l^2 + m^2 + n^2 = \cos^2\phi_{\text{Ax}}(\sin^2\theta_{\text{Ax}} + \cos^2\theta_{\text{Ax}}) + \sin^2\phi_{\text{Ax}} = 1$$

Table 11.2.2 summarises the conversion equations between the axis angles and the direction cosines as specified above.

It is really required to relate the wheel steer and camber angle changes to the suspension bump rather than just to the wheel centre bump. The suspension bump is the vertical position of the lowest point of the wheel, in the contact patch, the wheel considered geometrically as of constant radius, effectively the loaded radius R_{L}. With wheel bump, the contact point height is

$$Z_{\text{CP}} = Z_{\text{W}} - R_{\text{L}}\cos\gamma$$

The static (initial) value is

$$Z_{\text{CP0}} = Z_{\text{W0}} - R_{\text{L}}\cos\gamma_0$$

If the ground plane is the datum level plane, then this is zero. The actual suspension bump is

$$z_{\text{S}} = Z_{\text{CP}} - Z_{\text{CP0}} = (Z_{\text{W}} - Z_{\text{W0}}) + R_{\text{L}}(\cos\gamma_0 - \cos\gamma)$$

or

$$z_{\text{S}} = z_{\text{W}} + R_{\text{L}}(\cos\gamma_0 - \cos\gamma)$$

For a small camber angle, the wheel centre may therefore be taken as having the same vertical position change as the suspension bump, but preferably the effect of camber variation would be included, particularly for large camber angle changes (e.g. for a swing axle or other short transverse arm).

Table 11.2.2 Pivot axis angle and direction cosine conversion equations

$$l = -\cos\phi_{\text{Ax}}\cos\theta_{\text{Ax}}$$

$$m = -\cos\phi_{\text{Ax}}\sin\theta_{\text{Ax}}$$

$$n = \sin\phi_{\text{Ax}}$$

$$\phi_{\text{Ax}} = \text{asin}\,(n)$$

$$\theta_{\text{Ax}} = \text{atan}\left(\frac{m}{l}\right) = \frac{\pi}{2} - \psi_{\text{Ax}}$$

$$\psi_{\text{Ax}} = \text{atan}\left(\frac{l}{m}\right) = \frac{\pi}{2} - \theta_{\text{Ax}}$$

Considering a simple trailing arm, with the wheel also rotationally fixed relative to the arm, as the arm rises the lowest point of the wheel is not a fixed point of the wheel. Hence, there is no very simple direct solution for the lowest point in general. Further analysis of the wheel bottom point is given in Section 5.8.

Consider the vertical plane transverse to the pivot axis, Figure 11.2.1(c), which is not the same as the plane of the wheel axis. Given the initial dimensions, the radius from the pivot axis to the contact patch point can be established, as can the angle of this line. However, this is complicated by any out-of-horizontal inclination of the pivot axis. The required arm bump angle for any suspension bump can then be solved explicitly. With the datum plane at ground level, in the initial position,

$$Z_{W0} = R_L \cos \gamma_0$$

Numerically, one method is to project the pivot axis into the longitudinal vertical plane of the wheel centre, giving the pitch centre point E_P, and into the transverse vertical plane giving point E_S, the swing centre point. The perpendicular may be dropped onto the pivot axis from the wheel centre. The length of the perpendicular follows, as does the angle.

Given the arm angular position angle θ_A, to calculate the wheel centre bump,

$$z_W = L_A(\sin \theta_A - \sin \theta_{A0})\cos \phi_{Ax}$$

The suspension bump is

$$z_S = z_W + R_L(\cos \gamma_0 - \cos \gamma)$$

This gives the suspension bump as

$$z_S = L_A(\sin \theta_A - \sin \theta_{A0})\cos \phi_{Ax} + R_L(\cos \gamma_0 - \cos \gamma)$$

Direct solution would require an algebraic relationship between the camber angle γ and the arm angle θ_A, a relationship which is not known accurately in advance.

Geometrically, then, it is much more convenient to approximate the suspension bump as the wheel centre bump, although this is not precise. This limitation is easily overcome in a numerical solution.

11.3 Wheel Axis Geometry

Wheel axis geometry was analysed in Chapter 5. The main results are summarised for convenient reference in Table 11.3.1.

Table 11.3.1 Steer and camber angle conversion equations

Equations
$l = \cos \gamma \sin \delta$
$m = -\cos \gamma \cos \delta$
$n = \sin \gamma$
$\gamma = \operatorname{asin} n$
$\delta = \operatorname{atan}\left(-\dfrac{l}{m}\right)$

11.4 The Trailing Arm

The pure trailing arm has its pivot axis exactly perpendicular to the vehicle centre plane. In general, there is wheel static camber and toe, but for simplicity initially consider these to be zero. The wheel axis is then parallel to the pivot axis. Bump then has zero effect on steer and camber angles (Table 11.4.1). The initial bump scrub and bump scrub variation coefficients are zero, and the geometric roll centre is at ground level.

There is a variation of the pitch arm radius, Figure 11.4.1. The arm angle is given by

$$\sin\theta_A = \frac{Z_W - H_{Ax}}{L_A}$$

The initial, static, value of the arm angle is given by

$$\sin\theta_{A0} = \frac{Z_{W0} - H_{Ax0}}{L_A}$$

and is generally non-zero. The simplest geometry occurs when the initial angle is zero, but in practice this may give excessive braking anti-rise (Chapter 10).

Practical pure trailing arms often include some static toe and camber. These angles are always small, and can be expected to have only a limited effect on the bump coefficients. However, this really needs to be demonstrated rather than just claimed. Therefore it is desirable to obtain explicit expressions for them. This is also a useful demonstration of method.

Consider a pure trailing arm with static camber angle γ_0. For simplicity, the pivot axis is taken as at the same initial height as the wheel centre. For a passenger car the static camber would be 1 or 2 degrees, but for a racing saloon with rear trailing arm it would possibly be beyond $-3°$ (-0.05 rad). By inspection, when the trailing arm is hypothetically rotated up to an arm angle of 90°, the camber angle has become zero and a toe-out angle of $\delta = \gamma_0$ has appeared. When the arm is lowered to $-90°$, the camber has again become zero, but the steer toe-out angle has become $\delta = -\gamma_0$. Prospectively, the relationships are

$$\delta = \gamma_0 \sin\theta_A$$
$$\gamma = \gamma_0 \cos\theta_A$$

Expanding the sine and cosine series for the arm angle out to a maximum of the second-order term gives

$$\delta = \gamma_0 \,\theta_A$$
$$\gamma = \gamma_0 (1 - \tfrac{1}{2}\,\theta_A^2)$$

For small arm angles, the suspension or wheel bump is related to the arm angle simply by

$$\theta_A = \frac{z_S}{R_P}$$

Table 11.4.1 Coefficients for a pure trailing arm, no static wheel angles, so all zero

$\varepsilon_{BS1} = 0$
$\varepsilon_{BS2} = 0$
$\varepsilon_{BC1} = 0$
$\varepsilon_{BC2} = 0$
$\varepsilon_{BScd0} = 0$
$\varepsilon_{BScd1} = 0$

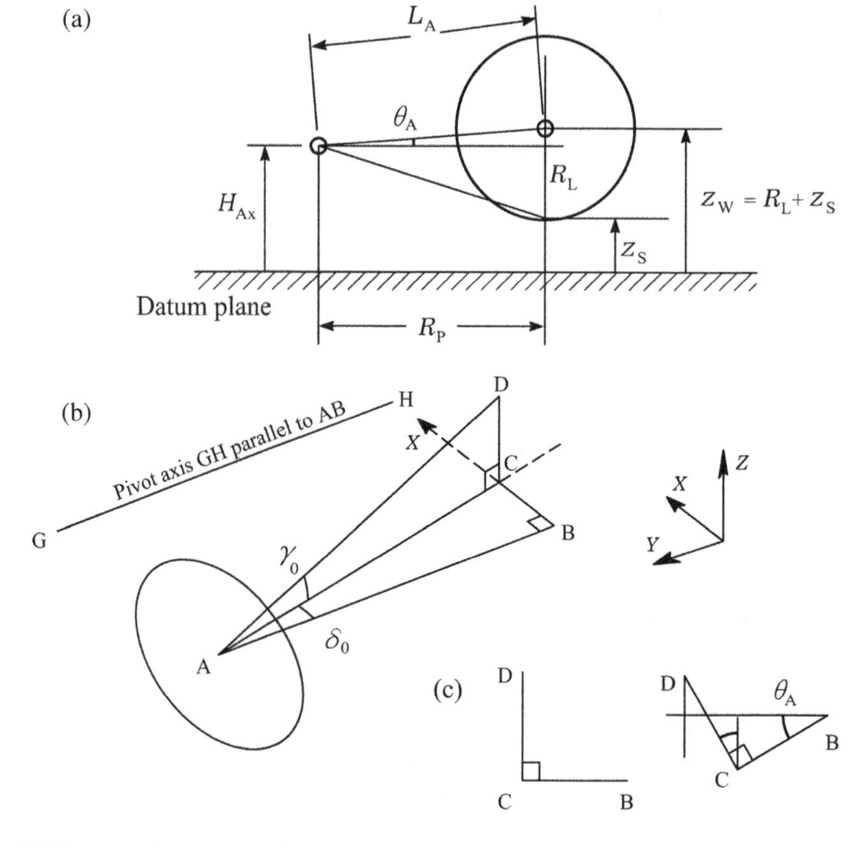

Figure 11.4.1 Pure trailing arm: (a) side view, showing arm length L_A, pitch arm radius R_P; (b) wheel axis with static steer and camber angles; (c) rotation of point D about centre at B.

so the wheel angles become

$$\delta = \frac{\gamma_0}{R_P} z_S$$

$$\gamma = \gamma_0 - \frac{\gamma_0}{2R_P^2} z_S^2$$

Therefore the static camber angle has introduced linear bump steer and quadratic bump camber coefficients:

$$\varepsilon_{BS1} = \frac{\gamma_0}{R_P}$$

$$\varepsilon_{BC2} = -\frac{\gamma_0}{2R_P^2}$$

Of these, the former is more significant. With a static camber of −0.05 rad and a pitch arm radius of 0.45 m, the result is a linear bump steer coefficient of −0.11 rad/m, which is −0.64 deg/dm (deg/decimetre) or −0.16 deg/inch. This is not negligible.

Consider now zero static camber, but a static toe-out angle of δ_0. At an arm angle of plus 90°, this produces zero steer angle and a camber angle of $\gamma = -\delta_0$. At an arm angle of −90° again the steer angle has gone to zero, but the camber angle becomes $\gamma = +\delta_0$. Prospectively, then,

$$\delta = \delta_0 \cos\theta_A$$
$$\gamma = -\delta_0 \sin\theta_A$$

For small angles, as before, this becomes

$$\delta = \delta_0 - \frac{\delta_0}{2R_P^2} z_S^2$$
$$\gamma = -\frac{\delta_0}{R_P} z_S$$

The result is therefore quadratic bump steer and linear bump camber coefficients

$$\varepsilon_{BS2} = -\frac{\delta_0}{2R_P^2}$$
$$\varepsilon_{BC1} = -\frac{\delta_0}{R_P}$$

Because the static toe angles are likely to be small in all cases, and such angles are often introduced with the intention that the toe angle will go even closer to zero when actually running, these effects are likely to be small. For example, with $\delta_0 = -0.02$ rad and $R_P = 0.45$ m we have $\varepsilon_{BS2} = 0.05$ rad/m^2 = 0.028 deg/dm^2.

By these simple expressions, summarised in Table 11.4.2, the effect of static wheel toe and camber angles are easily evaluated rather than just assumed to be insignificant.

A more thorough analysis of the above geometry, now to be given, confirms that the expressions given are good approximations, and demonstrates a useful principle. Figure 11.4.1(b) shows the wheel axis geometry with the actual pivot axis GH.

Now as far as changes of steer and camber angles are concerned, rotations about alternative parallel axes are entirely equivalent, adding only a pure translation. Rotation about the real axis is of course necessary for evaluation of actual movement, such as bump displacements. In the case of the pure trailing arm, the real pivot axis is parallel to the Y axis, and parallel to the line AB. Therefore, for investigation of steer and

Table 11.4.2 Coefficients for a pure trailing arm, with static wheel angles (δ_0, γ_0)

$$\varepsilon_{BS1} = \frac{\gamma_0}{R_P}$$
$$\varepsilon_{BS2} = -\frac{\delta_0}{2R_P^2}$$
$$\varepsilon_{BC1} = -\frac{\delta_0}{R_P}$$
$$\varepsilon_{BC2} = -\frac{\gamma_0}{2R_P^2}$$

The bump scrub coefficients remain zero.

camber angles only, simply consider the wheel and wheel axis to rotate about the line AB. Positive arm angle θ_A is a positive bump position, with a lowering of point C relative to B, and a lowering and forward movement of point D, the second point of the wheel axis AD. It is changes of the position of the line AD that are of interest, with point D rotating about B as seen in Figure 11.4.1(c). As established earlier,

$$\mathrm{AD} = L$$
$$\mathrm{BC} = L\cos\gamma_0 \sin\delta_0$$
$$\mathrm{CD} = L\sin\gamma_0$$

After rotation, the value of X_D, measured from the line AB, is

$$X_D = (L\cos\gamma_0 \sin\delta_0)\cos\theta_A + (L\sin\gamma_0)\sin\theta_A$$

The consequent steer angle is then given by

$$\sin\delta = \frac{X_D}{L} = \cos\gamma_0 \sin\delta_0 \cos\theta_A + \sin\gamma_0 \sin\theta_A$$

Now introducing a small-angle approximation for the steer and camber (including $\cos\gamma_0 = 1$), and expanding the arm angle to the second term of its series,

$$\delta = \delta_0(1 - \tfrac{1}{2}\theta_A^2) + \gamma_0\theta_A$$

Inserting the small-angle approximation

$$\theta_A = \frac{z_S}{R_P}$$

gives the toe-out steer angle as

$$\delta = \delta_0 + \frac{\gamma_0}{R_P}z_S - \frac{\delta_0}{2R_P^2}z_S^2$$

which is in agreement with the bump steer coefficients obtained earlier (Table 11.4.2).

Similarly, for camber after rotation (see again Figure 11.4.2 (b)), relative to the line AB, the height of point D is

$$Z_D = -(L\cos\gamma_0 \sin\delta_0)\sin\theta_A + (L\sin\gamma_0)\cos\theta_A$$

and the sine of the camber angle is given correctly by

$$\sin\gamma = \frac{Z_D}{L} = -\cos\gamma_0 \sin\delta_0 \sin\theta_A + \sin\gamma_0 \cos\theta_A$$

Using the small-angle approximations for δ and γ, and expanding the arm angle to the quadratic term, gives

$$\gamma = -\delta_0\theta_A + \gamma_0(1 - \tfrac{1}{2}\theta_A^2)$$

Now substituting the linear approximation for the arm angle in terms of the bump position gives

$$\gamma = \gamma_0 - \frac{\delta_0}{R_P}z_S - \frac{\gamma_0}{2R_P^2}z_S^2$$

again in agreement with the semi-intuitive expressions of Table 11.4.2.

Table 11.4.3 Pure trailing arm static toe and camber angle bump steer effects

$\sin\delta = \cos\gamma_0 \sin\delta_0 \cos\theta_A + \sin\gamma_0 \sin\theta_A$
$\sin\gamma = -\cos\gamma_0 \sin\delta_0 \sin\theta_A + \sin\gamma_0 \cos\theta_A$

This also illustrates the use of rotation about the most convenient parallel axis for investigating steer and camber angles.

The accurate equations for the angles are collected in Table 11.4.3. These are easy enough to use, but even for this extremely simple case are less revealing than the simpler approximate expressions of Table 11.4.2.

11.5 The Sloped-Axis Trailing Arm

The pure trailing arm has a geometric roll centre at ground level, which may be regarded as too low. One way to raise it is to incline (i.e. 'slope') the pivot axis in rear view, out of the horizontal plane, Figure 11.5.1, introducing the appropriate bump scrub coefficient. Also, a sloped axis introduces a bump

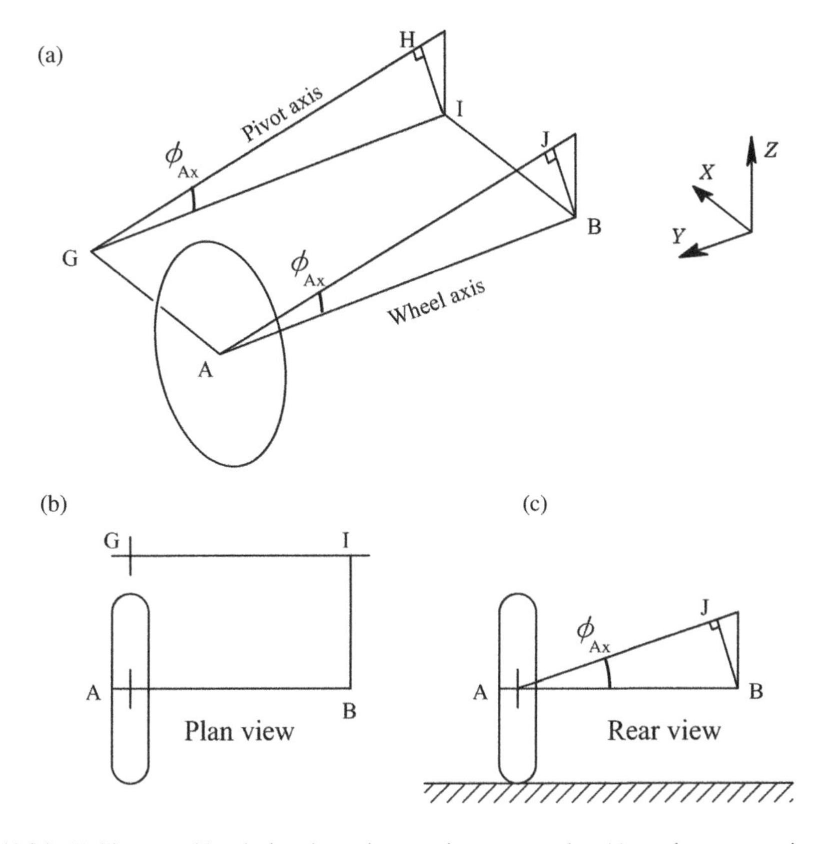

Figure 11.5.1 Trailing arm with a sloping pivot axis, no static steer or camber: (a) rear three-quarter view; (b) plan view; (c) rear view.

steer effect, which may be desired. For simplicity, here, the wheel is considered to have zero static toe and camber angles. The pivot axis remains perpendicular to the vehicle centreline as seen in plan view, but is sloped at the angle ϕ_{Ax} in rear view, with a positive angle considered to be with the pivot axis rising as it moves from the wheel towards the vehicle centreline.

To investigate the changes of wheel axis angle, consider rotation of this axis about the line AJ parallel to the real pivot axis GH. The wheel centre A does not move. The second wheel axis point B moves around the axis AJ, in a circle around the point J. To obtain convenient approximate equations, the slope of the pivot axis is considered to be small. This is not too bad an approximation as practical values are likely to be less than 10°. Positive rotation of the arm, positive wheel bump, moves point B initially backwards relative to point J, away from point I.

The basic line lengths are:

$$\text{AB} = L$$
$$\text{BJ} = L \sin \phi_{Ax}$$

Relative to the line AB, with X forwards and Z upwards, approximately,

$$X_B = -L \sin \phi_{Ax} \sin \theta_A$$
$$Z_B = L \sin \phi_{Ax}(1 - \cos \theta_A)$$

The resulting bump steer and camber angles are given by

$$\sin \delta = \frac{X_B}{L} = -\sin \phi_{Ax} \sin \theta_A$$

$$\sin \gamma = \frac{Z_B}{L} = \sin \phi_{Ax}(1 - \cos \theta_A)$$

With the usual small-angle approximations, as in previous sections, the resulting angles may be expressed as

$$\delta = -\frac{\phi_{Ax}}{R_P} z_S$$

$$\gamma = \frac{\phi_{Ax}}{2R_P^2} z_S^2$$

Table 11.5.1 Coefficients for a sloped-axis trailing arm, with static wheel angles

$$\varepsilon_{BS1} = -\frac{\phi_{Ax} - \gamma_0}{R_P}$$

$$\varepsilon_{BS2} = -\frac{\delta_0}{2R_P^2}$$

$$\varepsilon_{BC1} = -\frac{\delta_0}{R_P}$$

$$\varepsilon_{BC2} = \frac{\phi_{Ax} - \gamma_0}{2R_P^2}$$

$$\varepsilon_{BScd0,Y} = \tan \phi_{Ax}$$

$$\varepsilon_{BScd1,Y} = 0$$

so the result is a linear bump steer coefficient and a quadratic bump camber coefficient:

$$\varepsilon_{\mathrm{BS1}} = -\frac{\phi_{\mathrm{Ax}}}{R_{\mathrm{P}}}$$

$$\varepsilon_{\mathrm{BC2}} = +\frac{\phi_{\mathrm{Ax}}}{2R_{\mathrm{P}}^2}$$

This is a similar result to the analysis for wheel static camber on the trailing arm, as in Table 11.4.2, but with inverted signs. This is as would be expected, as the angle changes are basically a property of the lack of parallelism of the wheel axis and the pivot axis. Therefore the sloped-axis trailing arm, for a reasonably small inclination angle, has the properties of Table 11.5.1. The bump scrub coefficients included in the table follow easily by inspection of the rear view.

11.6 The Semi-Trailing Arm

The semi-trailing arm has its pivot axis swept in plan view, at the sweep angle ψ_{Ax}, considered positive when the pivot axis is angled rearwards whilst passing towards the centreline, as GH in Figure 11.6.1. For simplicity, in the initial analysis, the pivot axis is considered to be in the horizontal plane, that is, at zero slope angle at wheel centre height. Also, the wheel initial toe and camber angles are taken as zero.

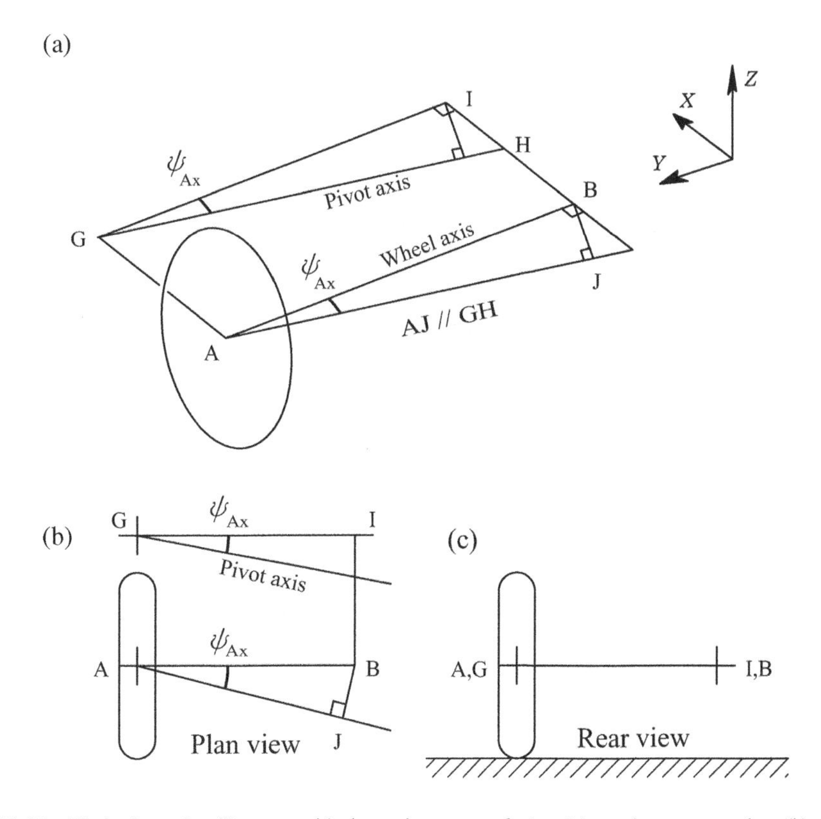

Figure 11.6.1 The basic semi-trailing arm, with pivot axis sweep angle ψ_{Ax}: (a) rear three-quarter view; (b) plan view; (c) rear view.

As the suspension arm rotates, for angle analysis the wheel axis AB may be considered to rotate about the axis AJ parallel to the real pivot axis GH, with B rotating about J. The basic dimensions are

$$\mathrm{AB} = L$$

$$\mathrm{BJ} = L \sin \psi_{\mathrm{Ax}}$$

Relative to the line AJ, point B moves in a circle, with coordinates, relative to AJ, of

$$X_{\mathrm{B}} = -L \sin \psi_{\mathrm{Ax}}(1 - \cos \theta_{\mathrm{A}})$$

$$Z_{\mathrm{B}} = -L \sin \psi_{\mathrm{Ax}} \sin \theta_{\mathrm{A}}$$

The bump steer and camber angles are given by

$$\sin \delta = \frac{X_{\mathrm{B}}}{L} = -\sin \psi_{\mathrm{Ax}}(1 - \cos \theta_{\mathrm{A}})$$

$$\sin \gamma = \frac{Z_{\mathrm{B}}}{L} = -\sin \psi_{\mathrm{Ax}} \sin \theta_{\mathrm{A}}$$

Using the usual small-angle approximations, as in previous sections, the angles may be expressed approximately as

$$\delta = -\psi_{\mathrm{Ax}} \tfrac{1}{2} \theta_{\mathrm{A}}^2$$

$$\gamma = -\psi_{\mathrm{Ax}} \theta_{\mathrm{A}}$$

and in terms of the suspension bump as

$$\delta = -\frac{\psi_{\mathrm{Ax}}}{2R_{\mathrm{P}}^2} z_{\mathrm{S}}^2$$

$$\gamma = -\frac{\psi_{\mathrm{Ax}}}{R_{\mathrm{P}}} z_{\mathrm{S}}$$

so the effects introduced by the axis sweep angle are a quadratic bump steer coefficient and a linear bump camber coefficient:

$$\varepsilon_{\mathrm{BS2}} = -\frac{\psi_{\mathrm{Ax}}}{2R_{\mathrm{P}}^2}$$

$$\varepsilon_{\mathrm{BC1}} = -\frac{\psi_{\mathrm{Ax}}}{R_{\mathrm{P}}}$$

The small-angle approximations on the pivot axis angle may not be so good in this case, as sweep angles up to 30° may be used, but the results of this simple analysis give an excellent insight into the effects of pivot axis sweep – specifically, a second-order bump steer and a first-order bump camber.

As discussed in Chapter 8, the lateral bump scrub for this suspension can be found simply by projecting the real pivot axis into the transverse vertical plane of the wheel centres, giving the swing centre point E_{S} at height H_{S}. The bottom point of the wheel rotates about this point, in rear view, so the lateral bump scrub rate coefficient is

$$\varepsilon_{\mathrm{BScd0,Y}} = \frac{H_{\mathrm{S}}}{R_{\mathrm{S}}} = \frac{H_{\mathrm{S}}}{R_{\mathrm{P}}} \tan \psi_{\mathrm{Ax}}$$

Table 11.6.1 Coefficients for a semi-trailing arm with sloped axis and static wheel angles

$$\varepsilon_{\mathrm{BS1}} = -\frac{\phi_{\mathrm{Ax}} - \gamma_0}{R_{\mathrm{P}}}$$

$$\varepsilon_{\mathrm{BS2}} = -\frac{\psi_{\mathrm{Ax}} + \delta_0}{2R_{\mathrm{P}}^2}$$

$$\varepsilon_{\mathrm{BC1}} = -\frac{\psi_{\mathrm{Ax}} + \delta_0}{R_{\mathrm{P}}}$$

$$\varepsilon_{\mathrm{BC2}} = \frac{\phi_{\mathrm{Ax}} - \gamma_0}{2R_{\mathrm{P}}^2}$$

$$\varepsilon_{\mathrm{BScd0,Y}} = \frac{H_{\mathrm{S}}}{R_{\mathrm{S}}} = \frac{H_{\mathrm{S}} \tan \psi_{\mathrm{Ax}}}{R_{\mathrm{P}}}$$

$$\varepsilon_{\mathrm{BScd1,Y}} = -\frac{1}{R_{\mathrm{S}}} = -\frac{\tan \psi_{\mathrm{Ax}}}{R_{\mathrm{P}}}$$

where $L_{\mathrm{P}} = L_{\mathrm{S}} \tan \psi_{\mathrm{Ax}}$ is the pitch arm length, Figure 11.0.1. Therefore, approximately,

$$\varepsilon_{\mathrm{BScd0,Y}} = \frac{H_{\mathrm{S}}}{R_{\mathrm{P}}} \psi_{\mathrm{Ax}}$$

and the lateral bump scrub rate variation coefficient is

$$\varepsilon_{\mathrm{BScld,Y}} = \frac{1}{R_{\mathrm{S}}} = \frac{\psi_{\mathrm{Ax}}}{R_{\mathrm{P}}}$$

These results are summarised in Table 11.6.1, including superposed terms for wheel static toe and camber.

11.7 The Low-Pivot Semi-Trailing Arm

A pivot height different from the wheel centre height would normally be introduced to change the longitudinal properties, such as the anti-lift or anti-rise.

From the principle of the parallel axis rotation, it is evident that a lowering of the pivot axis should have no effect on the bump steer or bump camber angles.

However, this may be a confusing point, because when the pivot axis is at a different height from the wheel centre, then the second-order bump steer ($\varepsilon_{\mathrm{BS2}}$) of the semi-trailing arm seems to become asymmetrical about the static bump position. In effect this would introduce a first-order bump steer

$$\varepsilon_{\mathrm{BS1}} = (R_{\mathrm{L}} - H_{\mathrm{Ax}})\varepsilon_{\mathrm{BS2}}$$

This is typically a small effect, but not really negligible; for example, $L_{\mathrm{P}} = 0.450$ m, $R_{\mathrm{L}} = 0.300$ m, $H_{\mathrm{Ax}} = 0.150$ m, $\psi_{\mathrm{Ax}} = 0.30$ rad, $\varepsilon_{\mathrm{BS2}} = -0.741$ rad/m^2, $\varepsilon_{\mathrm{BS1}} = -0.111$ rad/m $= -0.64$ deg/dm. However, the use of this term presupposes that the wheel static steer and camber angles are specified when the wheel centre is at the same height as the pivot axis. If these are specified at the static position, as they would normally be, then the effects are automatically accounted for, and compensated, with no resulting first-order effect other than that from the wheel axis angles. When the wheel angles are specified at the level arm position, at the static position there are extra wheel axis angles that offset the above effect.

The bump scrub may be affected considerably, of course, but the equations of Table 11.6.1 continue to apply.

Varying the axis height at constant arm radius will vary the static arm length, which would change the relationship between the arm angle and suspension bump, so having some small, but non-zero, effect on the coefficients. Numerical results do show some small effects due to axis height.

11.8 The Transverse Arm

The basic transverse arm, equivalent to the simple swing axle, has a pivot axis parallel to the vehicle centreline, at or close to the wheel centre height. For a true swing axle, the swing arm length is somewhat less than half of the track, wheel lateral location being by the driveshafts which are pivoted at the sides of the final drive unit, Figure 11.8.1. An extra arm is used to control the longitudinal position of the wheel, hence also completing definition of the pivot axis. The pitch arm length and radius are infinite, and therefore unsuitable for use in some equations.

In the absence of wheel static steer and camber, it is evident that the wheel camber angle is

$$\gamma = -\theta_A$$

and the steer angle remains zero. Therefore, for this simple case, nominally,

$$\varepsilon_{BC1} = -\frac{1}{R_S}$$

The lateral bump scrub rate is

$$\varepsilon_{BScd0,Y} = \frac{H_{Ax}}{R_S}$$

and the lateral bump scrub rate variation is

$$\varepsilon_{BScd1,Y} = -\frac{1}{R_S}$$

The introduction of static steer and camber angles has some interesting effects at large arm angles, approaching 90°, beyond the practical range but of interest for geometry and for accurate analytic expressions, because at an arm angle of 90° the properties are discontinuous.

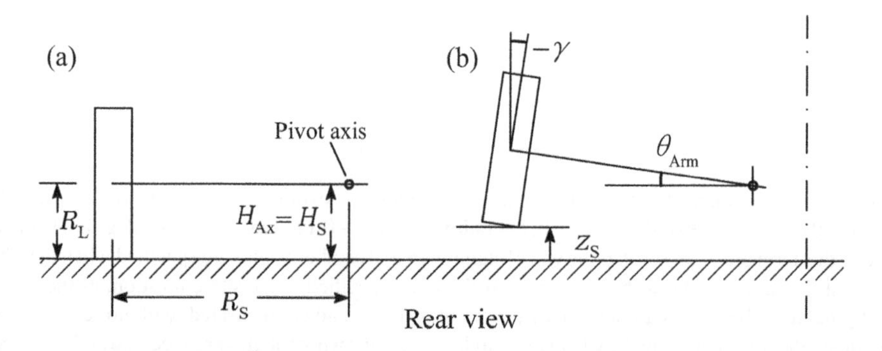

Figure 11.8.1 The simple transverse arm (swing axle), rear view.

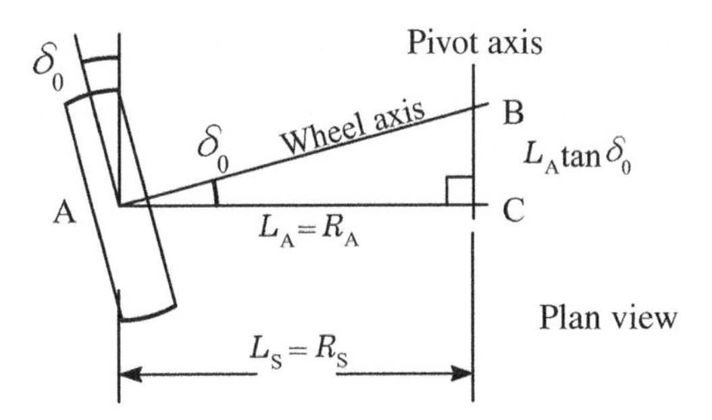

Figure 11.8.2 The swing axle with static toe angle, plan view.

Introducing a static camber angle γ_0, within the practical range of operation this simply adds γ_0 to the bump camber. Introducing a positive static toe-out angle δ_0, at an arm angle of 90° ($\pi/2$ rad) the camber angle becomes ($\pi/2 - \delta_0$) and the steer angle is zero. If δ_0 is negative (i.e. a toe-in), then the camber at 90° arm angle becomes $-(\pi/2 - \delta_0)$. At $-\pi/2$ arm angle, the camber angle becomes $-(\pi/2 - \delta_0)$ and the steer angle is zero. In fact in these extreme arm positions the result depends on an infinitesimal toe-in or toe-out.

Figure 11.8.2 shows the plan view of a swing axle with static toe angle. The wheel axis is AB, and line AC is the perpendicular to the pivot axis. When the arm moves in bump, the wheel centre A moves around C, and so towards C in the plan view. Therefore the steer angle is given by

$$\tan\delta = \frac{L_A \tan\delta_0}{L_A \cos\theta_A} = \frac{\tan\delta_0}{\cos\theta_A}$$

As the arm angle approaches 90°, the tangent of the steer angle goes to infinity, and the steer angle tends to sign(δ_0) times $\pi/2$ (90°). Considering practical cases, for a small arm angle the approximation may be made that

$$\tan\delta \approx \frac{\tan\delta_0}{(1 - \frac{1}{2}\theta_A^2)} \approx \tan\delta_0(1 + \tfrac{1}{2}\theta_A^2)$$

This then gives

$$\delta \approx \delta_0 + \frac{\delta_0}{2R_A^2} z_S^2$$

so introducing a quadratic bump steer coefficient

$$\varepsilon_{BS2} = \frac{\delta_0}{2R_A^2}$$

Because the static steer angle is likely to be small, the value of this is not a very significant effect in practice. Within the practical range,

$$\delta \approx \delta_0$$
$$\gamma = \gamma_0 - \frac{1}{R_S} z_S$$

Table 11.8.1 summarises the results.

Table 11.8.1 Coefficients for a simple transverse arm, with static wheel angles

$$\varepsilon_{\mathrm{BS1}} = 0$$
$$\varepsilon_{\mathrm{BS2}} = \frac{\delta_0}{2R_{\mathrm{A}}^2} \approx 0$$
$$\varepsilon_{\mathrm{BC1}} = -\frac{1}{R_{\mathrm{S}}}$$
$$\varepsilon_{\mathrm{BC2}} = 0$$
$$\varepsilon_{\mathrm{BSc0d,Y}} = \frac{H_{\mathrm{Ax}}}{R_{\mathrm{S}}}$$
$$\varepsilon_{\mathrm{BScd1,Y}} = -\frac{1}{R_{\mathrm{S}}}$$

11.9 The Sloped-Axis Transverse Arm

The transverse-arm pivot axis may be sloped out of the horizontal plane, at angle ϕ_{Ax} whilst remaining parallel to the centreline in plan view, giving the sloped-axis transverse arm. Typically this is done to obtain anti-squat and anti-rise characteristics, or bump steer, to be analysed here. For consistency with trailing and semi-trailing arms, positive slope angle ϕ_{Ax} will be considered to be when the axis rises towards the rear of the vehicle. Practical swing axle design may use positive or negative values of pivot axis inclination.

Figure 11.9.1 illustrates this case. For practical values of axis inclination, the effect on camber is small, but the effect on steer is significant. In Figure 11.9.1(a), when the wheel bump is $z_{\mathrm{W}} \approx z_{\mathrm{S}}$, the centre of the wheel will move forwards by $z_{\mathrm{S}} \tan \phi_{\mathrm{Ax}}$ with a resulting bump toe angle (toe-out as positive) of

$$\delta = -\frac{z_{\mathrm{S}} \tan \phi_{\mathrm{Ax}}}{R_{\mathrm{S}}}$$

with, therefore, a first bump steer coefficient

$$\varepsilon_{\mathrm{BS1}} = -\frac{\tan \phi_{\mathrm{Ax}}}{R_{\mathrm{S}}} \approx -\frac{\phi_{\mathrm{Ax}}}{R_{\mathrm{S}}}$$

Table 11.9.1 summarises the results for this type of rigid-arm suspension.

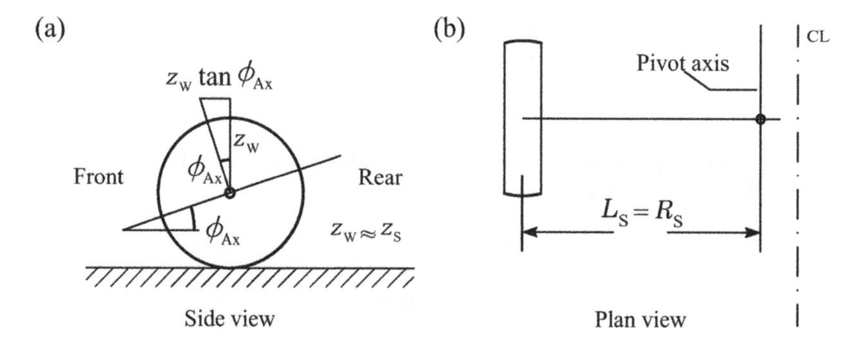

Figure 11.9.1 The transverse arm with sloped axis: (a) side view, (b) plan view.

Table 11.9.1 Coefficients for a sloped-axis transverse arm, with static wheel angles

$$\varepsilon_{\text{BS1}} = -\frac{\phi_{\text{Ax}}}{R_{\text{S}}}$$

$$\varepsilon_{\text{BS2}} = 0$$

$$\varepsilon_{\text{BC1}} = -\frac{1}{R_{\text{S}}}$$

$$\varepsilon_{\text{BC2}} = 0$$

$$\varepsilon_{\text{BScd0,Y}} = \frac{H_{\text{Ax}}}{R_{\text{S}}}$$

$$\varepsilon_{\text{BScd1,Y}} = -\frac{1}{R_{\text{S}}}$$

As a matter of interest, in 1940 Henry Ford was granted a patent for a swing axle explicitly featuring a slope angle to give bump toe-out, Figure 11.9.2. With a large load on the vehicle, or in riding bumps, the large first bump camber coefficient of the swing axle causes rather a large camber angle with associated side forces, which cause excessive tyre wear. This can be offset by introducing bump toe-out (negative ε_{BS1}) to reduce the net side force to zero. The relevant tyre characteristics are the cornering stiffness C_α relating the cornering force to the slip angle, and the camber stiffness C_γ relating the side force to the camber angle. For a suspension bump z_{S}, the first-order bump steer and camber angles are

$$\delta = \varepsilon_{\text{BS1}} z_{\text{S}}$$

$$\gamma = \varepsilon_{\text{BC1}} z_{\text{S}}$$

The consequent tyre side force is

$$\begin{aligned} F_{\text{Y}} &= C_\alpha \delta + C_\gamma \gamma \\ &= C_\alpha \varepsilon_{\text{BS1}} z_{\text{S}} + C_\gamma \varepsilon_{\text{BC1}} z_{\text{S}} \end{aligned}$$

To eliminate the bump side force requires

$$\frac{\varepsilon_{\text{BS1}}}{\varepsilon_{\text{BC1}}} = -\frac{C_\gamma}{C_\alpha}$$

Now the ratio of the tyre coefficients is mainly a characteristic of the tyre carcase construction. In the days of the Ford patent, bias-ply tyres were the norm, with a ratio of about 0.15. Nowadays, radial-ply tyres are normal for passenger cars, with a ratio of about 0.05. A swing axle has a first bump camber coefficient of about −1.7 rad/m (−10 deg/dm), so for bias-ply tyres the required bump steer coefficient is about plus 1.5 deg/dm, and for radial ply tyres about 0.5 deg/dm. Using, from Table 11.9.1,

$$\phi_{\text{Ax}} = -\varepsilon_{\text{BS1}} R_{\text{S}}$$

with a swing arm radius of about 0.6 m, the required axis slope is −9° for bias ply tyres and −3° for radial ply tyres. The negative slope means that the axis should be down towards the rear, in agreement with Figure 11.9.2.

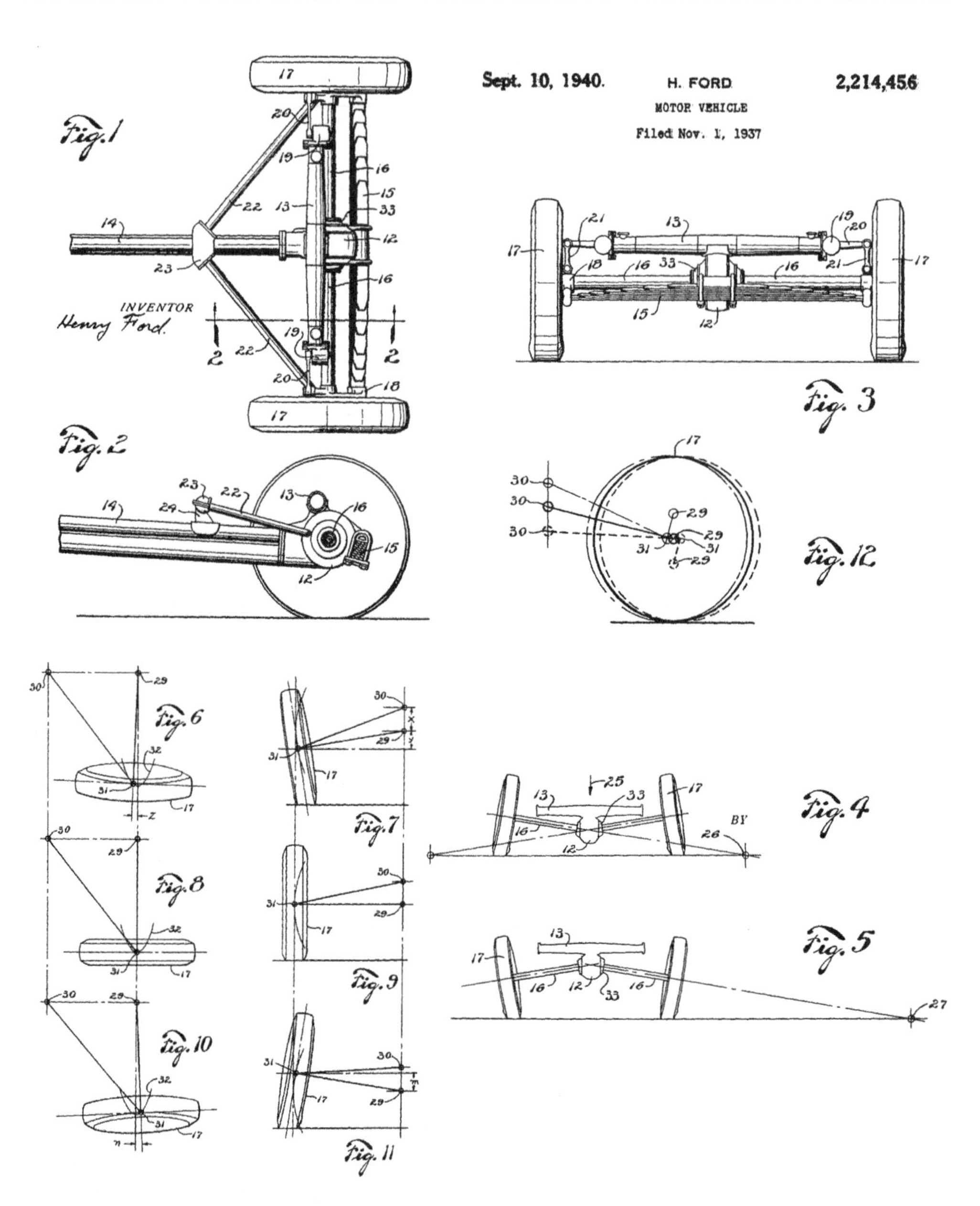

Figure 11.9.2 A swing axle with bump toe-out to compensate for side forces resulting from bump camber action (Ford, 1937).

11.10 The Semi-Transverse Arm

Analogous to the semi-trailing arm, the semi-transverse arm has a pivot axis in the horizontal plane, but angled in plan view, the pivot axis being fairly closely but not exactly parallel to the *X* axis. For consistency with semi-trailing arms, the angle of the axis can be measured alternatively from the negative *Y*-axis

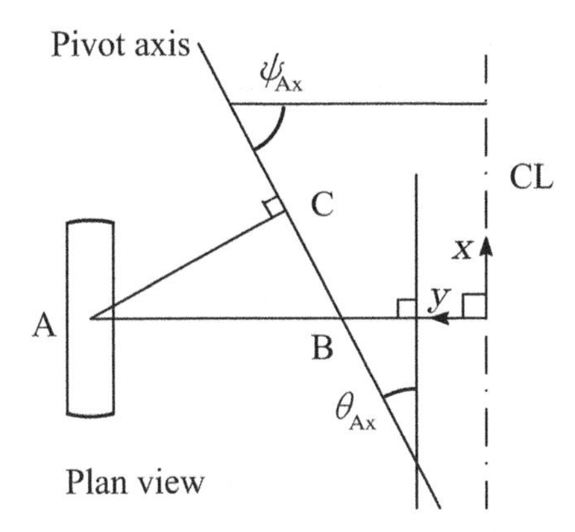

Figure 11.10.1 Semi-transverse arm in plan view, axis sweep angle ψ_{Ax} from the Y axis, axis yaw angle θ_{Ax} from the X axis.

direction (left wheel), giving the sweep angle ψ_{Ax}. However, as in Figures 11.0.1 and 11.10.1, the actual deviation from a simple transverse arm is the complementary angle, the axis yaw angle, which will be denoted by θ_{Ax}:

$$\theta_{Ax} = \frac{\pi}{2} - \psi_{Ax}$$

This angling of the axis has several effects. In Figure 11.10.1, dropping a perpendicular from the wheel centre to the pivot axis, it may be seen that the wheel centre A moves in an arc about point C, the foot of the perpendicular.

The radius of the arc of point A is the arm length L_A. As the wheel bumps or droops, the wheel centre A moves, in plan view, towards C, by a distance p given by

$$p = L_A(1 - \cos\theta_A) = L_A \tfrac{1}{2}\left(\frac{z_S}{R_A}\right)^2 \approx \frac{1}{2R_A} z_S^2$$

The movement of the wheel centre in the longitudinal X direction is

$$X_A = p \sin\theta_{Ax} \approx \frac{\theta_{Ax}}{2R_A} z_S^2$$

There is therefore a bump steer angle, approximately

$$\delta = -\frac{X_A}{L_{AB}} = -\frac{X_A}{R_A} = -\frac{\theta_{Ax}}{2R_A^2} z_S^2$$

consequently giving a second bump steer coefficient

$$\varepsilon_{BS2} = -\frac{\theta_{Ax}}{2R_A^2}$$

The pitch and roll force characteristics are solved in the usual way, projecting the axis into the relevant planes to obtain the swing centre E_S and pitch centre E_P points. Table 11.10.1 summarises the coefficients for the semi-transverse arm.

Table 11.10.1 Coefficients for a semi-transverse arm, including axis slope and static wheel angles

$$\varepsilon_{\mathrm{BS1}} = -\frac{\phi_{\mathrm{Ax}}}{R_{\mathrm{A}}}$$

$$\varepsilon_{\mathrm{BS2}} = -\frac{(\theta_{\mathrm{Ax}} - \delta_0)}{2R_{\mathrm{A}}^2}$$

$$\varepsilon_{\mathrm{BC1}} = -\frac{1}{R_{\mathrm{S}}}$$

$$\varepsilon_{\mathrm{BC2}} = 0$$

$$\varepsilon_{\mathrm{BScd0,Y}} = \frac{H_{\mathrm{S}}}{R_{\mathrm{S}}}$$

$$\varepsilon_{\mathrm{BScd1,Y}} = -\frac{1}{R_{\mathrm{S}}}$$

11.11 The Low-Pivot Semi-Transverse Arm

A low pivot axis on a transverse arm is typically used to reduce the height of the roll centre (e.g. the low-pivot swing axle). The effect on bump steer is similar to the effect on the semi-trailing arm. The existing second-order bump steer, if any, is no longer symmetrical about the static position, so some first-order bump steer is introduced:

$$\varepsilon_{\mathrm{BS1}} = (R_{\mathrm{L}} - H_{\mathrm{Ax}})\varepsilon_{\mathrm{BS2}}$$

However, this again assumes that the wheel static steer and camber are specified at the horizontal arm position. Normally they would be specified at the static position. At the horizontal arm position, these angles are different, introducing additional effects, and compensating for the above term.

The bump scrub coefficients continue as in Table 11.10.1 with appropriate evaluation of the swing centre height H_{S}.

11.12 General Case Numerical Solution

The computer numerical solution of the rigid arm is quite easily performed, given certain standard three-dimensional coordinate geometry routines discussed in a subsequent chapter. The basic problem is specified by the static position in vehicle coordinates, Figure 11.12.1, by

(1) the coordinates of a point A on the pivot axis $(x_{\mathrm{A}}, y_{\mathrm{A}}, z_{\mathrm{A}})$,
(2) the direction cosines of the pivot axis $(l_{\mathrm{A}}, m_{\mathrm{A}}, n_{\mathrm{A}})$,
(3) the coordinates of a point W on the wheel axis $(x_{\mathrm{W}}, y_{\mathrm{W}}, z_{\mathrm{W}})$,
(4) the direction cosines of the wheel axis $(l_{\mathrm{W}}, m_{\mathrm{W}}, n_{\mathrm{W}})$.

Other information required includes the track, and the wheel centre position, if this is not already known to be point W.

Two distinct analyses are required, static (the initial position) and displacement (arm rotated, leading to bump coefficients).

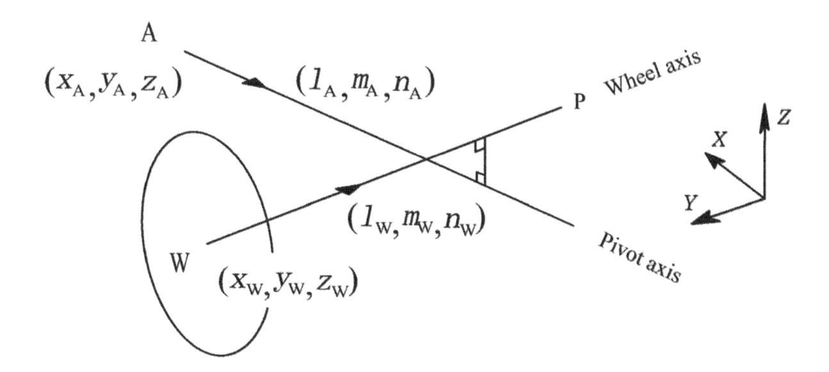

Figure 11.12.1 Pivot axis through A and wheel axis through W, each axis defined by a point and direction cosines. P is a methodological second point on the wheel axis.

In a static analysis of the suspension it is desired to obtain:

(1) the pitch centre, pitch arm radius and length;
(2) the swing centre, swing arm radius and length;
(3) the bump scrub rate coefficient;
(4) the roll centre height.

The pitch centre is obtained by intersecting the pivot axis with the vertical longitudinal plane of the wheel–ground contact point, effectively of the wheel centre. The swing centre is obtained by intersecting the pivot axis with the vertical transverse plane of the two wheel centres of the axle. The intersection of a line with a plane requires standard routines discussed in Chapter 15. In special cases, one or other of these points may be at infinity (e.g. a pure trailing arm or a pure transverse arm).

The bump scrub rate coefficient $\varepsilon_{\mathrm{BScd0}}$ follows as H_S/R_S in the general case, and the geometric roll centre height as $\frac{1}{2}T\varepsilon_{\mathrm{BScd0}}$.

In the displacement analysis, for a rotation of the arm, about the pivot axis, by some specified angle, it is required to determine

(1) the wheel steer angle,
(2) the wheel camber angle,
(3) the suspension bump,
(4) the local scrub rate,

with a series of such results through the suspension range, a quadratic polynomial curve fit then giving the bump coefficients. Interchange of the wheel axis angle specification (δ, γ) and the wheel axis direction cosines (l_W, m_W, n_W) is required, as given in Table 11.3.1.

The underlying geometrical problem is to rotate one line, the wheel axis, about another line, the pivot axis. The easiest way to do this is to use two points on the wheel axis, and rotate each point about the pivot axis, the new point positions then defining the new wheel axis. The first point on the axis can be the point W as originally specified – typically but not essentially the wheel centre. The second point can be any reasonable well-separated point, P, say at separation L from point W. Then the initial position of point P is

$$x_P = x_W + Ll_W$$
$$y_P = y_W + Lm_W$$
$$z_P = z_W + Ln_W$$

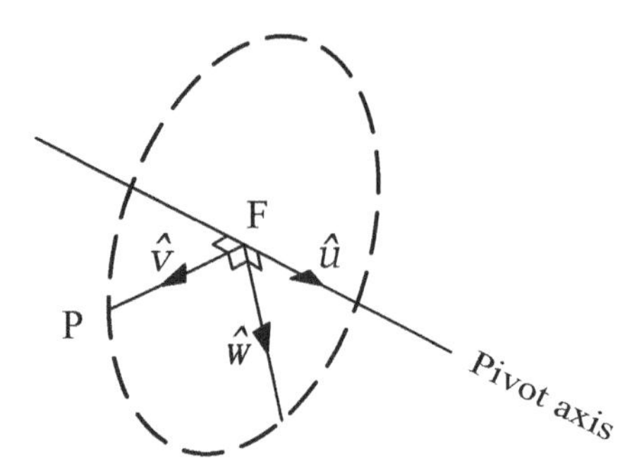

Figure 11.12.2 Generation of local coordinate system for rotation of a point about an axis by a specified angle.

At the scale of vehicle suspensions, it is convenient to use a separation of 1 metre. The direction cosines of the new axis then follow easily, and thence the wheel steer and camber angles for that bump position.

The residual problem, then, is the rotation of one point about a line by a specified angle. This is easy in two dimensions, so one approach is to specify a local coordinate system with the circular motion of the point in a local coordinate plane with perpendicular unit vectors ($\hat{\boldsymbol{v}}$, $\hat{\boldsymbol{w}}$) in the plane perpendicular to the axis $\hat{\boldsymbol{u}}$, Figure 11.12.2. The axis unit vector is already known. Drop a perpendicular from the point P onto the axis, with foot F. The length PF is the arc radius R for that point. One in-plane unit vector can be determined as being along FP and of unit length. The third unit vector, also in the plane of rotation, can then be determined by a vector cross product of the two known unit vectors,

$$\hat{\boldsymbol{w}} = (l_{\mathrm{w}}, m_{\mathrm{w}}, n_{\mathrm{w}}) = \hat{\boldsymbol{u}} \times \hat{\boldsymbol{v}} = \begin{vmatrix} \hat{\boldsymbol{i}} & \hat{\boldsymbol{j}} & \hat{\boldsymbol{k}} \\ l_{\mathrm{u}} & m_{\mathrm{u}} & n_{\mathrm{u}} \\ l_{\mathrm{v}} & m_{\mathrm{v}} & n_{\mathrm{v}} \end{vmatrix}$$

where the determinant is to be expanded with the usual sign alternations.

For a right-hand rotation by angle θ about the axis direction, the coordinates of the new position of P are then given by

$$x_{\mathrm{P}} = x_{\mathrm{F}} + l_{\mathrm{v}} \cos\theta + l_{\mathrm{w}} \sin\theta$$

$$y_{\mathrm{P}} = y_{\mathrm{F}} + m_{\mathrm{v}} \cos\theta + m_{\mathrm{w}} \sin\theta$$

$$z_{\mathrm{P}} = z_{\mathrm{F}} + n_{\mathrm{v}} \cos\theta + n_{\mathrm{w}} \sin\theta$$

with P rotating in the ($\hat{\boldsymbol{v}}$, $\hat{\boldsymbol{w}}$) plane as desired. Hence a new point P is determined.

By doing this for any two distinct points on the wheel axis, the new wheel axis is found, and the steer and camber angles follow.

In practice, the above is quite easily programmed, given the necessary three-dimensional routines, and is accurate.

11.13 Comparison of Solutions

Table 11.13.1 shows how the accurate numerical solution compares with the approximate algebraic expressions from the earlier Tables 11.6.1 (semi-trailing arm) and 11.10.1 (semi-transverse arm). The approximate expressions are evidently good enough for mental imagery of the effects, and even for general design use.

Table 11.13.1 Suspension: rigid-arm analysis, by Rigid Arm.f90 program

```
Arm type = Pure trailing arm, zero steer and camber

Pivot theta       (deg)    =    90.000
Pivot psi         (deg)    =     0.000
Pivot phi         (deg)    =     0.000
Wheel steer       (deg)    =     0.000
Wheel camber      (deg)    =     0.000
Pitch arm length  (m)      =     0.450

                               numerical analytic
epsBS1      (deg/dm)       =     0.000       0.000
epsBS2      (deg/dm2)      =     0.000       0.000
epsBC1      (deg/dm)       =     0.000       0.000
epsBC2      (deg/dm2)      =     0.000       0.000

Arm type = Trailing arm with sloped axis

Pivot theta       (deg)    =    90.000
Pivot psi         (deg)    =     0.000
Pivot phi         (deg)    =     5.000
Wheel steer       (deg)    =     0.000
Wheel camber      (deg)    =     0.000
Pitch arm length  (m)      =     0.450

                               numerical analytic
epsBS1      (deg/dm)       =    -1.114      -1.111
epsBS2      (deg/dm2)      =     0.000       0.000
epsBC1      (deg/dm)       =     0.000       0.000
epsBC2      (deg/dm2)      =     0.125       0.123

Arm type = Pure trailing arm, with steer and camber

Pivot theta       (deg)    =    90.000
Pivot psi         (deg)    =     0.000
Pivot phi         (deg)    =     0.000
Wheel steer       (deg)    =     2.000
Wheel camber      (deg)    =     2.000
Pitch arm length  (m)      =     0.450

                               numerical analytic
epsBS1      (deg/dm)       =     0.444       0.444
epsBS2      (deg/dm2)      =    -0.050      -0.049
epsBC1      (deg/dm)       =    -0.444      -0.444
epsBC2      (deg/dm2)      =    -0.050      -0.049

Arm type = Semi-trailing arm, zero steer and camber

Pivot theta       (deg)    =    72.000
Pivot psi         (deg)    =    18.000
Pivot phi         (deg)    =     0.000
Wheel steer       (deg)    =     0.000
Wheel camber      (deg)    =     0.000
Pitch arm length  (m)      =     0.450
```

(continued)

Table 11.13.1 (*Continued*)

			numerical	analytic
epsBS1	(deg/dm)	=	0.000	0.000
epsBS2	(deg/dm2)	=	-0.465	-0.444
epsBC1	(deg/dm)	=	-4.139	-4.000
epsBC2	(deg/dm2)	=	0.000	0.000

Arm type = Semi-trailing arm, with sloped axis

Pivot theta	(deg)	=	72.000
Pivot psi	(deg)	=	18.000
Pivot phi	(deg)	=	5.000
Wheel steer	(deg)	=	0.000
Wheel camber	(deg)	=	0.000
Pitch arm length	(m)	=	0.450

			numerical	analytic
epsBS1	(deg/dm)	=	-1.171	-1.111
epsBS2	(deg/dm2)	=	-0.461	-0.444
epsBC1	(deg/dm)	=	-4.139	-4.000
epsBC2	(deg/dm2)	=	0.146	0.123

Arm type = Semi-trailing arm, with steer and camber

Pivot theta	(deg)	=	72.000
Pivot psi	(deg)	=	18.000
Pivot phi	(deg)	=	0.000
Wheel steer	(deg)	=	2.000
Wheel camber	(deg)	=	2.000
Pitch arm length	(m)	=	0.450

			numerical	analytic
epsBS1	(deg/dm)	=	0.441	0.444
epsBS2	(deg/dm2)	=	-0.510	-0.494
epsBC1	(deg/dm)	=	-4.581	-4.444
epsBC2	(deg/dm2)	=	-0.049	-0.049

Arm type = Pure transverse arm, zero steer and camber

Pivot theta	(deg)	=	0.000
Pivot psi	(deg)	=	90.000
Pivot phi	(deg)	=	0.000
Wheel steer	(deg)	=	0.000
Wheel camber	(deg)	=	0.000
Arm radius	(m)	=	0.600

			numerical	analytic
epsBS1	(deg/dm)	=	0.000	0.000
epsBS2	(deg/dm2)	=	0.000	0.000
epsBC1	(deg/dm)	=	-9.596	-9.549
epsBC2	(deg/dm2)	=	0.000	0.000

Table 11.13.1 (*Continued*)

```
Arm type  = Transverse arm, with sloped axis

Pivot theta    (deg)  =      0.000
Pivot psi      (deg)  =     90.000
Pivot phi      (deg)  =      5.000
Wheel steer    (deg)  =      0.000
Wheel camber   (deg)  =      0.000
Arm radius      (m)   =      0.600

                               numerical  analytic
epsBS1     (deg/dm)      =      -0.848      -0.833
epsBS2     (deg/dm2)     =       0.000       0.000
epsBC1     (deg/dm)      =      -9.596      -9.549
epsBC2     (deg/dm2)     =       0.000       0.000

Arm type  = Pure transverse arm, with steer and camber

Pivot theta    (deg)   =     0.000
Pivot psi      (deg)   =    90.000
Pivot phi      (deg)   =     0.000
Wheel steer    (deg)   =     2.000
Wheel camber   (deg)   =     2.000
Arm radius      (m)    =     0.600

                               numerical  analytic
epsBS1     (deg/dm)      =      -0.012       0.000
epsBS2     (deg/dm2)     =       0.029       0.000
epsBC1     (deg/dm)      =      -9.590      -9.549
epsBC2     (deg/dm2)     =       0.000       0.000

Arm type = Semi-transverse arm, zero steer and camber

Pivot theta   (deg)  =    15.000
Pivot psi     (deg)  =    75.000
Pivot phi     (deg)  =     0.000
Wheel steer   (deg)  =     0.000
Wheel camber  (deg)  =     0.000
Arm radius     (m)   =     0.600
                            numerical  analytic
epsBS1     (deg/dm)     =     0.000       0.000
epsBS2     (deg/dm2)    =    -0.220      -0.208
epsBC1     (deg/dm)     =    -9.593      -9.549
epsBC2     (deg/dm2)    =     0.000       0.000

Arm type = Semi-transverse arm, with sloped axis

Pivot theta    (deg)  =    15.000
Pivot psi      (deg)  =    75.000
Pivot phi      (deg)  =     5.000
Wheel steer    (deg)  =     0.000
Wheel camber   (deg)  =     0.000
Arm radius      (m)   =     0.600
```

(*continued*)

Table 11.13.1 (*Continued*)

			numerical	analytic
epsBS1	(deg/dm)	=	-0.878	-0.833
epsBS2	(deg/dm2)	=	-0.221	-0.208
epsBC1	(deg/dm)	=	-9.593	-9.549
epsBC2	(deg/dm2)	=	0.000	0.000
Arm type = Semi-transverse arm, with steer and camber				
Pivot theta	(deg)	=	15.000	
Pivot psi	(deg)	=	75.000	
Pivot phi	(deg)	=	0.000	
Wheel steer	(deg)	=	2.000	
Wheel camber	(deg)	=	2.000	
Arm radius	(m)	=	0.600	
			numerical	analytic
epsBS1	(deg/dm)	=	0.080	0.000
epsBS2	(deg/dm2)	=	-0.193	-0.208
epsBC1	(deg/dm)	=	-9.678	-9.549
epsBC2	(deg/dm2)	=	-0.002	0.000

11.14 The Steered Single Arm

The single-arm suspension has sometimes been used at the front with steering. Two examples are the Dubonnet type, Figure 1.5.2 and the long transverse arm, Figure 1.6.3. The bump steer depends on the steering linkage, as described in Chapter 6, but in both cases accurate steering is possible. However, the steer axis (kingpin) inclination angle and steer axis (kingpin) caster angle, which affect steering feel, are of importance. The dependence of these on suspension bump can be represented by quadratic expressions in the usual way:

$$\theta_{\text{KI}} = \theta_{\text{KI0}} + \varepsilon_{\text{BKI1}} z_{\text{S}} + \varepsilon_{\text{BKI2}} z_{\text{S}}^2$$
$$\theta_{\text{KC}} = \theta_{\text{KC0}} + \varepsilon_{\text{BKC1}} z_{\text{S}} + \varepsilon_{\text{BKC2}} z_{\text{S}}^2$$

The kingpin inclination angle is generally taken as positive when it slopes inwards at the top, that is, analogous to the wheel camber angle but of reversed sign. The caster angle is positive when the axis slopes to the rear as it rises. The kingpin axis is fixed relative to the single arm, so the bump kingpin inclination coefficients are essentially the same as the bump camber coefficients.

The kingpin axis is approximately vertical, Figure 11.14.1, specified typically by two points, the ball joints, easily giving the parametric form of the steer axis line. Numerical solution of the moved axis then follows by the same principles as for the wheel axis. It is apparent that in the case of the transverse arm there will be large changes of inclination angle, which is undesirable, with small changes of caster angle. In contrast, the trailing and semi-trailing arms have small changes of kingpin inclination angle but large changes of caster angle:

$$\varepsilon_{\text{BKI1}} = \frac{\cos\theta_{\text{Ax}}}{R_{\text{A}}}$$

$$\varepsilon_{\text{BKC1}} = -\frac{\sin\theta_{\text{Ax}}}{R_{\text{A}}} = -\frac{\cos\psi_{\text{Ax}}}{R_{\text{A}}}$$

This large caster angle variation of the steered trailing arm, over 10° in each direction, is a serious difficulty, and contributed to the early demise of the type in favour of the double-wishbone suspension. The trailing arm has severe brake dive with the caster change because the arm adds to the body pitch angle effect. The leading arm, with extreme anti-dive, is less of a problem in this respect.

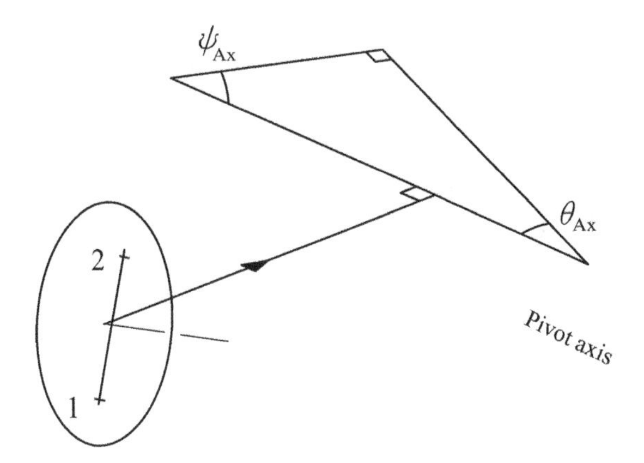

Figure 11.14.1 The single-arm suspension with a kingpin axis for steering.

The steered transverse arm was more successful, and was used for many years on small rear-engined family saloons. To minimise the kingpin inclination variation and the camber change, the arm was made long, with the pivot axis close to the vehicle centreline, Figure 1.6.3.

11.15 Bump Scrub

In the case of a rigid arm with zero slope of the pivot axis, and no wheel angles, the bump scrub is particularly simple to analyse. In Figure 11.15.1(a), which is the view along the pivot axis, the distance from that axis to the central contact point is L_C (not L_A to the wheel centre), and the angle of this radius from the horizontal is θ_C. The radius R_C is slightly different from R_A because of the wheel camber, but this small difference is neglected here.

In the plan view of Figure 11.15.1(b), for a locked-wheel vehicle, the resulting total scrub will be perpendicular to the pivot axis, along direction AC. For a (small) suspension bump z_S, the total scrub is

$$s = L_C \, d\theta \sin\theta_C = z_S \tan\theta_C$$

The total locked-wheel scrub rate coefficient is therefore

$$\varepsilon_{BScd0,T} = \tan\theta_C = \frac{H_{Ax}}{R_C} \approx \frac{H_{Ax}}{R_A}$$

This total locked scrub resolves into directions along the prospectively rolling wheel and perpendicular to it. Normally the small wheel steer angle is neglected, so that the scrub can be resolved more simply into vehicle X (longitudinal) and Y (lateral) components. The longitudinal scrub coefficient, relevant to braking, is

$$\varepsilon_{BScd0,X} = -\varepsilon_{BScd0,T} \sin\theta_{Ax} = -\frac{H_{Ax}\sin\theta_{Ax}}{R_A}$$

but

$$R_P \sin\theta_{Ax} = R_A$$

so

$$\varepsilon_{BScd0,X} = -\frac{H_{Ax}}{R_P}$$

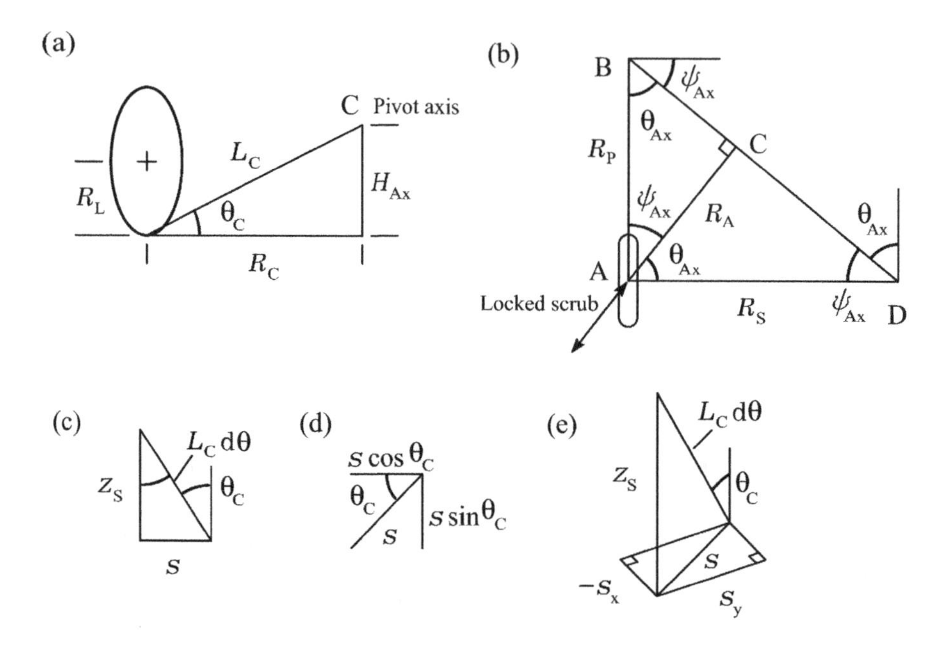

Figure 11.15.1 Bump scrub of the rigid arm without pivot axis slope.

Similarly, the Y component of the scrub gives a lateral scrub rate coefficient

$$\varepsilon_{\text{BScd0,Y}} = \varepsilon_{\text{BSc0,T}} \cos\theta_{\text{Ax}}$$

giving

$$\varepsilon_{\text{BScd0,Y}} = \frac{H_{\text{Ax}}}{R_{\text{S}}}$$

This lateral component is the one relevant to handling and the roll centre height, and the result here agrees with the simple calculation normally used.

The case of a pivot axis with slope is more complex, Figure 11.15.2. With an arm angular motion $d\theta$, the tangential displacement of the contact point of the locked wheel is

$$t = L_{\text{C}}\, d\theta$$

which occurs in the sloping plane perpendicular to the axis. This resolves into

$$s_1 = t \sin\theta_{\text{C}}$$

in the horizontal plane and

$$s_2 = t \cos\theta_{\text{C}}$$

in the sloping plane. This latter part further resolves into

$$s_3 = t \cos\theta_{\text{C}} \sin\phi_{\text{Ax}}$$

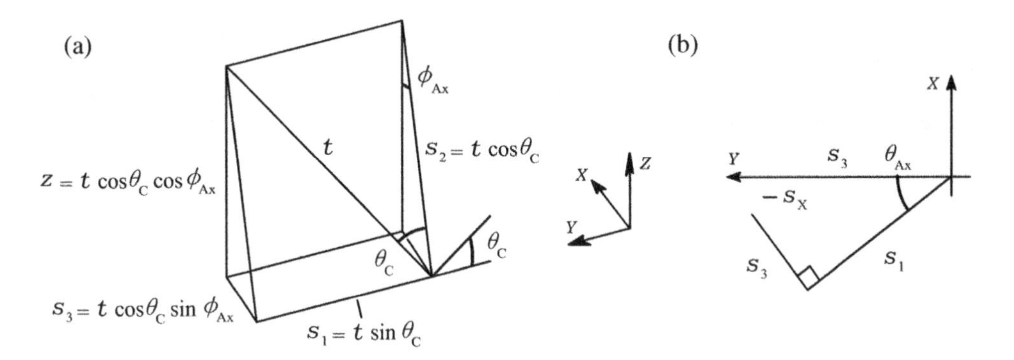

Figure 11.15.2 Bump scrub geometry for a rigid arm with pivot axis slope.

in the horizontal plane and

$$z_S = t\cos\theta_C\cos\phi_{Ax}$$

vertically. The Y component of the scrub is

$$\begin{aligned} s_Y &= s_1\cos\theta_{Ax} + s_3\sin\theta_{Ax} \\ &= t\sin\theta_C\cos\theta_{Ax} + t\cos\theta_C\sin\phi_{Ax}\sin\theta_{Ax} \end{aligned}$$

The lateral scrub rate is therefore

$$\varepsilon_{BScd0,Y} = \frac{s_Y}{z_S} = \frac{\tan\theta_C\cos\theta_{Ax}}{\cos\phi_{Ax}} + \frac{\sin\phi_{Ax}\sin\theta_{Ax}}{\cos\phi_{Ax}}$$

Now

$$\frac{H_{Ax}}{R_S} = \frac{H_{Ax}}{R_A}\frac{R_A}{R_S} = \frac{\tan\theta_C}{\cos\phi_{Ax}}\cos\theta_{Ax}$$

and

$$\sin\theta_{Ax} = \frac{L_{CD}}{R_S}$$

so

$$\begin{aligned} \varepsilon_{BScd0,Y} &= \frac{H_{Ax}}{R_S} + \frac{L_{CD}\tan\phi_{Ax}}{R_S} \\ &= \frac{H_{Ax}}{R_S} + \frac{H_S - H_{Ax}}{R_S} \\ &= \frac{H_S}{R_S} \end{aligned}$$

where H_S is the height of the swing centre, so the conventional simple expression is applicable even when the axis is sloped.

Of course, the argument for this property in the case of the force roll centre is very easily established, because the jacking force must be such that the moment of the lateral force about the axis is zero.

References

Ford, H. (1937). Motor vehicle. US Patent 2,214,456, filed 1 Nov 1937, granted 10 Sep 1940.

——————— // ———————

12

Double-Arm Suspensions

12.1 Introduction

This chapter deals with suspensions having location by two arms rather than the single-rigid-arm type of the previous chapter. The most common form of double-arm suspension is the double-transverse-arm suspension. Another type is the double-trailing-arm type. Example diagrams of these and others are given in Chapter 1. Strut-and-arm suspensions are also considered here, as variants of the double-arm type.

Double-transverse-arm suspensions are also known as double-lateral-arm, double-wishbone or double A-arm. The terms 'wishbone' and 'A-arm' are effectively interchangeable, at least as far as geometric properties are concerned – these names merely reflect the appearance of the arm, arising from the method of construction.

The common strut and transverse-arm suspension can, for geometric analysis, be considered as effectively a double-transverse-arm type in which the upper arm has more or less infinite length. This places some limitations on the geometric design possibilities, as will be shown.

The earliest cars had rigid front axles which suffered from steering geometry problems, as described in Chapter 1. To minimise this, the front springs had to be made stiff, to limit suspension movement. Independent front suspension, with its superior steering geometry, was introduced to allow soft springs and hence better ride quality whilst retaining good steering properties. The earliest independent types were the Dubonnet single leading or trailing arm, and the double transverse arm. The Dubonnet type soon fell from favour. The early double transverse-arm type had equal-length arms, parallel and level. Soon, the arms were made unequal in length, and placed at various angles to the horizontal in front and side view. Also, one or both arm pivot axes were sometimes angled in plan view. These changes to the details were made to obtain certain geometric properties giving desired dynamic properties. This chapter relates the geometric configuration to the consequent coefficients.

With double-arm suspensions, the steering properties, such as the bump steer coefficients, are dependent on the details of the steering mechanism, as already discussed to some extent in Chapter 5. This can be separated out from the basic properties of the arm geometry. Therefore the arm geometry is used to control the roll centre, the bump camber properties, the anti-dive and the bump caster. Specifically, the lateral coefficients to be controlled are:

(1) the lateral bump scrub rate $\varepsilon_{\text{BScd0,Y}}$ giving the roll centre height;
(2) the lateral bump scrub rate variation $\varepsilon_{\text{BScd1,Y}}$ giving the roll centre movement;
(3) the linear bump camber coefficient ε_{BC1};
(4) the quadratic bump camber coefficient ε_{BC2}.

Suspension Geometry and Computation J. C. Dixon

The longitudinal coefficients to be controlled are:

(1) the longitudinal bump scrub rate $\varepsilon_{\mathrm{BScd0,X}}$ giving the anti-dive;
(2) the longitudinal bump scrub rate variation $\varepsilon_{\mathrm{BScd1,X}}$ giving anti-dive variation;
(3) the linear bump caster coefficient $\varepsilon_{\mathrm{Bcas1}}$;
(4) the quadratic bump caster coefficient $\varepsilon_{\mathrm{Bcas2}}$.

There are therefore eight coefficients to be controlled. In the analysis process, the geometry is specified and the coefficients are deduced. In the design process, the values of the coefficients are specified, and the required geometry is deduced. The appropriate steering can then be added. In practice, of course, there may be some iteration in this process.

To control the eight variables as desired, the system must have at least eight suitable degrees of design freedom, four in the transverse properties and four in the longitudinal ones. Basically, these are the two arm lengths and the two arm angles to the horizontal. As seen in front view, this gives the first four factors, and in side view the second four. Strictly speaking, every coordinate affects every coefficient, but in practice the individual coefficients can be quite closely related to particular individual geometric features.

12.2 Configurations

Figure 12.2.1 shows a range of basic possibilities for the double-arm suspension. The pivot axis of each arm may be at almost any angle in plan view, although always approximately horizontal. Also, the arm lengths may vary considerably. The double transverse arm may have the pivot axes substantially parallel with the vehicle centreline, as shown, but to obtain certain characteristics one or other arm axis may be inclined in side view, or even more so in plan view.

The double-trailing-arm suspension seen in Figure 12.2.1(b) is simply a particular type of double arm with the pivot axes transverse to the vehicle centreline. Figure 12.2.1(c) shows a rarer type, in which the

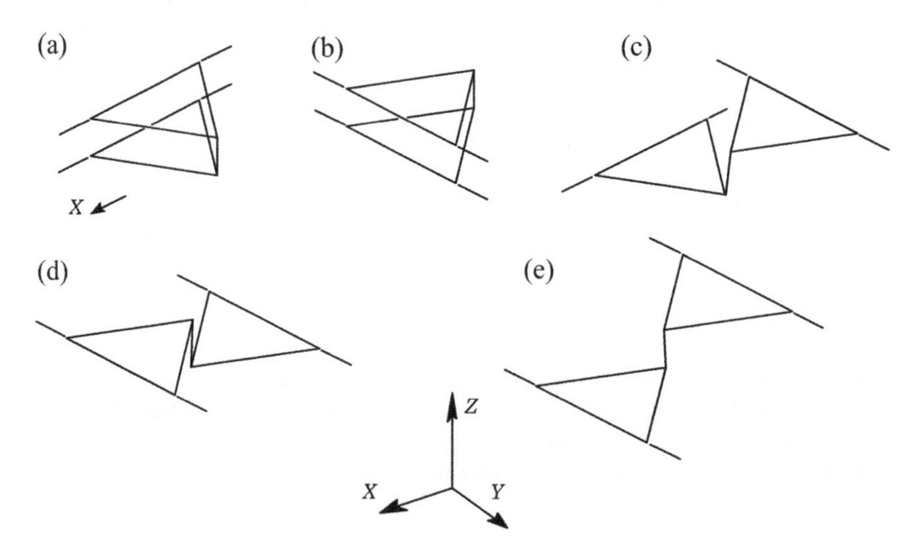

Figure 12.2.1 Basic layouts of double-arm suspension: (a) double transverse; (b) double trailing; (c) lower transverse, upper leading; (d) lower leading, upper trailing; (e) lower trailing, upper leading.

lower arm is a conventional transverse arm, but the upper arm is leading and has its axis transverse. This is geometrically, in front view, somewhat like a strut. This is a particularly extreme case of crossed axes. In Figure 12.2.1(d,e) are seen types with one leading and one trailing arm. These are unusual, but have the advantage that there is a Watt's linkage effect which minimises the fore and aft movement of the wheel centre, so these configurations are sometimes used for driven rear independent suspensions.

12.3 Arm Lengths and Angles

The three-dimensional system can be approached as two separate problems, each in two dimensions only, one problem being the front (or rear) view controlling the transverse properties, the other being the side view controlling the longitudinal properties. The correct analysis should really be done in a complete form in three dimensions, but the two-dimensional approximations are useful for human understanding and for initial design purposes.

The plane of an arm is defined as the geometric plane passing through the arm pivot axis and the centre of the ball joint at the arm outer end.

In the vehicle end view, seeing the transverse vertical plane of the unsteered wheel centres, the plane of the arm intersects this vertical plane to give a line of the arm. The actual pivot axis intersects the vertical plane at the inner point of the equivalent two-dimensional arm. It is less clear exactly where the outer point of the equivalent arm should be taken to be. Usually, the front-view position of the ball joint is used. This may not agree exactly with the plane of the arm, but is usually good enough. The result of this process is an equivalent front view, with the arms having specific lengths and angles to the horizontal, in the static position, Figure 12.3.1(a). The two arm lengths and two angles provide four degrees of design freedom, and prove to be good variables in controlling the desired transverse geometrical properties. However, it may be more convenient to think in terms of the average arm length and the length ratio or difference, and of the average angle and the convergence angle, as will be seen. Also, the equations are more readily expressed in terms of the 'shortness' of the links, the reciprocal of their length.

In side view, Figure 12.3.1(b), the plane of the arm may be intersected with the vertical longitudinal plane of the unsteered wheel centre along the side of the vehicle to give the line of the arm. The pivot axis of an arm intersects this plane at the point at one end of the equivalent two-dimensional arm. Often this particular point is at infinity. Nevertheless, its angular position to the horizontal from the wheel is important. Again, a point must be chosen to represent the ball joint centre, and often this will be the position of the ball joint as seen in the side view, although again this may not agree exactly with the line of the arm found from the intersection of the planes. This discrepancy is usually small. The result of this process is an equivalent side view, with the arms having specific lengths and angles to the horizontal, in the static position, Figure 12.3.1(b). These two lateral-view arm lengths and two angles provide four further degrees of design freedom, and prove to be good variables in controlling the desired longitudinal geometrical properties. However, again, it may be more convenient to think in terms of the average length

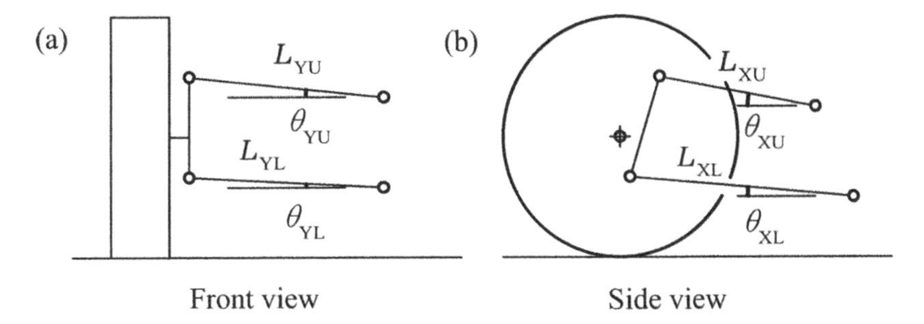

Figure 12.3.1 Double transverse arms, equivalent 2D mechanism: (a) front view; (b) side view.

and the length ratio or difference, or link shortness and shortness difference, and of the average angle and the convergence angle (i.e. angle difference).

It will be apparent from Figure 12.3.1 that the conceptual geometrical properties are related to the lateral-view arm lengths and angles by a similar process to the way that the lateral properties are related to the transverse arms and angles. The lateral properties will be examined first, the longitudinal properties then following easily.

12.4 Equal Arm Length

The earliest double-transverse-arm suspensions, seen in front view, had arms that were simply of equal length and horizontal when in the static position, Figure 12.4.1. The basic properties are easily found by inspection, and are shown in Table 12.4.1. The camber is unchanging. The initial bump scrub rate is zero, implying that the roll centre is initially at ground level. The bump scrub rate variation and associated roll centre movement depend on the lateral arm length L_Y.

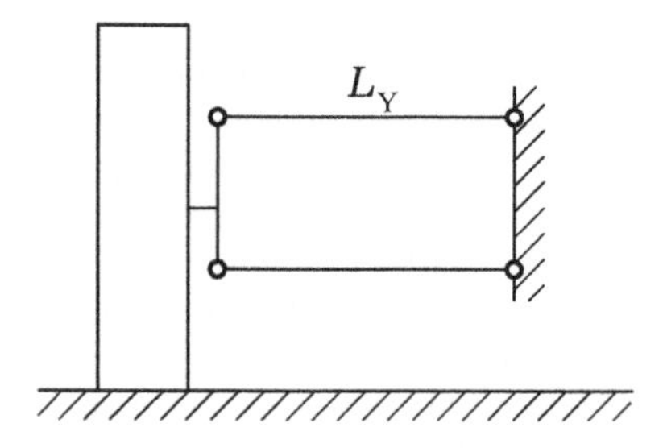

Figure 12.4.1 Parallel equal-length arms.

Table 12.4.1 Lateral properties of parallel horizontal double transverse arms

$$\varepsilon_{\mathrm{BScd0,Y}} = 0$$

$$\varepsilon_{\mathrm{BScd1,Y}} = -\frac{1}{L_Y}$$

$$\varepsilon_{\mathrm{BC1}} = 0$$

$$\varepsilon_{\mathrm{BC2}} = 0$$

12.5 Equally-Angled Arms

If the arms are angled in the static position, whilst remaining parallel, as in Figure 12.5.1, then the camber is still unchanging with bump. The arm angle is defined to be positive when the ball joint is higher than the

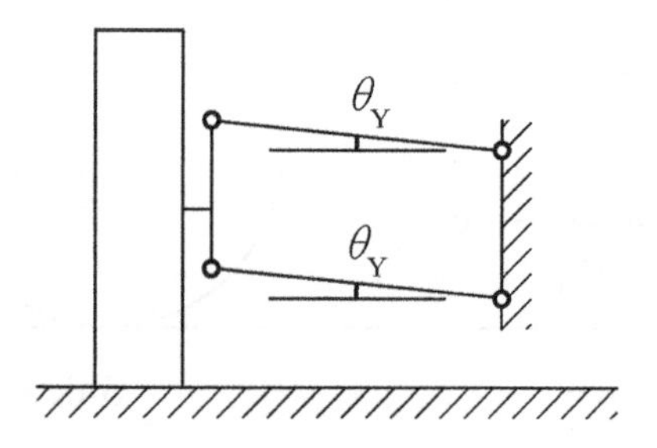

Figure 12.5.1 Angled parallel equal-length arms.

pivot axis, so increasing positively with suspension bump. With the non-zero static arm angle, the initial bump scrub rate is also no longer zero, and becomes

$$\varepsilon_{\mathrm{BScd0,Y}} = -\tan\theta_{\mathrm{Y}} \approx -\theta_{\mathrm{Y}}$$

The bump scrub rate variation is as before, depending on the lateral arm length.

12.6 Converging Arms

If the arms are at different angles in the static position, as was shown in Figure 12.3.1(a), having values θ_{YU} and θ_{YL} for the upper and lower arms respectively, then the arms converge, or diverge. The arm angle difference is

$$\theta_{\mathrm{YD}} = \theta_{\mathrm{YU}} - \theta_{\mathrm{YL}}$$

Usually this convergence is towards the body side of the wheel, giving positive θ_{YD}, but this is not absolutely invariable. The camber now changes in bump. There is a swing centre no longer at infinity. Consider the simple symmetrical case in which the average angle of the arms is zero, such that the upper arm is inclined at $\theta_{\mathrm{YU}} = +\frac{1}{2}\theta_{\mathrm{YD}}$ and the lower arm at $\theta_{\mathrm{YU}} = -\frac{1}{2}\theta_{\mathrm{YD}}$, Figure 12.6.1(a). The arms are of equal length L_{Y}. The height difference of the outer ball joints is H_{BJD}.

The divergence angle is given, approximately, in radians, by

$$\theta_{\mathrm{YD}} = \frac{H_{\mathrm{BJD}}}{R_{\mathrm{S}}}$$

so the swing arm radius is approximately

$$R_{\mathrm{S}} = \frac{H_{\mathrm{BJD}}}{\theta_{\mathrm{YD}}}$$

Figure 12.6.1(b) shows the lateral movement of the ball joints for a bump z_{S}. The resulting bump camber angle is

$$\gamma_{\mathrm{B}} = -\frac{2z_{\mathrm{S}}\tan\frac{1}{2}\theta_{\mathrm{YD}}}{H_{\mathrm{BJD}}} \approx -\frac{z_{\mathrm{S}}\,\theta_{\mathrm{YD}}}{H_{\mathrm{BJD}}}$$

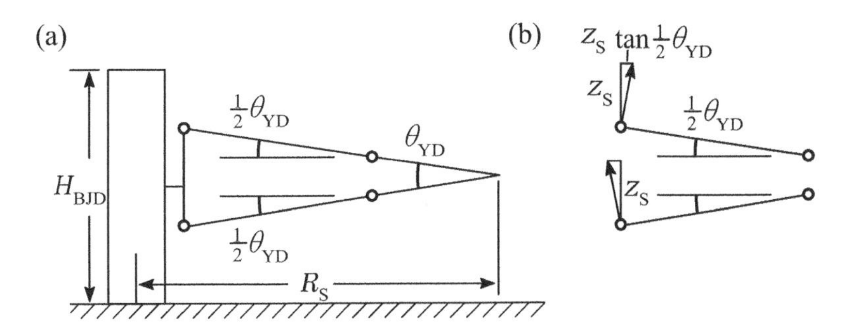

Figure 12.6.1 Converging arms.

Substituting for θ_{YD}, then

$$\gamma_B = -\frac{z_S}{R_S}$$

and the consequent linear bump camber coefficient is

$$\varepsilon_{BC1} = -\frac{1}{R_S} \approx -\frac{\theta_{YD}}{H_{BJD}}$$

Therefore the linear bump camber coefficient may be adjusted by control of the convergence angle of the arms, or, amounting to the same thing, by the length of the swing arm. If the convergence angle is negative, then the arms converge outwards, the swing arm radius is negative, and the linear bump camber coefficient is positive.

For unequal arm angles from horizontal, and small angles, continue to take the difference of the arm angles.

In Figure 12.6.1(a), it is apparent that converging the arms will also affect the bump scrub rate (roll centre height), but subsequently this may be changed independently as desired by adding the same slope to both arms.

12.7 Arm Length Difference

To examine the basic effect of arm length difference, consider the simple case of horizontal arms, as in Figure 12.7.1. It is usual for the upper arm to be the shorter one. Figure 12.7.1(b) shows the bumped position, with associated lateral movement of the ball joints because of the arm inclinations. Then

$$y_U = L_{YU}(1 - \cos\theta_{YU})$$

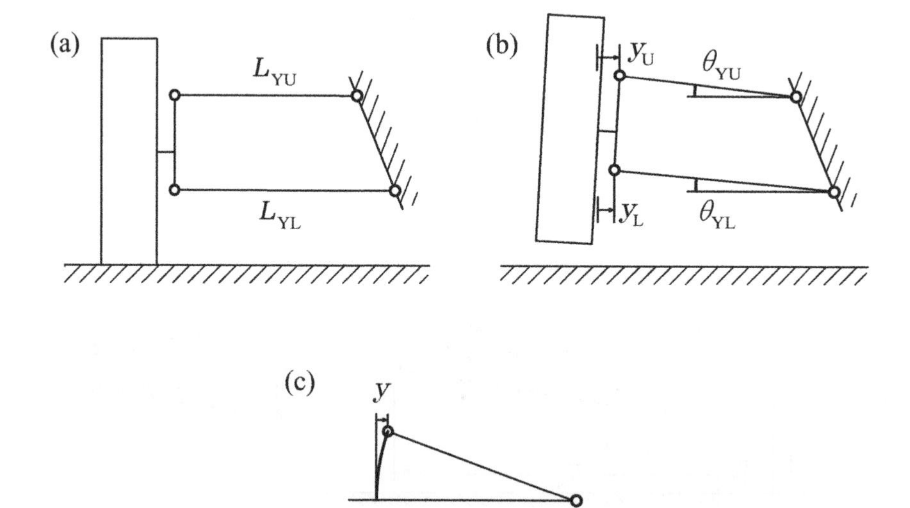

Figure 12.7.1 Double transverse arms with arm length difference.

Approximating the arm angle by

$$\theta_{YU} = \frac{z_S}{L_{YU}}$$

and using

$$\cos\theta_{YU} \approx 1 - \tfrac{1}{2}\theta_{YU}^2 = 1 - \frac{z_S^2}{2L_{YU}^2}$$

then

$$y_U = \frac{z_S^2}{2L_{YU}} \quad \text{and} \quad y_L = \frac{z_S^2}{2L_{YL}}$$

Consequently, the bump camber angle is

$$\begin{aligned}\gamma_B &= -\frac{y_U - y_L}{H_{BJD}} \\ &= -\frac{1}{2H_{BJD}}\left(\frac{1}{L_{YU}} - \frac{1}{L_{YL}}\right)z_S^2\end{aligned}$$

In terms of the arm shortness $S = 1/L$, this is

$$\gamma_B = -\frac{1}{2H_{BJD}}(S_{YU} - S_{YL})z_S^2$$

Using the arm shortness difference

$$S_{YD} = S_{YU} - S_{YL}$$

the bump camber angle is simply expressed as

$$\gamma_B = -\frac{S_{YD}}{2H_{BJD}}z_S^2$$

Hence, the arm length difference causes a quadratic bump camber coefficient

$$\varepsilon_{BC2} = -\frac{S_{YD}}{2H_{BJD}}$$

and the arm length difference may be used to control this aspect of the geometry as required. A negative quadratic bump camber coefficient is usually desired, in which case S_{YD} must be positive, requiring the upper arm to be the shorter one.

12.8 General Solution

The foregoing equations are a useful guide to the effect of various geometrical changes, but with arm convergence the bump scrub effects are not handled accurately enough for design purposes, so an improved analysis will now be given here. However, the resulting equations are more complex, and the

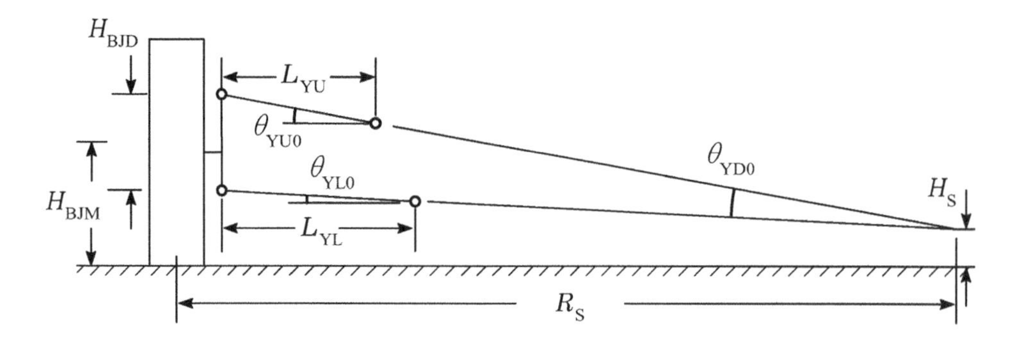

Figure 12.8.1 Double transverse arms, general case, front view of right wheel or rear view of left wheel.

effects of design changes are not as clear as in the previous very simple equations. Figure 12.8.1 shows the geometry, which now simultaneously includes different arm lengths, different arm angles, the ball joint static vertical spacing H_{BJD}, and the mean static ball joint height H_{BJM}.

The individual static-position heights of the upper and lower outer ball joints are H_{BJU} and H_{BJL}. Table 12.8.1 summarises the definitions and relationships for related variables. The values of H_{BJU}, H_{BJL}, H_{BJS}, H_{BJD}, H_{BJM} and f_{BJH} are constant for a given design. Their use simplifies subsequent equations.

The front-view equivalent link lengths are L_{YU} and L_{YL}. For convenience in the following equations the link shortnesses are

$$S_{YU} = \frac{1}{L_{YU}}; \qquad S_{YL} = \frac{1}{L_{YL}}$$

Table 12.8.1 Outer ball joint height relationships

The sum and difference are

$$H_{BJS} = H_{BJU} + H_{BJL}$$
$$H_{BJD} = H_{BJU} - H_{BJL}$$

The individual ball joint heights are

$$H_{BJU} = \tfrac{1}{2}(H_{BJS} + H_{BJD})$$
$$H_{BJL} = \tfrac{1}{2}(H_{BJS} - H_{BJD})$$

The mean outer ball joint static height is

$$H_{BJM} = \tfrac{1}{2}H_{BJS}$$

The ball joint height factor is the ratio of the sum to the difference:

$$f_{BJH} = \frac{H_{BJS}}{H_{BJD}}$$

which, more explicitly, is

$$f_{BJH} = \frac{H_{BJU} + H_{BJL}}{H_{BJU} - H_{BJL}}$$

Note that f_{BJH} is *not* the ratio of the height of the individual ball joints.

Table 12.8.2 Arm angle relationships

The static arm angle sum and difference are

$$\theta_{YS0} = \theta_{YU0} + \theta_{YL0}$$
$$\theta_{YD0} = \theta_{YU0} - \theta_{YL0}$$

At suspension bump position z_S, the link angles become

$$\theta_{YU} = \theta_{YU0} + S_{YU} z_S$$
$$\theta_{YL} = \theta_{YL0} + S_{YL} z_S$$

The sum and difference of these angles are

$$\theta_{YS} = \theta_{YU} + \theta_{YL}$$
$$\theta_{YD} = \theta_{YU} - \theta_{YL}$$

The individual angles are

$$\theta_{YU} = \tfrac{1}{2}(\theta_{YS} - \theta_{YD})$$
$$\theta_{YL} = \tfrac{1}{2}(\theta_{YS} - \theta_{YD})$$

By substitution, the sum and difference are

$$\theta_{YS} = \theta_{YS0} + S_{YS} z_S$$
$$\theta_{YD} = \theta_{YD0} + S_{YD} z_S$$

The shortness sum and difference are

$$S_{YS} = S_{YU} + S_{YL}$$
$$S_{YD} = S_{YU} - S_{YL}$$

The initial (static position) link angles are θ_{YU0} and θ_{YL0}, positive as shown in the figure, clockwise. These arm angles are defined positive such that they increase in bump. Table 12.8.2 summarises useful arm angle equations.

In Figure 12.8.1, the long triangle gives, approximately,

$$H_{BJD} = R_S \theta_{YD}$$

so the swing arm radius is

$$R_S = \frac{H_{BJD}}{\theta_{YD}} = \frac{H_{BJD}}{\theta_{YD0} + S_{YD} z_S}$$

The linear bump camber coefficient is

$$\varepsilon_{BC1} = -\frac{1}{R_{S0}} = -\frac{\theta_{YD0}}{H_{BJD}}$$

As before, the quadratic bump camber coefficient is

$$\varepsilon_{BC2} = -\frac{S_{YD}}{2H_{BJD}}$$

Table 12.8.3 Double transverse arms, summary of two-dimensional lateral equations

$\varepsilon_{\mathrm{BScd0,Y}} = \frac{1}{2}(f_{\mathrm{BJH}}\,\theta_{\mathrm{YD0}} - \theta_{\mathrm{YS0}})$	(12.8.1)
$\varepsilon_{\mathrm{BScd1,Y}} = \frac{1}{2}(f_{\mathrm{BJH}}S_{\mathrm{YD}} - S_{\mathrm{YS}})$	(12.8.2)
$\varepsilon_{\mathrm{BC1}} = -\dfrac{1}{R_{\mathrm{S}}} = -\dfrac{\theta_{\mathrm{YD0}}}{H_{\mathrm{BJD}}}$	(12.8.3)
$\varepsilon_{\mathrm{BC2}} = -\dfrac{S_{\mathrm{YD}}}{2H_{\mathrm{BJD}}}$	(12.8.4)

The height of the swing centre is

$$H_{\mathrm{S}} = H_{\mathrm{BJM}} - \tfrac{1}{2}\,\theta_{\mathrm{YS}}R_{\mathrm{S}} = H_{\mathrm{BJM}} - \tfrac{1}{2}H_{\mathrm{BJD}}\frac{\theta_{\mathrm{YS}}}{\theta_{\mathrm{YD}}}$$

This gives

$$H_{\mathrm{S}} = H_{\mathrm{BJM}} - \frac{1}{2}H_{\mathrm{BJD}}\left(\frac{\theta_{\mathrm{YS0}} + S_{\mathrm{YS}}z_{\mathrm{S}}}{\theta_{\mathrm{YD0}} + S_{\mathrm{YD}}z_{\mathrm{S}}}\right)$$

The local lateral bump scrub rate (i.e. at the local bump position z_{S}) is

$$\begin{aligned}
\varepsilon_{\mathrm{BScd,Y}} &= \frac{H_{\mathrm{S}}}{R_{\mathrm{S}}} = \frac{H_{\mathrm{BJM}}}{R_{\mathrm{S}}} - \frac{1}{2}\,\theta_{\mathrm{YS}} \\
&= \tfrac{1}{2}f_{\mathrm{BJH}}(\theta_{\mathrm{YD0}} + S_{\mathrm{YD}}z_{\mathrm{S}}) - \tfrac{1}{2}(\theta_{\mathrm{YS0}} + S_{\mathrm{YS}}z_{\mathrm{S}}) \\
&= \left\{\tfrac{1}{2}f_{\mathrm{BJH}}\,\theta_{\mathrm{YD0}} - \tfrac{1}{2}\theta_{\mathrm{YS0}}\right\} + \left\{\tfrac{1}{2}f_{\mathrm{BJH}}S_{\mathrm{YD}} - \tfrac{1}{2}S_{\mathrm{YS}}\right\}z_{\mathrm{S}}
\end{aligned}$$

However, the bump scrub rate is

$$\varepsilon_{\mathrm{BScd,Y}} = \varepsilon_{\mathrm{BScd0,Y}} + \varepsilon_{\mathrm{BScd1,Y}}z_{\mathrm{S}}$$

so the two bump scrub rate coefficients are solved, as summarised with earlier results in Table 12.8.3.

12.9 Design Process

Given a design specification for the coefficients, the equations of Table 12.8.3 must be solved for the front-view geometry. One possible design sequence is:

(1) Use $\varepsilon_{\mathrm{BC1}}$ to give θ_{YD0};
(2) Use $\varepsilon_{\mathrm{BScd0,Y}}$ to give θ_{YS0};
(3) Use $\varepsilon_{\mathrm{BC2}}$ to give S_{YD};
(4) Use $\varepsilon_{\mathrm{BScd1,Y}}$ to give S_{YS}.

In more detail:

(12.8.3): $$\theta_{\mathrm{YD0}} = -H_{\mathrm{BJD}}\varepsilon_{\mathrm{BC1}}$$

(12.8.1): $$\begin{aligned}
\theta_{\mathrm{YS0}} &= f_{\mathrm{BJH}}\,\theta_{\mathrm{YD0}} - 2\varepsilon_{\mathrm{BScd0,Y}} = -2(H_{\mathrm{BJM}}\varepsilon_{\mathrm{BC1}} + \varepsilon_{\mathrm{BScd0,Y}}) \\
\theta_{\mathrm{YU0}} &= \tfrac{1}{2}(\theta_{\mathrm{YS0}} + \theta_{\mathrm{YD0}}) \\
\theta_{\mathrm{YL0}} &= \tfrac{1}{2}(\theta_{\mathrm{YS0}} - \theta_{\mathrm{YD0}})
\end{aligned}$$

which completes solution of the arm angles.

The link lengths require the other two equations from Table 12.8.3:

(12.8.4): $S_{YD} = -2H_{BJD}\varepsilon_{BC2}$

(12.8.2): $S_{YS} = f_{BJH}S_{YD} - 2\varepsilon_{BScd1,Y} = -2(H_{BJS}\varepsilon_{BC2} + \varepsilon_{BScd1,Y})$

The actual arm shortnesses and lengths are then

$$S_{YU} = \tfrac{1}{2}(S_{YS} + S_{YD})$$

$$S_{YL} = \tfrac{1}{2}(S_{YS} - S_{YD})$$

and the actual arm lengths are just the reciprocals of the shortnesses. This completes the solution of the equivalent link lengths in front view.

Table 12.9.1 Double-wishbone design example

Double Transverse Arm Analysis - Analytic:

Data	Design Requirement:	Arm angles	Arm lengths
HBJU = 0.39000 m	eBScd0 = 0.10000 - - -	thYD = 7.21927 deg	SYS = 8.60000 -/m
HBJL = 0.21000 m	eBScd1 = -2.50000 -/m	thYS = 12.60507 deg	SYD = 1.08000 -/m
HBJS = 0.60000 m	eBC1 = -0.70000 rad/m	thYU = 9.91217 deg	LYU = 0.20661 m
HBJM = 0.30000 m	eBC2 = -3.00000 rad/m2	thYL = 2.69290 deg	LYL = 0.26596 m
HBJD = 0.18000 m			
fBJH = 3.33333 - - -			

Table 12.9.1 gives example numerical values, in which, from the specified coefficients, the equations of this section are used to obtain the geometry. The equations of Table 12.8.3 can then be used to obtain the coefficients of the proposed geometry, as a check, to compare the specified coefficients with the calculated ones for the derived design. This check shows that the equations are consistent. However, it does not prove them physically accurate, because of the modelling approximations, for example, the neglect of the lateral position of the ball joints, and small-angle approximations. Nevertheless, these equations are useful for design purposes, and the design may subsequently be refined by accurate numerical computer analysis in three dimensions.

12.10 Numerical Solution in Two Dimensions

Given a two-dimensional equivalent front-view linkage mechanism, as in Figure 12.10.1, there are several possible solution methods for a precise solution. In general, there is no explicit solution for a specified bump of the wheel-to-road contact point E. Therefore the practical approach is to move one arm (e.g. the lower arm AB) by an angle, and solve all points from that, including the suspension bump $z_S = z_E$ and scrub $s = \Delta y_E$. Using a suitable range of arm angles, quadratic curves can be fitted to $\gamma(z_S)$ and $s(z_S)$ to obtain the bump camber and bump scrub coefficients, and thence the bump scrub rate coefficients.

The actual geometrical solution may proceed in several ways. The simple length-and-angle approach begins by an initial position analysis to obtain the static position lengths and angles, including θ_{AB0}, θ_{BD0}, and θ_{EBF} (which is unchanging). Also needed are the unchanging lengths L_{AB}, L_{BD}, L_{CD} and L_{EB}. The lower arm is displaced by angle ϕ_{AB}. The procedure for a given displacement is given in Table 12.10.1.

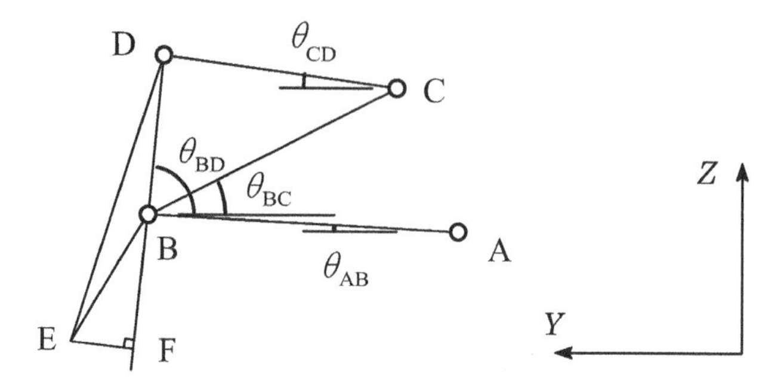

Figure 12.10.1 Double transverse arm, equivalent linkage mechanism for numerical solution.

Solving for at least three positions, over a suitable range of lower arm angle to give a realistic suspension bump range, fit quadratic curves to $\gamma(z_S)$ and $s(z_S)$ giving the bump camber coefficients ε_{BC1} and ε_{BC2}, and the lateral bump scrub coefficients $\varepsilon_{BSc1,Y}$ and $\varepsilon_{BSc2,Y}$. Then, also, the bump scrub rate coefficients are derived from the bump scrub coefficients by

$$\varepsilon_{BScd0,Y} = \varepsilon_{BSc1,Y}$$
$$\varepsilon_{BScd1,Y} = 2\varepsilon_{BSc2,Y}$$

as explained in Section 7.9.

Other solution methods are possible, and may be preferable, particularly if a good library of two-dimensional geometry subroutines is available.

Given points B and C, and lengths L_{BD} and L_{CD}, then upper the ball joint point D may be found by the intersection of two circles. If two circles do intersect then there are, in general, two intersections, and the solution of this uses a quadratic equation, producing two solutions, so the correct one must be selected.

In the static analysis, a point F, the foot of a perpendicular from point E onto DB extended, may be formed, giving lengths L_{BF} and L_{FE}. In the displaced position, point F follows easily from DB produced by a simple factor, and FE may be stepped along the perpendicular.

Table 12.10.1 Numerical solution of double transverse arms, two-dimensional front view for one position

(1) Specify lower arm angular displacement ϕ_{AB}.
(2) $\theta_{AB} = \theta_{AB0} + \phi_{AB}$.
(3) Calculate y_B and z_B from y_A, z_A, θ_{AB} and L_{AB}.
(4) Calculate L_{BC}.
(5) Obtain angle θ_{BC} using atan.
(6) Use cosine rule for angle θ_{CBD}, in triangle CBD, all sides known,
(7) $\theta_{BD} = \theta_{BC} + \theta_{CBD}$.
(8) The bump camber angle is $\gamma_B = \theta_{BD} - \theta_{BD0}$.
(9) The angular position of EB is $\theta_{EB} = \theta_{EB0} + \gamma_B$.
(10) Obtain y_E and z_E from y_B, z_B, θ_{EB} and L_{EB}.
(11) Suspension bump is $z_S = z_E$.
(12) Suspension scrub is $s = y_E - y_{E0}$.

Table 12.10.2 Comparison of coefficients obtained

```
Data:
                                Y            Z

Upper arm outer BJ  =       0.6100       0.4200  m
Lower arm outer BJ  =       0.6300       0.1800  m

Upper arm length    =      0.2000 m
Lower arm length    =      0.3000 m
Upper arm angle     =      8.0000 deg
Lower arm angle     =      1.0000 deg

HBJS  =  0.6000 m
HBJM  =  0.3000 m
HBJD  =  0.2400 m
fBJH  =  2.5000 - - -

Coeff     Analytic     Numerical

eBScd0     0.0742       0.0713       - - -
eBScd1    -2.0833      -1.9830       -/m
eBC1      -0.5091      -0.4896       rad/m
eBC2      -3.4722      -3.1541       rad/m2
```

These alternative methods are not particularly superior in two dimensions, but that style of approach is good in the more complex three-dimensional analysis. All the numerical methods are mathematically very accurate, if implemented correctly, but there may still be approximations in the conversion to an equivalent two-dimensional problem.

Table 12.10.2 compares the results obtained by the algebraic and numerical methods. The average discrepancy in this case is only about 6%. It may be concluded that the algebraic method is useful, but not precise because of the modelling assumptions, for example neglect of the ball joint lateral position from the wheel centre plane, a detail which is easily included in the numerical model. The algebraic model could be made more accurate, of course, but then the clarity of the equations would be reduced.

12.11 Pitch

As in the case of the lateral properties, which can be analysed approximately in two dimensions using an equivalent front view, the pitch properties, namely bump scrub (anti-dive, etc.) and bump caster variation, can be analysed approximately in a two-dimensional equivalent side view, as in Figure 12.11.1. The rear effective pivot points of the arms are the points at which the pivot axes penetrate the longitudinal vertical plane of the wheel centre.

The mathematical analysis of the side view follows very much along the lines of the front-view analysis, but with changes of notation and some changes of sign. The variables shown are the ball joint height difference H_{BJD}, the mean ball joint height H_{BJM}, the upper arm equivalent longitudinal length L_{XU} and angle θ_{XU}, the latter positive as shown in the figure and increasing in bump, and the lower arm length L_{XL} and angle θ_{XL}.

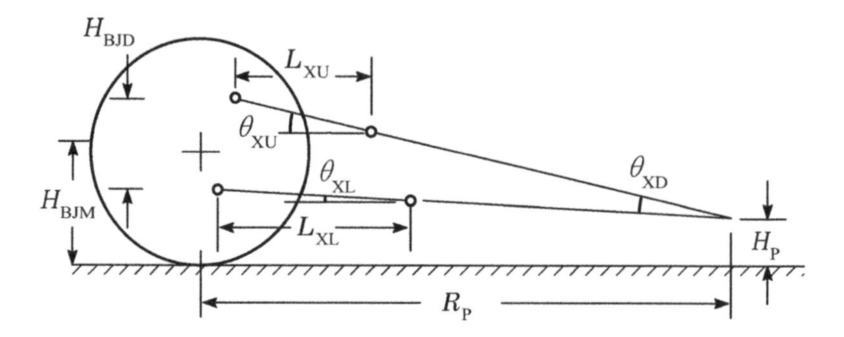

Figure 12.11.1 Equivalent two-dimensional side view of double transverse arms.

The steer axis caster angle is positive when the steer axis slopes backwards as it rises, which is the opposite of the camber angle when comparing Figures 12.8.1 and 12.11.1. The effective arm length is now the pitch arm radius R_P instead of the swing arm L_S. The linear bump caster coefficient becomes

$$\varepsilon_{Bcas1} = \frac{1}{R_P} = \frac{\theta_{XD0}}{H_{BJD}}$$

which has different signs from the equivalent expression for the transverse variables. The second bump caster coefficient also loses the minus sign:

$$\varepsilon_{Bcas2} = \frac{S_{XD}}{2H_{BJD}}$$

The resulting analysis equations are summarised in Table 12.11.1, which has sign changes in the last two equations, in which ε_{Bcas} appears instead of ε_{BC}.

The pitch design problem begins with a specification of the desired longitudinal coefficients, with use of the equations of Table 12.11.1 to solve the geometry. Again, this is substantially the same as for the transverse analysis:

(12.11.3): $\theta_{XD0} = H_{BJD}\varepsilon_{Bcas1}$

(12.11.1): $\theta_{XS0} = f_{BJH}\,\theta_{XD0} - 2\varepsilon_{BScd0,X} = 2\left(H_{BJM}\varepsilon_{Bcas1} - \varepsilon_{BScd0,X}\right)$

with

$$\theta_{XU0} = \tfrac{1}{2}(\theta_{XS0} + \theta_{XD0})$$

$$\theta_{XL0} = \tfrac{1}{2}(\theta_{XS0} - \theta_{XD0})$$

The two-dimensional effective link lengths are solved by

(12.11.4): $S_{XD} = 2H_{BJD}\varepsilon_{Bcas2}$

(12.11.2): $S_{XS} = f_{BJH}S_{XD} - 2\varepsilon_{BScd1,X}$

with

$$S_{XU} = \tfrac{1}{2}(S_{XS} + S_{XD})$$

$$S_{XL} = \tfrac{1}{2}(S_{XS} - S_{XD})$$

and the lengths follow immediately as reciprocals.

Table 12.11.1 Double transverse arms, summary of two-dimensional longitudinal equations

$\varepsilon_{\mathrm{BScd0,X}} = \frac{1}{2}(f_{\mathrm{BJH}}\,\theta_{\mathrm{XD0}} - \theta_{\mathrm{XS0}})$	(12.11.1)
$\varepsilon_{\mathrm{BScd1,X}} = \frac{1}{2}(f_{\mathrm{BJH}} S_{\mathrm{XD}} - S_{\mathrm{XS}})$	(12.11.2)
$\varepsilon_{\mathrm{Bcas1}} = \dfrac{1}{R_{\mathrm{P}}} = \dfrac{\theta_{\mathrm{XD0}}}{H_{\mathrm{BJD}}}$	(12.11.3)
$\varepsilon_{\mathrm{Bcas2}} = \dfrac{S_{\mathrm{XD}}}{2H_{\mathrm{BJD}}}$	(12.11.4)

In a practical design it may be that the secondary coefficients are set to zero,

$$\varepsilon_{\mathrm{BScd1,X}} = 0$$
$$\varepsilon_{\mathrm{Bcas2}} = 0$$

in which case

$$S_{\mathrm{XD}} = 0; \qquad S_{\mathrm{XS}} = 0$$
$$L_{\mathrm{XU}} = \infty; \qquad L_{\mathrm{XL}} = \infty$$

The infinite lengths in side view mean that the pivot axes are parallel to the vehicle centre plane. The angles are still evaluated as above, and the axes will generally be inclined to the horizontal.

As an alternative to the analytical solution in pitch, a two-dimensional numerical solution is a possibility, Figure 12.11.2, similar in general terms to the front view. The lower arm is rotated from the static position and the consequent caster angle, longitudinal scrub and suspension bump are solved and curve fitted for coefficients. The difference is that the ground contact point is the point initially directly below the wheel centre, and for longitudinal scrub analysis the wheel must be treated as locked in rotation. The angle FBE will normally be positive as shown, in the other direction from in the front view, and the perpendicular EF is also probably in the other direction. The suspension bump evaluation method is to determine the wheel centre bump and to subtract the wheel radius. This illustrates a limitation of the two-dimensional modelling, as the side-view bump calculation does not allow for variations of camber angle. In general, the solution proceeds by the same principles, with the same possible methods.

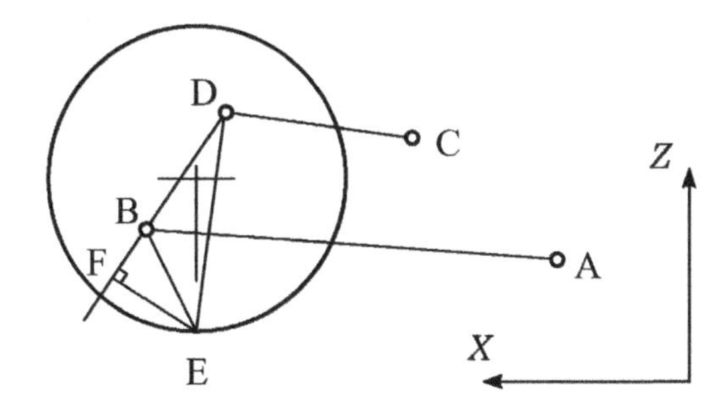

Figure 12.11.2 Side view of double transverse arms for two-dimensional numerical analysis.

In the case of front anti-lift traction analysis, or anti-dive braking with driveshafts transmitting the torque, as for inboard brakes, then the effective longitudinal bump scrub of the wheel centre must be investigated, using similar methods.

12.12 Numerical Solution in Three Dimensions

The two-dimensional solutions are generally quite good, but not strictly correct, and ideally a three-dimensional solution should be used. For the general case, with pivot axes at various angles, algebraic analysis is hardly practical other than after reduction to approximate two-dimensional equivalents, so computer numerical analysis is the practical accurate solution.

As the two-dimensional analyses show, each suspension arm will have values for the lengths and initial angles in front and side views (i.e. L_{XL}, θ_{XL}, L_{XU}, θ_{XU}). Given the ball joint position, the four variables dictate the pivot points of the arm as seen in front-view and side-view planes, so the arm axis in general runs somewhat diagonally to give both equivalent arms a finite length. The arm geometry is defined by the ball joint coordinates and by the arm pivot axis. The physical arm itself may vary in shape and construction, possibly being bent. Also, the part of the pivot axis met by the arm may vary considerably, to meet packaging requirements or to obtain certain compliance characteristics.

The method of solution of the three-dimensional problem is given in Table 12.12.1. If the suspension position is required for some predetermined suspension bump then iteration must be used to find the corresponding arm angle.

In addition to the analysis of Table 12.12.1, it may be desired to make further analysis of connected components (e.g. pushrods, rockers, spring length, or damper length) using similar techniques.

Exploring the steps of Table 12.12.1 in more detail, rotation of the lower arm requires a local coordinate system with one axis along the arm pivot axis, one axis radially outwards to the initial ball joint position, and one axis perpendicular to those two, found by vector cross products, this bringing the arc of the ball joint into a plane with two of the axes, allowing a two-dimensional calculation of the rotation with subsequent conversion into the basic body axes. An alternative approach is to specify a height for the lower ball joint above the datum plane, from which the complete lower ball joint coordinates can be solved.

Solution of the upper ball joint involves satisfying the arc of the upper ball joint about the upper arm pivot axis and the spacing between the wheel carrier ball joints. The best approach is to use equations satisfying the three known lengths, two from known points on the upper arm pivot axis, one from the known position of the lower ball joint, selecting one of the two resulting solutions. This 'tripod' problem is a useful general purpose subroutine.

Table 12.12.1 Three-dimensional numerical solution of double transverse arms

(1) Rotate the lower arm about its axis, giving the lower ball joint position.
(2) Solve for the upper ball joint according to the upper arm and the wheel carrier.
(3) Solve for the track-rod end balljoint according to the steering arm, the rack end position and track-rod length.
(4) Solve for the wheel centre and suspension bump.
(5) Solve for the locked-wheel contact point scrub.
(6) From the wheel carrier position, solve the wheel axis direction cosines.
(7) Deduce the camber angle, caster angle, steer angle, lateral scrub, longitudinal scrub, and suspension bump.
(8) Do several positions through the suspension range, and curve fit quadratics to give the relevant coefficients.

Solution of the track-rod end/steering rod ball joint is another tripod problem, with known points at the wheel carrier ball joints and at the rack end ball joint, with three known lengths.

Next, the position of the wheel centre and locked-wheel contact patch points must be found. These could be treated as another tripod problem, based on the now known two-steer-axis ball joints and the steering-arm end ball joint, and three known unchanging lengths. However, certainly in the case of the contact point, it is possible that the point to be found could fall exactly in, or very close to, the plane of the known points, in which case the tripod method is likely to be troublesome. Instead, a better method is to perform an initial analysis dropping a perpendicular from the contact point into the plane of the known three base points, giving a foot point F and an out-of-plane perpendicular length. The coordinates of F are a linear function of the coordinates of the three plane-defining points, with factors that are unchanging and determined in the initial analysis. When the wheel upright has been moved, the new point F follows easily, and the contact point can be stepped out on the new perpendicular to the plane.

From the wheel centre, the lowest point of the wheel may now be deduced, giving the suspension bump.

Solution of the wheel axis requires solution of the new position of two points on that axis. The initial analysis will prepare the information for this. This can be approached as two tripod problems, or, better, one of the points can be that on the axis and in the plane of the known three points. The three fixed lengths to one point for a tripod solution are unchanging and determined at the initial analysis.

As so often with computer programming, the devil is in the detail. Here are described the general approach and principles. Chapter 15 describes some useful three-dimensional geometrical routines.

12.13 Steering

The properties of the steering have been described in general terms in Chapters 5 and 6. For a double-arm suspension, the resulting bump steer can be expressed in terms of the linear and quadratic coefficients as explained.

In two dimensions, it is also possible to actually calculate the ideal point coordinates for the rack end. In three dimensions, it is possible to calculate an ideal axis for the rack end. The suspension is moved in bump, producing three steering arm end positions, P_1, P_2 and P_3, with lower arm angles corresponding to suspension bump positions approximately -0.10 m, static, and $+0.1$ m. The wheel steer angle is held at zero (treating the rack as disconnected), and the three-dimensional coordinates of the steering arm end ball joint are determined for each position.

These three steering arm end points define a plane. The ideal axis is perpendicular to this plane, so the normal to the plane gives the direction cosines of the ideal axis. Now set up a coordinate system (u, v, w) with origin at P_2, Figure 12.13.1. Axis w runs normal to the plane. Axis u runs in the plane from P_2 towards P_1. The third axis v is in plane, perpendicular to the first two. This gives two axes in the plane of the three points. The ideal point is within the plane, being the centre of the arc through the three points, and so is easily solved in two-dimensional geometry. Convert the known point coordinates of P_1, P_2 and P_3 to (u, v) coordinates. The centre must lie on the perpendicular bisector of a chord, so

$$u_C = \tfrac{1}{2}u_1$$

The radii P_2C and P_3C are the same, so, simply by Pythagoras' theorem,

$$R^2 = u_C^2 + v_C^2 = (u_C - u_3)^2 + (v_C - v_3)^2$$

so

$$v_C = \frac{u_3^2 + v_3^2 - 2u_3u_C}{2v_3}$$

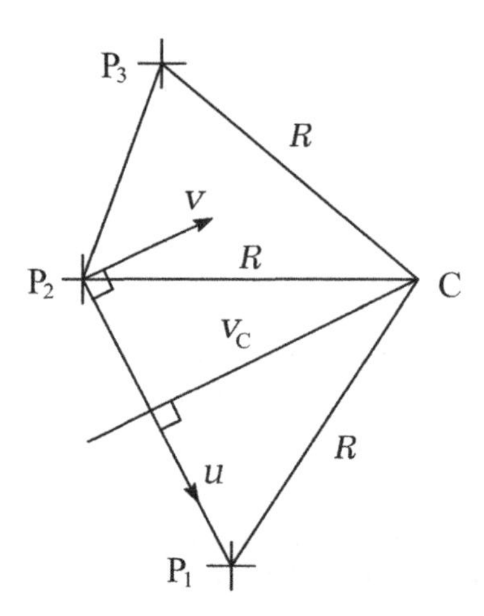

Figure 12.13.1 Determination of the ideal point in the plane of the three bumped steering arm end points.

and the arc radius R, the ideal track-rod length for a rack end at the ideal centre, then follows from the previous equation.

The coordinates of the ideal centre can then be converted to vehicle body coordinates. This point then, with the axis direction cosines, defines the ideal rack-end axis.

In the above investigation the arc of the steering arm end is not strictly in a perfect plane. Therefore, it may be preferred to obtain track-rod coordinates for more bump positions, and to fit a plane by statistical error minimisation techniques.

Also, if the two arm pivot axes are not parallel, then the ideal axis is not really a straight line. In particular, this occurs with the crossed axes of the one-leading-arm one-transverse-arm suspension, shown in Figure 12.2.1(c). In that case it may be worthwhile to obtain the ideal rack end position directly for various rack longitudinal positions.

The purpose of these investigations is, of course, to see the implications for rack length and ideal rack vertical position if it is moved longitudinally to alter the Ackermann factor.

12.14 Strut Analysis in Two Dimensions

The strut-and-arm suspension, commonly just called a strut suspension, typically uses a basically transverse lower arm, with the upper arm and wheel carrier (wheel upright) replaced by an integrated unit of wheel carrier, slider and spring–damper unit, acting on an upper trunnion where it connects to the inner part of the bodywork. The upper joint is a really a combination of the slider with a ball joint, or rubber bush, the latter allowing various angular movements. The combination acts geometrically like a trunnion, which is a slider passing right through a swivel joint (in two dimensions) or ball joint (in three dimensions), as seen in Figure 12.14.1. The steering action normally acts about a steering axis passing through the lower ball joint and the trunnion point. However, a separate steering axis could be provided, as used to be done on some double-transverse-arm suspensions.

The two-dimensional analytical methods applied to the double-transverse-arm suspension are easily adapted to the common strut suspension. The one issue to be resolved is how to find the equivalent upper arm. The plane of the equivalent upper arm is through the trunnion point and perpendicular to the strut slider centreline. Intersecting this with the transverse vertical plane gives the equivalent arm, as usual, but with length yet to be determined. Note that the strut slider is generally not aligned with the ball joint on the

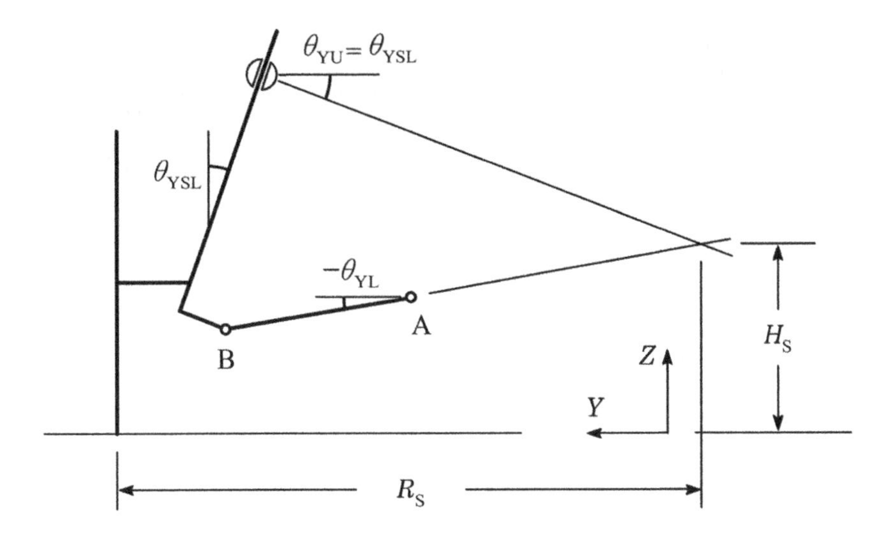

Figure 12.14.1 Strut-and-arm suspension, basic front-view geometry.

lower arm, so not coinciding with the steering axis which passes through the two joints, Figure 12.14.1. The lower arm angle is θ_{YL}, positive when the outer ball joint is higher than the axis, but usually negative in the static position.

The upper arm length can simply be taken effectively as infinite, that is, as having zero shortness. A slightly better equivalent is to take it to be equal to the swing arm length in front view and the pitch arm length in side view. In Figure 12.14.2, AB is the lower arm, E is the foot of the perpendicular from B onto the slider centreline, DE is the slider centreline, and BD is the steering axis through the lower ball joint B and trunnion centre D. Figure 12.14.2(b) shows the velocity diagram for a lower arm angular velocity Ω. The fixed points are A and C. The line ab has length value (units m/s)

$$\mathrm{ab} = \Omega L_{AB}$$

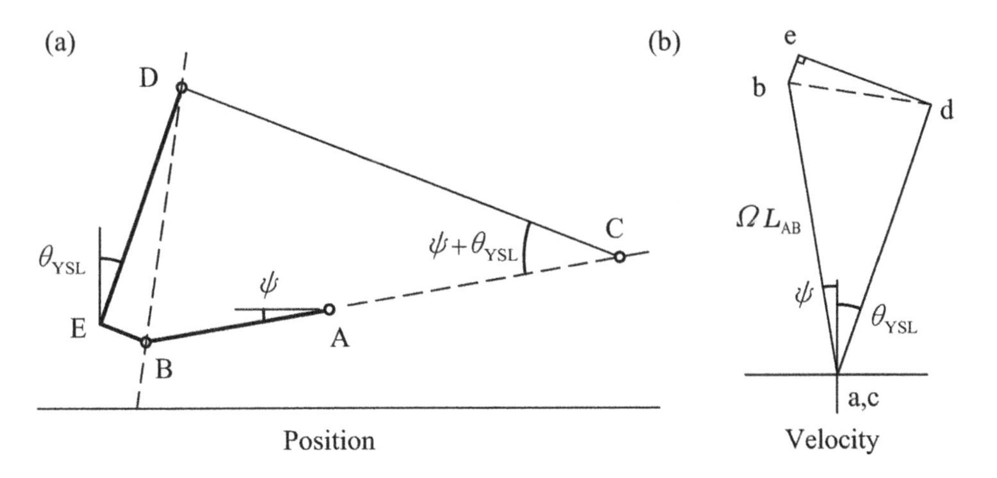

Figure 12.14.2 Basic strut-and-arm geometry for analysis of the effective upper arm length.

Table 12.14.1 Strut-and arm suspension, summary of two-dimensional lateral equations

$\varepsilon_{\mathrm{BScd0,Y}} = \frac{1}{2}(f_{\mathrm{BJH}}\,\theta_{\mathrm{YD0}} - \theta_{\mathrm{YS0}})$	(12.14.1)
$R_{\mathrm{S}} = \dfrac{H_{\mathrm{BJD}}}{\theta_{\mathrm{YD0}}}$	(12.14.2)
$\varepsilon_{\mathrm{BC1}} = -\dfrac{1}{R_{\mathrm{S}}}$	(12.14.3)
$S_{\mathrm{YU}} = \dfrac{1}{R_{\mathrm{S}}}$	(12.14.4)
$\varepsilon_{\mathrm{BScd1,Y}} = \frac{1}{2}(f_{\mathrm{BJH}}S_{\mathrm{YD}} - S_{\mathrm{YS}})$	(12.14.5)
$\varepsilon_{\mathrm{BC2}} = -\dfrac{S_{\mathrm{YD}}}{2H_{\mathrm{BJD}}}$	(12.14.6)

at the angle $\psi = -\theta_{\mathrm{YL}}$. The line cd is perpendicular to CD, and parallel to slider DE. Line bd is perpendicular to BD. Triangle cbd is similar to CBD. The radius of the movement arc of D, the velocity of D and the angular velocity of ED are related by

$$V_{\mathrm{D}} = \omega_{\mathrm{ED}} R_{\mathrm{D}}$$

The angular velocity is

$$\omega_{\mathrm{ED}} = \frac{\mathrm{ed}}{L_{\mathrm{ED}}}$$

The radius of the arc is therefore

$$R_{\mathrm{D}} = \frac{V_{\mathrm{D}}}{\omega_{\mathrm{ED}}} = \frac{\mathrm{cd}\, L_{\mathrm{ED}}}{\mathrm{ed}} = L_{\mathrm{CD}}$$

so the effective centre of the arc is at C, on the line BA, and the effective length of the upper arm is

$$L_{\mathrm{YU}} \approx R_{\mathrm{S}}$$

Adapting Table 12.8.3 with appropriate modifications for the upper arm gives Table 12.14.1 for the strut-and-arm suspension lateral properties.

Note that $S_{\mathrm{YD}} = S_{\mathrm{YU}} - S_{\mathrm{YL}}$ will inevitably be negative in practice, so $\varepsilon_{\mathrm{BC2}}$ will be positive. H_{BJD} is much larger than for a conventional double-wishbone suspension.

The longitudinal properties can be analysed in the same way, with similar adaptations to Table 12.11.1, giving Table 12.14.2.

Table 12.14.2 Strut-and-arm suspension, summary of two-dimensional longitudinal equations

$\varepsilon_{\mathrm{BScd0,X}} = \frac{1}{2}(f_{\mathrm{BJH}}\,\theta_{\mathrm{XD0}} - \theta_{\mathrm{XS0}})$	(12.14.7)
$R_{\mathrm{P}} = \dfrac{H_{\mathrm{BJD}}}{\theta_{\mathrm{XD0}}}$	(12.14.8)
$\varepsilon_{\mathrm{Bcas1}} = \dfrac{1}{R_{\mathrm{P}}}$	(12.14.9)
$S_{\mathrm{XU}} = \dfrac{1}{R_{\mathrm{P}}}$	(12.14.10)
$\varepsilon_{\mathrm{BScd1,X}} = \frac{1}{2}(f_{\mathrm{BJH}}S_{\mathrm{XD}} - S_{\mathrm{XS}})$	(12.14.11)
$\varepsilon_{\mathrm{Bcas2}} = \dfrac{S_{\mathrm{XD}}}{2H_{\mathrm{BJD}}}$	(12.14.12)

12.15 Strut Numerical Solution in Two Dimensions

A computer numerical analysis can be made of the strut-and-arm suspension in two dimensions, using a geometrically correct solution, without approximations for the upper arm. The technique required is different from the double-arm suspension. Figure 12.15.1 shows the geometry. AB is the lower arm, with pivot A and outer ball joint B, ED is the slider axis, at front-view slider initial angle θ_{YSL}, with E the foot of the perpendicular from B. Point F is the wheel contact point.

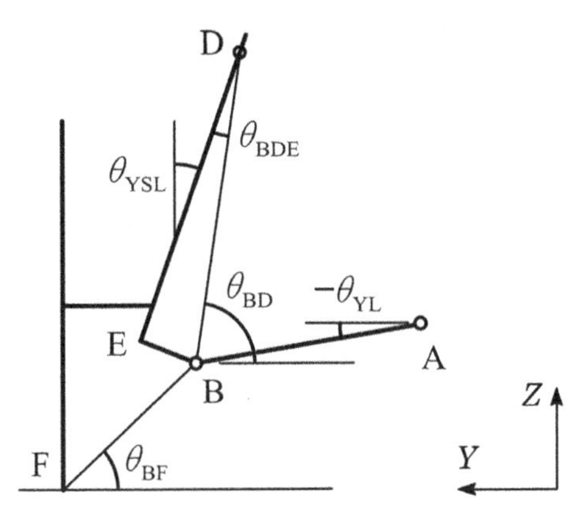

Figure 12.15.1 Strut-and-arm geometry for front-view two-dimensional numerical solution.

A preliminary analysis is made of some initial or unchanging lengths and angles, as in Table 12.15.1. Point B may be inboard or outboard of E, which must be accounted for, for example with a variable I_{BDE}, effectively representing a sign of angle θ_{BDE}. The solution for a given lower arm angle proceeds as in Table 12.15.2.

When implemented correctly, this is numerically very accurate, although still subject to the two-dimensional approximations from the real three-dimensional problem. The camber angle and scrub for a series of lower-arm angles can then be curve fitted against the bump to obtain the various coefficients. The results may be compared with the results of the analytical method, as in Table 12.15.3.

The comparison shows that the approximated analytic expressions are not very accurate, but do give a good qualitative estimate and understanding of the consequences of design changes.

Table 12.15.1 Strut-and-arm two-dimensional front view, numerical analysis initial calculations

(1) θ_{YL0} lower arm initial angle
(2) L_{AB} lower arm length
(3) (y_E, z_E), hence L_{BE}
(4) $\theta_{ED0} = \pi/2 - \theta_{YSL}$
(5) L_{FB} to the wheel contact point
(6) θ_{FB0} initial angle of FB
(7) $I_{BDE} = +1$ or -1 as BD steeper than ED or not
(8) Initial y_{F0} (half track)

Table 12.15.2 Strut-and-arm two-dimensional front view, numerical solution, one position

(1) Specify lower arm angular displacement ϕ_{AB}.
(2) $\theta_{AB} = \theta_{AB0} + \phi_{AB}$.
(3) Calculate (y_B,z_B) from y_A, z_A, L_{AB} and θ_{AB}.
(4) Obtain L_{BD} (D fixed).
(5) Obtain angle θ_{BD} from coordinates of B and D (use atan2).
(6) Obtain $\theta_{BDE} = \text{asin}\,(L_{BE}/L_{BD})$.
(7) $\theta_{ED} = \theta_{BD} - I_{BDE}\,\theta_{BDE}$.
(8) The bump camber angle is $\gamma_B = \theta_{ED} - \theta_{ED0}$.
(9) Obtain $\theta_{FB} = \theta_{FB0} + \gamma_B$.
(10) Contact point scrub is $y_F = y_B + L_{FB}\cos\theta_{FB} - y_{F0}$.
(11) Contact point bump is $z_F = z_B - L_{FB}\sin\theta_{FB}$.

Table 12.15.3 Strut-and-arm comparison of two-dimensional analytic and numerical solutions

```
Strut-and-Arm, summary comparison table:

Data:
                                      Y          Z

Trunnion centre D (m)               0.482      0.710
Lower arm outer  B (m)              0.606      0.196
Lower arm length =   0.280 m
Lower arm angle  =  -1.000 deg
Track            =   1.400 m
Slider angle     = 16.000 deg

Results:
                        Analytic        Analytic      Numerical
                        LU = LS       LU = infinity

eBC1      rad/m          -0.577          -0.577         -0.499
eBC2      rad/m2          2.913           3.474          2.075
eBScd0    ---             0.131           0.131          0.111
eBScd1    m-1            -4.713          -4.933         -4.162
```

12.16 Strut Design Process

Application of the two-dimensional analytical equations of Table 12.14.1 to design of a strut-and-arm suspension is very revealing. Whereas the double-arm suspension is very adaptable and can be specified to give almost any desired coefficients, the strut-and-arm suspension is very limited because of the long equivalent upper arm. The arm shortness difference

$$S_{YD} = S_{YU} - S_{YL}$$

is negative, so the quadratic bump camber coefficient

$$\varepsilon_{BC2} = -\frac{S_{YD}}{2H_{BJD}}$$

is inevitably positive, whereas negative values are favoured for double-arm suspensions. Also, control of the bump scrub rate variation is effectively lost, this, although being desirably negative, is much too large in magnitude, so control of the roll centre movement is poor. Even the linear bump camber coefficient

$$\varepsilon_{BC1} = -\frac{1}{R_S} = -\frac{\theta_{YD0}}{H_{BJD}}$$

may be a problem, because the ball joint height difference H_{BJD} is larger than for the double-arm suspension, so to achieve a desirable swing arm length of about 1.5 m requires a very large arm angle difference θ_{YD0}. This indicates need of a very steeply inclined slider, which is impractical for other reasons – for example, the excessive steering axis inclination angle is detrimental. This could be overcome by using a separate steering axis, but at extra cost. The only coefficient not really compromised is the linear bump scrub variation, so at least the initial roll centre height can be fixed correctly. However, the lower arm may be set at a steeper angle than is otherwise desirable to give a higher initial roll centre, and to reduce the swing arm length.

Geometrically, then, the strut-and-arm is a poor suspension configuration, and is used rather because it has a good motion ratio for the spring and damper, spreads the loads into the body well, and is economic in production.

Some manufacturers have sought to combine some of the benefits of both systems by using a double-arm system for the geometry, with the spring and damper acting, with pivotal connection, down onto the top of the upright or onto the end of the top arm (see Figure 1.6.8).

In sports and racing cars, the adaptability of the double-arm suspension is highly valued, and the geometric limitations of the strut-and-arm are a significant problem. In some cases conversions are made, often retaining the existing lower arm, using a subframe on each side, mounted at the lower arm and at the strut top, providing the mounting points for a new, upper, arm. This will cause problems with the steering unless the rack is repositioned and the length is corrected. This is not necessary in the special case of a rack which is located immediately in line with the lower arm.

12.17 Strut Numerical Solution in Three Dimensions

The strut-and-arm suspension can be analysed numerically in three dimensions by similar methods to the double-arm suspension, with some detail changes because of the slider trunnion (compare Table 12.17.1 with Table 12.12.1).

Table 12.17.1 Outline three-dimensional numerical solution of strut-and-arm suspension

(1) Rotate the lower arm about its axis, giving the lower ball joint position.
(2) Calculate the length between the lower ball joint and the trunnion centre.
(3) For the wheel carrier, using the unchanging distance of the lower ball joint from the slider axis, calculate the slider length from the foot of the perpendicular (from the lower ball joint) to the trunnion centre.
(4) Use a tripod solution based on the lower ball joint, the trunnion centre and the rack-end ball joint, with the three known lengths (two unchanging). This solves the foot of the perpendicular, hence the slider (wheel carrier) position.
(5) Solve for the wheel centre, and thence the lowest wheel point and bump.
(6) Solve the position of the contact point of the locked wheel.
(7) From the wheel carrier position, solve the wheel axis direction cosines.
(8) Deduce the camber angle, caster angle, steer angle, lateral scrub, longitudinal scrub, and suspension bump.
(9) Analyse several positions through the suspension range, and curve fit a quadratic to give the relevant coefficients.

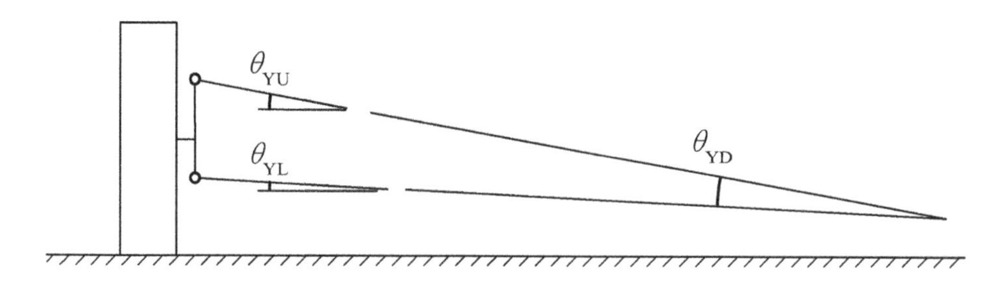

Figure 12.18.1 Modified double trailing arms, front view, effective arm length infinite.

12.18 Double Trailing Arms

The double-trailing-arm system, as shown in Figure 1.6.2, has long been used at the front of small passenger cars, often with transverse torsion bar springing, sometimes with a spring acting down onto the top arm.

In the common form of this suspension, the two pivot axes are parallel to one another, and perpendicular to the vehicle centre plane, which gives a very simple geometry. In front view the wheel simply moves vertically in bump. There is no camber and no lateral scrub. In side view, the trailing arms are parallel, giving zero caster change. The arms are of the same length, so the quadratic caster coefficient is also zero. There is quadratic longitudinal scrub.

A more interesting version is also possible, in which modifications are introduced to achieve desired coefficients. First, in front view, the arm axes may be inclined, Figure 12.18.1.

The lateral length factors are still

$$\begin{aligned} L_{YU} &= \infty; & L_{YL} &= \infty \\ S_{YU} &= 0; & S_{YL} &= 0 \\ S_{YS} &= 0; & S_{YD} &= 0 \end{aligned}$$

For double trailing arms, bump action does not alter the equivalent front-view arm angles, so

$$\theta_{YS} = \theta_{YS0} = \theta_{YU0} + \theta_{YL0}$$
$$\theta_{YD} = \theta_{YD0} = \theta_{YU0} - \theta_{YL0}$$

Table 12.8.3, previously presented for double transverse arms, is applicable for the lateral properties, inserting the above length factors, giving the lateral bump scrub and bump camber expressions seen in Table 12.18.1.

Because the arms are trailing (from the body) they lead from the wheel carrier, so in the terminology used for general transverse arms the lengths would be negative. To avoid this confusion, the side-view arm lengths are redefined as positive, as seen in Figure 12.18.2, which also shows positive arm angles, the arm angle increasing in bump. The lower arm shown has a greater angle, so the pitch arm length is positive, and the angle included at the pitch centre is positive, so being $-\theta_{XD0}$. Because of the redefinition of the arm lengths, Table 12.11.1 for transverse arms is not directly applicable.

The pitch arm radius is

$$R_P = \frac{H_{BJD}}{-\theta_{XD0}} = \frac{-H_{BJD}}{\theta_{XD0} + S_{XD} z_S}$$

The pitch centre height is

$$H_P = H_{BJM} + \tfrac{1}{2}\theta_{XS} R_P$$

Table 12.18.1 Double trailing arms, summary of equations

Lateral	
$\varepsilon_{\text{BScd0,Y}} = \frac{1}{2}(f_{\text{BJH}}\,\theta_{\text{YD0}} - \theta_{\text{YS0}})$	(12.18.1)
$\varepsilon_{\text{BScd1,Y}} = 0$	(12.18.2)
$\varepsilon_{\text{BC1}} = -\dfrac{1}{R_{\text{S}}} = -\dfrac{\theta_{\text{YD0}}}{H_{\text{BJD}}}$	(12.18.3)
$\varepsilon_{\text{BC2}} = 0$	(12.18.4)
Longitudinal	
$\varepsilon_{\text{BScd0,X}} = \frac{1}{2}(-f_{\text{BJH}}\,\theta_{\text{XD0}} + \theta_{\text{XS0}})$	(12.18.5)
$\varepsilon_{\text{BScd1,X}} = \frac{1}{2}(-f_{\text{BJH}}S_{\text{XD}} + S_{\text{XS}})$	(12.18.6)
$\varepsilon_{\text{Bcas1}} = \dfrac{1}{R_{\text{P}}} = -\dfrac{\theta_{\text{XD0}}}{H_{\text{BJD}}}$	(12.18.7)
$\varepsilon_{\text{Bcas2}} = -\dfrac{S_{\text{XD}}}{2H_{\text{BJD}}}$	(12.18.8)

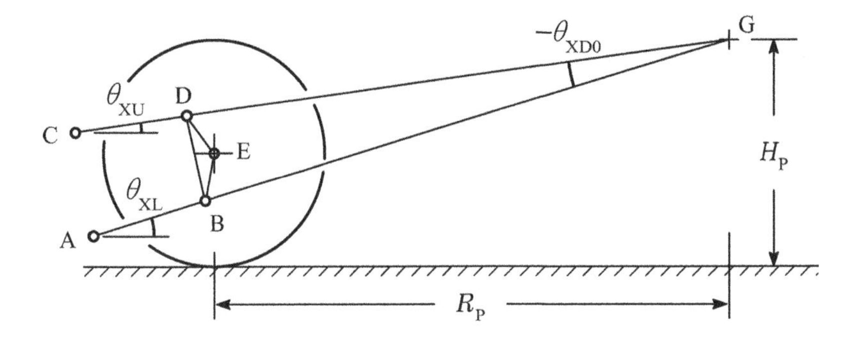

Figure 12.18.2 Modified double trailing arms, side view, arm lengths redefined as positive.

The local longitudinal bump scrub rate is

$$\varepsilon_{\text{BScd,X}} = \frac{H_{\text{P}}}{R_{\text{P}}} = \frac{H_{\text{BJM}}}{R_{\text{P}}} + \tfrac{1}{2}\,\theta_{\text{XS}}$$

$$= -\frac{H_{\text{BJM}}(\theta_{\text{XD0}} + S_{\text{XD}}z_{\text{S}})}{H_{\text{BJD}}} + \tfrac{1}{2}(\theta_{\text{XS0}} + S_{\text{XS}}z_{\text{S}})$$

Collecting terms then gives the longitudinal bump scrub coefficients shown in Table 12.18.1. The results are similar to Table 12.11.1 but with various sign changes.

The bump caster coefficients are derived as previously, with the results as in the table.

By this means the double-trailing-arm suspension may be given most of the desirable coefficients, although it continues to lack bump scrub variation (roll centre movement control) and quadratic bump camber.

12.19 Five-Link Suspension

A degenerate case of location by two arms (wishbones) and a steering rack occurs when each arm is exploded into two separate links, making five separate links in all, as seen for example in Figure 1.10.9.

This is claimed to offer greater design freedom to obtain some compliance properties, particularly for a driven wheel. Geometrically, the present problem is to solve for the wheel carrier position as it moves in bump. In general, analytic methods are not easily applied, because the equations are non-linear and unwieldy, so an iterative numerical solution is used.

The complete wheel carrier position is specified by six coordinates, say (x, y, z), of a given point, possibly the wheel centre, and angles (γ, δ, θ) of camber, steer and pitch. From these, the position of each of the six ball joints on the carrier can be calculated. The lengths of the spaces between these points and the corresponding points on the body can also be calculated, and compared with the link lengths. The position variables are then iteratively adjusted to reduce the link length discrepancies to nominally zero.

Normally the vertical position variable z would be specified, leaving five variables to determine. This can be done cyclically, making a series of passes through the set of variables, each variable being improved in turn. An increment of the variable is tested, to measure the change in each link space. These are the influence coefficients for that variable. The variable can then be changed to minimise the total discrepancy.

Alternatively, all five variables can be adjusted simultaneously. The influence coefficients of all five variables are determined together, which then gives five simultaneous equations for the new position variables.

The values of the influence coefficients found vary somewhat with the carrier position, so there is a need to re-evaluate them at each step. The second method above is slower per cycle, but becomes strongly convergent when close to the solution.

The bump properties can be investigated in the usual way, by solving for the carrier position at a series of suspension bumps and then curve fitting quadratic equations. The five-link suspension has all the design variability of a double-transverse arm, and so can be given any geometric properties desired in practice.

The method of solution is similar to that for a rigid axle, which is discussed in some detail in Chapter 13.

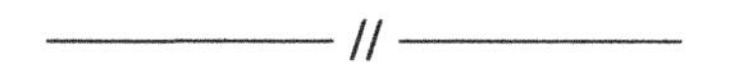

13

Rigid Axles

13.1 Introduction

This chapter deals with rigid axles having rigid link location. A rigid axle is simply defined as one on which the two wheels are physically connected without relative change of position other than a very small amount of axle compliance. Trailing-twist axles are excluded. Chapter 1 gives diagrams of many examples of rigid axles. Leaf-spring located axles are typically dealt with by considering an equivalent link system.

Three basic forms of geometric analysis can be used on a rigid axle: pivot-point analysis; link analysis; and computer numerical analysis. The purpose of such analysis is to reveal the suspension geometric characteristics in a similar way to those of an independent suspension. However, the variables used are different.

13.2 Example Configuration

Figure 13.2.1 shows a general four-link axle in simple line drawing form, similar to the real design shown in Figure 1.7.3. Lateral location is achieved by the plan-view inclination of the links. In some cases the two upper links meet at the axle, giving a single lateral location point there shared by the two links. More often, and more practical in construction, the two links may be brought close together but still separate.

In many cases there is single longitudinal upper link, with lateral location provided by a specific-purpose mechanism. Typically this is laid out horizontally just behind the axle, and is formed of either a simple Panhard rod or a Watt's linkage.

In Figure 13.2.1 it is apparent that the wide-spread lower links are likely to be a dominant factor in controlling the important steer angle of the axle. However, this also depends on the spacing at the front of the links.

13.3 Axle Variables

For an independent suspension, the fundamental independent variable is the individual suspension bump. For a rigid axle, there are two basic independent variables. These could be considered to be the suspension bump values at each of the two wheels, z_{SBL} and z_{SBR}, but more often the complete axle variables are used, namely the axle suspension heave z_{AS} and the axle suspension roll angle ϕ_S. The axle bump is measured at the centre of the axle, so, for small angles, the axle suspension heave (symmetrical double bump) and the

Suspension Geometry and Computation J. C. Dixon

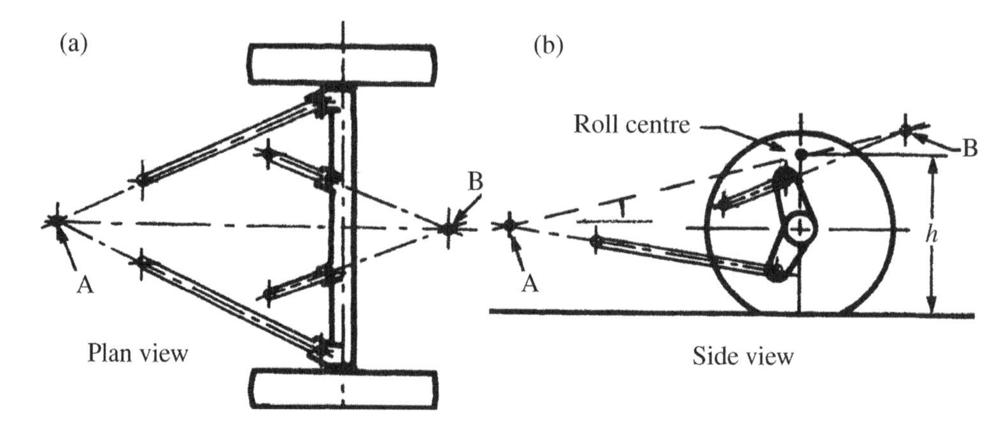

Figure 13.2.1 A general four-link axle.

axle suspension roll are given by

$$\begin{aligned} z_{AS} &= \tfrac{1}{2}(z_{SBL} + z_{SBR}) \\ \phi_S &= \phi_B - \phi_A = \frac{z_{SBR} - z_{SBL}}{L_{Ax}} \end{aligned} \tag{13.3.1}$$

where L_{Ax} is the length of the axle between the wheel centres, ϕ_B is the body roll angle and ϕ_A is the axle roll angle on the tyres. Correspondingly, the individual wheel suspension bumps left and right are

$$\begin{aligned} z_{SBL} &= z_{AS} - \tfrac{1}{2}L_{Ax}\phi_S \\ z_{SBR} &= z_{AS} + \tfrac{1}{2}L_{Ax}\phi_S \end{aligned} \tag{13.3.2}$$

The geometric properties, such as the roll steer coefficients, of the axle are usually expressed in terms of the suspension heave and roll, but could also be expressed in terms of the two single-wheel bumps.

A rigid body in three dimensions has six degrees of freedom, these being three displacements and three angles (expressed, for example, as the Euler angles). In the case of an axle these can be taken as the surge (X), the sway (Y), the heave (Z), the roll angle (ϕ), the pitch/caster angle (θ) and the steer/yaw/heading angle (δ). For an axle, the independent variables are the heave (Z position) and the roll angle, leaving four variables to be determined, corresponding to the remaining four degrees of freedom and the four links constraining the axle.

The displacements are straightforward, in the (X, Y, Z) directions, and performed first. The angles need to be acted on in a specific order. In general vehicle engineering the Euler angles are normally used, which are in the order heading angle, pitch angle and roll angle. In the case of an axle, the roll angle is an independent variable, and best taken first, followed by the steer angle and then the pitch angle. Note that the axle heave steer angle is not the same as a rigid axle with steered wheels, which has independent wheel steer rotation relative to the axle. The latter also involves the steering mechanism.

The roll angle, taken first, is about an axis parallel to the longitudinal reference X axis, through the axle centre. The steer angle is then about the vertical Z axis. The pitch angle is last, performed most conveniently about the axis line between the wheel centres.

Figure 13.3.1 shows the axle centreline AB (the line between the two wheel centres) with roll and steer angles applied in that order. In the context of a computer analysis, the angles must be related to the coordinates of the wheel centres A and B, with the angles taken in the correct order. The axle length

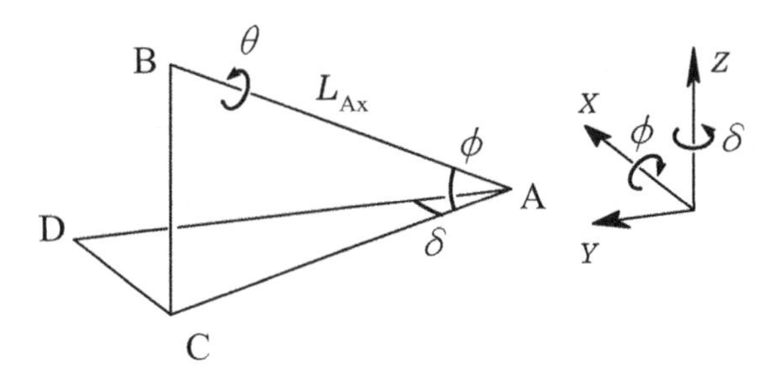

Figure 13.3.1 Roll and steer angles of an axle defined, taken in that order, AB being the line between the wheel centres, AD parallel to the Y axis, CD parallel to the X axis, BC parallel to the Z axis. Subsequent pitch is about the axle centreline AB.

between centres is $L_{Ax} = L_{AB}$. AD is parallel to the Y axis, CD is parallel to the X axis and BC is parallel to the Z axis. Then

$$L_{BC} = L_{Ax} \sin\phi = z_B - z_C$$

$$L_{CD} = L_{Ax} \cos\phi \sin\delta = -(x_C - x_D) = -(x_B - x_D)$$

$$L_{AD} = L_{Ax} \cos\phi \cos\delta = y_D - y_A = y_C - y_A = y_B - y_A$$

For given wheel centre coordinates, the angles may be deduced from

$$\sin\phi = \frac{z_B - z_C}{L_{Ax}}$$

and, with the steer angle positive for right-hand rotation about the Z axis,

$$\tan\delta = -\frac{x_C - x_D}{y_D - y_A} = -\frac{x_B - x_A}{y_B - y_A}$$

The axle pitch angle does not arise directly in the above, being applied subsequently about the axle centreline AB.

Table 13.3.1 summarises the geometric coefficients of an axle, expressed in terms of the suspension heave and roll, as usual. The single-wheel bump coefficients are easily derived from these, if desired, by considering a single-wheel bump as a combination of axle heave and roll (equations 13.3.2). This gives a further set of coefficients for each side separately, possibly with non-symmetrical values because of design features (e.g. a Panhard rod) or because of manufacturing tolerance asymmetry. The scrub values are only really of physical importance at the wheels, but the axle can be considered to have mean effective coefficients on its centreline.

In all cases there are, in principle, linear and quadratic coefficients (e.g. axle roll steer ε_{ARS0} and ε_{ARS1}) as for independent suspensions, although some of these may be rather small, particularly where the design intent is true lateral symmetry, and often neglected. Example values are given later. Even more terms could be considered, treating each coefficient as a polynomial.

Amongst the first six coefficients of Table 13.3.1, row 4 is of the greatest importance – heave steer and, particularly, roll steer. The surge, sway and pitch are mainly of significance for their influence on the effect found at the base of the wheels, reflected in the scrub coefficients. The longitudinal scrub can affect

Table 13.3.1 The main geometric coefficients of a rigid axle

			z_S heave	ϕ_S roll
(1)	X	surge	ε_{AHX}	ε_{ARX}
(2)	Y	sway	ε_{AHY}	ε_{ARY}
(3)	θ	pitch (caster)	ε_{AHP}	ε_{ARP}
(4)	δ	steer (yaw)	ε_{AHS}	ε_{ARS}
(5)	x	longitudinal scrub	$\varepsilon_{AHSc,X}$	$\varepsilon_{ARSc,X}$
(6)	y	lateral scrub	$\varepsilon_{AHSc,Y}$	$\varepsilon_{ARSc,Y}$

	Coefficient	Description	Base units	Common units
(1)	ε_{AHX}	Axle heave surge	m/m	— (non-dim.)
(2)	ε_{AHY}	Axle heave sway	m/m	— (non-dim.)
(3)	ε_{AHP}	Axle heave pitch (caster)	rad/m	deg/dm
(4)	ε_{AHS}	Axle heave steer	rad/m	deg/dm
(5)	$\varepsilon_{AHSc,X}$	Axle heave scrub longitudinal	m/m	— (non-dim.)
(6)	$\varepsilon_{AHSc,Y}$	Axle heave scrub lateral	m/m	— (non-dim.)
(7)	ε_{ARX}	Axle roll surge	m/rad	mm/deg
(8)	ε_{ARY}	Axle roll sway	m/rad	mm/deg
(9)	ε_{ARP}	Axle roll pitch (caster)	rad/rad	— (non-dim.)
(10)	ε_{ARS}	Axle roll steer	rad/rad	— (non-dim.)
(11)	$\varepsilon_{ARSc,X}$	Axle roll scrub longitudinal	m/rad	mm/deg
(12)	$\varepsilon_{ARSc,Y}$	Axle roll scrub lateral	m/rad	mm/deg

braking quality. The lateral roll scrub controls the roll centre height and its variation. For a small suspension roll angle ϕ_S,

$$z_{SBL} = +\tfrac{1}{2}\phi_S T$$

$$z_{SBR} = -\tfrac{1}{2}\phi_S T$$

The lateral scrub values (the actual scrub, not scrub rate) are

$$y_L = \varepsilon_{ARSc,YL}\phi_S$$

$$y_R = \varepsilon_{ARSc,YR}\phi_S$$

where the roll scrub coefficient has units m/rad. The roll centre height H is given by

$$H\phi_S = \tfrac{1}{2}(y_L + y_R)$$

so

$$H = \tfrac{1}{2}(\varepsilon_{ARSc,YR} + \varepsilon_{ARSc,YL})$$

The same value of the roll scrub coefficient can be expected on the two wheels, or the mean value can be taken, so

$$H = \varepsilon_{ARSc,Y}$$

This can also be understood directly by imagining the axle to rotate about the roll centre, displacing the bottom of the wheels to the side by $H\phi_S$, for small angles.

13.4 Pivot-Point Analysis

This method, previously described in Section 6.12, is described again here for convenient comparison with other methods. It is based on studying the support links to find two points A and B where forces are effectively exerted by the axle on the body, as in Figure 13.2.1. This assumes that the link lines do in fact meet, which in reality is not quite the case because of manufacturing tolerances when in the static position. Also, when the suspension is rolled, even in the ideal case the lines will no longer meet. The assumption of the locating points A and B is therefore a useful fiction, an engineering approximation. Kinematically, the line through points A and B forms a pivot axis for the axle relative to the body. Kinetically, the points A and B are also points at which the lateral forces can be transmitted from the axle to the body, giving the usual close relationship between the geometric, kinematic and kinetic properties. The line through AB penetrates the transverse vertical plane of the axle at the geometric roll centre, which in simple analysis is then also used as the force roll centre.

The axis AB also provides a way to investigate the roll steer characteristics of the axle. For certain purposes, the vehicle is perceived to roll about the roll axis, joining the roll centres of the front and rear suspensions. In the present context, however, a roll about such an axis consists of a roll about a horizontal longitudinal axis and a yaw angular displacement. Now, evidently, the yaw angle of the whole vehicle does not alter the suspension bump steer or roll steer angles. Therefore, to examine the roll steer, the body is considered to roll about a horizontal longitudinal axis, with a suspension roll angle ϕ_S. Whatever the height of this axis, the lateral displacement of the forward point A relative to the rear point B is

$$Y_{A/B} = -\phi_S(z_A - z_B)$$

This will create a steer angle of the axle, actually for a rear axle an understeer angle

$$\delta_{US} = \frac{Y_{A/B}}{X_{A/B}} = -\phi_S \frac{z_A - z_B}{x_A - x_B}$$

The corresponding linear roll understeer coefficient of a rear axle is then

$$\varepsilon_{RU1} = -\frac{z_A - z_B}{x_A - x_B} = \rho_A$$

where ρ_A is the inclination angle of the axle axis AB expressed in radians (not of the vehicle roll axis). The effect will be an understeer one when a rear axle has an axis lower at the front point, as shown in the diagram. For a front axle, understeer occurs for an axis pointing upwards at the front.

Suitable points A and B can be found for other axle link layouts. For example, if the lower links are parallel then the point A is at infinity, so AB is parallel to the bottom links. If the bottom link pair is replaced by a torque tube or similar system, then point A is the front ball joint. If transverse location is by a Panhard rod, then point B is the point at which the rod intersects the vertical central plane. A Watt's linkage is usually installed with the centre point of the small vertical member attached to the axle, and the outer ends of the links attached to the body (e.g. Figure 1.7.6).

A characteristic of the Panhard rod is that the roll centre rises for roll in one direction, and falls for the other, because of vertical motion of the point of connection to the body. For other axle lateral location systems it is similarly necessary to find the point where the line of action of the force intersects the central vertical longitudinal plane.

The 'pivot-point' method of analysis also gives some insight into the effect of suspension bump position on the suspension characteristics, and hence into the effect of additional vehicle load. By some analysis of the suspension details, it is possible, and often easy, to see how the points A and B will move vertically when the axle moves in heave (e.g. when the body lowers due to an extra load). In Figure 13.2.1 it is apparent that as the body lowers, the front point A will lower, and the rear point Bwill rise, for this particular design. This arrangement will tend to keep the roll centre at a fairly constant height. On the other hand, the inclination of the axis AB will increase with suspension bump and with load. The result will be an increased linear understeer from the rear axle, and of the linear understeer of the vehicle. This may be held to be an advantage of the system illustrated, in that a greater load on the rear of the vehicle has a partially compensating increase of linear understeer. However, the small vertical change of the roll centre height will not influence the increase of final oversteer due to the extra load in conjunction with the tyre characteristics. Arguably, then, it may be better to use an arrangement in which the roll centre is lowered as the load increases, giving preference to the more critical limit-state handling.

13.5 Link Analysis

The axle characteristics can usefully be investigated by direct analysis of the links, particularly with some simple configurations. Consider a rigid axle with a single-point upper lateral location and two lower parallel longitudinal members, along the lines of Figure 1.7.4 or 1.9.1, such that the lower longitudinal arms govern the steer angle of the axle, as illustrated here in Figure 13.5.1. The longitudinal links are characterised by their length, static angle to the horizontal, and spacing.

Consider then a body roll giving a suspension roll angle ϕ_S on such a system. Relative to the axle, the front of the left link will rise and the front of the right link will fall. In conjunction with the link static inclination in side view, this will give a linear roll steer effect. Also, the non-infinite length of the links may give a secondary steer effect.

The resulting understeer angle, δ_{US}, for a rear axle, will be given by

$$\delta_{US} = -\frac{dx_L - dx_R}{B}$$

where x is the forward movement of the axle point. For a small body roll angle, the left link front end rises, pushing the axle back, so

$$dx_L = -\tfrac{1}{2}B\phi_S\,\theta_{L0}$$

$$dx_R = +\tfrac{1}{2}B\phi_S\,\theta_{R0}$$

$$\delta_{US} = \tfrac{1}{2}\phi_S(\theta_{R0} + \theta_{L0})$$

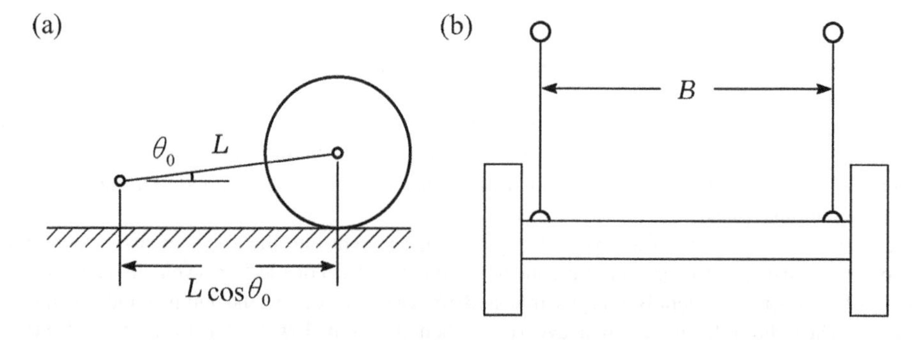

Figure 13.5.1 Rigid axle with steer governed by two parallel longitudinal links (other links not shown).

The consequent linear understeer coefficient is therefore

$$\varepsilon_{\mathrm{RU1}} = \tfrac{1}{2}(\theta_{\mathrm{R0}} + \theta_{\mathrm{L0}})$$

Considering now links of different lengths, (e.g. due to manufacturing tolerances),

$$\mathrm{d}x_{\mathrm{L}} = L_{\mathrm{L}}(1 - \cos\theta_{\mathrm{L}}) \approx L_{\mathrm{L}}(\tfrac{1}{2}\theta_{\mathrm{L}}^2) = L_{\mathrm{L}}\frac{1}{2}\left(\frac{\frac{1}{2}B\phi_{\mathrm{S}}}{L_{\mathrm{L}}}\right)^2$$

giving

$$\mathrm{d}x_{\mathrm{L}} = \frac{1}{8}\frac{B^2}{L_{\mathrm{L}}}\phi_{\mathrm{S}}^2$$

Then, in terms of the link shortnesses $S = 1/L$,

$$\delta_{\mathrm{US}} = \tfrac{1}{8}B(S_{\mathrm{R}} - S_{\mathrm{L}})\phi_{\mathrm{S}}^2$$

so the resulting quadratic understeer coefficient is

$$\varepsilon_{\mathrm{RU2}} = \tfrac{1}{8}B(S_{\mathrm{R}} - S_{\mathrm{L}})$$

Evidently, then, the first understeer coefficient arises from any difference in the static arm angles, as might occur due to manufacturing tolerances or to asymmetrical loading. The second understeer coefficient arises from differences of the arm lengths, which is just a matter of manufacturing tolerance, usually small.

Figure 13.5.2 shows a simple example axle in side view for two cases. The change of axle pitch angle can be analysed in a very similar way to the analysis of camber angle on an independent suspension. In the first case, for a bump $\mathrm{d}z$ the longitudinal displacement of the end of the link is

$$\mathrm{d}x = \mathrm{d}z\tan\theta \approx \mathrm{d}z\,\theta$$

Considering both links then, the change of pitch angle is

$$\mathrm{d}\theta = \frac{\mathrm{d}x_{\mathrm{U}} - \mathrm{d}x_{\mathrm{L}}}{H_{\mathrm{A}}} \approx \frac{\mathrm{d}z}{H_{\mathrm{A}}}(\theta_{\mathrm{U}} - \theta_{\mathrm{L}})$$

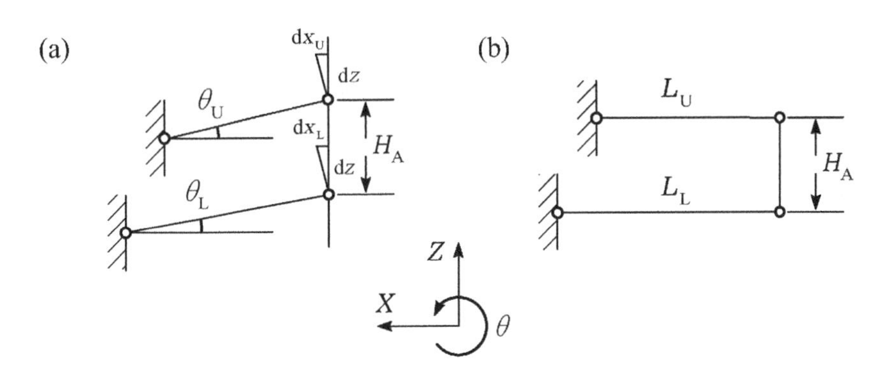

Figure 13.5.2 Example axle side view: (a) surge and pitch location by inclined links; (b) location by links of different length/shortnesses.

The result is a linear axle heave pitch coefficient

$$\varepsilon_{\mathrm{AHP1}} = \frac{\theta_{\mathrm{U}} - \theta_{\mathrm{L}}}{H_{\mathrm{A}}}$$

For links of different lengths, L_{U} and L_{L}, initially horizontal,

$$\mathrm{d}x = L\left(1 - \cos\left(\frac{\mathrm{d}z}{L}\right)\right) \approx L\tfrac{1}{2}\frac{\mathrm{d}z^2}{L^2} = \tfrac{1}{2}S\,\mathrm{d}z^2$$

The pitch angle change is then

$$\Delta\theta = \frac{S_{\mathrm{U}} - S_{\mathrm{L}}}{2H_{\mathrm{A}}}\mathrm{d}z^2$$

with a consequent quadratic axle heave pitch coefficient

$$\varepsilon_{\mathrm{AHP2}} = \frac{S_{\mathrm{U}} - S_{\mathrm{L}}}{2H_{\mathrm{A}}}$$

Further analysis along these lines may be applied to obtain the longitudinal scrub of the bottom of the wheel in bump, which is held to be significant in braking quality on rough roads, as any bump scrub at this point is likely to cause deterioration of the braking quality.

13.6 Equivalent Links

A Watt's linkage, designed correctly, produces an exactly straight line of its control point, over a well-defined range of movement, until the centre link reaches alignment with one of the lateral links. There are also other straight-line mechanisms that are sometimes used. Also, the Watt's linkage may be deliberately or accidentally imperfect, giving a very large radius of action. In a simple analytic or computer analysis, it may be convenient to represent such a mechanism by a simple equivalent link. The question arises, then, as to the accuracy of such an equivalent linkage. Such a situation typically arises where an axle has longitudinal location at each side by a Watt's linkage, but the only available software has conventional links. A very long conventional link can be used, but an actual length must be chosen. Consider a radius arm representing what is really a straight line. The equivalent link will be placed at the correct angle, so the linear effect should agree exactly, but there will be an unreal second-order effect. This would be zero if the link were infinitely long. The amount of second-order effect depends on the change of angle of the equivalent link. Representing the angle of the link by an 8-byte real number, the angle is of order 1 radian, so there will be no effect at all on the numerical solution if the angle change is less than about 10^{-15} radians. Considering a suspension displacement of 0.100 m, the actual length of the equivalent link needs to be 10^{14} m. In practice, it is easy to use a greater length, say 10^{20} m. With a real number range of 10^{308}, this still allows accurate computation. The criterion applied here is of course a rather rigorous one, and from the point of view of engineering accuracy, an equivalent link of say 1000 m would probably be adequate.

13.7 Numerical Solution

Whatever the configuration, a numerical analysis by computer can obtain the various characteristics of the axle, provide that suitable software is available, or written for the purpose, and small details, such as the effect of manufacturing tolerances of mounting points, are also easily investigated.

The basic rigid axle, as in Figure 13.5.1, is mounted on four links. This is the simplest form. Other more complex types, such as a Watt's linkage location, require a specialised program or the use of equivalent links, as discussed earlier.

In principle, this could be solved by writing equations for the link spaces in terms of the axle position variables, and setting these equal to the physical link lengths. In practice the equations would be unwieldy, and non-linear, and iteration would be required for a practical solution. In practice, then, there is really no way to avoid doing an iterative solution.

The basis of an iterative solution is that there is an existing position, somewhat in error, and a way to improve that estimated position. Basically this requires the following:

(1) a suitable set of variables representing the axle position;
(2) a way to calculate from those the coordinates of the link connection points;
(3) a calculation of the link length errors;
(4) a way to use the errors to improve the position.

In practice it is convenient to use a 'reference triangle' method, to be described here. Using the PointATinit and PointAT routines described in Chapter 15, a reference triangle can be defined, with all the axle points being calculated from that. Consider, then, an axle with points defined as follows:

$P_1, \ldots, P_4$	link connection points on the body;
$P_5, \ldots, P_8$	corresponding link points on the axle;
P_9	mid point of the axle (mean of P_{10} and P_{11});
P_{10}	centre of left wheel;
P_{11}	centre of right wheel;
P_{12}	bottom point of the left wheel;
P_{13}	bottom point of the right wheel.

The reference triangle is then, conceptually, rigidly attached to the axle. Various triangles may be used successfully, but the choice here, for three points forming a right-angled triangle, is

P_{T1} coincident with P_9, the axle centre point;
P_{T2} on the axle centreline (P_{10} to P_{11}), at a distance L_T from P_9 towards P_{10};
P_{T3} approximately above P_{T1}, distance L_T from P_9 and distance $\sqrt{2}L_T$ from P_{T2}, with x coordinate initially at the same coordinate as P_{T1}.

A suitable length for L_T is on the same scale as the suspension itself. A value of 1 m allows it to be omitted from some equations, for computational efficiency, but the equations are then no longer dimensionally correct. A value of 0.5 m is good. Another possibility is to use half of the axle length, bringing the second triangle point into coincidence with the left wheel centre.

The triangle is initially defined for the axle in its specified heaved and rolled position, the surge (X), sway (Y), steer angle (δ) and pitch angle (θ) yet to be determined. The axle must subsequently be moved in (X, Y, δ, θ) to satisfy the links without disturbing the heave and roll positions. This is facilitated by the choice and use of the triangle. Simply displacing the entire triangle in X or Y is not problematic. By choosing the first triangle point at the axle mid-point, the triangle can be rotated about Z in steer angle change without interfering with the heave or roll. Finally, by pitching the axle about the axle line $P_{10}P_{11}$ there is again no unwanted disturbance.

At each cycle of the iteration, the position of the axle points with link connections, $P_5, \ldots, P_8$, are calculated, using procedure PointAT, the link spaces are calculated, and hence the link length errors are known.

Four linear equations can be written for variation of the link space lengths with the axle position variables (X, Y, δ, θ). Denoting change of these variables as a 4-vector $\boldsymbol{X}$ gives a matrix equation

$$\Delta \boldsymbol{L} = \boldsymbol{C}\boldsymbol{X}$$

where $\boldsymbol{C}$ is a sensitivity matrix, actually varying significantly with the axle position. In more detail, with X as the variation of positions,

$$\Delta L_1 = C_{1,1}X_1 + C_{1,2}X_2 + C_{1,3}X_3 + C_{1,4}X_4$$

$$\Delta L_2 = C_{2,1}X_1 + C_{2,2}X_2 + C_{2,3}X_3 + C_{2,4}X_4$$

$$\Delta L_3 = C_{3,1}X_1 + C_{3,2}X_2 + C_{3,3}X_3 + C_{3,4}X_4$$

$$\Delta L_4 = C_{4,1}X_1 + C_{4,2}X_2 + C_{4,3}X_3 + C_{4,4}X_4$$

Now, ideally the known link space errors $E_1, \ldots, E_4$ will be eliminated, so the desired link space changes are just the negatives of the errors, giving the matrix equation

$$\boldsymbol{C}\boldsymbol{X} = -\boldsymbol{E}$$

Actually, this gives four simultaneous equations for $X_1, \ldots, X_4$, which when solved indicate the desired adjustment to the axle position. In practice the result is imperfect because the local sensitivity matrix varies with axle position, the problem not being truly linear. With revision of the sensitivity matrix at each cycle, quadratic convergence can be achieved to high accuracy.

The axle movements are conveniently applied to the axle by moving only the reference triangle as follows:

(1) Translate the complete triangle by the X and Y increments.
(2) Rotate the triangle by the steer angle correction (Z axis).
(3) Rotate the triangle in pitch about the axle wheel-to-wheel centreline.

Note that the axle iterative solution does not actually imply any particular sequence of angles defining the axle position (e.g. roll before steer before pitch). The final result of this process is simply the axle point position coordinates. The angle sequence arises subsequently when the coordinate position is interpreted as a combination of angular displacements.

Rotation of the triangle in steer is easy, Figure 13.7.1, with the Z coordinate playing no part. Given the initial (X_1, Y_1),

$$R = \sqrt{X_1^2 + Y_1^2}$$

$$\delta_1 = \operatorname{atan}\left(\frac{Y_1}{X_1}\right)$$

where the initial angle evaluation must obtain the correct quadrant (e.g. in Fortran use atan2). Then

$$X = R\cos(\delta_1 + \delta)$$
$$Y = R\sin(\delta_1 + \delta)$$

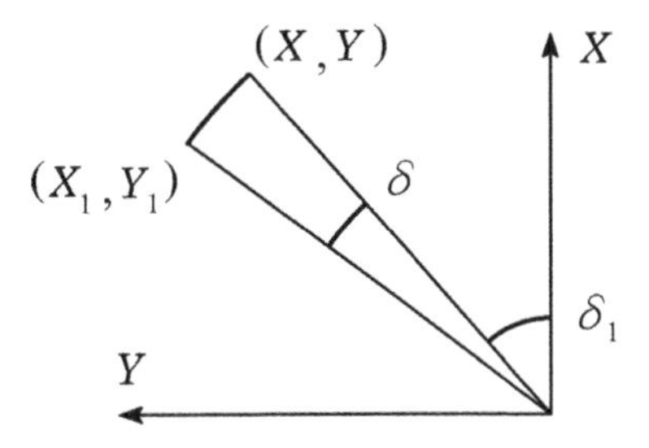

Figure 13.7.1 Steer rotation of a point.

Pitching of the axle is conveniently achieved by using X_{T3} as the variable, that is, the X coordinate of the upper point of the triangle, solving for a new triangle point having a known X coordinate ($X_{T3} + \Delta X_{T3}$) and known distance L_T from P_{T1} and $\sqrt{2}L_T$ from P_{T2}. This requires a routine to solve for the intersection of two spheres with a transverse vertical plane.

To calculate the pitch angle of the axle, calculate the pitch angle of the rigidly attached triangle. This is conveniently done by considering the plan-view length from P_{T1} to P_{T3},

$$L_{T13PV} = \sqrt{(X_{T3} - X_{T1})^2 + (Y_{T3} - Y_{T1})^2}$$

giving

$$\theta = \operatorname{asin}\left(\frac{L_{T13PV}}{L_T}\right)$$

This is the whole angle. The angle change due to heave and roll requires subtraction of the static pitch angle.

13.8 The Sensitivity Matrix

The key to successful iteration is the sensitivity matrix. For a given axle position, this is found analytically or numerically according to the variable. In principle, a small disturbance is made to the triangle/axle position and the changes of link space length are obtained. The sensitivity matrix is in fact

$$\boldsymbol{C} = \begin{bmatrix} \left(\frac{\partial L_1}{\partial X}\right) & \left(\frac{\partial L_1}{\partial Y}\right) & \left(\frac{\partial L_1}{\partial \delta}\right) & \left(\frac{\partial L_1}{\partial X_{T3}}\right) \\ \left(\frac{\partial L_2}{\partial X}\right) & \left(\frac{\partial L_2}{\partial Y}\right) & \left(\frac{\partial L_2}{\partial \delta}\right) & \left(\frac{\partial L_2}{\partial X_{T3}}\right) \\ \left(\frac{\partial L_3}{\partial X}\right) & \left(\frac{\partial L_3}{\partial Y}\right) & \left(\frac{\partial L_3}{\partial \delta}\right) & \left(\frac{\partial L_3}{\partial X_{T3}}\right) \\ \left(\frac{\partial L_4}{\partial X}\right) & \left(\frac{\partial L_4}{\partial Y}\right) & \left(\frac{\partial L_4}{\partial \delta}\right) & \left(\frac{\partial L_4}{\partial X_{T3}}\right) \end{bmatrix}$$

A small disturbance, ΔX say, therefore reveals the four sensitivities for the four link spaces to that variable.

For the simple longitudinal axle displacement (surge) X, the sensitivities are easily obtained analytically, according to the alignment of the particular link space. The change of link space is a dot product:

$$\Delta \boldsymbol{L} = -\Delta \boldsymbol{X} \cdot \hat{\boldsymbol{u}}$$

where $\hat{\boldsymbol{u}}$ is the unit vector along the link, actually from the vehicle body to the axle. Therefore the sensitivity is simply

$$\frac{\partial L_K}{\partial X} = -l_K$$

where l_K is the X-direction cosine of the link space (body to axle). Hence the first four sensitivities to X are easily determined.

Similarly, for axle lateral (sway) displacement Y, the link sensitivities are

$$\frac{\partial L_K}{\partial Y} = -m_K$$

where m is the Y-direction cosine.

For the sensitivity to the steer angle, a numerical approach is convenient. A small temporary axle test steer displacement δ_T is made to the triangle, the four link connection points are calculated, and the link space length changes evaluated. Alternatively, the movement of each link point could be calculated as was shown in Figure 13.7.1, but in terms of coordinate changes, the dot product of this taken with the link direction.

For the pitch sensitivity, again a numerical solution is required, using a small test increment ΔX_{T3} on triangle point 3, evaluating the link connection point positions and changes of space length.

In the case of the numerical evaluation of sensitivities, the question arises as to the size of increment to use. This is not highly critical, but in practice a linear displacement in metres or angular displacement in radians of 10^{-4} to 10^{-11} works well, using 8-byte variables, with 10^{-7} being near the centre of the range. Outside that range, convergence is inferior. The similarity of values for displacement and angle relates to the fact that a suspension has a scale of about 1 metre. Actual example values of sensitivity matrices are given in subsequent sections.

13.9 Results: Axle 1

The first example axle is shown in Figure 13.9.1. It is a conventional four-link type similar to that in Figure 13.2.1. Table 13.9.1 presents the numerical values for initialisation of the iteration process, beginning with the actual unmoved position coordinates. Also given are the values of the sensitivity matrix in the static position, where, for example, the first column is the response to X displacement of the axle, that is, the X direction cosines. Then the interpolation values f_2 and f_3 follow with the normal displacements of points relative to the reference triangle, as specified earlier.

Table 13.9.2 presents the successive iteration loop errors, where the quadratic convergence can be observed. Finally, the actual results for surge, sway, pitch angle and steer angle are given. The convergence error is sometimes exactly zero, but, for interest, 10^{-15} m is in any case only the diameter of the nucleus of an atom.

Table 13.9.1 Numerical solution, Axle 1: initialisation

```
Axle type: Standard 4-link

Initial position coordinates:
    K          x            y            z

    1     0.600000     0.300000     0.250000
    2     0.300000     0.200000     0.400000
    3     0.300000    -0.200000     0.400000
    4     0.600000    -0.300000     0.250000

    5     0.100000     0.500000     0.200000
    6     0.050000     0.070000     0.400000
    7     0.050000    -0.070000     0.400000
    8     0.100000    -0.500000     0.200000

    9     0.000000     0.000000     0.300000
   10     0.000000     0.680000     0.300000
   11     0.000000    -0.680000     0.300000

Link lengths (m) =    0.540833     0.281780     0.281780     0.540833

Static position sensitivities:

CS1(1,:)=     -0.924500     0.369800     0.184900     0.462250
CS1(2,:)=     -0.887217    -0.461353     0.073960    -0.025886
CS1(3,:)=     -0.887217     0.461353     0.018490     0.006472
CS1(4,:)=     -0.924500    -0.369800    -0.092450     0.231125

    K          f2           f3           dsp

    5     1.000000    -0.200000     0.100000
    6     0.140000     0.200000     0.050000
    7    -0.140000     0.200000     0.050000
    8    -1.000000    -0.200000     0.100000
    9     0.000000     0.000000     0.000000
   10     1.360000     0.000000     0.000000
   11    -1.360000     0.000000     0.000000

Initially rolled and heaved axle:

Axle heave      =    0.050000 m
Axle roll angle =    6.000000 deg
```

13.10 Results: Axle 2

The second example axle has a Panhard rod for lateral location, with a single top longitudinal link, as shown in Figure 13.10.1. Table 13.10.1 shows the static point coordinates, the interpolation factors and the final results. Table 13.10.2 shows the sensitivity matrix values for three positions – static, initially displaced and final – where significant differences may be seen. Also the distinctive values for the Panhard rod link are evident.

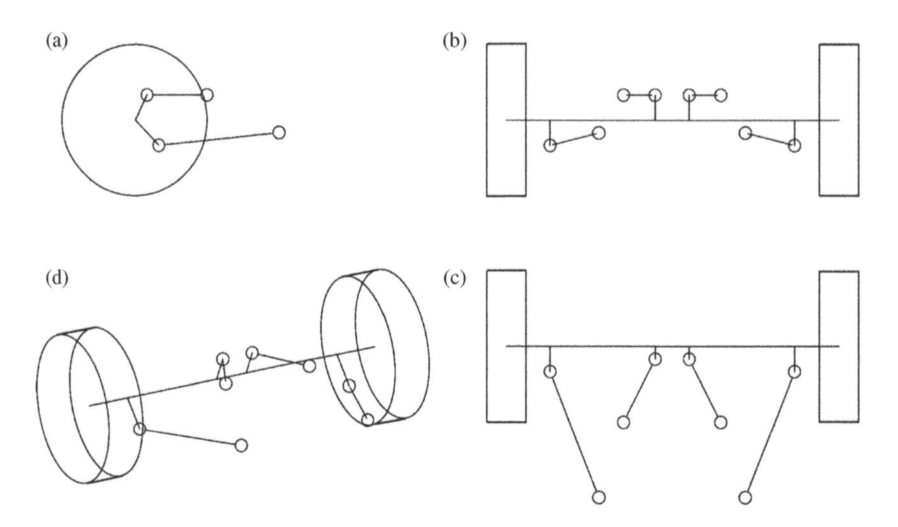

Figure 13.9.1 Example Axle 1, with convergent link location: (a) side view; (b) front view; (c) plan view; (d) axometric three-quarter view.

Table 13.9.2 Axle 1: iteration results

Iteration loop errors:

Loop	Link1	Link2	Link3	Link4	Total abs
0	0.317E-02	0.107E-01	-0.132E-02	-0.458E-02	0.198E-01
1	0.167E-03	0.395E-03	0.250E-03	0.106E-03	0.917E-03
2	0.162E-06	0.429E-06	0.299E-06	0.117E-06	0.101E-05
3	0.172E-12	0.496E-12	0.372E-12	0.134E-12	0.117E-11
4	-0.111E-15	-0.555E-16	0.555E-16	0.000E+00	0.222E-15

Values obtained:

Axle heave	= 0.050000 m
Axle roll angle	= 6.000000 deg
Axle surge	= 0.002335 m
Axle sway	= 0.011737 m
Axle pitch/caster	= 0.185575 deg
Axle steer	= -0.922967 deg

13.11 Coefficients

By solving for a range of points through the suspension range and fitting a quadratic curve, the basic coefficients may be obtained. This has been done here with 11 points for heave through a range −0.100 m to +0.100 m and for roll from −0.100 rad to +0.100 rad. The results are given in Table 13.11.1 for Axle 1 and Table 13.11.2. for Axle 2. As may be seen in both tables, for the pitch angle the quadratic fit is poor, and a much higher order of polynomial is required for a really good fit, but the pitch angle is relatively small. The quadratic fit for the important steer angle is good. Axle 1 is symmetrical, so heave has exactly zero

Table 13.10.1 Numerical solution, Axle 2: initialisation and iteration

```
Axle type: One top link, Panhard rod

Initial position coordinates:

    K         x                y                z

    1     0.600000         0.500000         0.250000
    2     0.300000         0.000000         0.500000
    3    -0.100000        -0.400000         0.350000
    4     0.600000        -0.500000         0.250000

    5     0.100000         0.350000         0.200000
    6     0.050000         0.000000         0.500000
    7    -0.100000         0.400000         0.320000
    8     0.100000        -0.350000         0.200000

    9     0.000000         0.000000         0.300000
   10     0.000000         0.680000         0.300000
   11     0.000000        -0.680000         0.300000

Link lengths (m) =   0.524404   0.250000   0.800562   0.524404

    K         f2               f3               dsp

    5     0.700000        -0.200000         0.100000
    6     0.000000         0.400000         0.050000
    7     0.800000         0.040000        -0.100000
    8    -0.700000        -0.200000         0.100000
    9     0.000000         0.000000         0.000000
   10     1.360000         0.000000         0.000000
   11    -1.360000         0.000000         0.000000

Initially rolled and heaved axle:

Axle heave          =     0.050000 m
Axle roll angle     =     6.000000 deg

Iteration loop errors:

Loop    Link1        Link2        Link3        Link4        Total abs
0    -0.345E-02   0.559E-02   -0.246E-02   0.253E-02    0.140E-01
1     0.726E-04   0.811E-04   -0.122E-04   0.732E-04    0.239E-03
2     0.111E-07   0.259E-07    0.339E-08   0.115E-07    0.519E-07
3    -0.111E-15   0.167E-14    0.111E-15   0.000E+00    0.189E-14
4    -0.111E-15   0.000E+00    0.000E+00   0.000E+00    0.111E-15

Values obtained:

Axle heave          =     0.050000 m
Axle roll angle     =     6.000000 deg
Axle surge          =     0.001558 m
Axle sway           =     0.003950 m
Axle pitch/caster   =     0.088394 deg
Axle steer          =     0.811626 deg
```

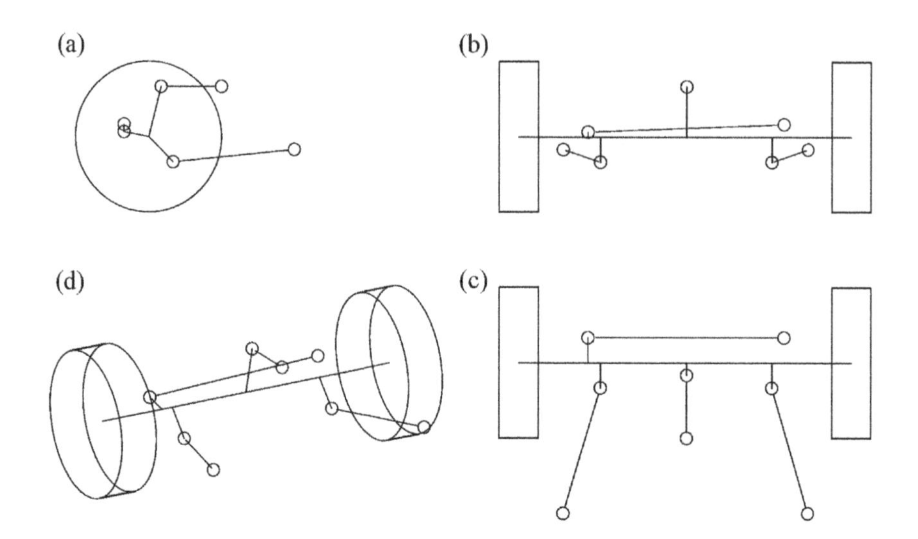

Figure 13.10.1 Example Axle 2, with Panhard rod lateral location: (a) side view; (b) front view; (c) plan view; (d) axometric three-quarter view.

steer and sway effects, but some heave–steer and heave–sway effects may be seen on Axle 2 because of the Panhard rod lateral location.

Table 13.11.3 presents the quadratic scrub coefficients found for the two axles. Axle 1 is exactly symmetrical, so heave causes zero lateral scrub. There is significant longitudinal scrub which could be designed out by adjusting the lateral view link angles. Roll causes 0.0356 m/rad linear longitudinal scrub,

Table 13.10.2 Axle 2 sensitivity matrices

```
Static position sensitivity matrix:

CS1(1,:)=    -0.953463   -0.286039    0.190693    0.333712
CS1(2,:)=    -1.000000    0.000000   -0.114416    0.000000
CS1(3,:)=     0.000000    0.999298    0.003814   -0.038139
CS1(4,:)=    -0.953463    0.286039   -0.066742    0.116799

Starting sensitivity matrix:

CS(1:) =     -0.953463   -0.269762    0.172102    0.316961
CS(2:) =     -1.000000   -0.083623   -0.411128   -0.024538
CS(3:) =      0.000000    0.993949   -0.005466   -0.099701
CS(4:) =     -0.953463    0.309628    0.209821   -0.289556

Final sensitivity matrix:

CS(1:) =     -0.963421   -0.259312    0.178379    0.322722
CS(2:) =     -0.979243   -0.064430   -0.411390   -0.023097
CS(3:) =     -0.004605    0.996852   -0.004667   -0.103098
CS(4:) =     -0.944616    0.320211    0.212084   -0.284411
```

Table 13.11.1 Basic heave and roll coefficients for Axle 1

Quadratic fit coefficients:

	C0	C1	C2	Q fit
Heave:				
Surge	-0.000065	-0.054334	1.497510	0.968790
Sway	0.000000	0.000000	0.000000	------
Pitch	0.008868	0.496882	1.345168	0.765415
Steer	0.000000	0.000000	0.000000	------
Roll:				
Surge	0.000000	0.000000	0.101814	0.999431
Sway	0.000000	0.095631	0.000000	0.999765
Pitch	-0.000020	0.000000	0.025374	0.840332
Steer	0.000000	-0.052557	0.000000	0.997766

closely associated with the roll steer which without any pitch effect would give $0.05256 \times 1.36/2 = 0.0357$ m/rad, in good agreement. The linear lateral scrub coefficient in roll is 0.395 m/rad, giving the roll centre height.

For Axle 2, with the Panhard rod, the longitudinal heave scrub happens to be similar to Axle 1, but the quadratic coefficients are much greater, and there is considerable asymmetry. There is some lateral heave effect. Roll causes an approximately antisymmetrical linear longitudinal scrub. The linear lateral scrub with roll averages 0.361 m/rad, giving the roll centre height.

Table 13.11.2 Basic heave and roll coefficients for Axle 2

Quadratic fit coefficient matrix:

	C0	C1	C2	Q fit
Heave:				
Surge	-0.000060	-0.070559	1.354386	0.973089
Sway	0.000028	0.041053	-0.751124	0.986469
Pitch	0.005208	0.403181	0.373187	0.784783
Steer	0.000043	0.040367	-0.728536	0.989236
Roll:				
Surge	-0.000001	0.000120	0.152865	0.993947
Sway	0.000000	0.060982	0.113727	0.999640
Pitch	-0.000010	0.000138	0.013249	0.815758
Steer	-0.000002	0.260935	0.108252	0.999043

Table 13.11.3 Scrub coefficients for Axles 1 and 2

Axle 1 Scrub coefficients:				
Heave (units m/m = - - -):				
Left X	=	0.000015	-0.196137	0.022234
Right X	=	0.000015	-0.196137	0.022234
Left Y	=	0.000000	0.000000	0.000000
Right Y	=	0.000000	0.000000	0.000000
Roll (units m/rad):				
Left X	=	0.000000	0.035612	0.322117
Right X	=	0.000000	-0.035612	0.322117
Left Y	=	0.000000	0.395191	-0.340655
Right Y	=	0.000000	0.395191	0.340655
Axle 2 Scrub coefficients:				
Heave (units m/m = - - -):				
Left X	=	-0.000018	-0.194421	0.832917
Right X	=	0.000040	-0.139523	-0.157875
Left Y	=	0.000021	0.041400	-0.749998
Right Y	=	0.000017	0.041117	-0.745286
Roll (units m/rad):				
Left X	=	0.000000	-0.175865	0.257751
Right X	=	-0.000001	0.177716	0.404189
Left Y	=	-0.000001	0.360889	-0.248543
Right Y	=	0.000000	0.361160	0.477119

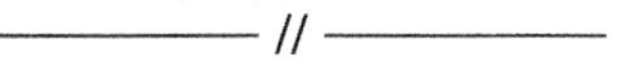

14

Installation Ratios

14.1 Introduction

The dynamic behaviour of a vehicle depends on the on the effect of the spring as seen at the wheel, commonly called the 'wheel rate' or 'effective spring stiffness', and on the damping coefficient effective at the wheel. However, the spring and damper are not installed at this point, but elsewhere on the suspension. The effective spring and damper characteristic at the wheel must therefore be related to the characteristics of the actual spring and damper. On a double-wishbone suspension they often operate about half-way out on the bottom suspension arm. On a racing car, often they operate through a linkage, including various forms of intermediate rocker. Basically the effect of the spring and damper depends on the ratio of velocities spring-to-wheel and damper-to-wheel when the wheel is displaced in its bump action. In general this velocity ratio, also known as the 'motion ratio' or 'installation ratio', is not a constant, but varies with the wheel bump position. Analysis of the effect of installation of the spring and damper therefore involves:

(1) evaluation of the relative motion of the spring, damper and wheel,
(2) consideration of the implications of this ratio.

Evaluation of the motion ratio is not too critical, especially in the early stages of design, so approximate methods, within 1 or 2 per cent, are of some value, and algebraic methods have the benefit of giving more insight into a design than do numerical methods. Hence the following sections describe algebraic or drawing methods for common suspensions. These methods may also be used as the basis of computer programs, as an alternative to the common finite increment of displacement method, or to the numerical velocity-diagram method.

The main principle involved is the geometry of the lever, or rocker. Rigid-arm suspensions, such as trailing arms or swing axles, follow on the same principle. Double wishbones require a different kind of analysis, to relate the wheel motion to the motion of the end ball joint of the relevant arm, usually the lower one, which is carrying the spring or damper, giving the first factor of the total motion ratio. Once this is known, normal rocker analysis will give the other factor. For struts, a slightly different analysis is required.

14.2 Motion Ratio

The main notation required here is as in Table 14.2.1. Note that the subscript K is used for the spring because the subscript S has already been allocated to 'suspension'.

At a given suspension bump position z_S from static ride height, the spring compression is x_K relative to static. A small further suspension bump motion δz_S results in a corresponding further spring compression

Table 14.2.1 Main nomenclature for motion ratios

L_D	m	damper length
L_K	m	spring length
R_D	—	damper motion ratio
R_K	—	spring motion ratio
V_D	m/s	damper compression velocity
V_K	m/s	spring compression velocity
V_P	m/s	velocity of intermediate pushrod
V_S	m/s	suspension bump velocity (wheel)
x_D	m	damper compression from static
x_K	m	spring compression from static
z_S	m	suspension bump

δx_K, Figure 14.2.1. The ratio of these is the displacement motion ratio for the spring at the suspension bump position z_S. This is denoted $R_{K/S}$ (spring relative to suspension bump), generally abbreviated to R_K:

$$R_{K/S} = \frac{\delta x_K}{\delta z_S}$$

Note that this is the spring compression increment divided by the suspension bump increment, not vice versa. It is the suspension bump movement that is considered to be the reference movement. This value of R_K will be independent of z_S and δz_S only if the system is linear, that is, if the motion ratio R_K is constant, which is generally not true. Therefore, in general the above expression is usable only for small δz_S. With a suitable computer program to analyse the suspension geometry, the motion ratio may be evaluated in the above way, with δz_S being given a suitable small value (e.g. 0.1 mm), depending on the precision with which the suspension position can be calculated. It is the actual suspension position that is solved for each of the adjacent conditions.

Mathematically, the motion ratio value is really the above ratio as δz_S tends to zero. The result, in the limit, is the derivative

$$R_K = \frac{dx_K}{dz_S}$$

This provides a possible means of evaluation if an explicit algebraic expression is available for spring compression x_K as a function of suspension bump displacement z_S. For example, if it is known that, over

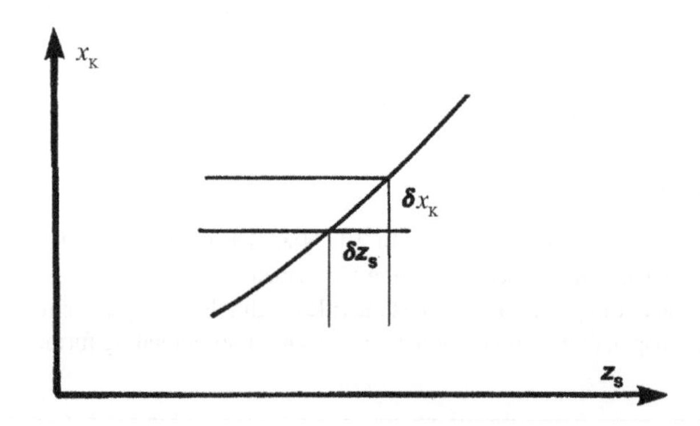

Figure 14.2.1 Spring compression displacement v. suspension bump displacement.

the range of interest, the spring compression, perhaps by an empirical curve fit, is

$$x_K = A_1 z_S + A_2 z_S^2$$

then

$$R_K = \frac{dx_K}{dz_S} = A_1 + 2A_2 z_S$$

The derivative definition of R_K may be extended by inserting a time element:

$$R_K = \frac{dx_K}{dt} \cdot \frac{dt}{dz_S}$$

The terms on the right-hand side are a velocity and a reciprocal velocity, so R_K may be written as a ratio of velocities:

$$R_K = \frac{V_K}{V_S}$$

The spring motion ratio is therefore identically equal to the quotient of spring compression velocity over suspension compression (bump) velocity. Because the numerator and denominator are both velocities, the quotient is dimensionless, and correctly called a ratio.

The velocity ratio definition provides a practical means of evaluating R_K if the velocity diagram can be drawn. Frequently this is feasible, even easy, for a two-dimensional set-up, and may also be practical for a three-dimensional suspension layout if the parts can be treated sequentially as two-dimensional parts (e.g. a racing suspension with pushrod and rocker, in many cases). Even a three-dimensional velocity diagram can be 'drawn' by computer. In a direct velocity-diagram solution of the suspension, the single actual suspension position is used, and the velocities are obtained directly at the actual suspension position; there is no question of an increment of position. In a manual solution, the velocity diagram may indeed be solved by accurate drawing, but equally if not better by sketching and algebraic solution.

Considering the displacement x_P of an intermediate pushrod, the overall spring motion ratio R_K may be expressed as

$$R_K = \frac{\delta x_K}{\delta z_S} = \frac{\delta x_K}{\delta x_P} \cdot \frac{\delta x_P}{\delta z_S}$$

or, in the limit, it is the derivative product

$$R_K = \frac{dx_K}{dx_P} \cdot \frac{dx_P}{dz_S} = R_{K/P} \cdot R_{P/S}$$

so the overall motion ratio is the product of the sequential ratios. By induction, this obviously extends to any number of sequential ratios. In terms of the velocities V_K, V_P and V_S,

$$R_K = \frac{V_K}{V_S} = \frac{V_K}{V_P} \cdot \frac{V_P}{V_S}$$

or, more generally, for multiple ratios in sequence

$$R_K = \frac{V_K}{V_{P1}} \cdot \frac{V_{P1}}{V_{P2}} \cdot \ldots \cdot \frac{V_{P,N-1}}{V_{P,N}} \cdot \frac{V_{P,N}}{V_S}$$

Hence a suspension mechanism can be solved by separate analysis of the sequential parts. This may be especially valuable where a complex suspension can be treated as separate sub-mechanisms, each analysable in two-dimensional motion, and each analysed in the most convenient way for that particular part.

In general the spring or damper motion ratio is not uniform through the bump motion, it is a function of z_S, the bump position of the suspension:

$$R_K \equiv R_K(z_S)$$

For passenger cars the value may be approximately constant or may increase slightly with bump, in the latter case usually also reducing with droop. For racing cars a rather rapidly increasing velocity ratio may be used. Where applied to a spring, this gives a rising stiffness with bump, and is called a 'rising-rate' suspension. Applied to a damper, it is rising-rate damping. Obtaining rising rate by mechanism design is generally much easier and more controllable than doing so by manufacturing rising rate springs or position-dependent dampers. Often the spring and damper are fitted co-axially and have the same motion ratio. For rising rate, this conveniently gives the desirable increase of wheel damping coefficient with increasing wheel rate. However, a rising rate motion ratio has a different effect on springs from that on dampers, as discussed in Sections 14.14 and 14.15.

14.3 Displacement Method

One of the ways to obtain the motion ratio value for a given nominal suspension bump position is by analysis of a pair of slightly different suspension positions. If the position analysis is undertaken by a drawing method it is prone to inaccuracy because of the relatively small difference of positions. Hence, the drawing must be undertaken by an experienced draughtsman at a large scale. With less emphasis on accuracy, a wide spread of the two positions will give an average ratio over the movement which may be useful in some cases. Also a sequence of, say, eight to ten positions may be drawn throughout the bump motion, and the damper compression plotted as a graph against suspension position, with the curve smoothed through the points. This helps to reveal any errors. The spring motion ratio for any particular position is then the gradient $\mathrm{d}x_K/\mathrm{d}z_S$ of the curve at the particular suspension bump position.

Drawing methods were the traditional technique in bygone days, but are regarded as somewhat archaic nowadays because of the availability of computers. However, drawing methods may still be of some value, for example if the particular configuration cannot be handled by existing software. In that case, careful drawing or velocity diagram analysis may be more expedient than writing new software, unless the configuration is known to be one of lasting interest. Computer-aided drawing packages offer potentially enhanced accuracy, of course, and can even be used to solve simple installation ratio problems.

In some cases the position relationships may be found algebraically. In three dimensions the solutions are too unwieldy for hand solution in other than isolated cases, for example to check early results of a computer analysis. However, if the system can be treated in a two-dimensional sequence then an analytical expression for positions may be possible, with differentiation giving the velocity ratio.

On one occasion the author was asked to adjudicate between two commercial computer programs which gave substantially different results. A careful drawing solution clearly sided with one of the two programs, and also suggested the reason for the fault in the other program. Drawing methods may therefore still be of value, even in this computerised day and age.

The actual solution of a given suspension position for a given body position (heave, roll and pitch) and suspension bump due to road displacement and tyre compression requires two- and three-dimensional coordinate geometry (analytical geometry), discussed in later chapters.

14.4 Velocity Diagrams

A velocity diagram is a coordinate diagram on which any point corresponds to a given velocity, drawn to a particular velocity scale (e.g. 10 mm/s : 1 mm). Hence, any point (X, Y) on any component has a velocity

(V_X, V_Y) which can be plotted on the diagram at a point (V_X, V_Y) using a suitable scale; in practice this is done for a few special points, essentially those at the joints between components, that is, at the pivot points and at sliders, at the ball joints in a suspension. The distances between points in the velocity diagram represent velocity magnitudes. The position of one point relative to another represents the relative velocity. The velocity diagram can be said to be drawn in 'velocity space'. The velocity diagram is constructed sequentially, point by point, according to the particular mechanism. It may drawn accurately to a given scale, or, often more conveniently and more accurately, simply sketched and solved algebraically. Accurate drawing requires some numerical calculation anyway. A full description of the principles of velocity diagram construction appears in standard texts (see the references at the end of this chapter).

Velocity diagrams are usually two-dimensional (V_X, V_Y), corresponding to common two-dimensional mechanisms, which have point positions (X, Y). Hence, they can be drawn on a two-dimensional surface, such as a piece of paper. Solution may be by geometrical drawing constructions, or by sketching and algebra with coordinate geometry.

However, three-dimensional velocity diagrams are also possible conceptually, although difficult to draw by hand, but can be represented in a computer by three-dimensional velocity vectors (V_X, V_Y, V_Z) obtained algebraically. This can even be extended to three-dimensional acceleration diagrams.

Velocity diagrams and displacement methods for obtaining spring and damper motion ratios are independent methods, and may be used to provide a check on each other.

14.5 Computer Evaluation

Commercial and proprietary packages are available for analysing suspension geometry on computers. At least one program, by the author, includes design facilities to specify, for example, a given bump steer for which the computer will choose suitable link dimensions.

When such numerical analysis packages work, they generally work well and accurately. However, not all packages can handle all possible configurations, especially for the more esoteric types used in racing. Also the 'assembly logic' of some packages is not always reliable, so that in some configurations there is a sudden total failure, usually manifested by a 'square root of negative number' error (see Chapter 16).

A full three-dimensional analysis program to obtain the motion ratios is a considerable job to write well. Such a program is certainly very useful, but provides specific numbers rather than design insight, and is best considered as an adjunct to qualitative understanding and simple algebraic models rather than completely replacing them.

14.6 Mechanical Displacement

If the suspension to be analysed already exists, or it is viable to construct one, perhaps adapting some other design, then it may be useful to actually measure the displacement graph of spring position and damper position against bump position, using dial gauges. This graph may be compared with solutions achieved by drawing, analytical methods or computers. The results are likely to be disappointing in some cases. Fair agreement is usually obtained for passenger cars, but in the case of racing cars, with long links and stiff springs, the elasticity of suspension members and even of the chassis at the mounting points may give rise to substantial discrepancies. In that case, if it is accepted that the discrepancy is indeed due to compliance in the suspension, then a more elaborate suspension model may be required. Figure 14.6.1 shows such a model, in which the suspension linkage compliance K_L is included. Under dynamic conditions, the suspension bump deflection

$$z_S = z_W - z_B$$

is no longer related directly to the spring deflection by a simple geometric motion ratio. However, for a given static position there are particular loads and deflections, so there is still a definite relationship $x_K(z_S)$.

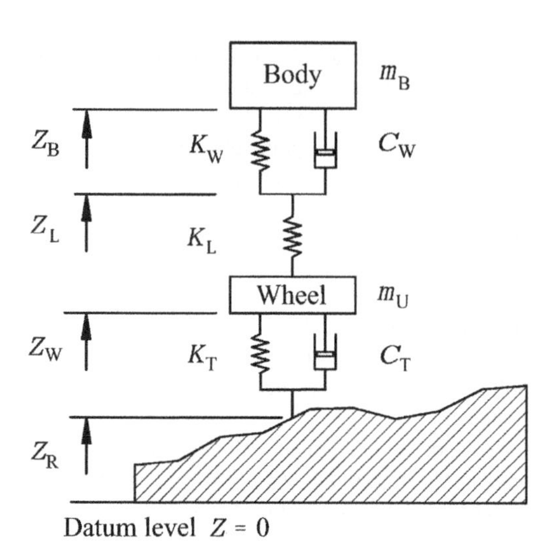

Figure 14.6.1 Heave-only suspension model including link compliance K_L, which could alternatively be placed above the K_W/C_W components.

Figure 14.6.1 assumes that the suspension spring and damper operate at the same point, as indeed is applicable for most racing cars, or else more than one suspension compliance must be included. In a static position, the damping forces are zero, and the link and suspension spring stiffnesses act in series.

The results of a motion ratio analysis must therefore be applied with some caution for more compliant suspensions. Even then, however, the motion ratio provides basic insight into the springs and dampers required, and into the design of a mechanism for rising rate.

14.7 The Rocker

The basic component in motion ratio evaluation is the rocker, or lever, which arises in principle, even if not explicitly as a rocker, in virtually all suspensions. In some cases, especially modern racing cars, a rocker is included as such to link the basic suspension to the spring and damper. By virtue of including a pushrod, or pullrod, and rocker, the spring and damper can be better positioned, vertically inside the body, or, as is common nowadays, laid down horizontally on top of the body or along the upper sides of the gearbox. This improves the aerodynamics by removing the spring–damper unit from the high-speed airflow, and the inclusion of a rocker in the system makes it very easy to change the motion ratio and rising rate simply by changing the rocker.

The function of a rocker, illustrated in Figure 14.7.1, can be specified by three aspects:

(1) the total rocker angular deviation between pushrods, θ_{RD};
(2) the rocker motion ratio, R_R;
(3) the rocker rising rate factor, f_R.

The rocker shown in Figure 14.7.1 is a completely general one. There are numerous simpler special cases. The angle θ_{RD} is the rocker deflection angle, the total angular difference between input and output, which will actually vary a little over the range of motion, and is specified at the normal ride height.

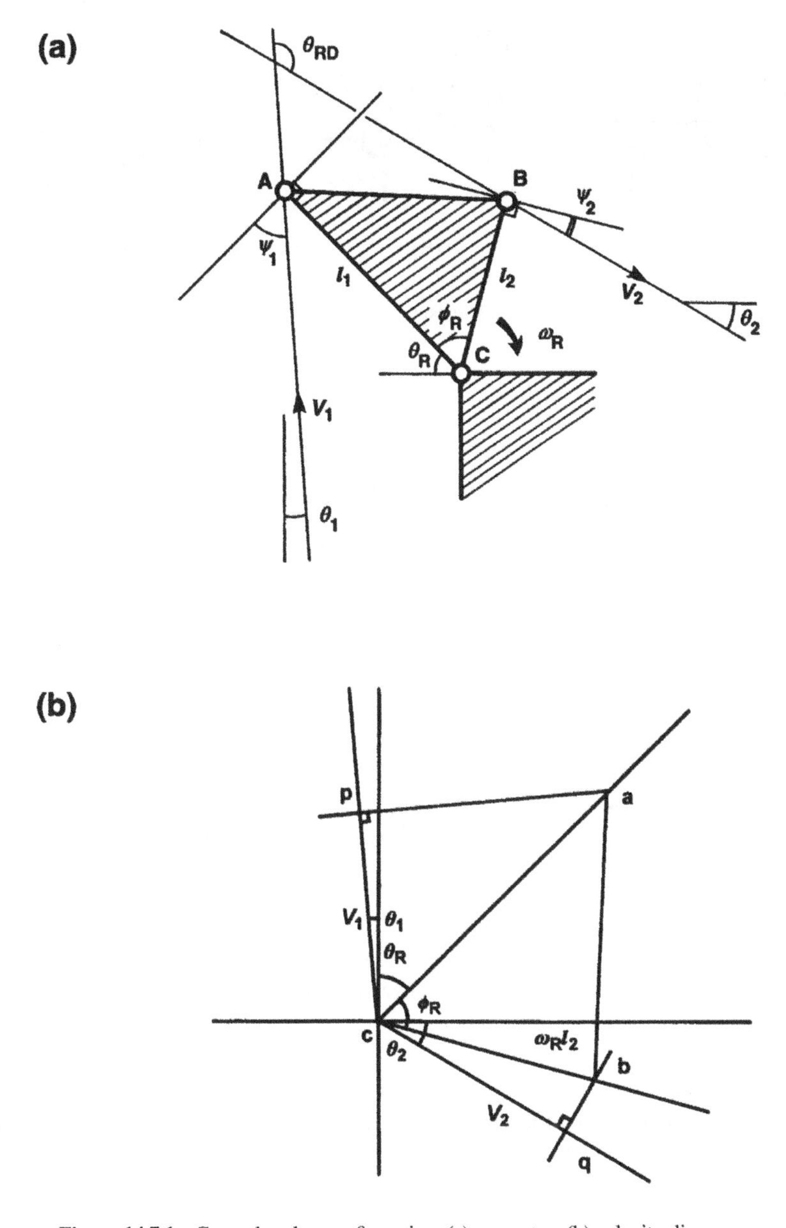

Figure 14.7.1 General rocker configuration: (a) geometry; (b) velocity diagram.

Figure 14.7.2 shows three examples of deflection angle θ_{RD} (0°, 90°, 180°). Intermediate deflection angles are equally possible.

The rocker motion ratio R_R, specified at the normal ride height, depends most obviously upon the arm lengths from the pivot axes to the input and output points, but also depends on the angular position of the input and output rods. Figures 14.7.3 and 14.7.4 illustrate some possible ways of achieving motion ratios V_2/V_1.

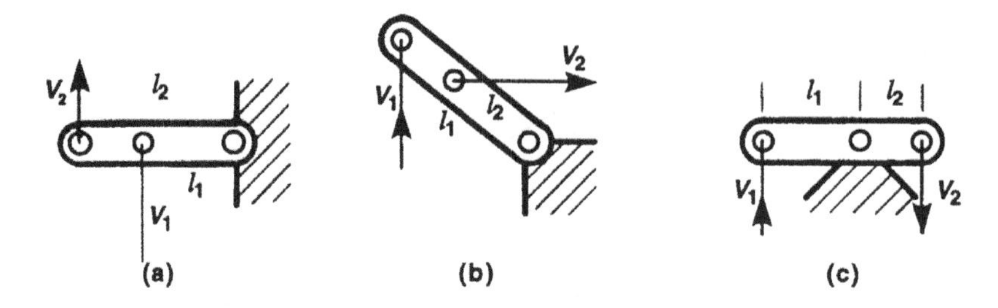

Figure 14.7.2 Simple rockers with various deflection angles.

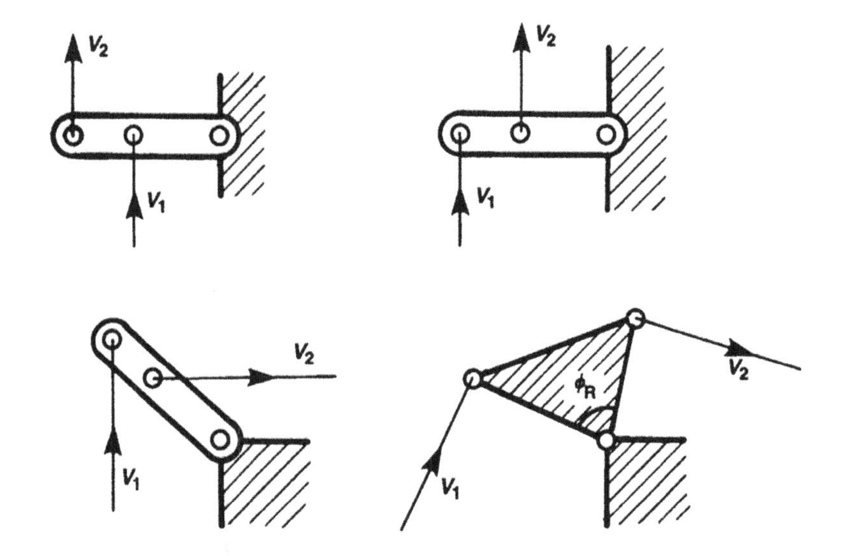

Figure 14.7.3 Simple rockers with various motion ratios.

The rising rate of a rocker is the proportional increase of motion ratio per unit of rocker rotation:

$$f_R = \frac{1}{R_R} \cdot \frac{dR_R}{d\theta_R}$$

To express this in terms of suspension motion at the wheel (i.e. as wheel rising rate), the geometry of the rest of the suspension must be known, so wheel rising rate is not necessarily a property of the rocker alone,

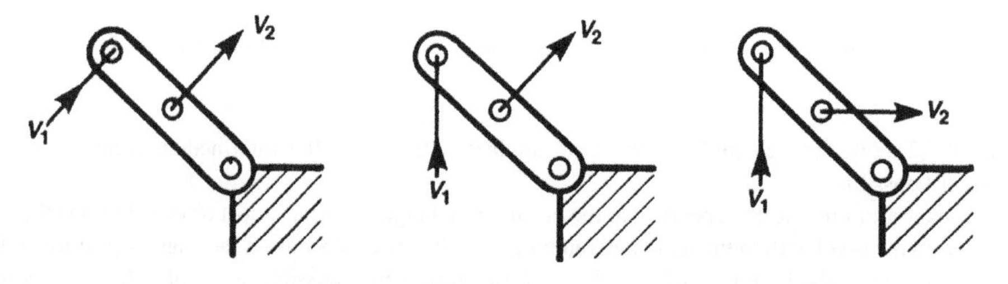

Figure 14.7.4 Simple rockers with various rising rates (zero in the first case).

but becomes so if the rest of the system is essentially linear. When a rising rate is desired, the design method commonly applied in practice is to guess at some dimensions and then to play with them on the computer. What is needed is a systematic design procedure.

The properties of the rocker may be deduced for special cases, as required, but in fact can be derived for the general rocker of Figure 14.7.1 quite easily, by virtue of an informed choice of representational parameters, arrived at by experience. It is assumed in the following analysis that the motion is planar. If the input or output rods are not parallel to the rocker plane, defined as a plane perpendicular to the axis of rotation of the rocker on the body, then the velocities in Figure 14.7.1 are the rocker plane velocities, related to the actual rod velocities by the cosine of the out-of-plane angle.

The parameters in Figure 14.7.1 are:

(1) the rocker included angle ϕ_R, positive when the output leads the input as shown;
(2) the input and output rocker arm lengths l_1 and l_2;
(3) the input and output rod offset angles ψ_1 and ψ_2, between the rod and the tangent perpendicular to the corresponding arm radius;
(4) the rocker position θ_R, from some appropriate datum, usually the normal ride position;
(5) the input and output rod velocities, in the rod directions, V_1 in and V_2 out;
(6) the rocker angular velocity ω_R.

Of the above, lengths l_1 and l_2 are constant, as is ϕ_R, these three parameters being the essence of the rocker geometry. Other parameters will vary, although ω_R is normally deemed to be some constant value for the purpose of analysis.

The performance parameters of the rocker, θ_{RD}, R_R and f_R, are derived as follows. The rocker deviation angle is

$$\theta_{RD} = \psi_1 + \phi_R + \psi_2$$

In practice, for design purposes, this is required in the form

$$\phi_R = \theta_{RD} - \psi_1 - \psi_2$$

Figure 14.7.1(b) is the velocity diagram for the rocker. At rocker angular speed ω_R, assumed for analysis, the tangential speed of point A is $\omega_R l_1$, perpendicular to line AC, so at angle θ_R from the vertical, giving point a representing V_A. Similarly, point b, representing V_B, is established at $\theta_R + \phi_R$.

The velocity of A can be resolved into components parallel and perpendicular to the input rod at angle θ_1 to the vertical. To do this, construct in the velocity diagram the line at θ_1 from the vertical, parallel to the input rod, and drop the perpendicular from a, giving point p. The length cp represents the velocity along the rod, whilst pa represents the tangential velocity of one end relative to the other. Similarly for the output, so the actual damper compression speed on the output is represented by cq. The rocker motion ratio R_R is given by

$$R_R = \frac{V_Q}{V_P} = \frac{\mathrm{cq}}{\mathrm{cp}}$$

Hence the velocity ratio of the rocker, defined by

$$R_R = \frac{V_2}{V_1}$$

is given by

$$R_R = \frac{\omega_R l_2(\theta_R + \phi_R - \theta_2)}{\omega_R l_1 \cos(\theta_R + \theta_1)}$$

This may be expressed more concisely as

$$R_R = \frac{l_2 \cos \psi_2}{l_1 \cos \psi_1}$$

This result may be summarised very simply by a physical interpretation. The denominator is the perpendicular distance from the axis to the line of action of the input link, Figure 14.7.1. The numerator is the perpendicular distance from the axis to the line of action of the output link.

As an alternative interpretation, using the rocker arm length motion ratio

$$R_{RL} = \frac{l_2}{l_1}$$

and the rod angle motion ratio

$$R_{R\psi} = \frac{\cos \psi_2}{\cos \psi_1}$$

the rocker motion ratio R_R may be expressed as

$$R_R = R_{RL} R_{R\psi}$$

The first of these new parameters, R_{RL}, is constant. The second, $R_{R\psi}$, varies as the rocker moves, and provides the basis of design for rising rate.

The input and output have tangential velocities, $V_{tan,in}$ and $V_{tan,out}$, one end relative to the other, of

$$V_{tan,in} = \omega_R l_1 \sin \psi_1$$

$$V_{tan,out} = \omega_R l_2 \sin \psi_2$$

These rotate the rods, thereby also altering the ψ angles, in a way which depends on the length of the input and output rods. This can be dealt with separately, and is, in any case, generally a much smaller effect than the basic rocker rising rate effect. Considering the input and output rods to remain parallel to their starting angle, and considering ψ to be positive as shown in Figure 14.7.1, at rocker angle position θ clockwise from normal ride height,

$$\psi_1 = \psi_{Z1} + \theta$$

$$\psi_2 = \psi_{Z2} - \theta$$

where ψ_{Z1} and ψ_{Z2} are the values of ψ at normal static ride height (zero deflection).

The rocker angle motion ratio factor is then

$$R_{R\psi} = \frac{\cos(\psi_{Z2} - \theta)}{\cos(\psi_{Z1} + \theta)}$$

Depending on the application, it may be convenient simply to think in terms of the motion ratios at two positions, say zero bump and some expected bump position. However, the mathematical rising rate factor, at zero bump defined earlier, was

$$f_R = \frac{1}{R_R} \frac{dR_R}{d\theta_R}$$

with

$$R_{R\psi} = R_{R\psi 0}(1 + f_R\,\theta)$$

The rocker angle motion ratio at θ is

$$R_{R\psi} = \frac{\cos(\psi_{Z2} - \theta)}{\cos(\psi_{Z1} - \theta)}$$

The rising rate factor f_R is given by

$$f_R = \frac{1}{\theta}\left(\frac{R_{R\psi}}{R_{R\psi,0}} - 1\right)$$

which, by considering infinitesimal θ, with substitution and condensation, becomes

$$f_R = \tan\psi_1 + \tan\psi_2$$

Hence, the rising rate is governed by the angles ψ_1 and ψ_2. Note that these are positive as defined in Figure 14.7.1 and the signs must be respected; also they may well change sign within the range of motion of the rocker. Physically, the reason for the rising rate is simply that the input moment arm is increasing, and the output one is decreasing.

At rocker deflection θ the motion ratio R_R is therefore estimated to be

$$R_R = R_{RL} R_{R\psi}(1 + f_R\,\theta)$$

This method of rising rate analysis is useful for a preliminary appraisal, giving a first estimate of the required values for ψ_1 and ψ_2 in combination, which then, in conjunction with the required rocker angular deflection θ_{RD}, gives the rocker included angle

$$\phi_R = \theta_{RD} - \psi_1 - \psi_2$$

Practically, because of the packaging problems of large rockers, fairly large rocker angular movements are used in practice, so once a first estimate has been made it is more practical to work with two or more actual rocker positions, for example normal ride height and a bump position, and to obtain two corresponding values of motion ratio.

The usual situation is that a given increase of ratio is required for a given rocker angular displacement, estimated from a given wheel bump motion, intermediate ratio and rocker input arm length. The required increase of rate may be shared between input and output. The required increase of rate on the input, say, is then known. Let the ratio be $r = R_2/R_1$. From the angle motion ratio

$$R_{R\psi} = \frac{\cos(\psi_{Z2} - \theta)}{\cos(\psi_{Z1} + \theta)}$$

may be obtained

$$r = \frac{\cos\psi_{Z1}}{\cos(\psi_{Z1} + \theta)}$$

This may be solved for ψ_{Z1} as

$$\tan\psi_{Z1} = \frac{1}{\sin\theta}\left(\cos\theta - \frac{1}{r}\right)$$

giving the correct initial angular position for the arm. In practice it may be difficult to obtain a low enough ratio with a practical rocker size, and ψ_{Z1} may be negative. This gives a rate which will fall slightly before then rising to the required value.

For design purposes, rocker design is a matter of juggling the lengths and angles according to the given input pushrod motion and the required rising rate. Two points to bear in mind are that basically it is the square of the motion ratio that controls the effect seen at the wheel, and that a motion ratio rising from normal ride height will also generally fall when below the normal ride height. Hence, large motion ratio changes must be designed with caution. Rising rate also causes an important additional stiffness term even with a linear spring (see Section 14.14).

14.8 The Rigid Arm

Rigid-arm suspensions, such as a trailing arm, have a single arm from a pivot axis, the wheel camber angle being rigidly fixed relative to the arm. It is possible, but unusual, to have steering with such a system. Rigid arms may be classified in various ways. From the geometric point of view, the important distinction is the angle, ψ_A, between the arm pivot and the vehicle centreline in plan view, Figure 14.8.1(a). Sometimes there is a non-zero angle of the arm pivot, ϕ_A, in rear view, Figure 14.8.1(b), as discussed in Chapter 11.

The basic classifications by ψ_A are:

(1) trailing arm (90°);
(2) semi-trailing arm (e.g. 70°);
(3) leading arm (90°);

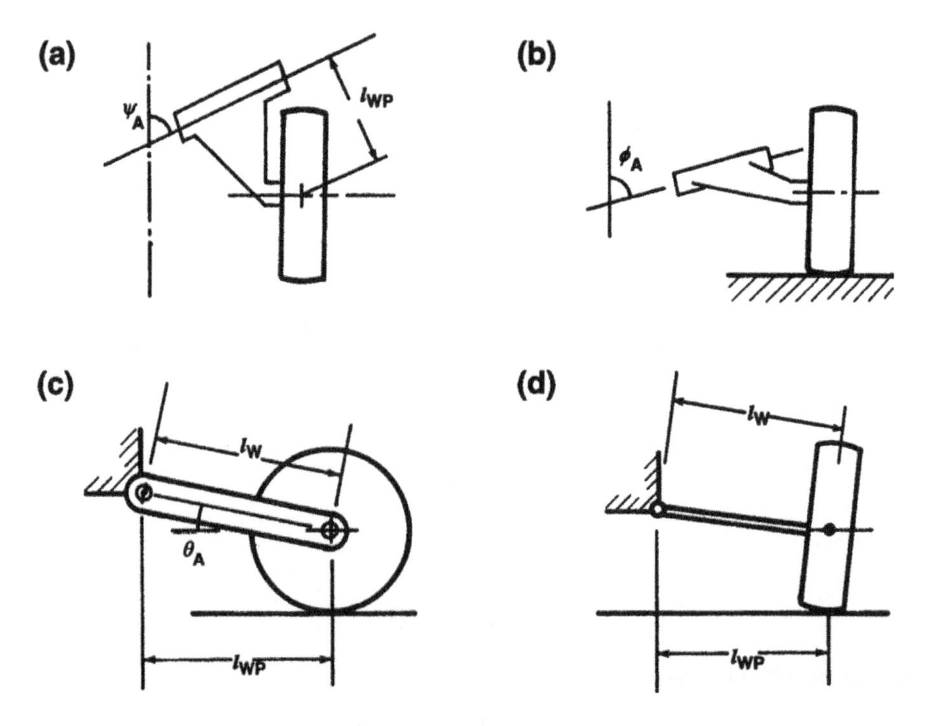

Figure 14.8.1 Rigid-arm suspensions: (a) plan view of semi-trailing arm; (b) rear view of semi-trailing arm; (c) side view of trailing arm; (d) rear view of swing axle.

(4) swing axle (0°);
(5) transverse arm (0°);
(6) semi-trailing swing axle (e.g. 45°).

The spring and damper usually act directly on the arm. In any case, it is necessary to obtain the relationship between the suspension bump velocity, that is, of the vertical wheel velocity at the contact patch, relative to the body, and the angular velocity of the arm. The radius of action of the wheel is l_{WP} in plan view, Figure 14.8.1(a). For an arm angular speed ω, the tangential speed of the wheel is ωl_{WP}, but this is not vertical in rear view, so the actual suspension wheel bump velocity V_S is

$$V_S = \omega_A l_{WP} \cos \phi_A$$

The velocity ratio of arm to wheel $R_{A/W}$ is therefore

$$R_{A/W} = \frac{\omega_A}{V_S} = \frac{1}{l_{WP} \cos \phi_A} \left[\text{rad s}^{-1}/\text{m} \cdot \text{s}^{-1}\right]$$

Note that the pivot axis plan angle θ_A does not appear in this expression. Neither does any influence of the angle of the arm in side view, because this has been incorporated by using the plan length l_{WP}, which may in fact vary somewhat through bump movement, because

$$l_{WP} = l_W \cos \theta_A$$

For any given bump position, θ_A follows, whence l_{WP} and $R_{A/W}$:

$$R_{A/W} = \frac{1}{l_W \cos \theta_A \cos \phi_A}$$

With some consideration, the above can be applied to any of the rigid-arm suspensions listed above.

The second part of the motion ratio then follows from the position of the spring or damper. Figure 14.8.2 shows the rigid arm in elevation viewed along the pivot axis. The spring may not be in the plane of the elevation, but will be close to it, with out-of-plane angle α_K.

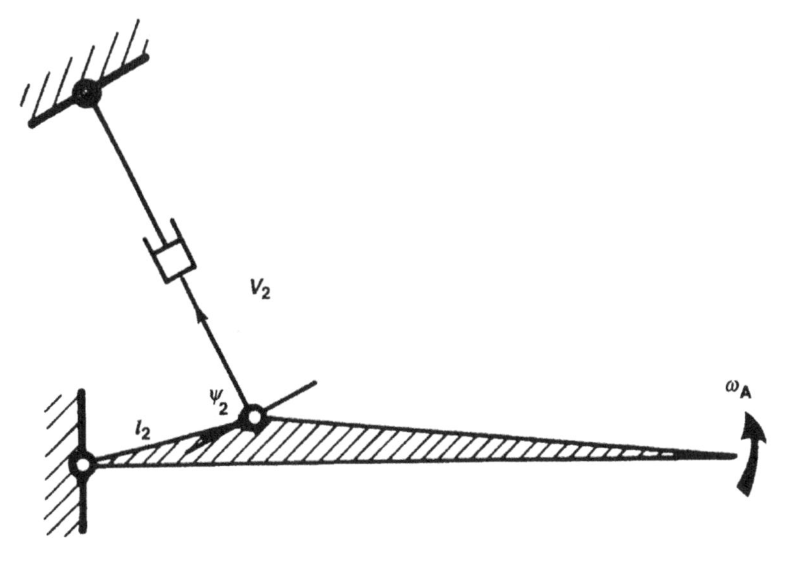

Figure 14.8.2 Rigid arm suspension with spring–damper unit, shown in elevation.

The rigid-arm analysis is now easily completed as for analysis of the output of the rocker:

$$V_2 = \omega_A l_2 \cos\psi_2$$

Allowing for the out-of-plane angle α_K, the spring compression velocity is

$$V_K \cos\alpha_K = V_2$$

so

$$V_K = l_2 \frac{\cos\psi_2}{\cos\alpha_K}\omega_A = l_2 \frac{\cos\psi_2}{\cos\alpha_K} \frac{V_B}{l_W \cos\theta_A \cos\phi_A}$$

Hence, the spring motion ratio R_K is

$$R_K = \frac{V_K}{V_S} = \frac{l_2}{l_W}\left(\frac{\cos\psi_2}{\cos\theta_A \cos\phi_A \cos\alpha_K}\right)$$

This is very similar to the expression for a simple rocker, but includes effects from the angles ϕ_A and α_K.

As in the case of the rocker, judicious choice of ψ_2 at zero bump will give a rising rate, or not, as desired, rising rate occurring for ψ_2 as shown, the angle reducing with bump action, increasing the effective damper moment arm. If anything, this is easier to design than an extra rocker because the angular motion of the suspension arm will generally be less than for a rocker, and the smaller angle makes a more linear progression possible. For example, a bump deflection z_s of, say, 100 mm on an effective arm length l, which may be as much as 1.3 m for a transverse arm or swing axle, gives an angular bump motion of the arm of about 4–5°. For a trailing arm of length 0.4 m, the angle is 15°.

14.9 Double Wishbones

The double wishbone or double A-arm suspension is a little more difficult to solve than the simple rigid arm. As before, it is necessary to establish a motion ratio between the suspension bump velocity and the angular velocity of the arm which operates the spring or damper, or operates the pushrod to the rocker for a racing car. Figure 14.9.1(a) shows the basic configuration.

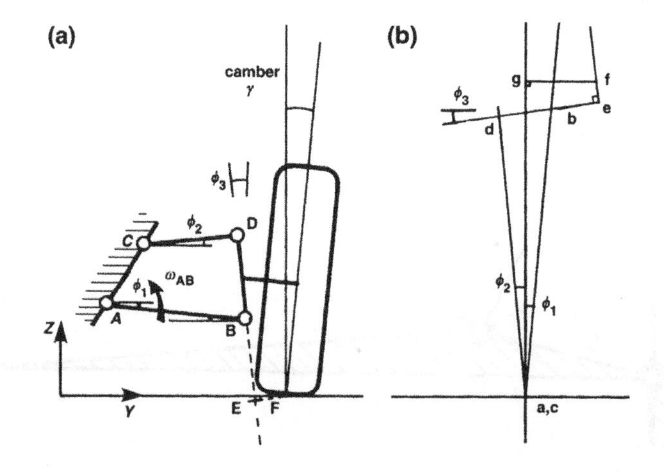

Figure 14.9.1 Double-wishbone suspension: (a) position diagram; (b) velocity diagram.

If the linear bump camber coefficient of the suspension is already known, then a particularly simple method is possible. The linear bump camber coefficient ε_{BC1} is the rate of change of wheel camber angle γ with suspension bump, arising from suspension geometry:

$$\varepsilon_{BC1} = \frac{d\gamma}{dz_S} \quad \text{(rad/m)}$$

For a suspension bump velocity V_S, and for a reasonably constant ε_{BC1}, usually an adequate approximation for the present purpose, the wheel camber angular velocity is

$$\frac{d\gamma}{dt} = \varepsilon_{BC1}\frac{dz_S}{dt} = \varepsilon_{BC1}V_S$$

If the bump camber rate is highly varying, then the local value should be used, allowing for the quadratic bump camber coefficient ε_{BC2}:

$$\gamma_{BC} = \varepsilon_{BC1}z_S + \varepsilon_{BC2}z_S^2$$

$$\varepsilon_{BC} = \frac{d\gamma}{dz_S} = \varepsilon_{BC1} + 2\varepsilon_{BC2}z_S$$

From Figure 14.9.1, the vertical velocity of B differs from the vertical velocity of F by the camber angular velocity multiplied by the lateral difference of position e, where

$$e = X_F - X_B$$

The vertical velocity of B is therefore

$$V_B = V_S - e\varepsilon_{BC}V_S = V_S(1 - e\varepsilon_{BC})$$

Hence, the motion ratio $R_{B/S}$ of ball joint B to suspension bump is

$$R_{B/S} = 1 - e\varepsilon_{BC}$$

Realistic values are $e = 0.1$ m and $\varepsilon_{BC} = 1$ rad/m, which will give $R_{B/S}$ a value of 0.9, a substantial deviation from 1.0 which should certainly be included in the analysis.

In the absence of prior information on the bump camber coefficient, a velocity diagram may be considered, as in Figure 14.9.1(b). This is more easily constructed by initially assuming an angular velocity ω_{AB} for the lower link, rather than a bump velocity of the wheel. The body is deemed to be stationary, so points A and C are fixed points, with zero velocity, and therefore appear as a and c at the origin of the velocity diagram. The tangential velocity of B relative to A is $\omega_{AB}l_{AB}$, and the line ab in the velocity diagram is perpendicular to link AB, the length of ab being the tangential velocity at the diagram scale. This establishes point B. Line cd is perpendicular to CD, and bd is perpendicular to BD; the intersection gives point d.

To obtain the velocity of F, at the bottom of the notionally rigid wheel, in the position diagram project line DB and drop perpendicular from F, giving E. In the velocity diagram, the rigid wheel with the wheel upright is solved by scaling. Hence,

$$\frac{\text{be}}{\text{db}} = \frac{\text{BE}}{\text{DB}}$$

giving point e. Draw the perpendicular from e. DBEF is a left turn, so dbef is also to the left. Use

$$\frac{\text{ef}}{\text{de}} = \frac{\text{EF}}{\text{DE}}$$

to give point f. Finally, drop the perpendicular from f to the vertical axis, giving point g.

This completes the velocity diagram to some convenient scale for some angular velocity ω_{AB} of the lower arm. This methodology may, of course, form the basis of a computer program where repeated analysis is desired.

The velocities of interest may now be read from the diagram:

(1) The vertical velocity of the point F, that is, the suspension bump velocity, V_S, is represented by ag ($V_S = V_{G/A}$).
(2) The wheel scrub (lateral) velocity V_{WS} is represented by fg.
(3) The tangential velocity of D relative to B, $V_{D/B}$, is represented by bd.
(4) The tangential velocity of D relative to C, $V_{D/C}$, is represented by cd.

Hence the following may be deduced:

(1) The motion ratio of the lower arm to suspension bump, in units rad s^{-1}/m s^{-1} = rad/m, is

$$R = \frac{\omega_{AB}}{V_S} = \frac{V_{B/A}}{l_{AB} V_S}$$

(2) The camber angular velocity

$$\frac{d\gamma}{dt} = \frac{V_{D/B}}{l_{DB}}$$

(3) The bump camber coefficient

$$\varepsilon_{BC} = \frac{1}{V_S}\frac{d\gamma}{dt} = \frac{V_{D/B}}{l_{DB} V_S} = -\frac{l_{db}}{l_{DB} l_{ag}}$$

(4) The basic roll centre height (unrolled)

$$h_{RC} = \frac{1}{2} T \frac{l_{fg}}{l_{ag}}$$

In the present context, it is the velocity ratio that is of interest. The lower arm may then be analysed as a rocker output for the spring or damper drive, as was the rigid arm, to give the overall motion ratio.

14.10 Struts

The strut suspension is the usual choice nowadays for the front of passenger cars. The use of a strut at the rear is a little unusual, but has featured in several cases.

The usual strut incorporates the damper into the body of the strut, and has a surrounding spring. An alternative design, the damper strut, has only the damper in the strut body, with the spring acting separately on one of the arms. Geometrical considerations are the same, although, of course, in the latter case it is the arm which must be analysed for the spring motion ratio.

Overall, the method of analysis is similar to that of the double-wishbone suspension. Figure 14.10.1(a) shows a strut suspension. This is in fact the simpler version where the strut axis passes through the ball joint at B.

If the bump camber coefficient is already available, then the vertical velocity V_{ZB} of B is given by

$$V_{ZB} = V_S(1 - e\varepsilon_{BC})$$

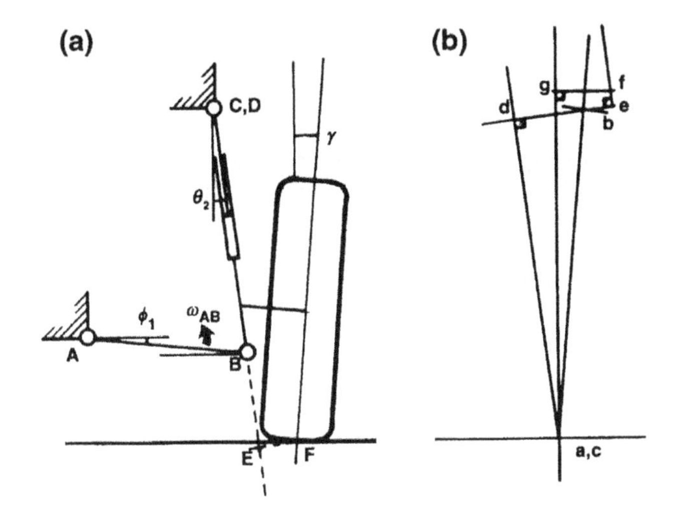

Figure 14.10.1 Strut suspension: (a) position diagram; (b) velocity diagram.

where

$$e = X_F - X_B$$

The tangential velocity of B is then

$$V_{B/A} = \frac{V_{ZB}}{\cos\phi_1}$$

and the damper compression velocity V_D is

$$V_D = V_{B/A}\cos(\phi_1 + \theta_2)$$

Hence,

$$R_D = \frac{V_D}{V_S} = (1 - e\varepsilon_{BC})\frac{\cos(\phi_1 + \theta_2)}{\cos\phi_1}$$

Realistic values may give a motion ratio below 0.9, in contrast to the naive expectation of a value close to 1.0.

The velocity diagram is shown in Figure 14.10.1(b). The body is stationary, so A and C are fixed points, appearing as a and c at the velocity diagram origin. Construction proceeds by assuming an angular velocity ω_{AB} for the bottom link:

$$V_{B/A} = \omega_{AB} l_1$$

perpendicular to AB, giving point b representing V_B in the velocity diagram. Velocity $V_{B/C}$ is the vector sum of longitudinal and tangential components, so construct a line through b perpendicular to CB and through c parallel to CB to intersect, giving point d. Point d represents the velocity of point D, which is a point notionally fixed to the lower part of the strut, and instantaneously coincident with the upper fixture point C. The damper compression velocity is represented by cd.

To obtain the suspension bump velocity, extend DB and drop the perpendicular from F to E. Use

$$\frac{\text{be}}{\text{bd}} = \frac{\text{BE}}{\text{BD}}; \quad \frac{\text{ef}}{\text{de}} = \frac{\text{EF}}{\text{DE}}$$

to give e and f, and drop the perpendicular from f onto the vertical axis to give g. Point f represents the motion of the base of the wheel, and g represents its vertical component.

From the velocity diagram can be obtained:

(1) the suspension bump velocity represented by ag;
(2) the wheel scrub velocity (fg);
(3) the tangential velocity of B relative to D (db).

Hence, the following may be calculated:

(1) the motion ratio R of the lower arm to suspension bump

$$R = \frac{\omega_{\text{AB}}}{V_{\text{S}}} = \frac{V_{\text{A/B}}}{l_{\text{AB}}\, V_{\text{A/G}}}$$

(2) the camber angular velocity

$$\frac{\text{d}\gamma}{\text{d}t} = \frac{V_{\text{B/D}}}{l_{\text{BD}}}$$

(3) the bump camber coefficient

$$\varepsilon_{\text{BC}} = \frac{\text{d}\gamma/\text{d}t}{V_{\text{S}}} = \frac{V_{\text{B/D}}}{l_{\text{BD}}\, V_{\text{S}}} = -\frac{l_{\text{bd}}}{l_{\text{BD}}\, l_{\text{ag}}} \left[\text{rad s}^{-1}/\text{m s}^{-1}\right]$$

(4) the basic roll centre height (unrolled)

$$h_{\text{RC}} = \frac{1}{2} T \frac{l_{\text{fg}}}{l_{\text{ag}}}$$

On front suspensions in particular, the damper axis is frequently aligned such that it does not pass through the ball joint at B, but rather inside or outside it. In that case the preceding analysis is still applicable, with the following provisos:

(1) The angle θ_2 used is that of the damper axis, not that of the steering axis CB.
(2) To obtain e and f, still use the steering axis line CB extended.

14.11 Pushrods and Pullrods

In formula racing cars in particular, it is normal practice nowadays to use double wishbones with a pushrod driving a rocker which operates the spring and damper. These can be solved readily by the processes already described, treating the motion ratio as the product of the sequence of ratios arising from the particular system. This generally involves:

(1) the relationship of wheel displacement to bottom arm angle;
(2) the relationship of bottom arm angle to pushrod displacement;
(3) the rocker ratio.

For a practical racing suspension, the wishbones are long and the pushrod angle may be quite low, giving a low velocity ratio. This gives proportionally larger pushrod forces, but the reduced stroke allows a compact rocker. Pullrods have been used in the past, but have now largely given way to pushrods. The pullrod analysis is very similar to that of the pushrod.

The rocker axis may be rotated to lay the damper along the vehicle. With the pushrod system this allows the front dampers to lie above the driver's legs, permitting the best aerodynamic shape for the front part of the vehicle.

In these more complex systems, when analysed as a series of two-dimensional sub-mechanisms, it may be necessary to incorporate some cosine factors to correct for out-of-plane motions, but these are usually quite small.

14.12 Solid Axles

Solid car axles, driven or undriven, may be located either by leaf springs, the Hotchkiss axle if driven ('live'), or, more often nowadays, by links. The springs and dampers may act on the axle itself, or on the links, as in Figure 14.12.1.

For bump analysis, if the spring or damper acts directly on the axle, as in Figure 14.12.2, above the wheel centreline, and the springs or dampers are vertical, then the motion ratio is very close to 1.0. However, the springs or dampers are sometimes angled, inwards at the top, in which case the bump motion ratio is $\cos\theta_K$ or $\cos\theta_D$.

If the springs act on the locating links, then the velocity diagram needs to be drawn, as in Figure 14.12.1. With this configuration, some rising rate can be achieved by angling the springs or dampers in side view. The vertical wheel velocity is found from the wheel centre, and its relationship to points B and C. The axle angular speed ω_A is

$$\omega_A = \frac{V_{B/C}}{l_{BC}}$$

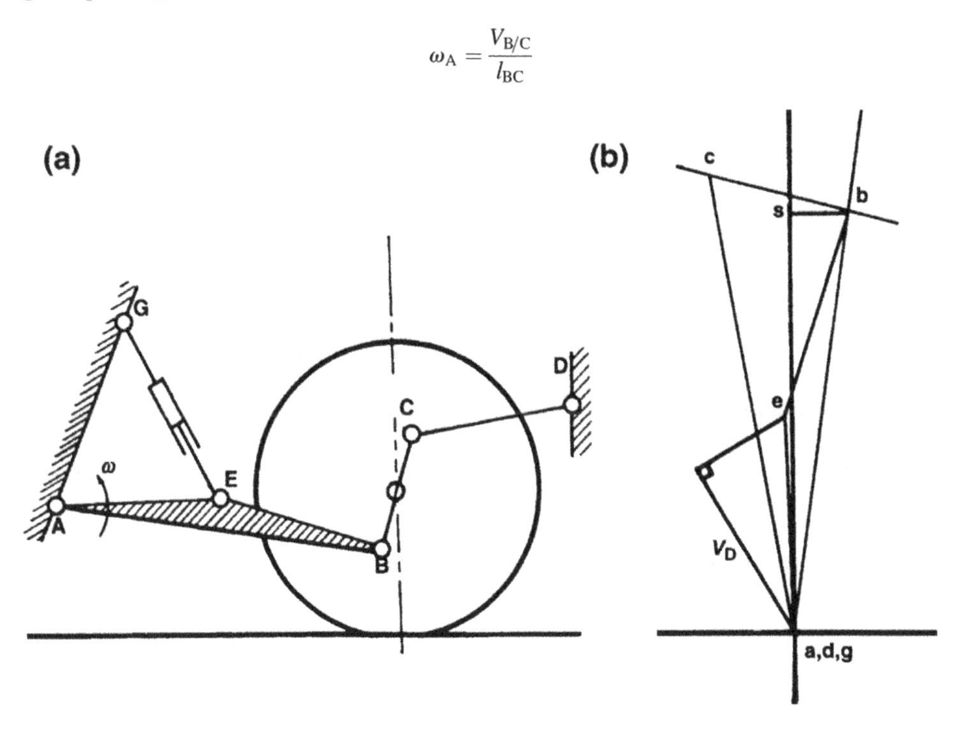

Figure 14.12.1 Example axle with link location, using a Watt's linkage at each end.

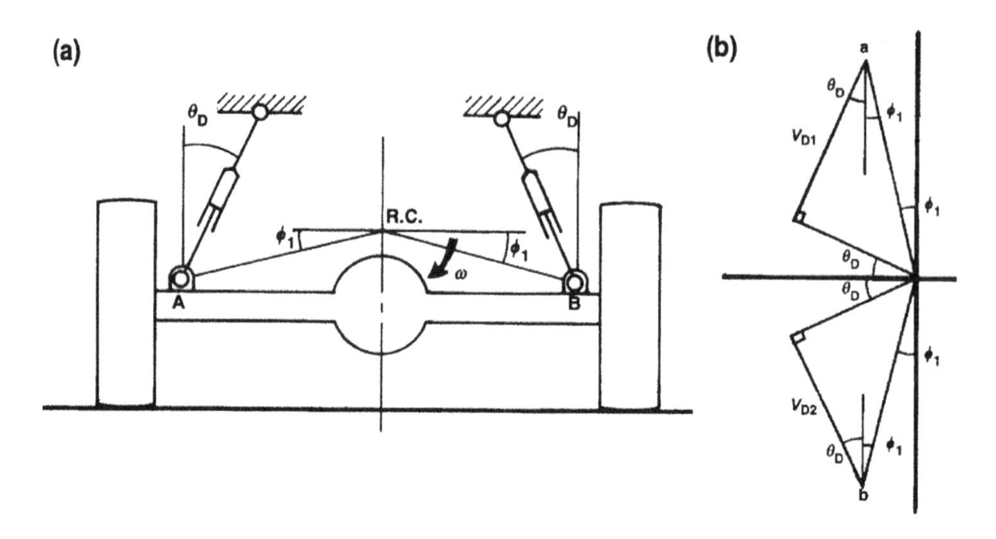

Figure 14.12.2 Inclined springs and dampers on an axle.

and the pitch/heave velocity ratio or coefficient ε_{APH} is

$$\varepsilon_{\text{APH}} = \frac{\omega_{\text{A}}}{V_{\text{S}}} \ [\text{rad/m}]$$

To analyse the roll stiffness or damping of the vehicle, for a simple axle, consider the vehicle to roll about the roll centre for the axle, which is the point of lateral location in the centre plane. For example, in the case of the use of a Panhard rod for lateral location, it is the point at which the rod pierces the longitudinal central vertical plane. The velocity diagram in Figure 14.12.2 shows that the damper velocity V_{D} is

$$V_{\text{D}} = \omega l_1 \cos(\phi_1 + \theta_K)$$

so the damper velocity ratio in roll is

$$R_{\text{K}\phi} = l_1 \cos(\phi_1 + \theta_K) \quad [\text{m s}^{-1}/\text{rad s}^{-1} = \text{m/rad}]$$

This will generally be substantially less than would be achieved with the springs acting directly at the wheels. This is because l_1 is typically only about 0.7 of half the track, to allow the springs to clear the wheels. This substantially reduces the roll stiffness, which may therefore be supplemented by an anti-roll bar. On the other hand, on a driven axle, the lower roll stiffness may be held to be an advantage, giving better traction during cornering.

For springs or dampers acting on the links instead of the axles, the above process gives the velocity of the end of the link (without θ_{K}). The link is then analysed to obtain the actual damper velocity.

14.13 The Effect of Motion Ratio on Inertia

Consider a suspension with a point mass at some position on the links, having a motion ratio R_{m} relative to the relevant suspension motion (wheel movement). Initially consider this motion ratio to be uniform, that is, not dependent on the suspension bump position. In this case,

$$R_{\text{m}} = \frac{\text{d}x_{\text{m}}}{\text{d}z_{\text{S}}} = \frac{x_{\text{m}}}{z_{\text{S}}} = \frac{\dot{x}_{\text{m}}}{\dot{z}_{\text{S}}} \qquad (\text{constant } R_{\text{m}})$$

where the movement of the mass is denoted by $x_{\rm m}$. This is not intended to suggest that the mass actually moves in the x-coordinate direction, but to indicate that it is not generally in the same z-direction as the suspension. The velocity of the mass is then

$$\dot{x}_{\rm m} = R_{\rm m}\dot{z}_{\rm S} = R_{\rm m}V_{\rm S}$$

and the acceleration of the mass is

$$\ddot{x}_{\rm m} = R_{\rm m}\,\ddot{z}_{\rm S} = R_{\rm m}A_{\rm S}$$

The force exerted on the mass to provide its acceleration is

$$F_{\rm m} = m\ddot{x}_{\rm m}$$

Neglecting friction, the corresponding force $F_{\rm Sm}$ at the suspension point, by virtual work is given by

$$F_{\rm Sm}\,\delta z_{\rm S} = F_{\rm m}\,\delta x_{\rm m}$$

giving

$$F_{\rm Sm} = \frac{{\rm d}x_{\rm m}}{{\rm d}z_{\rm S}}F_{\rm m} = R_{\rm m}F_{\rm m}$$

This result also follows naturally from the concept of a frictionless lever. The force required at the suspension is then

$$F_{\rm Sm} = R_{\rm m}F_{\rm m} = R_{\rm m}m\,\ddot{x}_{\rm m} = mR_{\rm m}^2\,\ddot{z}_{\rm S}$$

Therefore at a constant motion ratio the effective inertia, seen at the wheel, of a point mass, is just $mR_{\rm m}^2$.

Although perhaps slightly laboured, this development shows the assumptions in the theory behind this result.

The case of variable motion ratio is slightly more complicated. By definition, the motion ratio is, for linear or non-linear cases,

$$R_{\rm m} = \frac{{\rm d}x_{\rm m}}{{\rm d}z_{\rm S}} = \frac{\dot{x}_{\rm m}}{\dot{z}_{\rm S}}$$

The time rate of variation of the motion ratio can usefully be expressed in terms of its spatial variation and the suspension velocity:

$$\dot{R}_{\rm m} = \frac{{\rm d}R_{\rm m}}{{\rm d}t} = \frac{{\rm d}R_{\rm m}}{{\rm d}z_S}\cdot\frac{{\rm d}z_{\rm S}}{{\rm d}t} = R'_{\rm m}\dot{z}_{\rm S}$$

The velocity of the mass is simply

$$\dot{x}_{\rm m} = R_{\rm m}\dot{z}_{\rm S}$$

The acceleration of the mass in terms of $z_{\rm S}$ follows by the usual rules for differentiation of a product of two variables:

$$\ddot{x}_{\rm m} = R_{\rm m}\,\ddot{z}_{\rm S} + \dot{R}_{\rm m}\dot{z}_{\rm S}$$

The force at the suspension is then

$$F_{\rm Sm} = R_{\rm m}F_{\rm m} = R_{\rm m}m\ddot{x}_{\rm m} = mR_{\rm m}(R_{\rm m}\,\ddot{z}_{\rm S} + \dot{R}_{\rm m}\dot{z}_{\rm S})$$

giving

$$F_{\rm Sm} = m(R_{\rm m}^2 \ddot{z}_{\rm S} + R_{\rm m}\dot{R}_{\rm m}\dot{z}_{\rm S})$$

Substituting the result obtained earlier for the time derivative of $R_{\rm m}$ finally gives

$$F_{\rm Sm} = m(R_{\rm m}^2 \ddot{z}_{\rm S} + R_{\rm m}R'_{\rm m}\dot{z}_{\rm S}^2)$$

For a constant motion ratio, this reduces to the earlier simple expression deduced for that case, as it should. When the motion ratio is not constant, that is, actually varying with $z_{\rm S}$, then there is no simple equivalent inertia.

14.14 The Effect of Motion Ratio on Springs

The spring stiffness depends on a spatial derivative of spring length, so the effect of the motion ratio of the springs is in general different from the effect on inertia or damping coefficient. The spring may be linear (constant stiffness) or non-linear, and the motion ratio may be uniform in $z_{\rm S}$ or non-uniform. It is generally believed that the effect of motion ratio on effective spring stiffness as seen at the wheel is proportional to the spring motion ratio squared. However, this is true only for a constant motion ratio.

Consider first a linear spring with constant motion ratio $R_{\rm K}$ (recall that subscript K is used for springs because subscript S has already been allocated to 'suspension'). By definition, the motion ratio is

$$R_{\rm K} = \frac{{\rm d}x_{\rm K}}{{\rm d}z_{\rm S}} = \frac{\dot{x}_{\rm K}}{\dot{z}_{\rm S}}$$

The spring force is $F_{\rm K}$ and the spring stiffness is $K_{\rm K}$. By the principle of virtual work, or consideration of a frictionless lever,

$$F_{\rm K}\,\delta x_{\rm K} = F_{\rm SK}\,\delta z_{\rm S}$$

where $F_{\rm SK}$ is the effective spring force seen at the active suspension point (wheel motion). This gives

$$F_{\rm SK} = R_{\rm K}F_{\rm K}$$

The effective stiffness seen at the wheel, with constant $R_{\rm K}$, is

$$K_{\rm SK} = \frac{{\rm d}F_{\rm SK}}{{\rm d}z_{\rm S}} = R_{\rm K}\frac{{\rm d}F_{\rm K}}{{\rm d}z_{\rm S}} = R_{\rm K}\frac{{\rm d}F_{\rm K}}{{\rm d}x_{\rm K}}\frac{{\rm d}x_{\rm K}}{{\rm d}z_{\rm S}} = R_{\rm K}^2 K_{\rm K}$$

So, for this simple case, the effective stiffness simply depends on $R_{\rm K}^2$:

$$K_{\rm SK} = R_{\rm K}^2 K_{\rm K} \quad (\text{constant } R_{\rm K}) \tag{14.14.1}$$

Now consider a non-linear spring with a constant motion ratio:

$$\begin{aligned} K_{\rm SK} &= \frac{{\rm d}F_{\rm SK}}{{\rm d}z_{\rm S}} = \frac{{\rm d}F_{\rm SK}}{{\rm d}x_{\rm K}} \cdot \frac{{\rm d}x_{\rm K}}{{\rm d}z_{\rm S}} \\ &= \frac{\rm d}{{\rm d}x_{\rm K}}(R_{\rm K}F_{\rm K}) \cdot \frac{{\rm d}x_{\rm K}}{{\rm d}z_{\rm S}} \end{aligned}$$

But the motion ratio $R_{\rm K}$ is constant, so

$$K_{\rm SK} = R_{\rm K} \cdot \frac{{\rm d}F_{\rm K}}{{\rm d}x_{\rm K}} \cdot R_{\rm K} = R_{\rm K}^2 K_{\rm K}$$

Table 14.14.1 Example values of rising-rate stiffness effect

Parameter	Symbol	Units	Value
Suspension force	F_S	kN	4.000
Motion ratio (ref)	R_K	—	0.500
Spring force	F_K	kN	8.000
Spring stiffness	K_K	kN/m	128.0
Rising rate	dR_K/dz_S	m^{-1}	2.000
Basic stiffness	$K_K R_K^2$	kN/m	32.0
Rising rate stiffness	$F_K R'_K$	kN/m	16.0
Total stiffness	K_{SK}	kN/m	48.0

so the non-linearity of the spring does not alter the basic effective stiffness relationship, although, of course, the local value of the spring stiffness must be used.

Consider now a linear spring with varying motion ratio, $R'_K \neq 0$. The suspension force is as before:

$$F_{SK} = R_K F_K$$

where the local value of the motion ratio must be used, according to z_S. By the usual differentiation of a product,

$$K_{SK} = \frac{dF_{SK}}{dz_S} = \frac{d}{dz_S}(R_K F_K) = R_K \cdot \frac{dF_K}{dz_S} + \frac{dR_K}{dz_S} \cdot F_K$$

The first term is as in equation (14.14.1), but there is now also a new term. The total stiffness is

$$K_{SK} = R_K^2 K_K + F_K \frac{dR_K}{dz_S} \tag{14.14.2}$$

This additional second term is $F_K R'_K$. For the special case of constant R_K the derivative vanishes and the second term is zero, with equation (14.14.2) reducing to equation (14.14.1). The usual practical case is with a derivative somewhat positive, that is, a 'rising-rate' suspension. With a large spring force F_K, this can have a considerable effect on the stiffness seen at the wheel. By way of example values, see Table 14.14.1. In this example, the rising rate is only 2/m, about 0.05/inch, but the effect is substantial.

When a non-linear spring is used with a variable motion ratio, the theory is similar to the above. No specific conclusions can be drawn, although equation (14.14.2) continues to apply.

Changing the motion ratio of the springs does not allow the use of reduced-mass springs. Of course, a smaller motion ratio reduces the spring stroke required but correspondingly increases the force required from it. These factors are simply in the ratio and inverse ratio respectively of the motion ratio, and have no net effect. Also, by energy considerations, at least in the case of a linear spring it is fairly clear that an intermediate lever does not reduce the mass required to store a given amount of energy, the work done in bump action. This is confirmed by more detailed analysis.

14.15 The Effect of Motion Ratio on Dampers

It is generally believed that the effective damping coefficient is proportional to the damper motion ratio squared. This is only true for a linear damper, and may be seriously in error for non-linear cases as found on many real dampers. However, a varying motion ratio is of no special significance for dampers, unlike springs, only the local actual motion ratio value matters.

Consider a linear damper with a constant motion ratio. It is evident that

$$F_D = C_D V_D$$
$$V_D = R_D V_S$$
$$F_{SD} = R_D F_D$$
$$F_{SD} = R_D C_D V_D = R_D^2 C_D V_S$$
$$C_{SD} = R_D^2 C_D$$

so in this case the effective damping coefficient simply depends on the motion ratio squared.

Now consider a linear damper with a varying motion ratio,

$$F_D = C_D V_D$$
$$V_D = R_D V_S$$
$$F_{SD} = R_D F_D = R_D C_D V_D = R_D^2 C_D V_S$$

as before. Only the local value of the damper motion ratio R_D affects the damper; unlike the case of a spring, the gradient R'_D has no effect.

Consider now a non-linear damper exerting a damper force F_D related to the damper velocity V_D by

$$F_D = C_1 V_D^n$$

where the exponent n has a value 1 for a linear damper (notionally pure viscosity), but may vary in practice from zero (effective Coulomb damping) to 2 (pure fluid dynamic damping). At a damper motion ratio R_D and suspension bump speed V_S, the damper speed, positive in compression, is

$$V_D = R_D V_S$$

The actual damper force F_D is

$$F_D = C_1 (R_D V_S)^n$$

and the suspension damper force F_{SD} at the wheel is

$$F_{SD} = R_D F_D$$

Hence the effective damping force at the wheel is

$$F_{SD} = C_1 R_D^{1+n} V_S^n$$

The shape of the influence of V_S on F_{SD} is retained as V_S^n, but the actual coefficient is scaled by the damper coefficient ratio, which in this case is

$$R_{DC} = R_D^{1+n}$$

Some special cases to be considered are the following:

(1) Exponent $n = 0$, corresponding to dry Coulomb friction (old-fashioned snubbers) or to hydraulic dampers with a sudden-acting blow-off valve. In this case the damper velocity makes no difference, as long as it is moving, so the coefficient ratio and the force ratio are both equal to the velocity ratio:

$$R_{DC} = R_D \quad (n = 0)$$

(2) Exponent $n = 1$, corresponding to a linear damper:

$$R_{DC} = R_D^2 \quad (n = 1)$$

(3) Exponent $n = 2$, corresponding to a hydraulic damper with a valve that is fixed, (e.g. fully closed at low speed or fully open at high speed):

$$R_{DC} = R_D^3 \quad (n = 2)$$

(4) A representative intermediate case at damper speeds around the knee of the curve, with $n \approx 0.5$, giving

$$R_{DC} = R_D^{1.5} \quad (n = 0.5)$$

It is common for the motion ratio to be around 0.7, sometimes even as low as 0.4, from which it will be apparent that the damper coefficient ratio will vary very widely for different damper force–speed relationships.

Considered from a design aspect, the characteristic required at the wheel is the starting point, which is transformed by the motion ratio to the required damper characteristic. A low motion ratio therefore calls for a high damping coefficient for the damper itself. The use of a low or high motion ratio does not allow the use of a smaller or lighter damper. A low motion ratio reduces the stroke required, but correspondingly increases the damper force needed.

14.16 Velocity Diagrams in Three Dimensions

The three-dimensional velocity diagram is, in principle, a straightforward extension of the idea of the two-dimensional velocity diagram, but in practice is a good deal more difficult conceptually. All velocities have three components, e.g. $V_A = (V_{Ax}, V_{Ay}, V_{Az})$ for a point A, this velocity being 'plotted' in a three-dimensional velocity space. The method used is in effect geometrical constructions in a three-dimensional space. Actual drawing-instrument methods are not practical, although the results can be drawn as isometrics, and the solution must be by algebraic methods. The solution can often use existing three-dimensional geometry library routines. The alternative method of determining the velocities by a small displacement using an existing suspension position solver is generally easier, although imperfect in principle because of the increment size required for accuracy. The three-dimensional velocity diagram has the merit of being an exact method in principle.

In all cases, the position of the suspension must first be fully known. In the case of a double-transverse-arm suspension, proceed by specifying an angular velocity Ω of the bottom arm. The tangential velocity of the lower outer ball joint point A is then ΩR, with a direction perpendicular to the plane of the lower arm. The direction cosines (l, m, n) of the normal to the plane are therefore also the direction cosines of the velocity direction, so the velocity components can be deduced, giving

$$\bar{V}_A = (V_{Ax}, V_{Ay}, V_{Az}) = V_A(l, m, n)$$

The velocity of the upper outer ball joint B relative to the velocity point a for physical point A is just tangential. Therefore it is in a plane through velocity point a, perpendicular to the line from one ball joint to the other. The direction cosines of the normal for this plane are therefore those of the line AB. This gives a known plane in the velocity diagram through the known velocity point a. Also, the upper ball joint velocity must be perpendicular to the plane of the upper arm, giving a line from the velocity origin parallel to the normal to the plane. The intersection of the line and the plane in the velocity diagram gives point b, the velocity of upper ball joint point B.

To complete the solution of the wheel carrier/slider, the velocity of the steering-arm/track-rod end ball joint is required. Unlike in the case of two-dimensional velocity diagrams, in three dimensions a solid component does not have a similar velocity image, and the image is distorted. The clearest method is to drop a perpendicular from the steering-arm ball joint C onto the steer axis AB, giving a foot point F. Now a straight line in an ideally-rigid object does map into a straight-line image in the three-dimensional velocity diagram, so velocity point f can be calculated by simple proportion from velocity points a and b, in the same ratio as F lies on AB or AB extended. The velocity of C relative to F is purely tangential, and is therefore perpendicular to the plane ABCF, which has a normal with solvable direction cosines. Hence, the direction cosines of the line fc can be found, giving a line fc, so far of unknown length, from the known point f. Considering the rack end of the track rod, this point D has a known, specified velocity depending on the rack only, usually zero, giving velocity point d. Relative to D, the track-rod end C has a velocity perpendicular to the line CD, so it is in a plane in the velocity diagram perpendicular to line CD and containing the origin. The line from f intersects this plane at velocity point c for physical point C, completing the basic solution of the wheel upright. The three known velocities for points A, B and C can now be used to calculate angular velocities about the X, Y and Z axes, that is, the bump camber, bump caster and bump steer angular velocities.

To solve the scrub velocity at the contact point D, the wheel is treated as locked in rolling rotation, so the wheel and wheel carrier are considered to be a single rigid component. The contact point D is therefore a fixed point on the carrier. Drop a perpendicular from D to a new foot E on the ball joint line AB produced, solving e on ab by the same proportion as E on AB. Also, use the tangential velocity of C and the length of FC to obtain an angular velocity of the carrier about the axis AB. Apply this to DE to obtain a relative tangential velocity of D from E, which has the direction of the normal to the plane ABFED, hence solving the magnitude and direction of the velocity of D. This resulting velocity

$$V_{\rm D} = (V_{\rm Dx}, V_{\rm Dy}, V_{\rm Dz})$$

contains the longitudinal scrub velocity $V_{\rm Dx}$, the lateral scrub velocity $V_{\rm Dy}$, and the bump velocity $V_{\rm Dz}$. All the bump coefficients then follow easily; for example, the local longitudinal bump scrub rate coefficient is

$$\varepsilon_{\rm BScd,X} = \frac{V_{\rm Dx}}{V_{\rm Dz}}$$

and the local lateral bump scrub rate coefficient is

$$\varepsilon_{\rm BScd,Y} = \frac{V_{\rm Dy}}{V_{\rm Dz}}$$

Obtaining values for these over a range of suspension bump positions, the usual coefficients may be obtained, using

$$\varepsilon_{\rm BScd} = \varepsilon_{\rm BScd0} + \varepsilon_{\rm BScd1} z_{\rm S}$$

applied in longitudinal and transverse directions.

In principle, to find the suspension bump velocity, find the wheel centre velocity, and deduce the velocity of the lowest point of the wheel arc, allowing wheel rotation, that is, the lowest point of a circle on the wheel centre with radius equal to the wheel loaded radius. The wheel centre is a fixed point on the wheel carrier. The velocity of the lowest point is the velocity of the wheel centre plus the change in height due to camber change from wheel loaded radius, camber angle and camber angular velocity. Considering the wheel as free to rotate allows the bottom point to have an extra motion giving a tangential velocity component, which is purely horizontal. This leaves the lateral scrub component and the bump velocity unchanged. The result is therefore the same as above – simply the vertical component of the locked-wheel bottom point velocity, namely $V_{\rm Dz}$.

For a strut-and-arm suspension, again begin with a bottom arm angular velocity and the bottom ball joint point A. Obtain the velocity component directly towards the fixed trunnion point B. Point C on the slider, instantaneously coincident with B, has a velocity V_C directed along the slider, but with a component in the direction of AB equal to the component of V_A in that direction. This is because for two points on a rigid body the separation is constant, so the velocity components along the line of the points must be equal. Hence, vector dot products can be used to obtain V_C. This gives the velocity of two known points of the carrier/slider. The rack end D has a known, independently specified, velocity, probably zero. To solve for the steering-arm/track-rod end E, at the track-rod end this has a velocity in a plane perpendicular to DE from the velocity point d. Also, considering it as the steering-arm end, drop a perpendicular from E onto the line AC to foot G, solving g by proportion. The velocity $V_{E/G}$ is perpendicular the plane ACGE, giving a line in the velocity diagram from g, intersecting the plane from d at the desired point e. Hence, the carrier/slider has three known velocity points, and is solved. The velocity of the contact point can be solved as for the double-transverse-arm suspension.

Other suspension types can be solved using similar methods. The use of three-dimensional geometrical routines in the process of solution is apparent. In position analysis, the two main three-dimensional procedures are the 'pole-on-triangle' and the 'tripod'. Correspondingly, two main velocity diagram procedures can be used, rather than the *ad hoc* approach above.

The 'pole-on-triangle' velocity diagram procedure is required to solve the velocity of a point fixed relative to a base triangle which moves but is of unchanging shape and size. The first perpendicular required is that from P 'down' to F in the plane of the triangle. Point F has factors for its position relative to the three corners of the triangle. The velocity of F is found by applying the same factors to the corner velocities. Also drop the perpendicular from the point P_1 into the line P_2P_3 of the triangle to give a foot P_4. The velocity of P_4 may be found by proportion, using the interpolation factors of the physical point P_4 on P_2P_3. Next, calculate angular velocities about the rectangular axes aligned with P_2P_3 and P_1P_4. Use these angular velocities to calculate the velocity of P relative to the velocity of F. Chapter 15 gives further details.

The 'Tripod' position problem has an analogue in the tripod velocity problem, but whereas the first is the intersection of three spheres, the latter is the intersection of three planes in the velocity diagram, associated with three simultaneous linear equations. Consider rod 1, with base velocity (u_1, v_1, w_1) and rod centreline direction cosines (l_1, m_1, n_1). In the velocity diagram, the known base velocity gives a point, and the velocity of the other end of the rod must be in a plane through that point, the plane having a normal with direction cosines the same as the rod. The equation of that plane is simply

$$l_1u + m_1v + n_1w = l_1u_1 + m_1v_1 + n_1w_1$$

There are three such equations, so the unknowns (u, v, w) can be solved, this being the intersection point of the three planes. Tantalisingly, it is possible to solve explicitly using a direct line-and-plane method, using say rod 1 to give a plane, and rods 2 and 3 to give a line. This line is of course the intersection of the velocity planes of rods 2 and 3. The out-of-line velocities of the rod ends are independent because, instantaneously, they do not act along the rods, they only rotate them about a transverse axis. Taking a pair of rods, the total velocity of the joint point can be seen as one part perpendicular to the plane of the triangle, and one part in the plane. The latter part can be solved from the in-plane base velocities by two-dimensional methods, giving a velocity point. From there, the velocity of P is on a line perpendicular to the triangle, giving a line to intersect the plane of rod 1. This is, however, actually more difficult to implement than the general three-plane method.

14.17 Acceleration Diagrams

Acceleration diagrams are important in general mechanism design. In the case of vehicle suspensions, there are two applications. The first is a determination of the accurate accelerations of the suspension components to evaluate the acceleration forces. The second is for use as a derivative in the evaluation of velocity diagrams. Acceleration diagrams are quite easy in two dimensions for simple mechanisms such as pin-jointed ones (e.g. wishbones), and somewhat trickier for struts. In three dimensions they can become very tricky.

References

Morrison, J.L.M. and Crossland, B. (1970) *An Introduction to the Mechanics of Machines*, 2nd edn in SI units, London: Longman. ISBN 0582447291 and 0582447313. Contains material on velocity and acceleration diagrams.

Open University Course T235, *Engineering Mechanics: Solids*, contains detailed material on drawing velocity diagrams. The entire course (on basic mechanical engineering) is available free at http://intranet.open.ac.uk/open-content/. Click on 'Visit the OpenLearn website', enter 'LabSpace', click on topics 'Technology', scroll down to 'Engineering Mechanics: Solids' and click on 'Browse'. See Block 3 for velocity diagrams.

——— // ———

15

Computational Geometry in Three Dimensions

15.1 Introduction

The numerical or computational solution of a suspension geometry problem depends heavily on three-dimensional coordinate geometry, which just is a more complicated variant of the more familiar two-dimensional flat-plane paper-surface geometry. It is assumed here that the reader is familiar with basic two-dimensional coordinate geometry and has studied basic three-dimensional geometry and vector analysis. Presented here is an *aide-mémoire* of some useful reference material, and the mathematical solution of some problems particularly relevant to the solution of suspension geometry.

Although it seems physically that a point is simpler than a line which is simpler than a surface, this belies the problems of mathematical representation. A surface can be represented by a single equation. It takes two equations to represent a line (the intersection of two surfaces) and three equations to represent a point (the intersection of three surfaces). In fact, it is often most convenient to represent a line by three equations, in its parametric form.

In general, algebraic variables are in italic, to indicate that they represent a value. Vectors are indicated by bold type, but they are still variables so they are in bold italic. The point named P_1, in non-bold roman (upright) script has a position vector $\boldsymbol{P}_1$ in bold italic, which has a magnitude P_1, which is non-bold italic.

15.2 Coordinate Systems

Overwhelmingly, the coordinate system of greatest importance and use is the one of simple rectangular axes, typified by the usual (X, Y, Z) system. However, cylindrical coordinates are sometimes used, in effect just polar coordinates plus a length dimension. This can arise, for example, in the case of a suspension arm, which rotates about an axis at some arbitrary angle. The motion of the outer end of the arm is an arc which is not conveniently represented in the coordinate system of the vehicle body, but using cylindrical coordinates aligned with the axis the movement becomes simply a circular motion.

Sometimes linear non-rectangular coordinates can be used advantageously, as in the case of the PointIT routine (Section 15.17).

The distance between two points is given by Pythagoras' theorem in three dimensions, in rectangular coordinates, by

$$L_{\mathrm{AB}} = \sqrt{(x_{\mathrm{A}} - x_{\mathrm{B}})^2 + (y_{\mathrm{A}} - y_{\mathrm{B}})^2 + (z_{\mathrm{A}} - z_{\mathrm{B}})^2}$$

Suspension Geometry and Computation J. C. Dixon

This is invariably a positive number, possibly zero. The squares inside the root ensure that the kernel can never be negative.

The idea of a length, which is never negative, must be distinguished carefully from the idea of a displacement or position, which may be negative. In a coordinate system, the variables represent positions, not lengths, and can be negative, indicating a location in the other direction. This distinction between length and displacement also arises importantly in the case of the position of a point 'above' a plane. Given a specific plane normal, the point displacement from the plane could be positive 'up' or negative 'down', for example as in the PointAT routine (Sections 15.21 and 15.22).

15.3 Transformation of Coordinates

It is often convenient to retain rectangular coordinates but to shift the origin. This may be done to facilitate algebraic solution of problems, or to improve computational efficiency. Considering base coordinates (x, y, z), and local coordinates (u, v, w) which are parallel to (x, y, z) but displaced without rotation, placing the origin of (u, v, w) at $P_1 = (x_1, y_1, z_1)$ gives

$$\begin{aligned} u_1 &= 0; \quad & v_1 &= 0; \quad & w_1 &= 0 \\ u_2 &= x_2 - x_1; \quad & v_2 &= y_2 - y_1; \quad & w_2 &= z_2 - z_1 \end{aligned}$$

and so on.

Rotation of coordinates is less often required, but is easily achieved. In the base coordinates (x, y, z) the alternative coordinate system (u, v, w) has the same origin and is rotated such that the direction cosines of (u, v, w) in (x, y, z) are (l_u, m_u, n_u), (l_v, m_v, n_v) and (l_w, m_w, n_w), giving the transformation

$$u = l_u x + m_u y + n_u z$$
$$v = l_v x + m_v y + n_v z$$
$$w = l_w x + m_w y + n_w z$$

The direction cosines of the secondary axes are not fully independent (nine direction cosines but only three degrees of freedom), and compatibility must be maintained.

If the origin is simultaneously moved to (x_0, y_0, z_0) then the combined transformation is

$$u = l_u(x - x_0) + m_u(y - y_0) + n_u(z - z_0)$$
$$v = l_v(x - x_0) + m_v(y - y_0) + n_v(z - z_0)$$
$$w = l_w(x - x_0) + m_w(y - y_0) + n_w(z - z_0)$$

15.4 Direction Numbers and Cosines

Consider a straight line with any two distinct points upon it. By Pythagoras' theorem the length is

$$L_{12} = \sqrt{(x_2 - x_1)^2 + (y_2 - y_1)^2 + (z_2 - z_1)^2}$$

In rectangular coordinates, any set of three numbers, usually denoted (L, M, N), being

$$(L, M, N) = ((x_2 - x_1), (y_2 - y_1), (z_2 - z_1))$$

is called a set of the direction numbers for the line. Points P_1 and P_2 and the length L_{12} may vary, only the ratios of the direction numbers are significant. The direction numbers are sometimes also known as direction ratios, but this is not a good name as individually they are displacements.

In the special case that the length is unity, then the direction numbers have special values called direction cosines, denoted (l, m, n). Generally

$$l = \frac{L}{L_{12}}; \quad m = \frac{M}{L_{12}}; \quad n = \frac{N}{L_{12}}$$

and

$$l = \frac{x_2 - x_1}{L_{12}}; \quad m = \frac{y_2 - y_1}{L_{12}}; \quad n = \frac{z_2 - z_1}{L_{12}}$$

The direction cosines are not independent, having the property that their Euclidean length is unity:

$$l^2 + m^2 + n^2 = 1$$

The line makes three angles (α, β, γ), the direction angles, with the three coordinate axes (x, y, z) respectively, with

$$l = \cos\alpha; \quad m = \cos\beta; \quad n = \cos\gamma$$

The direction cosines are very useful for representing the direction of a line, or of a displacement, velocity, acceleration, force, etc. For example, a displacement with components

$$\boldsymbol{D} = (D_x, D_y, D_z) = (lD, mD, nD)$$

where $\boldsymbol{D}$ is the vector and D is its magnitude. In this case, it is evident that the components are obtained by using the cosine of the angles with the various axes.

A line may be considered to 'go' in either direction, that is, P_2 may be on either side of P_1, so the set of numbers $(-l, -m, -n)$ is an equally valid set of direction cosines, but with reversed direction.

The direction cosines are (1, 0, 0) for the x axis, (0, 1, 0) for the y axis and (0, 0, 1) for the z axis.

15.5 Vector Dot Product

This is also known as the vector scalar product, obviously because the result is a scalar value (non-vector) to be distinguished from the vector cross product. Given two vectors $\boldsymbol{A} = (u_A, v_A, w_A)$ and $\boldsymbol{B} = (u_B, v_B, w_B)$, the dot product is

$$P = \boldsymbol{A} \cdot \boldsymbol{B} = u_A u_B + v_A v_B + w_A w_B$$

That is, the matched pairs of components are multiplied and the resulting products are added. The result is a simple scalar value, without direction.

The great value of the dot product calculation is that the result is equal to the product of the vector magnitudes and the cosine of the angle between them (considered, if necessary, with a common starting point),

$$P = \boldsymbol{A} \cdot \boldsymbol{B} = AB\cos\theta$$

If the vectors $\boldsymbol{A}$ and $\boldsymbol{B}$ have units of length, then P has units of area.

In the special case when the base vectors are unit vectors, that is, vectors of unit length, $\hat{\boldsymbol{a}}$ and $\hat{\boldsymbol{b}}$, then

$$P = l_A l_B + m_A m_B + n_A n_B = \cos\theta$$

which provides a valuable method of obtaining the angle between two directions or lines.

In the event that the angle is a right angle (90°), then the cosine and the dot product must be zero, which can be a useful check on the perpendicularity of two lines.

A key application of the dot product is in obtaining components of displacements, velocities, accelerations, forces, etc., along any direction. Given a vector $\boldsymbol{A}$ with components (x_A, y_A, z_A) and a line with direction cosines (l, m, n), then the component of $\boldsymbol{A}$ in that direction is

$$A_{comp} = lx_A + my_A + nz_A$$

If the other direction of the line is considered, then all three direction cosines are reversed in sign, and the component changes sign but retains its magnitude.

Consider, for example, a moving rigid rod AB, with velocities at its ends

$$\begin{aligned} \boldsymbol{V}_A &= (u_A, v_A, w_A) \\ \boldsymbol{V}_B &= (u_B, v_B, w_B) \end{aligned}$$

The rod end positions are (x_A, y_A, z_A) and (x_B, y_B, z_B). Then the direction cosines of the rod, A to B, are

$$l = \frac{x_B - x_A}{L_{AB}}; \quad m = \frac{y_B - y_A}{L_{AB}}; \quad n = \frac{z_B - z_A}{L_{AB}}$$

The velocity component at A in the direction of the rod itself, that is, the axial velocity of the rod, is, by the dot product,

$$V_{A,axial} = lu_A + mv_A + nw_A$$

Similarly, at the other end,

$$V_{B,axial} = lu_B + mv_B + nw_B$$

For a rigid rod, these two axial components must be equal.

A good example application of the dot product may be found in Section 15.14.

15.6 Vector Cross Product

The vector cross product, denoted

$$\boldsymbol{P} = \boldsymbol{A} \times \boldsymbol{B}$$

is itself a vector, with magnitude

$$P = AB \sin\theta$$

where the angle θ is the angle between the two vectors, considered if necessary to be translated to a common starting point. If $\boldsymbol{A}$ and $\boldsymbol{B}$ have units of length, then $\boldsymbol{P}$ has units of area, and is in fact the area of a parallelogram defined by the two vectors, that is, twice the area of the triangle that they form. The direction of the resulting vector is perpendicular to both $\boldsymbol{A}$ and $\boldsymbol{B}$, being given by the right-hand rule. Where $\boldsymbol{A}$ is the right-hand thumb, and $\boldsymbol{B}$ is the first finger, then $\boldsymbol{P}$ is in the direction of the second finger. If $\boldsymbol{A}$ and $\boldsymbol{B}$ are taken in the other order, then the product is reversed in direction, that is,

$$\boldsymbol{B} \times \boldsymbol{A} = -\boldsymbol{A} \times \boldsymbol{B}$$

The product vector $\boldsymbol{A} \times \boldsymbol{B}$ has the property that it is parallel to the normal to the plane containing the vectors $\boldsymbol{A}$ and $\boldsymbol{B}$. This is probably its most important feature, as it can therefore be used to generate that normal vector.

The actual calculation can conveniently be summarised in a determinant. A generally non-unit vector $\boldsymbol{N}$ with the normal direction to the plane of $\boldsymbol{A}$ and $\boldsymbol{B}$ is given by the easily remembered form

$$N = \boldsymbol{A} \times \boldsymbol{B} = \begin{vmatrix} \hat{\boldsymbol{i}} & \hat{\boldsymbol{j}} & \hat{\boldsymbol{k}} \\ x_A & y_A & z_A \\ x_B & y_B & z_B \end{vmatrix}$$

where $(\hat{\boldsymbol{i}}, \hat{\boldsymbol{j}}, \hat{\boldsymbol{k}})$ with circumflex marks are unit vectors of the (x, y, z) axes respectively. Expanding the determinant in the usual way, with alternating signs,

$$\begin{aligned} N &= \hat{\boldsymbol{i}}(y_A z_B - y_B z_A) - \hat{\boldsymbol{j}}(x_A z_B - x_B z_A) + \hat{\boldsymbol{k}}(x_A y_B - x_B y_A) \\ &= \hat{\boldsymbol{i}} N_x + \hat{\boldsymbol{j}} N_y + \hat{\boldsymbol{k}} N_z \end{aligned}$$

This calculation of the vector normal to a plane is frequently required in suspension geometry analysis.

15.7 The Sine Rule

The sine rule is occasionally useful in suspension analysis, and applies to three-dimensional triangles as well as to two-dimensional ones, both being of necessity in a single plane, Figure 15.7.1. However, it is expressed in terms of the two-dimensional properties of the triangle rather than in direct three-dimensional terms. For a triangle with sides a, b and c, with opposing angles A, B and C,

$$\frac{\sin A}{a} = \frac{\sin B}{b} = \frac{\sin C}{c}$$

It is easily proved by dropping a perpendicular. It can equally well be expressed in reciprocal form, $a/\sin A = b/\sin B = c/\sin C$.

In three dimensions, considering the cross product of various pairs of sides, in each case the magnitude of the product is equal to twice the area of the triangle. Hence,

$$L_{31} L_{12} \sin \theta_1 = L_{12} L_{23} \sin \theta_2 = L_{23} L_{31} \sin \theta_3 = 2S_T$$

Dividing by $L_{12} L_{23} L_{31}$ gives

$$\frac{\sin \theta_1}{L_{23}} = \frac{\sin \theta_2}{L_{31}} = \frac{\sin \theta_3}{L_{12}}$$

agreeing with the sine rule as expressed in two-dimensional terms.

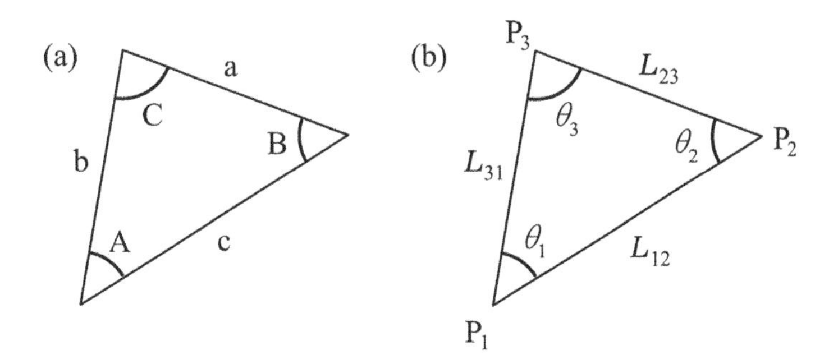

Figure 15.7.1 Triangles for the sine rule, alternative notations.

15.8 The Cosine Rule

The cosine rule is often useful, much more so than the sine rule. In Figure 15.8.1, the three sides of any planar triangle are related by

$$a^2 = b^2 + c^2 - 2bc\cos A$$

or

$$\cos A = \frac{b^2 + c^2 - a^2}{2bc}$$

In three-dimensional terms, the cosine of A is given by the dot product, in local coordinates based on P_1, as

$$(\boldsymbol{P}_2 - \boldsymbol{P}_1)\cdot(\boldsymbol{P}_3 - \boldsymbol{P}_1) = bc\cos A$$

with

$$\cos A = \frac{u_2u_3 + v_2v_3 + w_2w_3}{L_{31}L_{12}}$$

Actually,

$$\begin{aligned} b^2 + c^2 - a^2 &= (u_2^2 + v_2^2 + w_2^2) + (u_3^2 + v_3^2 + w_3^2) - (u_3 - u_2)^2 - (v_3 - v_2)^2 - (w_3 - w_2)^2 \\ &= 2(u_2u_3 + v_2v_3 + w_2w_3) \\ &= 2(P_2 - P_1)\cdot(P_3 - P_1) \\ &= 2bc\cos A \end{aligned}$$

That is,

$$\cos A = \frac{b^2 + c^2 - a}{2bc}$$

proving the cosine rule directly in three-dimensional coordinates.

The component $b\cos A$ is sometimes useful, being given by

$$b\cos A = \frac{b^2 + c^2 - a^2}{2c}$$

without use of the (computationally relatively slow) transcendental cosine function.

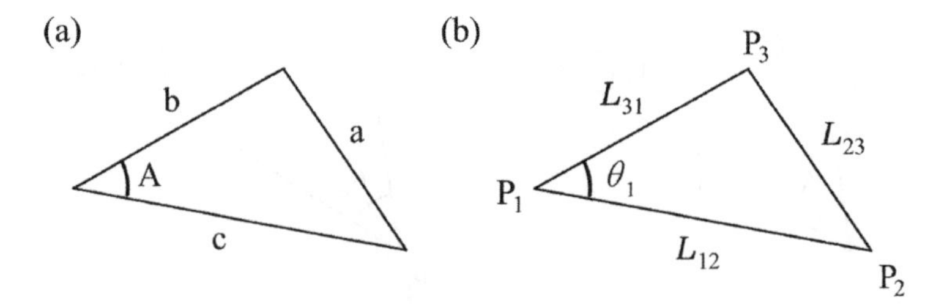

Figure 15.8.1 Triangles for the cosine rule, alternative notations.

15.9 Points

In geometry, a point is an ideal concept of location, with zero extension is space, being just a position. It is often defined as the intersection of two ideal geometric lines having no thickness, but this is applicable only in two dimensions. In three dimensions two lines do not meet, in general. The point could then be defined as the intersection of three planes, or of a line with a plane, or of a line with some other surface such as a spherical shell or a cone.

In practical three-dimensional coordinate geometry the point, having three degrees of freedom, is a position in space measured by three coordinates. These are the coordinates of the system in use, usually a rectangular coordinate system, so the point becomes the location (x, y, z). However, a point on a line is constrained in two degrees of freedom, and may be represented by a single parameter, for example the parameter of path length s in the parametric representation of a line.

15.10 Lines

The geometric definition of a point as the intersection of two lines (in two dimensions) may be regarded as somewhat circular, as a line is often defined as passing through two points, that is, two distinct points are sufficient to define a straight line. This is still true in three dimensions. A straight line can also be considered to be the intersection of two planes. In general, the intersection of two surfaces gives a line, not necessarily straight, for example a plane and a spherical shell gives a circle, and in some cases the line may be in distinct parts, as in the case of the intersection of a plane with a torus in some positions.

In the main, however, it is the straight line and circle that are of importance. There are several common and useful alternative forms of representation of a straight line.

(1) The standard form of a line between two points P_1 and P_2:

$$\frac{x-x_1}{x_2-x_1}=\frac{y-y_1}{y_2-y_1}=\frac{z-z_1}{z_2-z_1}$$

(2) The standard form of a line through P_1 with direction cosines (l, m, n):

$$\frac{x-x_1}{l}=\frac{y-y_1}{m}=\frac{z-z_1}{n}$$

This is also correct when direction numbers (L, M, N) are used instead, as these are in the same ratio to each other:

$$\frac{x-x_1}{L}=\frac{y-y_1}{M}=\frac{z-z_1}{N}$$

which is equivalent to the standard form.

(3) The parametric form of a line between two points:

$$x=x_1+f(x_2-x_1)$$
$$y=y_1+f(y_2-y_1)$$
$$z=z_1+f(z_2-z_1)$$

Here, the value of the parameter f controls the position of a point on the line. At P_1 we have $f=0$, whereas at P_2 we have $f=1$. The value of f may also be negative, with the point on P_2 to P_1 extended, or greater than 1 with the point on P_1 to P_2 extended. Calculating a length L_{12} for P_1 to P_2, it is apparent that the displacement of the point from P_1 is

$$s = f\,L_{12}$$

(4) The parametric form of a line with known direction cosines. This is a special case of the above in which a point P_2 is considered to be unit distance from P_1, that is, $L_{12} = 1$. The parameter f then becomes the actual path length from P_1, so, for example,

$$x = x_1 + s\frac{x_2 - x_1}{L_{12}} = x_1 + sl$$

where l is the direction cosine, giving in all

$$x = x_1 + l\,s$$
$$y = y_1 + m\,s$$
$$z = z_1 + n\,s$$

This parametric form is the most convenient form for general use.

15.11 Planes

The plane can be expressed in a single equation, but in various alternative ways.

(1) The general form is given by

$$Ax + By + Cz + D = 0$$

Here (A, B, C) are called the direction numbers, and are actually direction numbers of the normal to the plane. In the general form, the constant D is included in the left-hand side.

(2) The normal form is

$$lx + my + nz = p$$

Here (l, m, n) are the direction cosines of the normal to the plane. The variable p is the signed length of the normal from the origin to the plane in the direction (l, m, n). Note that, in contrast to the general form, p appears on the right-hand side of the equation. This normal form is the most useful form overall.

(3) The intercept form is written as

$$\frac{x}{a} + \frac{y}{b} + \frac{z}{c} = 1$$

The variables a, b and c are the intercepts of the plane on the x, y and z axes, respectively. In suspension analysis, this is not a very useful form. By multiplying the intercept form shown by abc the general form may easily be obtained.

(4) The final way is as a plane through three points. A plane passing through three points P_1, P_2 and P_3 having coordinates (x_1, y_1, z_1) *et cetera* has the equation given by the determinant

$$\begin{vmatrix} x - x_1 & y - y_1 & z - z_1 \\ x_2 - x_1 & y_2 - y_1 & z_2 - z_1 \\ x_3 - x_1 & y_3 - y_1 & z_3 - z_1 \end{vmatrix} = 0$$

This form evidently gives preference to the point P_1, which is treated somewhat as an origin of local coordinates, and two alternative forms could be written based on the other two points. In local coordinate variables based on P_1 the above determinant becomes particularly simple:

$$\begin{vmatrix} u & v & w \\ u_2 & v_2 & w_2 \\ u_3 & v_3 & w_3 \end{vmatrix} = 0$$

Expanding the determinant gives:

$$u(v_2w_3 - v_3w_2) - v(u_2w_3 - u_3w_2) + w(u_2v_3 - u_3v_2) = 0$$

Comparing this with the normal form

$$lx + my + nz = p$$

it is apparent that the length of the perpendicular from the origin to the plane is zero. This is correct in local coordinates based on P_1, which is a point at the origin and in the plane. The direction numbers of the plane are evidently

$$\begin{aligned} L &= v_2w_3 - v_3w_2 \\ M &= u_3w_2 - u_2w_3 \\ N &= u_2v_3 - u_3v_2 \end{aligned}$$

This is the same result as would be obtained by taking the vector cross product of $(\boldsymbol{P}_2 - \boldsymbol{P}_1)$ and $(\boldsymbol{P}_3 - \boldsymbol{P}_1)$ in that order, giving a vector normal to the plane. This vector can then be divided by its length to give direction cosines for the normal.

15.12 Spheres

Spheres, or perhaps more correctly geometric spherical shells, often arise in suspension analysis. Their representation is simple, but not always convenient for analysis because the equation is non-linear. For a radius R_1 on a centre point (x_1, y_1, z_1), the distance from the centre point, as simply given by Pythagoras' equation in three dimensions, is constant, giving

$$(x - x_1)^2 + (y - y_1)^2 + (z - z_1)^2 = R_1^2$$

This expands to

$$x^2 + y^2 + z^2 - 2x_1x - 2y_1y - 2z_1z = R_1^2 - (x_1^2 + y_1^2 + z_1^2)$$

where the right-hand side is constant.

The intersection of two spheres is an important practical problem. Where the intersection does occur, the result is generally a circle, although a single touching point is a possibility in principle. Subtracting the equations of two spheres gives the equation of a plane. This is called the radical plane of the two spheres. It is perpendicular to the line joining the centres, that is to say, the line of the centres is its normal. If the two spheres do not intersect, then the radical plane lies between the spheres. If they do intersect, then the circle of intersection lies in the radical plane. Considering two spheres with equations as above, that is,

$$x^2 + y^2 + z^2 - 2x_1x - 2y_1y - 2z_1z = c_1$$

$$x^2 + y^2 + z^2 - 2x_2x - 2y_2y - 2z_2z = c_2$$

subtraction gives the equation of the radical plane as

$$(x_2 - x_1)x + (y_2 - y_1)y + (z_2 - z_1)z = \frac{c_1 - c_2}{2}$$

Dividing this by the length L_{12} between the centres immediately gives the normal form of a plane. The direction cosines are evidently directed along the line of the centres, so the plane itself is perpendicular to that line, as would be expected by symmetry.

15.13 Circles

In two-dimensional geometry, circles are easily represented and, although non-linear, are highly analytic. In three dimensions, the circle can be inconvenient to represent because it may be at any orientation.

The circle is a line, so being the intersection of two surfaces, and requiring two simultaneous equations. One combination is the plane of the circle with the centre and a given radius. In effect, then, the circle is represented as the intersection of a plane with a sphere, giving a great circle of the sphere. In some cases, the circle arises naturally as the intersection of two spheres, probably of unequal radius, as in the case of a suspension arm producing the circular arc of the outer ball joint.

To represent the circle in this way we can use seven variables: the circle centre (x_C, y_C, z_C), the direction cosines of the normal to the plane of the circle (l, m, n) and the radius R. A point has only three degrees of freedom, so evidently these seven variables contain some redundancy. Actually it is not necessary to specify the centre, it may be convenient to specify any point on the axis, along with one point of the circle, $(x_A, y_A, z_A, l, m, n, x_1, y_1, z_1)$, using nine variables. The circle is then perceived as a section of a circular cylinder.

To actually deal with the circle numerically, it is really necessary to use local coordinates with the correct alignment. It is most convenient to have one given point on the circle itself (which, in practice, is usually available) along with the axis, if necessary dropping a perpendicular to establish the centre of the circle. The local coordinates (u, v, w) then have an origin at the centre of the circle (x_C, y_C, z_C). The third axis w points along the axis (l, m, n). The first axis u points from the centre towards the known point of the circle. Considering unit vectors $\hat{\boldsymbol{u}}$ and $\hat{\boldsymbol{w}}$ in these directions, the unit vector

$$\hat{\boldsymbol{v}} = \hat{\boldsymbol{w}} \times \hat{\boldsymbol{u}}$$

from the cross product gives the desired third axis, perpendicular to the other two, with $(\hat{\boldsymbol{u}}, \hat{\boldsymbol{v}}, \hat{\boldsymbol{w}})$ being a right-hand set. In more detail

$$\hat{\boldsymbol{v}} = \hat{\boldsymbol{w}} \times \hat{\boldsymbol{u}} = \begin{vmatrix} \hat{\boldsymbol{i}} & \hat{\boldsymbol{j}} & \hat{\boldsymbol{k}} \\ l_w & m_w & n_w \\ l_u & m_u & n_u \end{vmatrix}$$

so

$$\begin{aligned} l_v &= m_w n_u - m_u n_w \\ m_v &= l_u n_w - l_w n_u \\ n_v &= l_w m_u - l_u m_w \end{aligned}$$

A rotation of the arm or radius by angle θ from the first position, right-hand about $\hat{\boldsymbol{w}}$ then gives a local coordinate position

$$\begin{aligned} u &= R\cos\theta \\ v &= R\sin\theta \\ w &= 0 \end{aligned}$$

This position can then be converted into (x, y, z) coordinates as described in Section 15.3. In calculation of the last axis v by the vector cross product, the correct order of the vectors must be maintained to give a right-hand set of conventional notation.

15.14 Routine PointFPL2P

Purpose: 'Point at Foot of Perpendicular onto Line of 2 Points' – to obtain the point F at the foot of a perpendicular dropped from a specified point P onto a line defined by two points

Inputs: Coordinates of line points P_1 and P_2, and of point P, i.e. coordinates (x_1, y_1, z_1), (x_2, y_2, z_2), (x_P, y_P, z_P)

Outputs: The foot point coordinates (x_F, y_F, z_F), and the perpendicular length L_{PF}

Notes: None

Solution: The required point lies at some fraction f from P_1 towards P_2 (where the fraction f may also be negative or exceed 1). This point has

$$\begin{aligned} x &= x_1 + f(x_2 - x_1) = (1-f)x_1 + fx_2 \\ y &= y_1 + f(y_2 - y_1) = (1-f)y_1 + fy_2 \\ z &= z_1 + f(z_2 - z_1) = (1-f)z_1 + fz_2 \end{aligned} \tag{15.14.1}$$

The line from P to (x, y, z) must be perpendicular to the line P_1P_2. Therefore the dot product (equal to $\cos\theta$) must be zero, requiring

$$(x_F - x_P)(x_2 - x_1) + (y_F - y_P)(y_2 - y_1) + (z_F - z_P)(z_2 - z_1) = 0$$

By substitution, an equation is obtained for the fraction f:

$$\begin{aligned} (x_1 + f(x_2 - x_1))(x_2 - x_1) + (y_1 + f(y_2 - y_1))(y_2 - y_1) + (z_1 + f(z_2 - z_1))(z_2 - z_1) &= 0 \\ x_1(x_2 - x_1) + f(x_2 - x_1)^2 + y_1(y_2 - y_1) + f(y_2 - y_1)^2 + z_1(z_2 - z_1) + f(z_2 - z_1)^2 &= 0 \end{aligned}$$

with the result

$$f = \frac{(x_1x_2 + y_1y_2 + z_1z_2) - (x_1^2 + y_1^2 + z_1^2)}{L_{12}^2}$$

Substituting the value of f back into equations (15.14.1) gives the desired coordinates of the foot F. Also, the length of the perpendicular is

$$L_{PF} = \sqrt{(x_F - x_P)^2 + (y_F - y_P)^2 + (z_F - z_P)^2}$$

Comments: The method is robust, there is always a solution. There are no particular problems provided that the line specification points are not coincident, which should be checked early on, and in particular before the division by L_{12}^2.

15.15 Routine PointFPLPDC

Purpose: 'Point at the Foot of a Perpendicular onto a Line with Point and DCs' – to obtain the point at the foot of a perpendicular dropped from a point P onto a line defined by one point and direction cosines (parametric form)

Inputs: Line point P_1 coordinates (x_1, y_1, z_1), direction cosines (l, m, n) and point P coordinates (x_P, y_P, z_P)

Outputs: Foot point coordinates (x_F, y_F, z_F), perpendicular length L_{PF}

Notes: The direction cosines have the property $l^2 + m^2 + n^2 = 1$. This problem also illustrates a useful method of minimisation by derivatives.

Solution: A point S is at displacement s along the line from the reference point P_1. (The displacement s may be negative.) The point S is

$$\begin{aligned} x &= x_1 + ls \\ y &= y_1 + ms \\ z &= z_1 + ns \end{aligned} \tag{15.15.1}$$

The squared length of the prospective perpendicular line PS is

$$L_{PS}^2 = (x - x_P)^2 + (y - y_P)^2 + (z - z_P)^2$$

By substitution,

$$L_{PS}^2 = (x_1 + ls - x_P)^2 + (y_1 + ms - y_P)^2 + (z_1 + ns - z_P)^2$$

When S is the foot of the perpendicular, the length of the perpendicular is at a minimum. For an extremum the derivative is zero, so use

$$\frac{dL_{PS}^2}{ds} = 2(x_1 + ls - x_P)l + 2(y_1 + ms - y_P)m + 2(z_1 + ns - z_P)n = 0$$

Collecting terms,

$$(x_1 l + y_1 m + z_1 n) + s(l^2 + m^2 + n^2) - (x_P l + y_P m + z_P n) = 0$$

which gives the value of s as

$$s = (x_P - x_1)l + (y_P - y_1)m + (z_P - z_1)n \tag{15.15.2}$$

Substituting this value of s into equations (15.15.1) will give the desired coordinates of the foot F. The length of the perpendicular is then

$$L_{PF} = \sqrt{(x_F - x_P)^2 + (y_F - y_P)^2 + (z_F - z_P)^2}$$

Comments: There is always a solution, and the method is robust.

15.16 Routine PointITinit

Purpose: 'Point In a Triangle initialisation' – to obtain the factors for location of a point in the plane of a triangle, in terms of the triangle corner point coordinates. This is for later use by PointIT, after the triangle has moved, carrying the point P with it. The point P must be accurately in the plane of the triangle, but not necessarily within the bounding sides of the triangle.

Inputs: Coordinates of the three triangle-defining points P_1, P_2 and P_3, and of the in-plane point to be analysed P

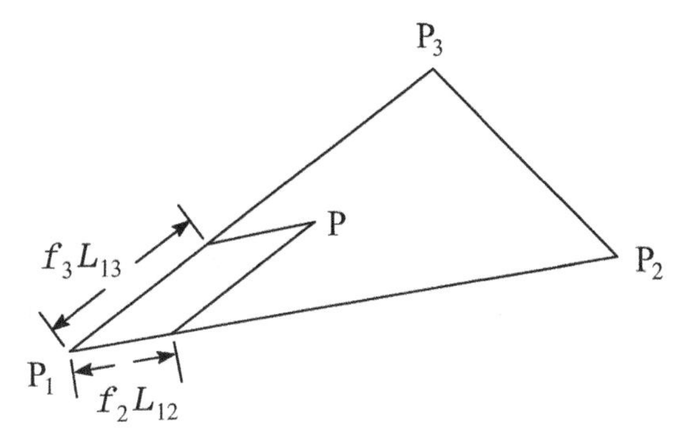

Figure 15.16.1 Factors f_2 and f_3 for PointITinit and PointIT.

Outputs: The factors f_2 and f_3, Figure 15.16.1

Notes: The coordinates of a point P in the plane of a triangle may be expressed symmetrically by

$$\begin{aligned} x_P &= f_1 x_1 + f_2 x_2 + f_3 x_3 \\ y_P &= f_1 y_1 + f_2 y_2 + f_3 y_3 \\ z_P &= f_1 z_1 + f_2 z_2 + f_3 z_3 \end{aligned} \qquad (15.16.1)$$

The additional factor f_1 is

$$f_1 = 1 - f_2 - f_3 \qquad (15.16.2)$$

Slightly more simply, using only two factors, it may be represented by

$$\begin{aligned} x_P &= x_1 + f_2(x_2 - x_1) + f_3(x_3 - x_1) \\ y_P &= y_1 + f_2(y_2 - y_1) + f_3(y_3 - y_1) \\ z_P &= z_1 + f_2(z_2 - z_1) + f_3(z_3 - z_1) \end{aligned} \qquad (15.16.3)$$

Using local coordinates $u = x - x_1$, $v = y - y_1$ and $w = z - z_1$,

$$\begin{aligned} u_P &= f_2 u_2 + f_3 u_3 \\ v_P &= f_2 v_2 + f_3 v_3 \\ w_P &= f_2 w_2 + f_3 w_3 \end{aligned} \qquad (15.16.4)$$

The local coordinates have origin at P_1. In effect, non-rectangular coordinate axes are used for the factors. The first coordinate axis is along P_1P_2. The second is along P_1P_3, not perpendicular to the first.

A vectorised version of this routine solving many points is useful.

Solution: Equations (15.16.1) give three simultaneous equations for f_1, f_2 and f_3. In view of the additional equation (15.16.2), this implies that the equations are over-specified. They are, however, consistent provided that P is in the plane of the triangle.

Use any two equations from (15.16.3) or (15.16.4) to solve for f_2 and f_3, for example,

$$\begin{aligned} f_2(x_2 - x_1) + f_3(x_3 - x_1) &= x_P - x_1 \\ f_2(y_2 - y_1) + f_3(y_3 - y_1) &= y_P - y_1 \end{aligned}$$

These two simultaneous equations may be solved in the usual way. The dividing determinant of the coefficients is

$$d = (x_2 - x_1)(y_3 - y_1) - (x_3 - x_1)(y_2 - y_1)$$

Then

$$f_2 = \frac{(x_P - x_1)(y_3 - y_1) - (y_P - y_1)(x_3 - x_1)}{d}$$

$$f_3 = \frac{(y_P - y_1)(x_2 - x_1) - (x_P - x_1)(y_2 - y_1)}{d}$$

The second f could alternatively be solved by use of one of the equations in (15.16.4).

Comments: Note that even for a well-conditioned problem (a 'good' triangle, with points nowhere near in line), data sensitivity may result in an attempt to divide by zero; that is, $d=0$ may occur, or d very small may occur, for example for a triangle parallel or nearly parallel to the xy plane. This is a data sensitivity.

The normal to the triangle has direction numbers given by the cross product $(\boldsymbol{P}_2 - \boldsymbol{P}_1) \times (\boldsymbol{P}_3 - \boldsymbol{P}_1)$. In local coordinates this is given by

$$\boldsymbol{N} = \begin{vmatrix} \hat{\boldsymbol{i}} & \hat{\boldsymbol{j}} & \hat{\boldsymbol{k}} \\ u_2 & v_2 & w_2 \\ u_3 & v_3 & w_3 \end{vmatrix}$$

so

$$\boldsymbol{N} = \hat{\boldsymbol{i}}(v_2 w_3 - v_3 w_2) + \hat{\boldsymbol{j}}(u_3 w_2 - u_2 w_3) + \hat{\boldsymbol{k}}(u_2 v_3 - u_3 v_2)$$

The third of these terms is

$$N_z = \hat{\boldsymbol{k}}\,\{(x_2 - x_1)(y_3 - y_1) - (x_3 - x_1)(y_2 - y_1)\} = \hat{\boldsymbol{k}} d$$

so, in the method used in the example, the divisor is the z component of the normal to the triangle. This gives the physical interpretation to the data sensitivity issue. $N_z = d$ will be zero, or very small, when the triangle normal is perpendicular, or nearly so, to the z axis.

Even for a well-conditioned problem with a good triangle shape (the three defining points not near to being on a straight line), the solution route must be correctly chosen. To avoid inaccuracy in critical cases, calculate the components of the normal and select the one of largest magnitude. This indicates which two equations from (15.16.3) will give the best solution route.

The triangle can be any shape, it does not require any identifiable direction, so it may be equilateral if desired.

15.17 Routine PointIT

Purpose: 'Point In a Triangle' – to obtain the point in a triangle given by the factors obtained by PointITinit. The triangle may have moved in any way, but must retain its shape.

Inputs: Coordinates of the three new triangle points P_1, P_2 and P_3, and the factors f_2, f_3

Outputs: The sought point P coordinates (x_P, y_P, z_P)

Notes: The new triangle must retain its shape (the side lengths must be unchanged). The points must be presented to PointITinit and PointIT in the same order. A vectorised version of this routine solving many points is useful.

Solution: The desired point is given simply by

$$\begin{aligned} x_P &= x_1 + f_2(x_2 - x_1) + f_3(x_3 - x_1) \\ y_P &= y_1 + f_2(y_2 - y_1) + f_3(y_3 - y_1) \\ z_P &= z_1 + f_2(z_2 - z_1) + f_3(z_3 - z_1) \end{aligned}$$

which is the same as (15.16.3).

Comments: This uses the non-rectangular coordinate system of the triangle sides. See Section 15.16 for further discussion. The symmetrical solution uses all three factors and three multiplications per variable. However, it is probably more accurate in most cases to use the equations above.

15.18 Routine PointFPT

Purpose: 'Point at the Foot of a Perpendicular into a Triangle' – to obtain the coordinates of the foot of a perpendicular from point P into the plane of a triangle $P_1P_2P_3$

Inputs: Coordinates of the triangle points P_1, P_2, P_3 and of the point P

Outputs: Coordinates of the foot point F, and the length L_{PF}

Notes: None

Comments: The three points of the triangle define a plane, and this routine is just a pseudonym for PointFPPl3P.

15.19 Routine Plane3P

Purpose: 'Plane of 3 Points' – convert to normal form (l, m, n, p)

Inputs: The coordinates of the three triangle points

Outputs: l, m, n, p – the direction cosines (l, m, n) of the normal to the triangle plane, and p, the length of the normal from the origin to the plane in the direction (l, m, n). Usually the sign of p is positive, but the set of four variables may have all signs reversed with valid results.

Notes: Arguably, one set (l, m, n, p) represents one surface of the plane, while $(-l, -m, -n, -p)$ represents the other surface, but this is not a significant issue here.

Solution: Consider any point in the plane, the point being at a distance L from the origin. Taking the dot product of the point position vector (x_P, y_P, z_P) with the plane normal (l, m, n) must give $p = L\cos\theta$, where θ is the angle between the position vector and the plane normal. Therefore, from the three points of the triangle there are three equations, and there is also the direction-cosine interrelationship:

$$\begin{aligned} lx_1 + my_1 + nz_1 &= p \\ lx_2 + my_2 + nz_2 &= p \\ lx_3 + my_3 + nz_3 &= p \\ l^2 + m^2 + n^2 &= 1 \end{aligned} \tag{15.19.1}$$

These are four equations for four unknowns, but the last equation is non-linear.

Practical solution can proceed as follows. Take the vector cross product

$$\boldsymbol{N} = (\boldsymbol{P}_2 - \boldsymbol{P}_1) \times (\boldsymbol{P}_3 - \boldsymbol{P}_1),$$

to give the standard normal to the plane. In local coordinates on point P_1 this is given by:

$$\boldsymbol{N} = \begin{vmatrix} \hat{\boldsymbol{i}} & \hat{\boldsymbol{j}} & \hat{\boldsymbol{k}} \\ u_2 & v_2 & w_2 \\ u_3 & v_3 & w_3 \end{vmatrix}$$

so

$$\boldsymbol{N} = \hat{\boldsymbol{i}}(v_2w_3 - v_3w_2) + \hat{\boldsymbol{j}}(u_3w_2 - u_2w_3) + \hat{\boldsymbol{k}}(u_2v_3 - u_3v_2)$$

Normalise this vector by its length to give a unit vector $\boldsymbol{N}$/len($\boldsymbol{N}$), which has the direction cosines (l, m, n) of the plane. Then use one equation from (15.19.1) to obtain p, completing the solution.

Comments: There are no particular problems, provided that the triangle is well defined (points separated and not in line).

15.20 Routine PointFP

Purpose: 'Point at Foot of Perpendicular' – to obtain the coordinates of the foot of a perpendicular from a specified point P onto a normal-form plane.

Inputs: The plane normal-form parameters (l, m, n, p) and the point P coordinates

Outputs: The coordinates of the foot point F of the perpendicular, and the length of the perpendicular

Notes: The direction cosines have the property $l^2 + m^2 + n^2 = 1$.

Solution: The perpendicular must have direction cosines (l, m, n). The length of the perpendicular is s from P in the direction (l, m, n). This point has the coordinates

$$\begin{aligned} x &= x_P + ls \\ y &= y_P + ms \\ z &= z_P + ns \end{aligned} \qquad (15.20.1)$$

The equation of the plane is

$$lx + my + nz = p$$

By substitution,

$$(x_P + ls)l + (y_P + ms)m + (z_P + ns)n = p$$

Collecting terms,

$$(x_P l + y_P m + z_P n) + s(l^2 + m^2 + n^2) = p$$

giving

$$s = p - (x_P l + y_P m + z_P n) \qquad (15.20.2)$$

Substituting this value of s into equations (15.20.1) gives the coordinates of the foot F of the perpendicular.

Comments: The solution is robust, with no particular problems. However, the direction of the normal is not standardised (see the next routine).

15.21 Routine PointFPPl3P

Purpose: 'Point at Foot of Perpendicular into a Plane of 3 Points' – to obtain the coordinates of the foot of a perpendicular from a specified point P onto a plane defined by three points. In practical suspension analysis, this is the usual form of definition of a plane, and this has the advantage of giving a standardised direction of the plane normal, essential for use in routines such as PointAT.

Inputs: Coordinates of the three points P_1, P_2 and P_3 defining the plane, and of the point P

Outputs: The coordinates of the foot point F of the perpendicular, and the displacement along the standard direction of perpendicular to give P from foot F

Notes: The direction cosines have the property $l^2 + m^2 + n^2 = 1$.

Solution: The perpendicular has direction cosines (l, m, n). The length of the perpendicular is s, possibly negative, from F to P in the direction (l, m, n). The direction of the normal is standardised in the positive direction of the cross product

$$\boldsymbol{N} = (\boldsymbol{P}_2 - \boldsymbol{P}_1) \times (\boldsymbol{P}_3 - \boldsymbol{P}_1)$$

It is convenient to use local coordinates (u, v, w) based on P_1, for example,

$$u_2 = x_2 - x_1; \quad v_2 = y_2 - y_1; \quad w_2 = z_2 - z_1$$

The direction numbers of the normal are

$$\begin{aligned} L_P &= v_2w_3 - v_3w_2 \\ M_P &= w_2u_3 - w_3u_2 \\ N_P &= u_2v_3 - u_3v_2 \end{aligned}$$

Dividing by the Euclidean length of this normal gives the direction cosines (l, m, n) of the plane normal.

For a displacement s from foot F to P in the positive normal direction, the position of F is given by

$$\begin{aligned} u_F &= u_P - ls \\ v_F &= v_P - ms \\ w_F &= w_P - ns \end{aligned} \tag{15.21.1}$$

The equation of the plane, which in local coordinates goes through the origin, is

$$\begin{aligned} lx + my + nz &= P \\ lu + mv + nw &= 0 \end{aligned}$$

By substitution,

$$l(u_P - ls) + m(v_P - ms) + n(w_P - ns) = 0$$

giving

$$s = lu_P + mv_P + nw_P \tag{15.21.2}$$

Substituting this value of s into equations (15.21.1) gives the coordinates of the foot F of the perpendicular.

Comments: The solution is robust, with no particular problems. The direction of the normal is standardised, which is very useful.

15.22 Routine PointATinit

Purpose: 'Point "Above" a Triangle initialisation' – to obtain factors for use by routine PointAT to determine the position of a point relative to a triangle, the point being in any relative location (above, below or in the plane of the triangle). The triangle may subsequently move in any way, taking the point with it, but the triangle must retain its shape.

Inputs: The coordinates of the defining points of the triangle in a known position, and the coordinates of the point to be analysed

Outputs: Factors f_2 and f_3 (for the foot of the perpendicular; see Figure 15.22.1), and the directed out-of plane distance s

Notes: A vectorised version of this routine solving many points is useful.

Solution: Drop the perpendicular from P into the plane of the triangle, using routine PointFPT, giving the foot point F and the signed out-of plane distance s. Depending on how routine PointFPT is written, the sign of s may need to be changed here. The foot F may then be analysed by routine PointITinit to give f_2 and f_3.

Taking the cross product of $\boldsymbol{P}_2 - \boldsymbol{P}_1$ and $\boldsymbol{P}_3 - \boldsymbol{P}_1$ in that order gives a vector perpendicular to the plane of the triangle. This is the positive direction for the out-of-plane displacement s.

Comments: There is one new problem beyond the use of PointFPT and PointATinit, which is the sign of s, which should be positive in the direction of the defined normal. This also requires that the normal

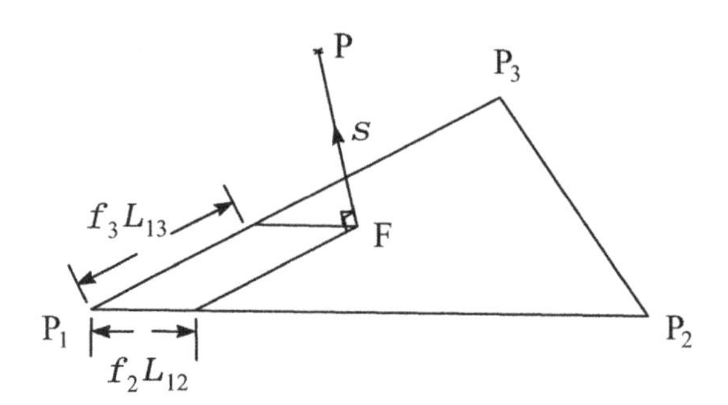

Figure 15.22.1 Factors f_2 and f_3, and out-of-plane displacement s for PointATinit and PointAT.

vector be calculated in the same way in PointATinit and PointAT, and that the triangle defining points are presented to these routines in the same order.

15.23 Routine PointAT

Purpose: 'Point "Above" a Triangle' – to obtain the position of a point 'above' (i.e. relative to) a triangle (above, below or exactly in the plane), using the factors obtained in an initial position by PointATinit

Inputs: The triangle point coordinates and the factors from PointATinit analysing the initial position

Outputs: The coordinates of P

Notes: This problem/routine is also sometimes known as 'Pole-on-a triangle', 'Rod-on-a-triangle' or, consequently, just 'Perch'. A vectorised version of this routine solving many points is useful.

Solution: Use the provided factors f_2 and f_3, Figure 15.22.1, to obtain a point in the plane of the triangle, using routine PointIT. This is the foot F of the perpendicular from the point, so calculate the normal to the plane by

$$\boldsymbol{N} = (\boldsymbol{P}_2 - \boldsymbol{P}_1) \times (\boldsymbol{P}_3 - \boldsymbol{P}_1)$$

Move from F along this normal by a distance s (possibly negative), reaching P.

Comments: The order of vectors in the calculation of $\boldsymbol{N}$ must be the same as in PointATinit, or the out-of-plane displacement will be reversed. There are no particular problems provided that the defining triangle is good (points not in line). The points must be presented to PointATinit and PointAT in the same order.

15.24 Routine Points3S

Purpose: To obtain the two points at the intersection of three spheres

Inputs: For each of three spheres, the coordinates of the centre and the radius

Outputs: The integer number of distinct real solutions, and the coordinates of the solution points

Notes: This is also known as the 'Tripod' problem or routine. If intersection does occur, the two solution points are symmetrical about the plane of the centres. That is, the line from one solution to the other is normal to that plane, and the mid-point of the line is a point in the plane, that is, the solutions are equidistant from the plane. Various solution methods are possible, only one of which is given here.

Solution: The equation of a sphere has the basic form

$$(x - x_1)^2 + (y - y_1)^2 + (z - z_1)^2 = R_1^2$$

The equations for the three spheres may be expanded to

$$\begin{aligned}
x^2+y^2+z^2-2x_1x-2y_1y-2z_1z &= R_1^2-(x_1^2+y_1^2+z_1^2)\\
x^2+y^2+z^2-2x_2x-2y_2y-2z_2z &= R_2^2-(x_{21}^2+y_2^2+z_2^2)\\
x^2+y^2+z^2-2x_3x-2y_3y-2z_3z &= R_3^2-(x_3^2+y_3^2+z_3^2)
\end{aligned} \tag{15.24.1}$$

These are three independent non-linear simultaneous equations for the intersection point (x, y, z). Subtracting any two of these equations removes the unknown non-linear terms, giving a linear equation. If this is done in three pairs, the result is three linear equations. Unfortunately, such an approach fails, because the three resulting equations are no longer independent.

The physical interpretation of this is instructive. Subtracting the equations of a pair of spheres gives the equation of their radical plane. If the spheres intersect then the circle of intersection lies in the radical plane. In the context of three spheres, any pair of the three radical planes intersect in a line perpendicular to the plane of the centres. This line passes through the sphere intersection solution points sought, and so all three pairs of radical planes have the same intersecting line, and so are not independent. This applies geometrically and algebraically. In a numerical context, the radical planes may be very slightly skewed due to numerical imprecision, in which case their intersection is a single point. However, this is not a useful point because although it is on the desired line of the solutions it is positioned along the line in an effectively unpredictable way, depending on the minutiae of the numerical calculation. A different solution strategy must therefore be sought.

One good method of solution is to obtain only two radical planes, say for S_1–S_2 and S_1–S_3, and to intersect these for a line through the solution points. Then intersect this line with any of the sphere equations, say S_1, to obtain the two points.

Using local coordinates based on the centre of sphere 1, $(u=x-x_1$, etc), the three sphere equations become:

$$u^2+v^2+w^2=R_1^2=c_1 \tag{15.24.2}$$

$$u^2+v^2+w^2-2u_2u-2v_2v-2w_2w=R_2^2-(u_2^2+v_2^2+w_2^2)=c_2 \tag{15.24.3}$$

$$u^2+v^2+w^2-2u_3u-2v_3v-2w_3w=R_3^2-(u_3^2+v_3^2+w_3^2)=c_3 \tag{15.24.4}$$

The radical planes S_1-S_2 and S_1-S_3, by subtraction, are

$$u_2u+v_2v+w_2w=\tfrac{1}{2}(c_1-c_2)=c_4 \tag{15.24.5}$$

$$u_3u+v_3v+w_3w=\tfrac{1}{2}(c_1-c_3)=c_5 \tag{15.24.6}$$

These two equations may be used to eliminate two variables. Say eliminate w first, using $w_3(15.23.5)-w_2(15.23.6)$ to give

$$u(w_3u_2-w_2u_3)+v(w_3v_2-w_2v_3)=w_3c_4-w_2c_5$$

Now write

$$\begin{aligned}
a_1 &= w_3u_2-w_2u_3\\
a_2 &= w_3v_2-w_2v_3\\
a_3 &= w_3c_4-w_2c_5
\end{aligned}$$

Then using

$$a_4 = \frac{a_3}{a_2}; \qquad a_5 = \frac{-a_1}{a_2}$$

the relationship between v and u can be simply expressed as

$$v = a_4 + a_5 u \tag{15.24.7}$$

where the constants are known. Now using equation (15.23.5),

$$u_2 u + v_2(a_4 + a_5 u) + w_2 w = c_4$$
$$u(u_2 + a_5 v_2) + w_2 w = c_4 - a_4 v_2$$

This gives a simple linear equation for $w(u)$ with known coefficients:

$$w = b_1 + b_2 u \tag{15.24.8}$$

Equations (15.24.7) and (15.24.8) can be substituted into (15.24.2) to give a quadratic equation in u only, so the problem is effectively solved, producing potentially two values for u, with corresponding values for v and w then solved by the subsequent equations. If the quadratic equation produces no real solutions, then the spheres do not have the required intersection (possibly two of them do meet). If the quadratic equation produces one real solution (i.e. coincident real solutions of the same value), then the intersections are coincident. This would occur, for example, when two spheres just touch at a single point, and the third sphere intersects the first two, passing exactly through the touching point, or, seen in another way, two spheres intersect in a circle with the third sphere just touching the circle. In the case of suspension analysis, the expected result would be two separate real solutions, indicating the expected two possible physical intersections. Given the two u solution values, u_A and u_B, then equations (15.19.7) and (15.19.8) give the other coordinates, providing the complete solutions (u_A, v_A, w_A) and (u_B, v_B, w_B).

Comments: One of these two solutions will be the desired one. The other corresponds to a geometrically correct but physically inappropriate assembly of the components. The correct solution must subsequently be selected carefully if problems are to be avoided, but this is not the province of routine Points3S, which lacks the necessary extra information to select the desired solution from the two candidate points.

As with some other routines, it will be noticed here that there are several possible solution paths, with possible data sensitivities according to the alignments, even for a basically well-conditioned problem. The crucial point is the order of elimination of the variables, which should be based on the avoidance of divisors of small magnitude (i.e. select the solution route with the largest magnitude divisors).

15.25 Routine Points2SHP

Purpose: 'Points at the intersection of 2 Spheres and a Horizontal Plane'

Inputs: Coordinates of the centres of the two spheres and corresponding radii, and the z value of the horizontal plane z_{HP}

Outputs: The integer number of real solutions found and the solution point coordinates

Notes: The solution of an independent suspension conveniently begins with a position of the lower outer ball joint. This could be done by specifying the rotation of the lower arm, but is easily done by specifying the height of the ball joint, which is approximately related to the suspension bump value, then solving for the x and y coordinates according to the arm geometry.

Solution: The equations to be satisfied are two spheres and a horizontal plane:

$$(x-x_1)^2+(y-y_1)^2+(z-z_1)^2=R_1^2 \tag{15.25.1}$$

$$(x-x_2)^2+(y-y_2)^2+(z-z_2)^2=R_2^2 \tag{15.25.2}$$

$$z=z_{\mathrm{HP}} \tag{15.25.3}$$

In local coordinates, $u=x-x_1$, etc., and substituting $z=z_{\mathrm{HP}}$,

$$u^2+v^2+w^2=R_1^2$$

$$u^2+v^2+w^2-2u_2u-2v_2v-2w_2w=R_2^2-(u_2^2+v_2^2+w_2^2)$$

$$w=w_{\mathrm{HP}}=z_{\mathrm{HP}}-z_1$$

There are potentially two circles of intersection of the spheres with the plane, with equations given by substituting for w:

$$u^2+v^2=R_1^2-w_{\mathrm{HP}}^2=c_1$$

$$u^2+v^2-2u_2u-2v_2v=R_2^2-(u_2^2+v_2^2+w_2^2)-w_{\mathrm{HP}}^2+2w_2w_{\mathrm{HP}}=c_2$$

Subtraction of these two circle equations gives the equation of the radical line:

$$2u_2u+2v_2v=c_1-c_2$$

Choosing to eliminate u first,

$$v=-\frac{u_2}{v_2}u+\frac{c_1-c_2}{2v_2} \tag{15.25.4}$$

where division by v_2 is required. If the absolute value of u_2 exceeds that of v_2, it is better to eliminate u first, with

$$u=-\frac{v_2}{u_2}v+\frac{c_1-c_2}{2u_2} \tag{15.25.5}$$

Substituting one of these two options into (15.24.1) or (15.24.2) gives a quadratic equation in one variable. The other then follows from (15.24.4) or (15.24.5).

Comments: The physical interpretation of this is that substituting the constant plane height into the sphere equations gives the equations of two circles in the plane (prospectively). The intersection of the two circles is a simpler two-dimensional problem, easily solved. Equations (15.20.4) and (15.20.5) are alternative equations for the radical line of the two circles. If this line is exactly or nearly parallel with the v axis (y axis) then solve for v first because $v(u)$ is badly conditioned. In this case, eliminate u, with (15.20.5), solving for v, because $\mathrm{abs}(u_2)>\mathrm{abs}(v_2)$.

15.26 Routine Point3Pl

Purpose: 'Point in 3 Planes' – to obtain the intersection point of three planes, each plane being specified by any one point in the plane and the direction cosines of the normal to the plane.

Inputs: For each plane, the coordinates of one point in it, and the direction cosines of its normal, for example (x_1,y_1,z_1) and (l_1, m_1, n_1)

Outputs: The coordinates of the one intersection point (x_P, y_P, z_P)

Notes: This routine is required for three-dimensional velocity diagram analysis of the Points3S ('Tripod') problem.

Solution: The solution point must lie in each plane, so satisfying simultaneously, by dot products,

$$l_1 x_P + m_1 y_P + n_1 z_P = l_1 x_1 + m_1 y_1 + n_1 z_1 = c_1$$

$$l_2 x_P + m_2 y_P + n_2 z_P = l_2 x_2 + m_2 y_2 + n_2 z_2 = c_2$$

$$l_3 x_P + m_3 y_P + n_3 z_P = l_3 x_3 + m_3 y_3 + n_3 z_3 = c_3$$

These are three simultaneous linear equations in x_P, y_P and z_P, conveniently solved by use of a general simultaneous equation solver (e.g. Routine SimEqns).

Comments: In the case of a suspension, the practical problem would always be well conditioned, but data sensitivities could easily arise due to critical alignments of the planes with the axes. A Gaussian elimination solver with pivoting will correctly negotiate such problems, and produce a good solution. Alternatively, use an *ad hoc* solver by the usual methods, with the order of elimination of the variables chosen to avoid small divisors.

15.27 Routine 'PointLP'

Purpose: 'Point of a Line and a Plane' – to obtain the coordinates of the point of intersection of a general parametric-form line and a normal-form plane

Inputs: The parametric variables of the line, and the normal form variables of the plane

Outputs: The coordinates of the intersection point

Notes: None

Solution: At the line parametric point S, the line coordinates are

$$\begin{aligned} x &= x_L + l_L s \\ y &= y_L + m_L s \\ z &= z_L + n_L s \end{aligned} \tag{15.27.1}$$

The equation of the plane is

$$l_P x + m_P y + n_P z = p_P$$

By substitution,

$$l_P(x_L + l_L s) + m_P(y_L + m_L s) + n_P(z_L + n_L s) = p_P$$

Collecting terms and making s the subject,

$$s = \frac{p_P - (x_L l_P + y_L m_P + z_L n_P)}{l_P l_L + m_P m_L + n_P n_L} \tag{15.27.2}$$

The divisor will be recognised as the dot product of the plane normal and the line direction. Substituting the resulting value of s into equations (15.26.1) gives the desired coordinates.

Comments: The solution is not completely robust. The dot product could cause a division by zero. Physically, if the dot product is zero then the plane normal and the line are perpendicular, in which case the line and plane will not meet, that is, will meet at infinity. If the defining point of the line is in the plane, the value of s is zero, which is not problematic. If the point is in the plane and the dot product is zero then the numerical solution for s is 0/0, but 0 would usually be an acceptable value. Therefore evaluate the numerator first. If this is not zero, evaluate the denominator. If this is zero, return an indicator of this situation. One way to handle such infinities is homogeneous coordinates.

15.28 Routine Point3SV

Purpose: To obtain the velocity of a point positioned by three rods (the position of which is found by Points3S)

Inputs: The coordinates of the three locating points, the coordinates of the apex point, and the velocity components of the three base points

Outputs: The velocity components of the apex point (u_P, v_P, w_P)

Notes: None

Solution: In the 3D velocity diagram, each base point has a given velocity point. Relative to the base point, the other-end velocity is tangential only. Relative to the base velocity point, then, the solution point must be in a plane with a normal given by the rod axis direction cosines. The solution velocity is the therefore the intersection in the velocity diagram of the three such planes, one for each rod.

The first rod has base coordinates (x_1, y_1, z_1). The apex position is (x_P, y_P, z_P). From these, the first rod direction cosines are easily determined. Similarly for the others.

For each rod, the base point velocity, for example (u_1, v_1, w_1), and the corresponding direction cosines, for example (l_1, m_1, n_1), give a plane in the three-dimensional velocity diagram with an equation such as

$$l_1 u + m_1 v + n_1 w = l_1 u_1 + m_1 v_1 + n_1 w_1 = c_1$$

This gives three simultaneous linear equations in the velocities (u, v, w). The intersection of the three planes can be solved by the use of routine Point3Pl.

Comments: Routine Point3Pl will obtain the intersection point, called with the correct arguments. Although velocities are being solved for, this requires a geometric construction in 'velocity space', that is, in the three-dimensional velocity diagram.

15.29 Routine PointITV

Purpose: 'Velocity of a Point In a Triangle' – to obtain the velocity of a point in the plane of a triangle, given the triangle corner velocity components and the point position factors f_2 and f_3

Inputs: The corner velocity components for the triangle, for example (u_1, v_1, w_1), and the point position factors f_2 and f_3 (found by PointITinit)

Outputs: The point velocity components

Notes: The triangle must retain its shape and order of defining points, but may move in any way. The velocities of the corners of the triangle must be compatible, with the side lengths remaining constant. The variables (u, v, w) here are velocities, not local positions.

Solution: By differentiation to give velocities from positions, it is evident that the velocity of the point is given in terms of the triangle corner velocities by the same factors as is its position in terms of the corner positions. The required factors are found by PointITinit, so

$$\begin{aligned}
u_P &= u_1 + f_2(u_2 - u_1) + f_3(u_3 - u_1) \\
v_P &= v_1 + f_2(v_2 - v_1) + f_3(v_3 - v_1) \\
w_P &= w_1 + f_2(w_2 - w_1) + f_3(w_3 - w_1)
\end{aligned}$$

Comments: The solution is straightforward, other than that caution is required to provide only compatible velocities of the three base points. The velocities of two ends of a rigid rod must have equal components along the rod. This longitudinal velocity is the dot product of the complete velocity with the rod direction cosines, so for two ends A and B the condition is

$$lu_A + mv_A + nw_A = lu_B + mv_B + nw_B$$

15.30 Routine PointATV

Purpose: 'Velocity of Point Above Triangle' – to calculate the velocity of a point with a position specified relative to a rigid triangle, above, below or exactly in the plane of the triangle

Inputs: The coordinates of the triangle corners, the factors f_2 and f_3 and the out-of-plane displacement s of the point P (from routine PointATinit), and the velocity components of each defining corner of the triangle (which velocities must be compatible)

Outputs: The velocity components of the point

Notes: This routine is used for three-dimensional velocity diagram analysis for points with position defined by PointAT.

Solution: The coordinates of P_1, P_2 and P_3 are given. Also the factors f_2 and f_3 for F and the out-of-plane displacement s for P. The foot F of the perpendicular from P has coordinates solved by routine PointIT, using f_2 and f_3. The coordinates of P are solved by routine PointAT. Therefore all positions are known.

The velocity of F is calculated by routine PointITV, using the factors f_2 and f_3.

It remains to calculate the velocity of P relative to F. Length L_{PF} is constant, so the relative velocity is purely tangential. Drop a perpendicular from P_1 onto line P_2P_3 with foot P_4. The velocity $V_{P/F}$ of P relative to F is parallel to the plane, with the components shown, V_{T1} being parallel to line P_1P_4 and V_{T2} being parallel to line P_2P_3, Figure 15.30.1. The velocity of P_4 can be found from the velocities of P_2 and P_3 by linear interpolation or extrapolation according to the position factor of P_4 along P_2P_3 (P_4 may be outside P_1P_3, with $f_4 < 0$ or $f_4 > 1$).

Calculate the unit normal to the plane for F towards P. By dot products with this unit vector, the normal (out-of-plane) velocity components at each point, V_{N1}, V_{N2}, V_{N3} and V_{N4}, can be determined, as seen in the figure.

The angular velocity of the triangle may be considered in three components. The first is the in-plane rotation, about an axis perpendicular to the plane, which has no effect on $V_{P/F}$, so will not be

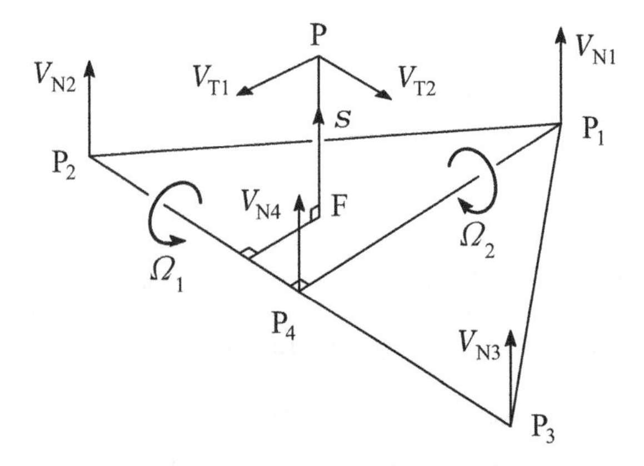

Figure 15.30.1 Tangential velocity components V_{T1} and V_{T2} for the point P relative to F, derived from the out-of plane velocities V_{N1} etc. at each corner.

considered further. The right-hand angular velocity about any axis parallel to the line P_2P_3 is

$$\Omega_1 = \frac{V_{N1} - V_{N4}}{L_{14}}$$

The first tangential velocity component is then

$$V_{T1} = \Omega_1 L_{FP} = (V_{N1} - V_{N4}) \frac{L_{FP}}{L_{14}}$$

Now, to obtain the second tangential component, the angular velocity about an axis parallel to P_4P_1 is

$$\Omega_2 = \frac{V_{N2} - V_{N3}}{L_{23}}$$

giving

$$V_{T2} = \Omega_2 L_{FP} = (V_{N2} - V_{N3}) \frac{L_{FP}}{L_{23}}$$

The velocity of P now follows by the addition of the two tangential components to V_F.

Comments: The solution is basically robust, with divisions by lengths that must be non-zero in practical cases. The solution is not symmetrical in the use of the three points of the triangle, and the in-plane perpendicular could equally well be dropped from one of the other corners. The end result for V_P should be the same.

The solution is straightforward in concept, but more detailed than other routines here, and care is needed to ensure a correct implementation. In practice, the accuracy is very good. A point to be careful about is that the corner input velocities must be compatible (the base triangle is rigid).

When the point P is safely out-of-plane, routine Point3SV could be used. However, this is unsatisfactory for in-plane points, and is really intended for configurations in which the point position is solved by Points3S. When the triangle is rigid, routine PointATV here is the appropriate solution method.

15.31 Rotations

The rotational position of a solid body can be represented and solved by the use of points on the body with vector analysis, and this is generally considered to be the practical method, using routines such as those described above. It is particularly suitable for mechanism analysis, such as that for suspensions, which tends naturally to deal with specific points on each member. An interesting alternative is the use of quaternions. These are particularly suitable for use on isolated solid bodies, such as aircraft or spacecraft, but could possibly be used with advantage for some three-dimensional mechanism or suspension analyses. Appendix D gives an explanation of the properties of quaternions and their use.

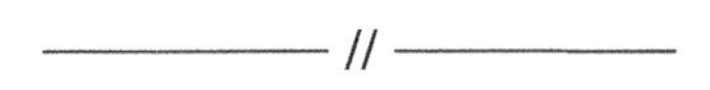

16

Programming Considerations

16.1 Introduction

This chapter deals with some particular issues in programming that bear upon the solution of suspension geometry. Some of these issues are fairly specific to suspensions (e.g. the 'assembly' problem of selecting one of two or more prospective solutions for a point), while others are more general (e.g. generally good programming practice). Three-dimensional geometrical programs are of sufficient complexity that the best programming practices are really needed to achieve reliable success.

16.2 The RASER Value

The RASER value for software is a reliability, accuracy, speed and ease of use rating. The reliability of a subroutine is primarily its ability to survive without crashing in any data environment, and depends on its failure modes analysis. Accuracy can be divided into accuracy for 'normal' cases and accuracy for difficult cases. Speed again can be divided into speed on normal cases and speed for difficult cases. Ease of use of a subroutine basically depends on its argument list.

The idea behind a RASER value is to draw attention to the quality of a piece of software. Just being conscious of the concept will concentrate a programmer's mind, the idea that his program might actually be quantitatively scored even more so. Such a scoring system works basically as follows. Program quality is on a scale from 0 to 1. A notionally perfect program scores 1.000000, but in practice no program is above some improvement. The maximum total score available is in fact 0.999999.

The first 9 is for reliability and ease of use. Reliability means not crashing, whatever the data values. With bad ('impossible') data, the correct error state IE values must be returned (Section 16.3). Ease of use depends on a suitable routine name, and a considered arrangement of the argument list (Section 16.9).

The second 9 is for accuracy on typical cases (i.e. the easy ones). The third 9 is for speed on such cases.

The fourth 9 is for accuracy on the occasional particularly difficult cases. The fifth 9 is for speed on such cases. Difficult cases include problems with bad conditioning and ones with data sensitivity problems.

The sixth 9 is for code compactness. Actually, this was once very important, in the early days of small computer memories, but is not so important nowadays for numerical software, which tends to be small anyway. However, there is much obese software around, so a low score is a possibility here.

Evidently, a score of 0.911111 beats a score of 0.899999. Reliability is more important than any other aspect.

These topics are dealt with in more detail in subsequent sections.

16.3 Failure Modes Analysis

Failure modes analysis was developed to improve the overall performance and reliability of large commercial and military systems, these being sufficiently unpredictable due to complexity or adverse influence that complete reliability could not be assured. Then, instead of simply accepting occasional failure and its consequences, the idea of design for operation under conditions of partial failure was introduced. This was by no means new, in fact, as engineering design had included it in one form or another for many years, including, for example, plastic failure design techniques.

The basic idea behind failure modes analysis is to accept that failure might or will sometimes occur, and to anticipate the likely forms of such failure and to be prepared for them, that is, to be aware of the prospective failure modes and to ameliorate the damage. This concept is readily applicable to computer software design, with fruitful results.

Program or subroutine outcomes may be classified as follows:

(1) a program 'crash';
(2) solution claimed good, is good;
(3) solution claimed good, is bad;
(4) solution failure stated, but actually good solution possible;
(5) solution failed, failure admitted.

To some extent this depends on the problem. A very difficult problem may require an inordinate amount of time, in which case it may be appropriate to abandon the attempt. It is widely accepted that the worst outcome is (3), a claimed good solution that is actually bad. Result (4), failing to find an achievable solution, is not ideal, but may be acceptable. The program may work well on normal problems, but it may not work on some particularly difficult cases. If the difficult case are rare, the program may still be useful. And, of course, it may simply not be known how to solve the difficult cases. The program may also fail by doing the computations but losing accuracy in some way. These issues are dealt with in subsequent sections.

It is usual for a numerical program to include at the end of its argument list an integer variable IE or Ierr, which is used to indicate its 'failure status', or, in modern sales parlance, its 'success status'. IE = 0 indicates no undue problems and a claimed good result, with other values giving an indication of less satisfactory outcomes. Typically a routine will have an output code along these lines:

```
IE = 0 Good result.
IE = 1 Data outside acceptable range.
IE = 2 There is no solution.
IE = 3 Accuracy lost.
IE = 4 SQRT unexpected negative argument.
IE = 5 ASIN or ACOS argument out of range.
IE = 6 LOG argument out of range.
IE = 7 Division by zero.
IE = 8 Failed to converge in iteration.
IE = 9 Points must not be in line.
IE = 10 Problem in lower subroutine call.
IE = 11 . . .
```

16.4 Reliability

Reliability in a broad sense is the ability of a program to exit gracefully whilst not making unreasonable claims about its results. In a narrower sense, as dealt with in this section, reliability is the ability of a program to keep working, not crashing, regardless of the data presented to it. This means anticipating any problems and dealing with them in advance, rather than being a passive subject of difficulties. The main problems in this regard are IE items 4–7 in the list of the previous section.

In a simple program, IE items 4–6 will cause a program termination – a 'crash'. In many languages, so will item 7. In Fortran, item 7 gives a result of infinity, which is a mixed blessing, at best. If the program terminates, then data may be lost, for example several minutes of typing in suspension coordinates, which is invariably undesirable. This problem is easily solved. Whenever one of these, or any other 'risky' routine, is to be used, first test the argument for acceptability. To a neophyte programmer, this may seem a nuisance, but experienced programmers know that in the end taking the trouble to do this saves a good deal of time and frustration. The number of occasions on which it is needed is not very great.

For example, the code fragment, all in real numbers

```
x=sqrt(a**3-d**3+f-g)
```

is risky, for two reasons. Even if the data supplied are good, it is hard to anticipate the outcome of the calculation of the argument value. Of course, there may be additional information available to the programmer to indicate that this value 'cannot' be negative. However, the arguments supplied may be outside the allowed range. If these have not been checked at the outset, the argument may be negative in defiance of the 'known-to-be-positive' value. The solution is simple:

```
xsq = a**3-d**3+f-g
If (xsq<0) Goto 14
x = sqrt(xsq)
```

where label 14 is in an error block at the end of the routine, such as

```
14 Pause 'Negative Sqrt argument.' ; Stop
```

The failure is anticipated, and dealt with in a civilised manner, according to the programmers wishes. Similarly,

```
If (a<1 .or. a>1) Goto 15
x = asin(a)
. . .
If (b<0) Goto 16
x = log10(b)
. . .
If (c==0) Goto 17
x=(p+q)/c
. . .
! Error block:
14 IE=4 ; Return
15 IE=5 ; Return
16 IE=6 ; Return
17 IE=7 ; Return
```

16.5 Bad Conditioning

If equations are badly conditioned then they cannot be solved accurately by any method. Typically this arises with simultaneous equations where, say, there are two equations each representing one line. If the lines are at a very small angle to each other then the accuracy of the solved point is poor in the direction along the lines. This is a distinct problem from general 'data sensitivity', which can occur even if the problem is basically well conditioned. In short, if the problem is badly conditioned then there is no way to ensure accurate results, compared with a well-conditioned case. If the problem is well conditioned, then accurate results are possible, but not necessarily by one particular method of solution. The quality of the solution will depend on the method used, which must be selected correctly according to the particular data.

Consider the problem of two linear simultaneous equations in variables (u, v):

$$l_1 u + m_1 v = p_1 \tag{16.5.1}$$

$$l_2 u + m_2 v = p_2 \tag{16.5.2}$$

Physically, this arises, for example, in the two-dimensional solution of the velocity of the joint of two rods of known length and position and specified base velocities. These equations are badly conditioned if they are almost the same equation, that is, if the coefficients are in almost exactly the same ratios. Just multiplying one equation by a constant does not change the conditioning. If $l_1 = l_2$ and $m_1 = m_2$ and p_1 is similar in value to p_2, this would be parallel lines at a close spacing, actually not meeting. Now if, say, l_2 is changed slightly, the lines will meet at a very small angle. This is bad conditioning. The constants may all be different, but nearly the same, and it is still bad conditioning. Consider the case when the lines are roughly parallel with the x axis, and meet at a small angle. A small change in the constant p for one equation will make a large difference in the x value of the solution. More precise calculations will not necessarily help – if the actual physical data are of limited accuracy then the true value of x will actually be known only within wide limits.

To solve the equations, we can eliminate one variable, say u, by using l_2(16.5.1) – l_1(16.5.2) to give

$$u(l_2 l_1 - l_1 l_2) + v(l_2 m_1 - l_1 m_2) = l_2 p_1 - l_1 p_2$$

giving the solution as

$$v = \frac{l_2 p_1 - l_1 p_2}{l_2 m_1 - l_1 m_2} \tag{16.5.3}$$

or solve for u by the same method as for v, using elimination of v by $m_2(1) - m_1(2)$ to give

$$u = \frac{m_1 p_2 - m_2 p_1}{l_2 m_1 - l_1 m_2} \tag{16.5.4}$$

In either (16.5.3) or (16.5.4) it is apparent that if the matching coefficients are similar in value then the numerator and denominator will be small, and both will be subject to inaccuracy due to subtractive cancellation. The problem is badly conditioned, and there is little that can be done to help, short of using quadratic precision numbers, and although that would reduce the problem of the numerical errors, the problem of the large effect of small changes in the data would remain. Fortunately, in practical suspension analysis the geometrical problem must be basically well conditioned or the suspension itself will be ineffective as a structure, with hopelessly bad location, say due to bad triangulation.

16.6 Data Sensitivity

Consider again the problem of two linear simultaneous equations in variables (u,v):

$$l_1u + m_1v = p_1 \tag{16.6.1}$$

$$l_2u + m_2v = p_2 \tag{16.6.2}$$

Solving for v first gave

$$v = \frac{l_2p_1 - l_1p_2}{l_2m_1 - l_1m_2} \tag{16.6.3}$$

There are now three ways to solve for the other variable, u. We can substitute for the known v value into equation (16.6.1) or (16.6.2), giving

$$u = \frac{p_1 - m_1v}{l_1} \tag{16.6.4}$$

or

$$u = \frac{p_2 - m_2v}{l_2} \tag{16.6.5}$$

or solve for u by the same method as for v, using elimination of v by m_2(16.6.1) $-$ m_1(16.6.2) to give

$$u = \frac{m_1p_2 - m_2p_1}{l_2m_1 - l_1m_2} \tag{16.6.6}$$

The divisor in equations (16.6.3) and (16.6.6) is

$$d = l_2m_1 - l_1m_2 \tag{16.6.7}$$

Consider now a well-conditioned case. Then the divisor d is then not unreasonably small, and equation (16.6.3) will evaluate v accurately. However, it may be that, say, the direction cosine l_1 is small, in which case equation (16.6.4) will be near to 0/0, and inaccurate. In fact, l_1 may actually be zero, and equation (16.6.4) may be incapable of solving the problem at all, even though the problem is well conditioned. This is a data sensitivity. It could arise physically simply because one of the rods is parallel to an axis; for example, in this case rod 1 parallel to the v axis, giving $m_1 = 1$ and $l_1 = 0$. The difficulty is purely due to the method of solution attempted, it is not fundamental to the problem. Using instead equation (16.6.5) may give a good solution, or, for other data values, may instead be a problem. Equation (16.6.6) depends on the value of d rather than on the individual alignments, so depends only on the conditioning. To avoid unnecessary problems, in this case it is probably best to evaluate d once and to use equations (16.6.3) and (16.6.6).

If problems arise in a basically well-conditioned problem, it is due to the use of a coordinate system that aligns in a troublesome way with the physical configuration. In the above problem, using equations (16.6.3) and (16.6.6) as the method is equivalent to using rotated coordinates so that the quality of the solution depends only on the conditioning; that is, a well-conditioned problem cannot then have any 'bad alignments'.

A related general rule of alternative evaluations is to use the one with the largest divisor, thereby minimising the risk of attempting 0/0 or something close to it.

In a problem of two linear simultaneous equations, the basic physical analogue is the intersection of two lines in a plane. For three linear equations, it is the intersection of three planes in a three-dimensional space. Similar problems to the above may arise in such cases, and may be dealt with in similar ways, basically by avoiding small divisors as far as possible, choosing a solution route with the largest ones. The solution route may therefore have to be determined according to the particular data values.

Another good example of data sensitivity is the problem of deducing the interpolation factor f for a point P known to lie on a line P_1P_2, with all coordinates known. The value of f is given by

$$f = \frac{x_P - x_1}{x_2 - x_1} = \frac{y_P - y_1}{y_2 - y_1} = \frac{z_P - z_1}{z_2 - z_1} \tag{16.6.9}$$

Any one of these may give the required result, in general, but any one of them could easily fail because the line P_1P_2 happens to be parallel to one of the axes, causing a division by zero. There are two good solutions. As usual, one method is to avoid a small divisor, so, assuming here actual separation of the points P_1 and P_2,

```
If (abs(x2-x1) > abs(y2-y1) .and. abs(x2-x1) > abs(z2-z1)) then
                f = (xp-x1)/(x2-x1)
elseif (abs(y2-y1) > abs(z2-z1)) then
                f = (yp-y1)/(y2-y1)
else
                f = (zp-z1)/(z2-z1)
endif
```

will select the best equation to use, giving reliability and the best accuracy. Admittedly, however, it is rather bulky. Another possibility is to use

```
f = Len1P/Len12
```

which is coordinate independent. This will work unless L_{12} is zero, but that case is not useful anyway. Whether this is a better option really depends on whether the relevant lengths are already known or not.

16.7 Accuracy

A 'blunder' in a calculation is when there is a wrong operation performed or a wrong number used. A blunder would be, for example, a program error such as a plus sign where there should be a minus sign, or the wrong variable name used, or an incorrect solution chosen, as in Section 16.10. When there is a blunder, the answer is completely wrong, becoming meaningless. In contrast, mathematically or statistically speaking, an 'error' in a calculation is a small inaccuracy due to limited precision of the numbers, or the accumulation of small errors by successive numerical processes.

Setting aside gross inaccuracies due to blunders, which are up to the programmer to eliminate by care and testing, the accuracy of the computational results depends on the inherent precision of the variable representation used and on the skill of the programmer in avoiding inaccuracy-introducing processes.

In the early days, the representation of real numbers was far from standardised. Many early computer languages represented the real numbers by a 4-byte word, that is, $4 \times 8 = 32$ bit representation. This is a scientific representation with several of the binary bits used for the exponent value, the remainder for the 'mantissa' or 'characteristic', the actual sequence of digits. In decimal terms, a 4-byte number has a precision of about 7 decimal digits. Some specialised scientific computers used a 60-bit representation with greater precision.

Fortran introduced a 'double precision' real variable usually of 64 bits. This was not completely standardised at the time because of the interests of the various hardware manufacturers. However, this is now commonly available in Fortran and C and other languages.

The Basic language varies considerably from one implementation to another. Many Basics have had 4-byte real numbers, some have had 5-byte reals, and more recently some have had 8-byte real numbers. Basic also introduced the 'numeric' data type of 8 bytes storage (64 bits), which is used for both integers and reals. One bit is used to indicate whether the number being represented is a real or an integer, so the precision of the reals is reduced by one bit compared with a true 64-bit real representation. This is simpler for beginning programmers, who no longer have to bother about the distinction between integer and real numbers, but is inefficient, and most scientific programmers prefer to keep a clear distinction, particularly because the two types of number do different jobs.

The modern IEEE real number standard gives 53-bit precision, the other bits being used for the sign and the exponent value. The 53 bits give about 16 decimal digits of precision, so 8-byte real numbers are held to a remarkably close figure. On a bar 1 metre long, this is 10^{-16} m, which may be compared with the diameter of an atom, about 0.1 nm (10^{-10} m), or the diameter of an atomic nucleus, which is about 10^{-15} m. Such a precise representation evidently does not really have much physical meaning, and is there for a different reason, to ensure that the ultimate result of a long calculation with accumulating inaccuracies can still result in a useful final value. It has been said with justification that to an engineer the precision is 3 useful decimal digits with 13 guard digits.

Quad-precision numbers use a 16-byte representation, and have about 35 decimal digits of precision.

The amount of calculation in a suspension is such that with 8-byte real numbers a final precision of about 1 part in 10^{12} can normally be achieved. This is far more accurate than has physical significance. If the results are worse than this then either the problem is a particularly awkward one, due to poor conditioning, or, more likely in practice, the problem has a data sensitivity and the program is solving it by a poor method. Methods that work around data sensitivities by selecting the solution route with the largest divisors produce the most accurate results, as described in Section 16.6. Because of the remarkable precision of the 8-byte real numbers now in common use, data sensitivities often have only a limited impact, but care in avoiding such problems is still the sensible approach.

16.8 Speed

In the early days of computing, the speed and efficiency of software were of paramount importance. Computers were slow, and very unreliable because of the thermionic valves. Programs had to run fast to go to completion before a valve failure. The introduction of transistors greatly improved reliability, and large-scale integration has subsequently produced personal computers of amazing speed and memory capacity. The analysis of one position of a suspension can now be done in microseconds, so there is no longer any need for a program to be efficient, it will seem to run instantaneously anyway. It must be said that in a professional context, the logical thing to do is to minimise the programmer time, not the computer time. Another way to view this is that the programmer's time should be spent on making the program reliable and accurate, rather than fast.

Nevertheless, it remains true that it is very satisfying to most programmers to write an efficient (i.e. fast) program. As long as this is not to the detriment of reliability and accuracy, there is probably no harm in this, and if the routine is one that will be used numerous times, it may be beneficial. However, starting writing a program with speed as the main factor is likely to lead to problems. It is best to write the program initially for clarity, to be sure that it really does do the job correctly. Later, when it is working, it could be polished up for speed. There are two rules of optimisation sometimes quoted:

(1) Do not optimise for speed.
(2) If you must do it, do not do it yet.

The reason for this is that a really speed-optimised program may be unclear in operation, causing issues of reliability and maintenance.

Most programs have 'hot spots' – parts where the greater part of the work is done, usually in the innermost loops. The best way to improve the performance of an existing program is to polish up the efficiency of the hot spots only, and leave the rest as simple and clear as possible.

Such tactical optimisation is very secondary to strategic optimisation, which is to choose an efficient algorithm in the first place. In the case of suspension geometry, the strategic optimisation is to use true analytic geometric solutions where possible, rather than iterative solutions, which are generally much slower.

16.9 Ease of Use

Ease of use depends on a suitable routine name, and a considered arrangement of the argument list. There is no absolutely correct way to do this, but long ago it was found that the sensible arrangement is to have the incoming data first with the results later, generally to have integers before reals, but to put control variables at the end. In individual cases, judgement is needed. For example, an integer carrying the value of the declared size of an array might reasonably be placed just before the array name, rather than collected together with other integers away from the array name.

16.10 The Assembly Problem

Some routines return more solutions than apply physically to the problem, a difficulty that can certainly arise in the case of suspension geometry analysis. A good example is the routine Points3S, solving the position at the apex of three rods of known length and known base positions (Section 15.24). This is frequently required for suspension geometry, and there is only one valid solution for the actual suspension. However, the routine returns two equally valid geometrical solutions. The problem is to choose the one that is appropriate to the physical problem.

As a specific example, consider the solution of the upper ball joint, based on the upper arm pivot points, P_4 and P_5, and the lower ball joint P_3. The three rod lengths are known. Routine Points3S solves the geometric problem using a quadratic equation which gives two solutions. This is geometrically correct – there are indeed two points that satisfy the equations. One point is the desired one. The other solution found is one that is on the opposite side of the plane of the base points, Figure 16.10.1. The routine calling Points3S and using its results has the problem of selecting the desired solution in some way. In this case there are several possibilities. One is to choose the solution that is nearest to the original (static) position, by calculating two lengths. Actually, there is no need to calculate the actual lengths, the two square root operations can be avoided by comparing the squares of the lengths. Denoting the static positions as (xs,ys,zs) and the moved position as (x,y,z) this would be implemented as follows:

```
Call Points3S(. . . ,xa,ya,za,xb,yb,zb . . .)
len1sq=(xa-xs(6))**2+(ya-ys(6))**2+(za-zs(6))**2
len2sq=(xb-xs(6))**2+(yb-ys(6))**2+(zb-zs(6))**2
If(len1sq<len2sq)then
        x(6)=xa ; y(6)=ya ; z(6)=za
else
        x(6)=xb ; y(6)=yb ; z(6)=zb
endif
```

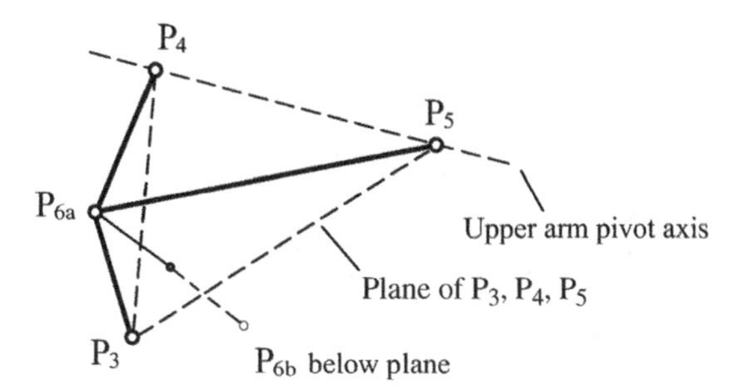

Figure 16.10.1 Two solutions by Points3S for the upper ball joint in the case of double-transverse arms.

This usually works satisfactorily, but consideration should be given to the possibility, in general geometry at least, that in an extreme displacement, at very large bump, this could fail, because the wrong point becomes closer.

Considering Figure 16.10.1 again, a good alternative could be to apply a simpler criterion based on a single component of the position. For transverse arms, the longitudinal (x) component of position varies little, and would probably be a troublesome choice. Because of the lateral inclination of the plane of the base points, the lateral (y) component of the two options is larger for the desired case, but again this is under threat in the case of a large bump displacement. However, the vertical (z) component makes a good discriminator, and is not bump sensitive, as the upper solution in the static position remains the upper solution throughout the motion range. This is therefore the preferred method in this case, and can be implemented easily as follows:

```
Call Points3S(. . . ,xa,ya,za,xb,yb,zb . . .)
If(za>zb)then
        x(6)=xa ; y(6)=ya ; z(6)=za
else
        x(6)=xb ; y(6)=yb ; z(6)=zb
endif
```

This is simpler, which is usually an advantage, even easier to understand, and faster than the previous method. However, it does require geometric insight into the problem.

A suitable method can be applied in each case, using an appropriate coordinate for the particular case, as desired. In the case of the solution of the steering-arm end, Figure 16.10.2, the two solution options are strongly discriminated by the x coordinate. For that case, with the x coordinate positive forwards, with a rear steering arm, as usual, then:

```
Call Points3S(. . . ,xa,ya,za,xb,yb,zb . . .)
If(xa<xb)then
        x(8)=xa ; y(8)=ya ; z(8)=za
else
        x(8)=xb ; y(8)=yb ; z(8)=zb
endif
```

If the steering arm and rack are forward of the steering axis, then the selection criterion must be reversed. Obviously, in each case the geometry must be considered carefully to select the most suitable method.

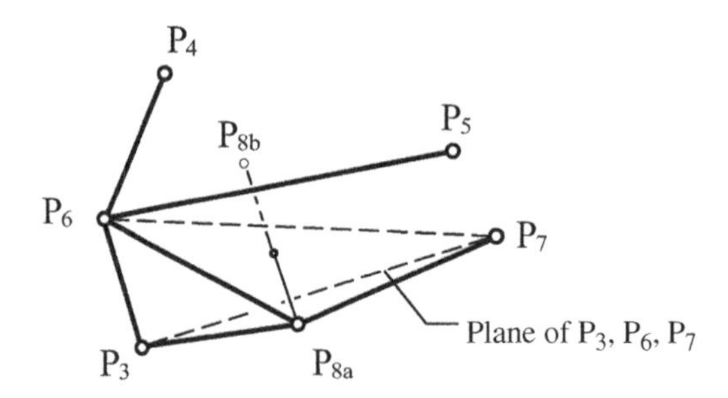

Figure 16.10.2 Two solutions by Points3S for the steering-arm end ball joint (track-rod outer end) in the case of double-transverse arms.

If an incorrect choice is made (i.e. an assembly blunder), then there may be various consequences. One possibility is a completion of the computational analysis with largely nonsensical numerical values resulting, but the routine returning IE = 0 (i.e. claiming a good solution), the worst possible outcome as discussed earlier. Another likely outcome, bad, but probably not as bad as the last event, is that at some subsequent stage the geometry will become impossible. There will be other links that fail to meet, not being long enough for the unphysical problem now presented, typically manifesting itself as a call to a square root with a negative kernel, or a quadratic equation without any real roots (i.e. no physical solution). At this point the routine should abandon the computation and return a suitable IE value to some higher program. However, the routine will not have any way of knowing what has really gone wrong, and the programmer – or worse, a subsequent user – will be left wasting a good deal of time trying to sort it out. Therefore, the 'assembly' issue is one that must be attended to carefully by the programmer.

16.11 Checksums

For hundreds of years, checksums have been used in scientific and financial calculations. They are highly applicable to numerical programming. Generally, a routine is required to solve one or more equations, so on completion the equations can be checked for satisfaction. As a simplistic example, a routine to solve the equation

$$ax^3 + 3b^2 = c/x$$

would have a section at the end along the lines of

```
    e = ax^3 + 3*b^2 - c/x
    If(abs(e) > tol) Goto 15
    Return
    . . .
    15 IE=2 ; Return
or  15 Pause 'Accuracy lost.' ; Stop
```

where `tol` is a tolerance, possibly 10^{-6}, depending on the particular case.

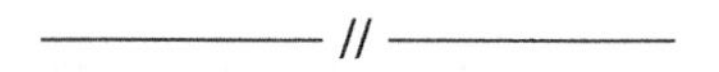

17

Iteration

To go beyond is as wrong as to fall short. (Confucius XI-XV-3)

17.1 Introduction

The English verb 'to iterate' comes from the Latin verb *itero*, to repeat, and the noun *iteratio*, meaning repetition. However, in the context of computing it has a sharply different meaning from simple repetition, the latter being simply 'looping'. A *loop* is just a series of instructions used in sequence, repeated several times. Iteration is a special case of repetition in which a variable, the accurate value of which is sought, is progressively refined in value from an initial estimate. Hence, in computational use:

> *to iterate* (verb): to seek a value by successively improving approximations.
> *iterate* (noun): the variable being refined.
> *iteration* (noun): the process of iterating the iterate.

Iteration, then, is a method in which successive improvements are made to one or more variables, and in which the correct solution is only approached asymptotically, the exact result being reached only after an infinite number of loops. In practice, a finite number of cycles are used, on the basis that the latest solution has become near enough, or is exact to within the precision of the computer number representation. Basic explanations of iterative methods may be found in many texts, for example Shoup (1979, 1984).

Iteration is known as an indirect method (although there is a process called direct iteration), to be contrasted with direct methods, which use a single sequence of calculations to arrive at the solution, usually mathematically exact in principle (i.e. algebraically exact). However, iteration is sometimes more accurate than a direct method because it can avoid cumulative numerical error. Iterative improvement (polishing) can sometimes be applied with advantage to a solution initially obtained by a direct method.

An important advantage of iterative methods is that error is not propagated, that is, an inaccuracy at one stage does not place a limit on the accuracy of the final result. However, there can be problems of lack of convergence. This means that iterative processes sometimes lack reliability. Even for an iteration that is convergent in principle, whether the iteration actually converges on the desired solution or not depends on the initial estimate. Practical experience of iterative methods shows that great care is required to obtain reliable results. Also, iteration tends to be a slow process, because of the repetition of the loop, usually much slower than a direct method. However, in many cases there is no direct method. Then there is no option – iteration must be made to work or there will be no result at all.

In the context of vehicle suspension analysis, and related topics such as general mechanism design, the basic analysis can often be performed either by direct geometric solution or iteration. As described in this book, direct geometric solution requires a good understanding of the problem, more than is required by iteration, but both methods have been used to write useful programs. There are some situations where iterative solutions are unavoidable. Solution of the position geometry with a specified suspension bump position for a double-transverse-arm suspension is one good example. There is no direct solution, so the method must be iterative. Similarly for strut suspensions. Another good example is the five-link suspension (see Figure 1.10.9). This has no direct solution at all, so all position analysis must be iterative. This applies also to the rigid axle, with solution detailed in Chapter 13. Iteration, then, is a necessary process in some forms of suspension analysis. One aspect of practical suspension problems is that the iterative function is generally well behaved, which facilitates efficient solution, and directs the choice of methods.

This chapter, then, gives an overview of iterative methods which might be used for suspension analysis, discussing their application and relative merits.

17.2 Three Phases of Iteration

It is useful to think of an iteration as having three phases, Figure 17.2.1. In the first phase, the function may be analytic but rather erratic, so ambitious high-order iterative methods may have problems. In the second phase, the function has become smooth locally and high-order methods come into their own, with fast convergence. In the third phase, the numerical noise on the function value becomes significant, and numerical derivatives in particular become a problem.

A mathematical algebraic variable has infinite precision. A real number in a computer does not. The smallest increment in a numerical variable that can be added to that variable and make it different from the initial value is limited by the numerical precision of the number representation. The size of the minimum increment relative to the variable value x is called epsilon(x), and in some languages is available as an intrinsic function under that name. The minimum usable increment in x is given by

$$dx = x \times \text{epsilon}(x)$$

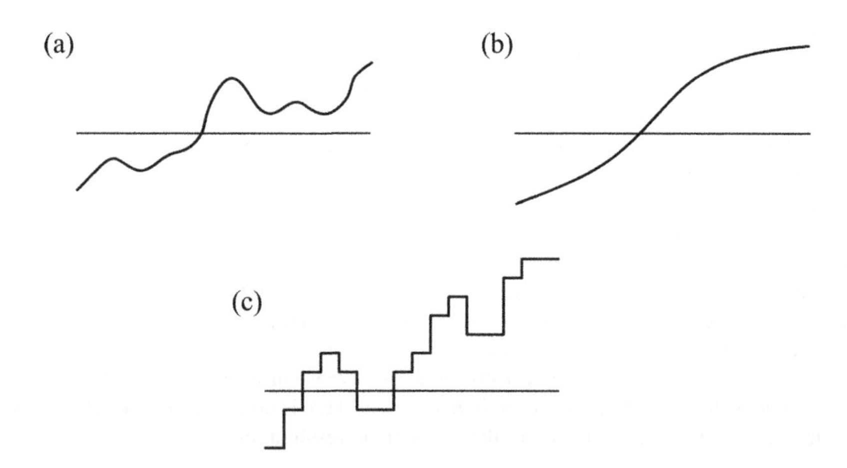

Figure 17.2.1 The three phases of iteration: (a) phase 1, wide range, with erratic analytical fluctuations; (b) phase 2, mid-range, smooth curve, fast iteration possible; (c) phase 3, micro-scale, numerical staircase, possibly erratic or large steps.

For an 8-byte real number, the precision is usually 53 bits, so a change of one part in 2^{53} is just the one last bit, the minimum change that can be registered, about 1 part in 10^{16}. Hence, any attempt to determine an iterate to better than 1 part in 2^{53} will not succeed. Actually, the situation is slightly more complicated than that, with the real number processor using a precision of 60 bits, having extra guard digits, the basic representation having a notional precision of 53 bits, and an epsilon corresponding to 52 bits ensuring a usable increment. For 4-byte real numbers the factor is usually 2^{23}. The useful accuracy in the iterated function f may be much worse, because it is the result of a series of calculations, possibly rather complicated geometrical ones, and subtractive cancellation may have reduced the meaningful accuracy to well below epsilon(f). Attempting to iterate the independent variable to a narrower range than corresponds to the uncertainty in $f(x)$ also will not give meaningful results.

In the case of polynomials, it is possible to estimate the accuracy with which the polynomial value can be calculated. However, this is not very practical for more complicated routines, although some testing and experience will indicate realistic values.

17.3 Convergence

Convergence analysis is the study of conditions under which an iterative solution will converge to a solution, and of methods of improving the rate of convergence and ultimate accuracy.

There are some similarities between the convergence of a numerical process and the convergence of a mathematical series. In the evaluation of various mathematical functions, series summations are used, as may be found in the handbooks. For example, the series for sin(x) is

$$\sin(x) = x - \tfrac{1}{6}x^3 + \tfrac{1}{120}x^5 - \cdots + (-1)^K \tfrac{1}{K!}x^K + \cdots$$

This series, or one similar to it (a Chebyshev economised series), is actually used by a computer to evaluate the sine function whenever it is used. A series is said to be convergent mathematically if its sum for an infinite number of terms is a finite value. However, a series may be mathematically convergent but effectively useless. For it to be useful, it must become accurate within a reasonable number of terms, so the terms must reduce in size rapidly. This is rapid convergence.

In an iterative solution to a problem, if the successive solution values become closer to a single value, this is convergence. However, the convergence must be usefully fast. It is often assumed that convergence will be to the correct solution, but it may not be so in some cases – that is, where there is more than one solution it is possible to converge on a solution, but the wrong one.

In an iterative solution, if the successive values become progressively further from the solution, this is divergence. Such divergence may be monotonic (going one way) or oscillatory (alternating sign of error). Exponential divergence is often recognised by the rapid appearance of very large unphysical values for variables. A possible solution is reformulation of the problem by some sort of inversion of the equations. Another possible corrective is to stabilise the process by taking the average of the new and old values. This can be refined further by using a variable iteration factor to select some fraction of the change to the new value. This is discussed in detail later.

In a successful iteration, in phase 2 of the process at least, the relative error e_K at step K will in general decrease as

$$e_{K+1} = A e_K^P$$

where P is called the order of convergence and A is called the asymptotic error constant.

Linear convergence has $P = 1$, and requires $-1 < A < 1$ for convergence, and small magnitude of A for rapid success. With $A = 0.1$ each iteration leaves a residual error 0.1 times that of the previous iteration. This adds one decimal digit of accuracy at each iteration. In a binary search, $A = 0.5$, adding one bit of accuracy at each iteration.

Some processes exhibit quadratic convergence, in which $P=2$, and the number of accurate digits doubles at each iteration. Then with $A=1$ and an initial relative error of 0.1 the successive errors would be $0.1, 0.01, 10^{-4}, 10^{-8}, 10^{-16}$, etc. Obviously such a process goes much more rapidly at the end than does a simple linear iteration. Newton's first iteration is an example of this.

Some methods have a cubic convergence, in which the number of accurate digits triples at each iteration. An example is the Laguerre iteration for a root of a polynomial.

The order of convergence need not be an integer. For example, the secant method has $P = (1+\sqrt{5})/2$ 1.618, the 'golden ratio'.

By using the appropriate number of derivatives, methods can be derived having asymptotic convergence of any required order. One method of deriving such formulae is by reversion of the Taylor series expansion. High-order methods can also use information from the function value at several points. However, higher-order methods involve more calculation per cycle. Higher-performance methods, then, use more than one function evaluation per loop, or use derivatives in addition to the basic function value. The correct comparison of the performance of alternative iterations can better be expressed as the order of convergence per total of function and derivative evaluations. For example, an iteration with quadratic convergence has order of convergence $P=2$ per loop, but if it uses two function evaluations per loop then the order of convergence is only $\sqrt{2}$ per evaluation.

Methods with a high order of convergence are faster in principle, but tend to be more critical – at higher risk of not converging at all – and generally have a smaller catchment area, requiring a better starting estimate. Hence, it is common in general computing practice to begin with a slow robust method and to polish with a high-order method. Linear convergence is rather slow in principle, but if truly linear it can easily be accelerated into quadratic convergence (Section 17.8) giving much improved performance. There may be little to be gained by going beyond that, because of the reduced reliability.

Because suspension problems generally involve geometrically smooth functions, high-performance, high-order iterations are attractive and practical. Also, the acceleration of linear iterations is a very useful method in such problems.

17.4 Binary Search

The basic binary search is mentioned here for comparison, but is not likely to be needed for a suspension. One useful feature is that it is very easily implemented, and with the speed of modern computers even this slow but safe method can do an adequate job. One of its merits is absolute reliability.

It is necessary to begin with a pair of iterate values that are known to bracket a zero. At each iteration, the mid-point is investigated. The new point and the old point with opposite function value sign are retained. Therefore, the two points at any stage bracket the zero, which is a valuable feature because the zero is then known to a definite limit of accuracy. The estimate of the actual zero is always the mid-point of the current range, and the maximum error is half of the range. The range and the maximum error halve at each iterative cycle, adding one bit of accuracy. Hence, about 3 cycles are need for each additional decimal digit (actually $\log_2 10 \approx 3.322$). Starting with no accuracy, for an 8-byte number representation having 53 significant bits it would take 53 cycles with 53 function evaluations to achieve full resolution. However, in view of the limited accuracy of the function, less may be all that can advantageously be used. Nevertheless, the figure of 53 cycles is a useful basis for comparison.

In computing, a real number is stored as a characteristic and an exponent. The degree of precision depends on the number of bits allocated to storing its characteristic. For example, a 4-byte ($4 \times 8 = 32$ bit) real variable could have 3 bytes allocated to the characteristic, one being a sign bit, leaving 23 value bits, discriminating 2^{23} separate values, approximately 7 decimal digits ($\log_{10} 2^{23}$). Double precision, often the standard nowadays, typically uses 8 bytes total with a 53-bit characteristic value (and one sign bit) giving about 16 significant decimal digits, extended precision 12 bytes with 85 bits and about 25 decimal digits, and quadratic precision 16 bytes with 117 bits and 35 significant decimal digits. When a binary search is used, the number of cycles required for full resolution is directly proportional to the number of bits in the characteristic. However, methods with quadratic convergence require a number of cycles, if all goes well, depending only on the logarithm of the number of bits. This means that the binary search shows up quite well in comparison when 4-byte numbers are in use, but for high precision the comparison leans towards the more refined methods.

The main strength of the binary search is in dealing with erratic functions. Suspension geometric functions are generally well behaved, so the binary search is not a likely choice. However, with the speed of modern hardware even a binary search could well be fast enough to be satisfactory. Realistically, though, most programmers would regard it as a matter of professional pride to use a faster method.

17.5 Linear Iterations

A linear iteration is one in which successive values of the iterate have errors in constant ratio (i.e., in the case of a successful iteration, an exponential decay of the error). In a prospectively linear iteration, and others, success is not assured and the following problems may to occur:

(1) monotonic divergence;
(2) oscillatory divergence;
(3) monotonic convergence too slow;
(4) oscillatory convergence too slow;
(5) oscillation between two stable incorrect values;
(6) convergence to an inaccurate value;
(7) convergence to a totally wrong value;
(8) a wild leap away from a root.

The first four problems are inherently algorithmic and may be dealt with by various software improvements. Cases (1) and (2), divergence, can often be cured with a fixed-point iteration by turning the equations 'inside out', using an alternative formulation. In principle, divergence can be fixed by a negative constant iteration factor, but this is not always easy to do in practice, and accelerated convergence (Section 17.8) is a more reliable cure. Problems (3) and (4), convergence too slow, can be fixed with an iteration factor, as sometimes may (5). Case (6) is likely to occur because of numerical imprecision in the real number representation (e.g. due to subtractive cancellation problems). In case (7), the value may be algebraically correct for the equations, but not the desired value (e.g. a wrong root), and may possibly be fixed by the selection of a more suitable initial value. The final case, (8), is due to alighting on a point of very small gradient, which in some methods results in a grossly excessive adjustment. The reason for this is typically that the numerically found gradient is faulty because of local numerical noise when approaching the final value.

The behaviour of standard linear iterations, without use of an extra iteration factor, can be analysed with advantage. The error ratio is the ratio of successive errors. During an actual iteration, the true solution is not known accurately, so the errors themselves are not known accurately either. However, the error ratio is closely approximated by the ratio of successive changes, because in a truly linear iteration the changes are directly proportional to the errors. The correction ratio is the iterative change in the desirable direction (opposite to the error) divided by the error at the start of the step, and is normally positive ($a = 1 - r$), for example 0.6 for a slow monotonic convergence. Parameter e in Table 17.5.1 is a fairly small number, such as 0.1.

Applying a fixed-value multiplier to the calculated change can solve most of the problems, and give much more rapid convergence. The ideal multiplier is

$$w = \frac{1}{1 - r}$$

Table 17.5.1 Linear iterative behaviour

Case	Error ratio r	Change/error $a = 1 - r$	Behaviour of standard iteration
(1)	+ infinity	−infinity	Extreme divergence
(2)	$+(1+e)$	$-e$	Monotonic divergence
(3)	$+1$	0	No improvement
(4)	$+(1-e)$	e	Slow monotonic convergence
(5)	$+0.4$	0.6	Monotonic convergence
(6)	$+e$	$(1-e)$	Fast monotonic convergence
(7)	$+0$	1	Perfect iteration in 1 step
(8)	$-e$	$(1+e)$	Fast oscillatory convergence
(9)	-0.4	1.4	Oscillatory convergence
(10)	$-(1-e)$	$(2-e)$	Slow oscillatory convergence
(11)	-1	2	2-point stable oscillation
(12)	$-(1+e)$	$(2+e)$	Slow oscillatory divergence
(13)	-2	3	Oscillatory divergence
(14)	−infinity	infinity	Extreme divergence

with

$$x_{K+1} = x_K + \left(\frac{1}{1-r}\right)\Delta x_K$$

In a truly linear iteration, this would give a jump directly to the correct solution.

Real iterations are not exactly linear, so unfortunately the iteration correction factor does not really give the immediate solution, but can still give considerable improvements. For example, a factor of 1.83 often works well in relaxation iterations. Also, the concept does lead to much improved 'accelerated' linear iterations (Section 17.8), using an adaptive iteration factor, with a new value calculated for each cycle.

17.6 Iterative Exits

A loop exit condition is the logical condition under which the work of the loop is deemed to have been completed, and control is to pass out of the loop. For example, the condition here is that the loop counter K equals 30:

```
K=0
Do
      K=K+1
      ...
      If(K==30)exit
enddo
```

For an iteration, rather than a simple loop, the exit condition may be quite complicated, but will typically look something like

```
If(abs(u1-u1s)<tolu .and. abs(v1-v1s)<tolv)exit
```

It is more efficient to write

```
If(abs(u1-u1s)>tolu .or. abs(v1-v1s)>tolv)cycle
Exit
```

because if the first test fails, as it probably will many times, then there is no need to make the second test. Note that when the .and. changes to .or. the direction of the comparisons must change. The following split exit condition is much more relaxed, because it will exit if only one variable is within tolerance, which is probably not acceptable:

```
If(abs(u1-u1s)<tolu)exit
If(abs(v1-v1s)<tolv)exit
```

An alternative good way is

```
If(abs(u1-u1s)>tolu)cycle
If(abs(v1-v1s)>tolv)cycle
Exit
```

If there are many variables with tolerances, they can be put at the end of the main loop thus:

```
DoKI:
Do KI=...
      ...
      ...
      Do KV=1,NV
            If(abs(dx(KV))>tol) cycle DoKI
      enddo
      exit
enddo DoKI
```

where it is necessary to use a named main loop to cycle the outer loop instead of just the test loop.

Fortran 90 places its exit conditions in an exit statement, as shown, which is powerful and flexible, as they can be placed at any point in the loop. Other languages (e.g. Basics) often have the exit conditions only at the ends, e.g. do while (condition) or loop until (condition). Really it is preferable to have all methods available.

In a bracketing iteration, such as a binary search, the tolerance can be applied to the errors. During a non-bracketing iteration, such as fixed-point iteration, the true value of the iterate is not known, and neither are the errors, so the tolerance cannot be applied to the errors. Instead, it must be applied to the changes of the iterate. If the convergence is fast (i.e. the error ratio r is small), this will work well. If the convergence is only creeping then the remaining error may greatly exceed the change, being $1/r$ times that, so the error would then greatly exceed the escape tolerance. Creeping iterations require acceleration by fixed or adaptive factors. Ideally, other than instant perfection, there is a small overshoot of the correction, with the corrections and errors reducing rapidly in magnitude whilst alternating in sign, thus effectively bracketing the solution.

The basis for ending an iteration is then a comparison between successive approximations for a root. The assumption is that the error can be made as small as desired simply by performing more iterations. This is not always possible. Let d_K represent the absolute value of the difference between successive approximations to the root:

$$d_K = \text{abs}(x_K - x_{K-1})$$

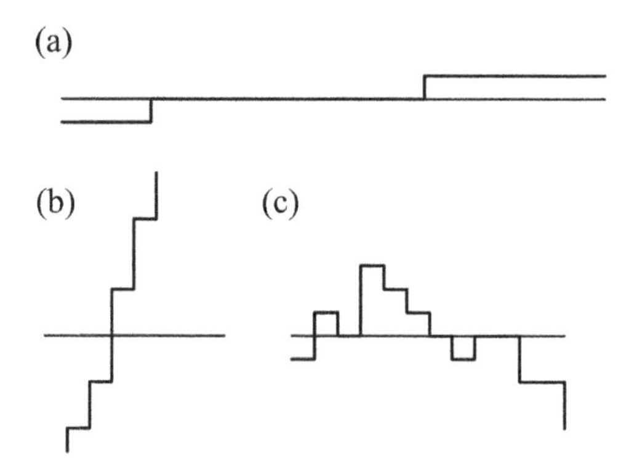

Figure 17.6.1 Three numerical staircase problems at small scale, x steps are $x \times$ epsilon(x): (a) very flat, many possible x values for a zero, small gradients; (b) very steep, possibly no true zero at all; (c) erratic 'noisy' function, possibly several disconnected zeros, possibly zero-value gradients.

The behaviour of d_K may be erratic for the first few iterations, during phase 1 (see Section 17.2), but if the process is convergent then d_K will decline considerably as the iteration proceeds through phase 2. In due course, phase 3 is entered, and in some cases it may be found that the calculated change to the iterate cannot consistently be reduced beyond a limiting value, no matter how many iterations are carried out. This phenomenon is generally caused by round-off errors in the computations, which give the 'erratic staircase' function shape at a small scale, Figure 17.6.1. For example, subtractive cancellation can cause wide flat areas with large steps between (i.e. a large-step staircase). Then the values for d_K may oscillate about some mean value, or perhaps suddenly leap to a much larger error, possibly infinite. No useful purpose is served by continuing computations beyond this point. The exit test should consider this. However, the test must allow some erratic behaviour of d_K, in particular that occurring in phase 1, because of the possibility of a poor starting value for the iterate.

There is much that can go wrong here, particularly in phase 3, depending on the type of iteration, with numerical derivatives being particularly risky. Again, according to the type of iteration and accuracy requirements, the solution to these problems may vary considerably. Two simple, indeed crude, but effective methods are:

(1) Simply perform a fixed number of iterations, the number found by experience and tests to be sufficient.
(2) Only seek sufficient accuracy. If this is much less than the ultimate numerical accuracy, which is often the case in engineering, then an easy exit tolerance can be used, avoiding all the problems of phase 3.

However, it is also often desired to achieve the best possible accuracy, and some more sophisticated control of the iteration is needed. First, for safety it is common to set a large maximum number of iterative cycles, such as 100 or 200, to catch any odd non-convergent failure cases, say entering a closed cycle. These should then exit as a failure state (e.g. IE = 8, 'Failed to converge in iteration').

Next, ideally the iteration should recognise that it has safely passed from phase 1 into phase 2. Depending on the particular problem being iterated, this may be done simply by requiring a certain minimum number of iterations, say 5, according to experience. Better is to detect a sequence of successively reducing correction magnitudes. This state can then be used to trigger accelerated convergence, if appropriate.

Finally, and most critically, the arrival at phase 3 must be detected and used to exit the iteration, or to start a phase 3 strategy to make the best of situations such as those in Figure 17.6.1, for example in (a) to find the centre of the zero band. Phase 3 can be considered to start when any of the following occur:

(1) Successive changes of the iterate are nearly down to $x \times$ epsilon(x).
(2) The function value changes are close to the limits on function accuracy, which is probably limited by numerical processes, subtractive cancellation, etc., much larger than epsilon(f).
(3) The changes to the iterate are small enough for the numerical $f(x)$ staircase form to be significant.
(4) The proposed change to the iterate is of an increased magnitude from the last cycle.

Generally, a good method is that once phase 2 is entered the changes are monitored, and accepted whilst they are sequentially smaller in magnitude. If an increase is found, then phase 3 has been reached, and if the last change was small enough then the iteration is effectively complete. This is sometimes called 'Garwick's device'. Garwick suggested that a relatively modest criterion of the form abs($x_{K+1} - x_K$) $< e$ should be used, and when this criterion has been satisfied then iteration should continue as long as the differences between consecutive x_K are diminishing. When finally we reach values for which the magnitude of the change increases then the last improving value of z should be accepted.

For the case of polynomials, see Adams (1967). See also Jenkins and Traub (1972) and Jenkins (1975).

17.7 Fixed-Point Iteration

Fixed-point iteration can be considered in two types, called here endogenous and exogenous. In the endogenous type, the base equation is simply rearranged into an iterative form. For a simple example, consider Wallis' equation (used by Wallis and Newton over 300 years ago)

$$y = x^3 - 2x - 5 = 0$$

As the basis of an iteration, this could be rearranged as

$$x = (2x + 5)^{1/3}$$

or as

$$x = \tfrac{1}{2}(x^3 - 5)$$

When two arrangements are possible, it seems invariable that one form is iteratively convergent and the other is divergent. In this case, the first form is the convergent one, having an error ratio of about 0.16 per cycle. The other formulation is wildly divergent, or, with a gloss, converges rapidly on infinity. The iteration to use would be

$$x_{K+1} = (2x_K + 5)^{1/3}$$

The form of function to be solved may not be known explicitly in algebraic form, but only as the result of a calculation, as a black box value. This is the effective situation with suspension iterations, which obviously are not amenable to an iteration of the above simple type.

Exogenous fixed-point iteration is using $x_{K+1} = x_K + Cf(x_K)$ where f is the function to be solved and C is a constant, the equation obviously being satisfied at $f = 0$ when x_K is a root. Here, $f(x)$ is not required in algebraic form, as it is not rearranged. Actually, almost any function F of f could prospectively be used for the iteration,

$$x_{K+1} = x_K + CF(f(x))$$

provided that the zero value at the root x is retained. The convergence properties depend on the function $F(f)$ and on the value chosen for the constant, which may also usefully be varied dynamically according to the progress of the iteration. Using the negative reciprocal of the gradient of f for C gives Newton's first iteration.

Fixed-point iteration on a complex variable can work quite well, but tends to be slow to converge, and will often oscillate between the conjugates. This may not be a problem as either one may be acceptable.

Fixed-point iterations are generally linear and may be accelerated with considerable performance gain, as described in Section 17.8. The problem of solving for Z_{LBJ} for a given suspension bump Z_{SB} can easily be expressed as a fixed-point iteration.

17.8 Accelerated Convergence

The simple linear iteration, as shown in Table 17.5.1, may be considerably improved in its rate of convergence by the introduction of a constant iteration factor, also known as an iteration speed factor, an acceleration factor, a stabilisation factor, a relaxation factor, etc., usually represented by w. The new value x_{K+1} taken is then calculated from the initially iterated new value x_{ITER} and the old value x_K:

$$x_{K+1} = x_K + w(x_{ITER} - x_K) \qquad (17.8.1)$$

Put simply, a fraction w of the proposed change is actually used. Evidently, an iteration factor of 0 will stop the iteration from advancing at all. A value less than 1 will slow down or stabilise the process. A factor of 1 gives simple standard iteration, whilst a value greater than 1 will accelerate and destabilise the process. For many problems, the optimum value has been investigated in detail. Unfortunately, the optimum may be impossible to calculate in advance.

Alternatively, an adaptive factor can be used. In adaptive iteration, a variable iteration factor, calculated anew for each iteration cycle, is used. The basic linear convergence, even creeping, divergent or oscillatory iterations, can by this means be converted into quadratic convergence.

If the error e in an iteration changes by a constant factor r in the successive estimates, this is a linear convergence, with exponential decay of the error, the successive errors being in a geometric progression. Consider an iteration with an unknown true value x, with three successive calculated non-accelerated iterate values numbered K, $K+1$ and $K+2$. The process therefore requires two new function evaluations per cycle. The three iterate values and the error ratio give:

$$x_K = x + e_K \qquad (17.8.2)$$

$$x_{K+1} = x + e_{K+1} \qquad (17.8.3)$$

$$x_{K+2} = x + e_{K+2} \qquad (17.8.4)$$

$$e_{K+2} = re_{K+1} \qquad (17.8.5)$$

Calculating $r \times$ (17.8.3) – (17.8.4) and using (17.8.5), which eliminates e, gives

$$rx_{K+1} - x_{K+2} = rx - x$$

A good estimate of the true x is therefore

$$x = \frac{x_{K+2} - rx_{K+1}}{1 - r} \qquad (17.8.6)$$

Hence, if a value can be obtained for the ratio r, a much better estimate of the true x is available. This method is applicable if r is fairly constant, and not too close to 1, that is, if the iteration is close to linear, in which case the ratio of errors can be calculated approximately from previous iterations, using the ratio of the changes:

$$\begin{aligned}\Delta_1 &= x_{K+1} - x_K \\ \Delta_2 &= x_{K+2} - x_{K+1}\end{aligned} \tag{17.8.7}$$

$$r = \frac{\Delta_2}{\Delta_1} = \frac{x_{K+2} - x_{K+1}}{x_{K+1} - x_K} \tag{17.8.8}$$

The enhanced solution will be exact if r really is constant, and good if r is changing slowly. For accuracy, the ratio should preferably be calculated from the proposed changes, not by subtraction of the complete iterated values, although this is not possible in the case of the typical suspension iteration for Z_{LBJ} given Z_{SB}. In summary,

$$r = \frac{\Delta_2}{\Delta_1}$$

$$x = \frac{x_{K+2} - r x_{K+1}}{1 - r} \tag{17.8.9}$$

By substitution, the above equations can be reformulated as

$$x = x_{K+2} + \left(\frac{r}{1-r}\right)(x_{K+2} - x_{K+1}) \tag{17.8.10}$$

or

$$x = x_{K+2} - \frac{(x_{K+1} - x_K)(x_{K+2} - x_{K+1})}{x_{K+2} - 2x_{K+1} + x_K} \tag{17.8.11}$$

and

$$x = \frac{x_{K+2} x_K - x_{K+1}^2}{x_{K+2} - 2x_{K+1} + x_K} \tag{17.8.12}$$

These forms show that the result is algebraically equivalent to Aitken's acceleration, and to Steffensen iteration. It is also easily shown that

$$x = x_K + \frac{\Delta_1^2}{\Delta_1 - \Delta_2} \tag{17.8.13}$$

$$x = x_{K+1} + \frac{\Delta_1 \Delta_2}{\Delta_1 - \Delta_2} \tag{17.8.14}$$

$$x = x_{K+2} + \frac{\Delta_2^2}{\Delta_1 - \Delta_2} \tag{17.8.15}$$

showing that the method is also the same as the delta-squared process. Actually, these final expressions of the principle may be the best ones to use, involving the least numerical inaccuracy. Of the above, equation (17.8.15) is the most accurate, unless, say, x_{K+2} is expanded and the increment is collected together and calculated as one item. The total increment on x_K may correspondingly be expressed in three ways:

$$\Delta x_K = \frac{\Delta_1^2}{\Delta_1 - \Delta_2} = \Delta_1 + \frac{\Delta_1 \Delta_2}{\Delta_1 - \Delta_2} = \Delta_1 + \Delta_2 + \frac{\Delta_2^2}{\Delta_1 - \Delta_2} \tag{17.8.16}$$

The division by a subtraction suggests some possible loss of accuracy in the increment, although this should not be serious unless the basic iteration is creeping with successive increments of almost the same size.

Accelerated convergence is excellent when it works, but it is not always cooperative. It requires a smooth underlying iteration with a constant error ratio, that is, a steady linear iteration. Iterations do not necessarily meet this criterion. In particular, when there are several variables involved in the loop, they interfere with each other and usually cause variable error ratios, so acceleration is likely to fail in such cases.

17.9 Higher Orders without Derivatives

It is possible to achieve much faster convergence than linear, typically quadratic per loop, by using data from more than one point. That is, instead of using a fixed-point type iteration, the information from the function value at several points is used simultaneously to give an improve iterate value. It is certainly reasonable to expect this to be more efficient than a normal fixed-point iteration, in which the old results are all discarded after a single use. In principle, all the test values of the function could be retained and used, say by fitting a polynomial to the N points and solving for a root. This is not a method commonly used in practice. However, it is common, and effective, to use the function value at two or three points. A higher-order iteration may alternatively be based on one point, using the function value and one or more derivative values at that point, as described in the next section.

The common methods based on the function values at two points are the secant iteration and *regula falsi*. These model the function as a straight line between the two points to predict the new iterate value. The usual three-point iteration is called Muller's method, which models the function as a parabola.

A secant is a straight line making two intersections with a curve, Figure 17.9.1, notionally cutting off a segment. The secant method is a widely used root iteration, possibly faster than Newton's method, with the difference that the analytic derivative is not used. Instead, two function values are used, with the function

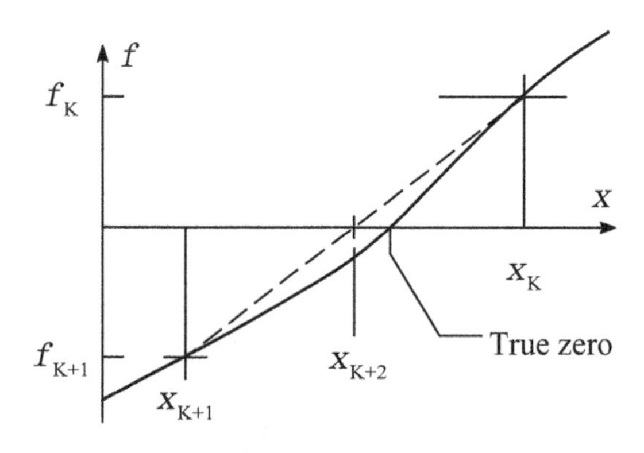

Figure 17.9.1 The secant method.

modelled as a straight line through these points to predict a zero. In effect then, this could be seen as similar to Newton's first iteration, but with a numerical derivative between the last two points, reducing the computational work per cycle.

With independent variable values x_K and x_{K+1}, with function values f_K and f_{K+1}, the gradient is

$$m = \frac{f_{K+1} - f_K}{x_{K+1} - x_K}$$

and the new iterate value is

$$x_{K+2} = x_{K+1} - \frac{f_{K+1}}{m}$$

Each cycle then uses one new function evaluation. The oldest point is discarded. This is the same as discarding the point with largest function value. Convergence is not quadratic, but it is super-linear, with convergence factor 1.618, the golden ratio. Care is needed with the completion criterion. Conte and De Boor (1972) describe a modified secant method in which lines of variable gradient are used.

Regula falsi, or 'false position', is a similar method of zero finding, being a modified secant method, in that the point discarded is the one with the same function sign as the new one. The two points used at each stage then have opposite function signs. By this means, the zero is kept bracketed, which gives certainty on the accuracy of the zero, and the new value is always interpolated, not extrapolated. The cost of this is somewhat slower convergence in most cases. This is a useful method if maximum accuracy is desired, as the bracketed zero facilitates phase 3 of the iteration.

Muller's method also uses only actual function evaluations, with no derivatives. This makes it particularly useful for some types of function. Three points are used. There are several variations of the method. In one method, given a root estimate x and an offset h, the function is evaluated at the three points $x - h$, x and $x + h$. A parabola is fitted to these points, and the two roots of the parabola are found. The one nearer to the original estimate is retained. The offset is reduced, and the process repeated, iterating down to an accurate root value. This clearly can work for real roots. Interestingly, it also works for complex roots.

In an alternative version, the three x values are not equally spaced. The three new values are the new one and the last two old ones nearest to the new one. By this means, only one new function evaluation is required per cycle, which proves to be more efficient. It is convenient to think of each new value as the current estimate, with the last two trailing values used to improve accuracy of the next estimate. Completion is deemed to occur when the change in the new estimate is small enough, and/or when the function value found is small enough.

Given any three values of the iterate (x_1, x_2, x_3) and the corresponding function values (f_1, f_2, f_3), the fitted quadratic gives three simultaneous equations in its coefficients a, b, and c, for example $f_1 = ax_1^2 + bx_1 + c$. The required solution for the quadratic coefficients is then, for example,

$$d = (x_1 - x_2)(x_1 - x_3)(x_2 - x_3)$$

$$a = \frac{(f_1 - f_2)(x_1 - x_3) - (f_1 - f_3)(x_1 - x_2)}{d}$$

$$b = \frac{(f_1 - f_3)(x_1^2 - x_2^2) - (f_1 - f_2)(x_1^2 - x_3^2)}{d}$$

$$c = f_1 - (ax_1^2 + bx_1)$$

There are two other obvious equations as alternatives or checks for c. The divisor d is symmetrical in the iterate values, and cannot be zero as long as the points are distinct. Alternatively, use

$$m_1 = \frac{f_2 - f_1}{x_2 - x_1}$$

$$m_2 = \frac{f_3 - f_2}{x_3 - x_2}$$

$$a = \frac{m_1 - m_2}{x_1 - x_3}$$

$$b = m_1 - a(x_1 + x_2)$$

$$c = f_1 - x_1(ax_1 + b)$$

With either set of the above equations,

$$x = \frac{2c}{-b - \text{sign}(b)\sqrt{b^2 - 4ac}}$$

The sign(b) is used to maximise the magnitude of the divisor, to give the nearer of the two prospective quadratic roots. The inverted arrangement of the standard quadratic solution is preferred, for accuracy. With the conventional expression, the nearer root uses a subtraction with subtractive cancellation accuracy loss. If a real root is expected then the root kernel should be checked to be positive.

In fact the method was invented by Muller for obtaining complex roots, and then found to work well for real roots too (see Muller, 1956). Efficient implementations are possible, particularly when using only one new function evaluation per iteration. The order of convergence is given as 1.84 on a single root and 1.23 on a double root. In the basic symmetrical method, there is some doubt over how to choose the reduced offset spacing, a problem avoided in the other method.

When approaching the limiting accuracy, Muller's method tends to suddenly increase the error, as do methods using a numerical derivative, so careful attention must be paid to timely exit from the loop. Conte and de Boor discuss the method in some detail.

These three methods, secant, *regula falsi* and Muller's method, all have fast convergence in phase 2, and do not need derivatives, so they obviously have potential application to suspension position iterations.

17.10 Newton's Iterations

Newton's iteration is also known as Newton–Raphson iteration. It was described by Newton in 1669, published in the book *Algebra* by Wallis in 1685, and again by Raphson in modified form in 1690. It is also known as Newton's first-order method, which is misleading because it has quadratic convergence. Successive x values are derived by modelling the function by its tangent. Hence, the function value f and derivative value f' are required at the current point, giving

$$x_{K+1} = x_K - \frac{f_K}{f'_K}$$

This explicit equation is due to Raphson. This method has quadratic convergence to the root if the function is well behaved locally, but is unsuitable for erratic functions. In other words, it is attractive when an analytical derivative is easily available, and fast but perhaps risky. A particular danger is in striking a point with $f' = 0$. It is at the other end of the spectrum from a binary search for a root, which is safe and steady. This method is in wide use.

The convergence factor of 2 (quadratic convergence) applies to one cycle, which requires evaluation of f and f'. The convergence factor per evaluation is therefore $\sqrt{2} = 1.414$, which is actually less than that for the secant method. Considering that the derivative is probably more easily evaluated than the function itself, possibly an effective convergence factor slightly better than $\sqrt{2}$ should be admitted.

A sufficient, but not necessary, condition for convergence of the Newton–Raphson method is that f' and f'' do not change sign in the interval $[x_1, r]$ and f' and f'' have the same sign (see Lapidus, 1962). These conditions are easily tested. Specifically, it is required for convergence that the initial error is $e < 2f'/f''$.

A possible alternative is the secant method, but the analytic derivative of Newton's method may be safer, particularly near to completion.

Schröder iteration is a Newton first-type iteration modified to account for a multiple root or root cluster. The explicit expression is

$$x_{K+1} = x_K - p\frac{f_K}{f'_K}$$

where p is an integer adjusting for the root multiplicity. This can be extremely effective, particularly on an isolated multiple root, when it can step almost exactly on to the root instead of giving only linear convergence. The way that this works is easily understood from a parabola in two dimensions touching the axis, a double root, where two steps of Newton's first iteration finds the root pair. A difficulty is that the root multiplicity is not normally known in advance. Schröder's original paper (Schröder, 1870) is in German; see also Rall (1966).

The following expression for an estimate of the multiplicity is due to Ostrowski (1973). Normal Newton–Raphson iteration is applied twice giving iterates x_1 and x_2 and x_3. Then

$$p = \left[\frac{1}{2} + \frac{x_1 - x_2}{x_3 - 2x_2 + x_1}\right]$$

where the Gaussian brackets indicate that the value must be not less than 1.

Halley's iteration, also known as Newton's second iteration, is sometimes called Newton's second-order method, which is misleading, because it has cubic convergence. This is an iteration which also uses a second derivative, giving faster convergence than Newton's iteration for well-behaved functions, preferably with analytic derivatives. The basic equation is

$$x_{K+1} = x_K - \frac{f}{f' - \left(\dfrac{ff''}{2f'}\right)}$$

where f, f' and f'' are evaluated at x_K. This may also be expressed as

$$x_{K+1} = x_K - \frac{2ff'}{2f'^2 - ff''}$$

Halley's iteration is also useful when there are multiple roots, as convergence of the first method is then only linear.

These iterations have potential application to suspension solution, but as they require derivatives it is not sufficient to have only the suspension positions. The derivative could be obtained by a position increment (i.e. use a numerical derivative), but a better way is to use a computed three-dimensional velocity diagram of the suspension to give the derivative, this being equivalent to an analytical derivative. Similarly, the second derivative could be obtained as a numerical derivative from a velocity solver, or as an analytical derivative from a computed three-dimensional acceleration diagram solution of the suspension.

17.11 Other Derivative Methods

In general, a greater convergence rate than quadratic may be achieved in favourable circumstances by the use of additional information. Such higher-order iterations typically require three items of function information. This may be found locally from f, f' and f'', giving several other iterations. The method of interpolation varies, but typically a quadratic equation (parabolic curve) is fitted to the local information and the roots of this are then found, the nearer one being selected as the new iteration point. The result is several interpolation formulae, which may usually be expressed as

$$x_{N+1} = x_N - C\frac{f}{f'}$$

where C is a variable 'correction factor' on the simple Newton iteration. For example, Halley's iteration is

$$x_{K+1} = x_K - \frac{2ff'}{2f'^2 - ff''} = x_K - \left(\frac{f}{f'}\right)\left(\frac{2f'^2}{2f'^2 - ff''}\right)$$

This can be expressed by its correction factor

$$C = \frac{2f'^2}{2f'^2 - ff''} = \frac{1}{1 - \dfrac{ff''}{2f'^2}}$$

When using the two derivatives at one point, as may be expected by dimensional analysis, the group ff''/f'^2 occurs frequently, and with an extra factor of 1/2 may usefully be called g:

$$g = \frac{ff''}{2f'^2}$$

The correction factor used in Halley's iteration is then simply

$$C = \frac{1}{1-g}$$

As another example, Olver's iteration, also cubically convergent, is

$$x_{K+1} = x_K - \frac{f}{f'} - \frac{1}{2}\frac{f''}{f'}\left(\frac{f}{f'}\right)^2$$

which has $C = 1 + g$.

Cauchy's iteration also uses the function value and its first two derivatives at one point. Its equation is

$$x_{K+1} = x_K - \frac{2f}{f' \pm \sqrt{f'^2 - 2ff''}}$$

with the sign of the root taken to maximise the absolute value of the denominator, going to the nearest prospective root.

Table 17.11.1 Single-point high-order iterations using f, f', and f''

Newton's iteration:	$C = 1$
Halley's iteration:	$C = \dfrac{1}{1-g}$
Olver's iteration:	$C = 1 + g$
Cauchy's iteration:	$C = \dfrac{2}{1+\sqrt{1-4g}}$
Laguerre's iteration (N is the polynomial order)	$C = \dfrac{N}{1+\sqrt{(N-1)(N-1-2Ng)}}$

Various such common iterations are characterised in Table 17.11.1 in terms of the correction factor g. When g goes to zero, as normally happens as a root is approached, then these expressions all tend to 1. If Laguerre's iteration is generalised by using an integer K instead of N, then setting $K = 1$ gives Newton's first-order iteration, and setting $K = 2$ gives Cauchy's iteration. When actually using the above expressions, it is usually better not to calculate g and substitute, but to expand the complete expression for Δx in the form most favourable for accurate calculation.

When Cauchy's iteration and Laguerre's iteration are expanded binomially in small g they tend to $1/(1-g)$ and ultimately to 1. With the square root there is extra calculation, but the possibility arises of a complex number result from an initially pure real root estimate. In some cases this may be regarded as an advantage, in that a complex root may be found without the need for an initially complex root estimate.

Other possibilities arises, apparently unnamed, for example:

$$C = \sqrt{1+2g}$$

which expands for small g into $1 + g$.

A method involving use of two points with two function values and the derivative at one of the points is straightforward, but it appears to be unnamed. It may have some advantages, particularly when it is desired to maintain root brackets. With two points, then instead one function value and two derivatives may be used, which is slightly easier. A quadratic equation is fitted to these values, with an easy solution of three simultaneous equations, and solved for the nearest root. It does not appear to be widely used, possibly because it is not well known, possibly because it requires rather more calculation than competing methods.

The function value and derivative value at two points is yet another possible method, requiring the fitting of a cubic, like a spline. Solution of the root of the cubic is, however, a less convenient aspect.

The potential application of these methods to suspension analysis is evidently similar to that of Halley's iteration.

17.12 Polynomial Roots

It is often required in scientific computing to find the roots of a polynomial. In practical scientific and engineering work, the coefficients are usually real, being calculated from the dynamical properties of the system, such as masses, angular inertias, stiffnesses, damping coefficients and so on, or being the result of a curve fit to experimental data. In the case of suspension geometry, a polynomial could be fitted to the curve of Z_{SB} against Z_{LBJ}, and used to obtain an estimate of the lower ball joint position for a given suspension bump. As a specific case, for an example double-transverse-arm suspension, such a fifth-order polynomial had coefficients as in Table 17.12.1. These coefficients are also of some interest in their own right, revealing some useful information about potential iteration issues on the true function rather than on the polynomial approximation. The gradient $C(1)$ is close to 1.0. so a simple fixed-point iteration is likely to be reasonably good.

Table 17.12.1 Polynomial fit coefficients: $Z_{SB} = C_0 + C_1 Z_{LBJ} + \ldots$

```
C(0) =  -0.213431204422253
C(1) =   1.06535142457698
C(2) =  -0.858178364024349
C(3) =   5.51778621761935
C(4) = -16.4240250458009
C(5) =  20.5603137932280
```

There have been many polynomial root finding programs. There are several fundamentally different methods, each with its strengths and weaknesses. Some methods find all the roots simultaneously, being based on an electrostatic analogy. These seem to be relatively inefficient, because methods that find only one root at a time have the advantage that they can use deflation to simplify the subsequent problem. However, simultaneous root finders do avoid the problems of deflation – loss of accuracy in subsequent roots.

In the case of an odd-order polynomial there must always be at least one real root, so it is common practice then to find that one real root first. For an even-order polynomial, the common scientific problem, the usual approach is to iterate on a single generally complex root, for example by Laguerre's method, Madsen's method, or the Jenkins–Traub method. An alternative is to extract a wholly real quadratic factor by Lin's method or Bairstow's method, although it is difficult to make these completely reliable with regard to convergence. Another possibility is to factor into two parts with real coefficients, say an eighth-order into two quartic factors, again by iteration, but again with some reliability problems.

Madsen's method uses Newton's first iteration on a complex root, but with multiple steps as proposed by Schröder (see above). This helps greatly with multiple or closely spaced roots (see Madsen, 1973, which includes a program in Algol).

Jenkins and Traub produced an elaborate program to find the roots of a polynomial aiming for good performance on difficult cases, for example when there are multiple roots or roots separate but close together. This uses various complex shifts. Also careful attention is given to estimating the prospective accuracy to give the best exit conditions (Jenkins and Traub, 1972; Jenkins, 1975). The real-coefficient program seems not to work correctly as published, but both programs repay study for the methods used.

Laguerre's method is an iterative method for finding the real or complex roots of a polynomial with real or complex coefficients. A program is given in Press et al. (1992). Convergence is not absolutely certain in all cases, but it is quite reliable in practice, and fast with cubic convergence to a single root. It usually takes only about four iterations for 8-byte (15-digit) real accuracy, so the efficiency is usually very good. Multiple roots have linear convergence. The underlying theory is quite simple, and involves an assumption that the desired root is at one distance from the estimate whilst all the others are at some other (much greater) distance. There is quite a lot of work in the loop, with complex-number evaluation of the polynomial and its first two derivatives, which are then combined to give a correction term to the root estimate. The Press et al. program includes a scheme for breaking out of a closed cycle that may occasionally occur.

Given a polynomial f of order N and a (generally complex) root estimate z, possibly starting with zero, the first and second polynomial derivatives are f' and f''. Evaluate first the two intermediate values $G = f'/f$ and $H = G^2 - f''/f$. Laguerre's correction to the root estimate is then given by

$$dz = -\frac{N}{G \pm \sqrt{(N-1)(NH - G^2)}}$$

where the sign is chosen to maximise the absolute value of the possibly complex denominator, naturally, to minimise the correction, so always trying to approach the nearest true root. Using 1 instead of N, the correction reduces to a Newton iteration, and using 2 instead of N gives a Cauchy iteration.

The equation above is easily modified for a root multiplicity M, but the improvement seems to be advantageous in practice only for a double root. Parlett's criterion is that, after three iterations, if a proposed increment exceeds 0.8 times the previous one then the root should be treated as multiple.

Bairstow's method is the extraction of a quadratic factor (x^2+ux+v) with real coefficients from a monic polynomial with real coefficients, using Newton's iteration extended to the two unknowns u and v simultaneously by partial derivatives. The starting values are often taken as zero, sometimes as $v=c_0/c_2$, $u=c_1/c_2$, where c counts from zero as the polynomial constant. If convergence fails, which is quite likely, other starting values may be successful. The method can be applied usefully to polynomials of order 3 or more, although in the case of the cubic it is slower than extracting a single real root. For cubics and quartics it can be applied analytically, whereas higher orders require synthetic division. The starting polynomial is divided by the candidate quartic factor to leave a remainder

$$R=R_1x+R_0$$

where the remainder coefficients are functions of the coefficients of the polynomial and of the putative factor. For a good factor these are zero, so an iteration is applied to reduce them. The analytic partial derivative of each is obtained with respect to u and v. This gives two simultaneous equations to solve for an ideal simultaneous correction given local linearity, with the solution

$$\Delta u=\frac{-R_1\dfrac{\partial R_2}{\partial v}+R_2\dfrac{\partial R_1}{\partial v}}{D};\quad \Delta v=\frac{R_1\dfrac{\partial R_2}{\partial u}-R_2\dfrac{\partial R_1}{\partial u}}{D}$$

where D is the Jacobian of the left-hand side of the two equations. The method may fail by (1) division by a zero denominator required, (2) divergence, (3) failure to converge due to oscillation, (4) convergence too slow. In practical application, the unreliability is a significant problem, and the method is only usable with a suitable initial estimate of the quadratic factor coefficients. When convergence occurs, excellent accuracy can be achieved. Therefore, in general, the method is at its best when used for polishing a quadratic factor found initially by other means. The zero denominator problem could possibly be dealt with in some cases by making an arbitrary small adjustment to u or v or both, to move away from the difficult spot.

Lin's method is iterative extraction of a quadratic factor with real coefficients from a polynomial with real coefficients, using synthetic division. Setting the remainder terms to zero gives two equations for new estimates of the factor coefficients. Lin's method tends to oscillate, and usually benefits from an iteration factor less than 1. Convergence is not reliable unless a good starting estimate is available, which limits the usefulness of the method. When convergence occurs, accuracy is excellent. For polynomials with real coefficients, as arise in many practical problems, quadratic factor extraction can be performed in all-real arithmetic. It is claimed to be particularly useful when there are two roots coincident or close together.

Lin's method is as follows. A polynomial of order N of the form

$$x^N+a_{N-1}x^{N-1}+\cdots+a_1x+a_0=0$$

can be expressed as a quadratic term times another term of order N–2 with a linear remainder:

$$(x^2+px+q)(x^{N-2}+b_{N-1}x^{N-3}+\cdots+b_3x+b_2)+(b_1x+b_0)=0$$

It is desired to reduce the remainder term to zero, when the quadratic expression will be a true factor. Assuming, then, that $b_1=0$ and $b_0=0$, and comparing coefficients,

$$b_{N-1}=a_{N-1}-p$$

$$b_{N-2}=a_{N-2}-pb_{N-1}-q$$

$$\ldots$$

$$b_{N-J}=a_{N-J}-pb_{N+1-J}-qb_{N+2-J}$$

$$\ldots$$

$$p = \frac{a_1 - qb_3}{b_2}$$

$$q = \frac{a_0}{b_2}$$

Note that the remainder coefficients are not actually used, being set to zero. The last two equations above are the new quadratic factor coefficients. Initial estimates must be made by some means for p and q. From these and the original a coefficients, b_{N-1} can be calculated. This allows b_{N-2} to be calculated, and so on. This is equivalent to the solution of two unknowns by direct iteration. In effect, synthetic division is used. The process is to be terminated in the usual way when the changes become sufficiently small. The process as described appears unreliable, convergence being problematic, this obviously depending on the initial estimates of the factor coefficients.

17.13 Testing

As test of a zero search for suspensions, an equivalent fifth-order polynomial for $Z_{SB}(Z_{LBJ})$ with coefficients as in Table 17.12.1 was used, producing sample loop results for a simple fixed-point iteration as in Table 17.13.1. The numbers of loops and evaluations for various methods vary somewhat with the

Table 17.13.1 Loop results from fixed-point iteration test on polynomial

```
Fixed Point Iteration, cf =  -1.00000000000000

      1  0.211396869688034        -0.111396869688034
      2  0.209960163628131         1.436706059902981E-003
      3  0.209998492473714        -3.832884558244154E-005
      4  0.209997479434359         1.013039354513801E-006
      5  0.209997506215864        -2.678150476387309E-008
      6  0.209997505507851         7.080122876867989E-010
      7  0.209997505526569        -1.871747201676044E-011
      8  0.209997505526074         4.948541576510479E-013
      9  0.209997505526087        -1.307287611496122E-014
     10  0.209997505526087         3.330669073875470E-016
     11  0.209997505526087         0.000000000000000E+000

Number of loop starts, evaluations = 11  12
Final x,v = 0.20999750552608684E+00 0.00000000000000000E+00

FPI Accelerated Convergence
      1  0.209978457157241        -0.111396869688034
      2  0.209997505524439        -1.955187930216540E-005
      3  0.209997505526087        -1.690925177655345E-012
Number of loop starts, evaluations = 4  9
Final x,v = 0.20999750552608684E+00 0.00000000000000000E+00

Secant Search, oldest point discarded:

      1  0.208025290425772        -2.023673967893436E-003
      2  0.209946718440993        -5.212926912367966E-005
      3  0.209997522583638         1.750849540349542E-008
      4  0.209997505525938        -1.527111770371903E-013
      5  0.209997505526087         2.775557561562891E-017
      6  0.209997505526087         0.000000000000000E+000
Number of loop starts, evaluations = 6  8
Final x,v = 0.20999750552608684E+00 0.00000000000000000E+00
```

Table 17.13.2 Comparison of methods on sample polynomial iteration

Method	Loops	Evaluations
Binary Search	53	55
Fixed Point Iteration	11	12
FPI Accelerated	4	9
Secant Search	6	8
Regula Falsi	7	9
Muller's Method	4	7
Newton	6	12
Halley	5	15
Olver	5	15
Cauchy	5	15
Laguerre	4	12

exact values used, but were typically in Table 17.13.2. All of the methods were capable of achieving the true zero. The most notable point is that the high-order methods do have fewer loops but the number of evaluations is not always greatly improved. Muller's method showed the fewest evaluations, but not greatly better than the accelerated fixed-point iteration, secant or *regula falsi*, and is more tricky to implement. The gradient of the function is quite close to 1, so the standard fixed-point iteration is quite good, and the acceleration advantages are not so great.

Tables 17.13.3–17.13.6 show the results of iterations applied to a true geometric solution of an example suspension to obtain Z_{LBJ} for a given Z_{SB}. The results can be seen to be similar to those for the polynomial. Even the simple fixed-point iteration would be satisfactory for most purposes. Actually, amongst these examples, Newton's method is potentially the fastest because the gradient changes slowly and really needs

Table 17.13.3 Fixed-point iteration

```
Test Fixed-Point Iteration on ZSB:

zs req = 1.000000000000000E-002

K          dz3                        z9                       err

1   0.0100000000000000      0.0102829990247792        0.28E-03
2   0.0097170009752208      0.0099914451983580       -0.86E-05
3   0.0097255557768628      0.0100002581161973        0.26E-06
4   0.0097252976606655      0.0099999922116523       -0.78E-08
5   0.0097253054490133      0.0100000002350037        0.24E-09
6   0.0097253052140096      0.0099999999929091       -0.71E-11
7   0.0097253052211005      0.0100000000002139        0.21E-12
8   0.0097253052208866      0.0099999999999936       -0.64E-14
9   0.0097253052208931      0.0100000000000004        0.43E-15
10  0.0097253052208926      0.0099999999999999       -0.97E-16
11  0.0097253052208927      0.0100000000000000       -0.42E-16
12  0.0097253052208928      0.0100000000000000        0.14E-16

Number of loop starts, evaluations = 13 13
dz3  =  9.725305220892787E-003
z(9) =  1.000000000000001E-002
err  =  0.14E-16
```

Table 17.13.4 Accelerated fixed-point iteration

```
Test Accelerated Fixed-Point Iteration on ZSB:

zs req = 1.000000000000000E-002

K          dz3                 z9                  er1        er2

1  0.0097253047610143  0.0099914451983580   0.28E-03  -0.86E-05
2  0.0097253052208929  0.0100000000142949  -0.47E-09   0.14E-10
3  0.0097253052208928  0.0100000000000000   0.18E-15  -0.42E-16

Number of loop starts, evaluations = 4 8
dz3  = 9.725305220892839E-003
z(9) = 1.000000000000007E-002
err  = 0.69E-16
```

Table 17.13.5 Secant iteration

```
Test Secant Iteration on ZSB:

zs req = 1.000000000000000E-002

K        dz3                  z9                  err

1  0.0097050902524657  0.0099791751548269   0.10+100
2  0.0097253431289573  0.0100000390518932  -0.21E-04
3  0.0097253052207387  0.0099999999998411   0.39E-07
4  0.0097253052208929  0.0100000000000002  -0.16E-12
5  0.0097253052208927  0.0100000000000001   0.21E-15
6  0.0097253052208926  0.0100000000000000   0.69E-16

Number of loop starts, evaluations = 7 9
dz3  = 9.725305220892617E-003
z(9) = 1.000000000000001E-002
err  = 0.14E-16
```

Table 17.13.6 Newton iteration

```
Newton's method with 3D velocity diagram for the derivative:

zs req = 1.000000000000000E-002

K          dz3                  z9                  err          slope

1  0.0100000000000000  0.0000000000000000  -0.10E-01  1.000000000000
2  0.0097253204527308  0.0102829990247792   0.28E-03  1.030287939502
3  0.0097253052208927  0.0100000156914399   0.16E-07  1.030173761288
4  0.0097253052208928  0.0100000000000000  -0.42E-16  1.030173754969
5  0.0097253052208928  0.0099999999999999  -0.69E-16  1.030173754969

Number of loop starts, evaluations = 5 8
dz3  =  9.725305220892773E-003
z(9) =  9.999999999999931E-003
err  = -0.69E-16
```

to be evaluated only once or twice, although a similar modification could be applied to the secant iteration. For those seeking a particularly efficient implementation, there are many possibilities here for performance improvements above the basic fixed-point iteration, applicable to more than just suspensions, of course.

References

Adams, D.A. (1967) A stopping criterion for polynomial root finding. *Communications of the ACM*, **10**(10), 655–658.

Conte S. D. and De Boor C. (1972) *Elementary Numerical Analysis: An Algorithmic Approach*, 2nd edn. New York: McGraw-Hill. A classic, well worth seeking out second hand.

Jenkins, M.A. (1975) Zeros of a real polynomial (Algorithm 493). *ACM Transactions on Mathematical Software*, **1**(2), 178–189.

Jenkins, M.A. and Traub, J.F. (1972) Zeros of a complex polynomial. *Communications of the ACM*, **15**(2), 97–99.

Lapidus L. (1962) *Digital Computation for Chemical Engineers*. New York: McGraw-Hill. Hard to find. Good explanations of general numerical methods. The chemical content is limited to the end-of-chapter examples.

Madsen, K. (1973) A root-finding algorithm based on Newton's method. *BIT*, **13**, 71–75.

Muller, D.E. (1956) A method for solving algebraic equations using an automatic computer. *Mathematical Tables and Other Aids to Computation*, **10**, 208–215.

Ostrowski, A.M. (1973) *Solution of Equations in Euclidean and Banach Space*. New York: Academic Press.

Press, W.H., Teukolsky, S.A., Vetterling, W.T. and Flannery, B.P. (1992) *Numerical Recipes in Fortran: The Art of Scientific Computing*. Cambridge: Cambridge Uuniversity Press, ISBN 052143064X. Also available in C and Basic. Well known reference, more-or-less essential, many programs, although no help on geometry.

Rall, L.B. (1966) Convergence of the Newton process to multiple solutions. *Numerische Mathematik*, **9**, 23–27.

Schröder, E. (1870) Über unendlich viele Algorithmen zur Auflösung der Gleichungen. *Mathematische Annalen* **2**, 317–365

Shoup T. E. (1979) *A Practical Guide to Computer Methods for Engineers*. Englewood Cliffs, NJ: Prentice Hall. ISBN 0136906516. Good explanations.

Shoup, T.E. (1984) *Applied Numerical Methods for the Microcomputer*. Englewood Cliffs, NJ: Prentice Hall. ISBN 0130414182. This is a development of Shoup (1979).

Of the books mentioned above, only Press et al. is in print. The others must be sought second hand, and can be found without too much difficulty apart from Lapidus. There are, of course, many books on numerical programming, but none that really help in the specific context of three-dimensional suspension geometry.

——— // ———

Appendix A

Nomenclature

Chapter 1 — Introduction - No Nomenclature

Chapter 2 — Road Geometry

A	m/s^2	acceleration
A_L	m/s^2	lateral acceleration in road plane
A_N	m/s^2	acceleration normal to the road
$F_{\rho N}$	N	normal pseudo-force in vehicle-fixed axes
g	m/s^2	gravitational field strength
h	m	ride height
J	m/s^3	jerk, dA/dt
$k_{\phi RS}$	rad/m	road torsion rate
L_W	m	wheelbase
M	Nm	moment
N_i	N	normal force at inner wheels
N_o	N	normal force at outer wheels
r	—	data point count
R	m	path radius of curvature
R_B	m	path banking radius of curvature
R_H	m	path horizontal radius of curvature
R_L	m	path lateral radius of curvature
R_N	m	path normal radius of curvature
R_V	m	path vertical radius of curvature
s	m	path length from a reference point
s_H	m	path length of projection into horizontal plane
T	m	axle track (tread)
W	N	weight force
w	m	ride-height sensor spacing
X	m	forward axis
x,y,z	m	vehicle or path-fixed axes
X,Y,Z	m	Earth-fixed coordinate axes
Y	m	axis to left

Suspension Geometry and Computation J. C. Dixon

Z	m	axis upwards
Z_R	m	road altitude above datum plane

Greek

β	rad	vehicle attitude angle
γ	rad	road camber angle
θ_{LN}	rad	wheelbase normal curvature angle
θ_R	rad	road longitudinal slope angle
κ	m^{-1}	path curvature
κ_B	m^{-1}	path banking curvature
κ_H	m^{-1}	path horizontal curvature
κ_L	m^{-1}	path lateral curvature
κ_N	m^{-1}	path normal curvature
κ_V	m^{-1}	path vertical curvature
ν	rad	path angle
σ	m	lateral position from road centreline
τ_C	$m^{-1} s^{-1}$	car (vehicle) turn-in
τ_P	m^{-2}	path turn-in
$\tau_{P,D}$	m^{-2}	driver-perceived path turn-in
ϕ_A	rad	axle roll angle (on tyres)
ϕ_B	rad	body roll angle
ϕ_R	rad	road lateral slope angle (bank, camber)
ϕ_S	rad	suspension roll angle
ϕ_{Sen}	rad	roll angle of sensor reference line
ϕ_κ	rad	angle of path curvature from horizontal
ψ	rad	vehicle heading angle

Subscripts

A	axle
B	body
C	car
H	horizontal
L	lateral
N	normal
P	path
R	road
V	vertical

Chapter 3 — Road Profiles

C	m	cosine component amplitude
f_N	Hz	natural frequency
f_R	Hz	frequency of road wave at vehicle speed V.
G	–	gradient
H	m	height
$k_{\phi RS}$	rad/m	road torsion
L_W	m	wheelbase
n_{SR}	cycles/m	road wave spatial frequency (see ω_{SR})
s	m	path length
s_H	m	path length in XY plane
S	m	sine component amplitude

T	m	vehicle track (tread), path spacing
T_R	s	period of a road wave at speed V
V	m/s	vehicle speed
X	m	longitudinal position, straight path length
Z	m	vertical position
Z_B	m	road banking (camber) height
Z_{B0}	m	road banking amplitude
Z_L	m	height of left path
Z_R	m	height of right path
Z_S	m	road centreline height
Z_{S0}	m	road centreline amplitude
Z_0	m	total component of a sinusoid

Greek

θ	rad	angle
θ_R	rad	road pitch angle
λ	m	sinusoid wave length
λ_R	m	road wave length
ϕ	rad	phase angle
ϕ_{RB}	rad	road banking angle
ϕ_{RT}	rad	road torsion angle between two axles
ω_R	rad/s	road fluctuation effective temporal frequency
ω_{SR}	rad/m	road fluctuation spatial frequency (see n_{SR})

Subscripts

B	antisymmetrical (banking)
L	left path
R	right path
S	symmetrical (centreline)
0	total amplitude (with phase angle)

Chapter 4 — Road Geometry

A	m/s^2	acceleration
a	m	front axle to centre of mass
b	m	rear axle to centre of mass
C_C	Ns/m	cushion damping coefficient
C_T	Ns/m	tyre damping coefficient total
C_W	Ns/m	suspension damping coefficient (total)
D_P	—	discomfort of passenger
D_T	—	discomfort of tyre
F_N	N	tyre normal force
g	m/s^2	gravity (ISO standard 9.80665 m/s^2)
h_T	m	tyre vertical deflection
k_ϕ	rad/m s^{-2}	roll gradient
K_C	N/m	cushion stiffness
K_T	N/m	tyre vertical stiffness total
K_W	N/m	suspension stiffness (wheel rate) total
L_{CP}	m	tyre contact patch length

m_B	kg	vehicle body mass
m_P	kg	passenger mass (effective on seat)
m_U	kg	unsprung mass total
R_U	m	tyre unloaded radius
R_L	m	tyre loaded radius
R_R	m	tyre rolling radius
R_{CE}	m	contact patch end radius
R_M	—	motion ratio
T	m	vehicle track (tread)
V	m/s	velocity
V_X	m/s	longitudinal velocity
Z	m	vertical position
z	m	ride position (from static)
Z_B	m	vehicle body vertical position
z_B	m	vehicle body ride position
z_C	m	cushion ride deflection (compression positive)
Z_P	m	passenger vertical position
z_P	m	passenger ride position
Z_R	m	road vertical position
z_S	m	suspension ride deflection (compression positive)
z_T	m	tyre ride deflection (compression positive)
Z_W	m	wheel vertical position

Greek

θ	rad	angle
θ_{CP}	rad	contact patch subtended angle
ϕ_A	rad	axle roll angle
ϕ_B	rad	body roll angle
ϕ_{BT}	rad	body torsion angle
ϕ_R	rad	road bank (roll) angle
ϕ_{RT}	rad	road torsion angle
ϕ_S	rad	suspension roll angle
Ω	rad/s	wheel angular spin speed

Subscripts

B	body (vehicle)
C	cushion
f	front
i	inner cornering side
L	left
o	outer cornering side
r	rear
R	right
s	spring
S	suspension
U	unsprung
W	wheel
0	static

Chapter 5 — Vehicle Steering

b	m	kingpin offset
c	m	caster offset
f	—	a factor
f_A	—	Ackermann factor
G	—	steering gear ratio
G_m	—	mean steering gear ratio
l	—	x-direction cosine
L	m	wheelbase, a length
L_f	m	front axle to centre of mass
L_{KP}	m	kingpin (steer axis) length lower to upper point
L_r	m	rear axle to centre of mass
m	—	y-direction cosine
n	—	z-direction cosine
p	m	a length (Figure 5.3.4)
q	m	a length (Figure 5.3.4)
R	m	path radius to centre of mass
R_f	m	path radius to centre of front axle
R_{OT}	m	offtracking
R_r	m	path radius to centre of rear axle
T	m	track (tread)
V	m/s	velocity
W_{KP}	m	kingpin (steer axis) diagonal horizontal spacing
x_{CT}	m	caster trail
X_{KC}	m	kingpin (steer axis) caster spacing $X_A - X_B$
X_{KCw}	m	kingpin (steer axis) caster spacing seen by steered wheel
y_{CO}	m	caster offset
Y_{KI}	m	kingpin (steer axis) inclination spacing ($Y_A - Y_B$, left wheel)
Y_{KIw}	m	kingpin inclination of steered wheel
z_S	m	suspension bump
z_{SJ}	m/rad	steer jacking
z_{SJA}	m/rad	steer jacking of an axle
Z_{KI}	m	kingpin (steer axis) vertical spacing
Z_{KP}	m	kingpin height lower to upper point

Greek

α	rad	tyre slip angle
β	rad	attitude angle
γ	rad	camber angle of road wheel
γ_S	rad	wheel camber angle due to steer
δ	rad	steer angle of road wheel
δ_K	rad	kinematic steer angle of road wheel
$\delta_{K,T}$	rad	kinematic steer angle of tractor unit
δ_{ref}	rad	reference steer angle of road wheel
δ_S	rad	steer angle of hand wheel
δ_5	rad	yaw angle at articulated fifth wheel
ε_{BCas1}	rad/m	linear bump caster coefficient
ε_{BCas2}	rad/m^2	quadratic bump caster coefficient
ε_{SC1}	—	linear steer camber coefficient (deg/deg)

ε_{SC2}	rad^{-1}	quadratic steer camber coefficient
ε_{SJ1}	m/rad	steer jacking linear coefficient
ε_{SJ2}	m/rad^2	steer jacking quadratic coefficient
ε_{SKC1}	—	steer kingpin caster linear coefficient (rad/rad)
ε_{SKC2}	rad^{-1}	steer kingpin caster quadratic coefficient (rad/rad^2)
ε_{SKI1}	—	steer kingpin inclination linear coefficient (rad/rad)
ε_{SKI2}	rad^{-1}	steer kingpin inclination quadratic coefficient (rad/rad^2)
λ	rad	Langensperger angle
λ_{f0}	rad	Langensperger angle, front, zero speed
θ_B	rad	body pitch angle
θ_{KC}	rad	kingpin (steer axis) caster angle
θ_{KC0}	rad	kingpin (steer axis) static straight caster angle
θ_{KCw}	rad	kingpin caster angle of steered wheel
$\theta_{KC,R}$	rad	kingpin (steer axis) caster angle relative to the road
θ_{KI}	rad	kingpin (steer axis) inclination angle
θ_{KIw}	rad	kingpin (steer axis) inclination angle of steered wheel
θ_{KP}	rad	kingpin (steer axis) total angle from vertical
θ_{StA}	rad	steering linkage Ackermann effect angle
$\theta_{StA,I}$	rad	ideal steering linkage Ackermann effect angle
θ_{TRGE}	rad	track rod geometry error
ϕ_{SJA}	rad	steer roll jacking of an axle
ψ_{KP}	rad	kingpin (steer axis) sweep angle
ψ_{KPw}	rad	kingpin (steer axis) sweep angle seen by steered wheel

Subscripts

f	front
KC	kingpin caster
KI	kingpin inclination
KP	kingpin (steer axis)
L	left
L	lower point of steer axis
r	rear
R	right
U	upper point of steer axis
0	initial, static, or at zero speed

Chapter 6 — Bump and Roll Steer

a	m	centre of mass to front axle
b	m	centre of mass to rear axle
e_H	m	rack height error
e_L	m	track-rod length error
k_{RUFV}	N^{-1}	variation of ε_{RU} with axle load
k_{RUZA}	m^{-1}	variation of ε_{RU} with axle suspension bump
K_A	N/m	axle suspension vertical stiffness
L_{AX}	m	steering-arm length perpendicular to track rod
L_{ITR}	m	ideal track-rod length
L_{TR}	m	track-rod length
T	m	axle track (tread)
z_A	m	axle mean suspension bump deflection
z_S	m	suspension bump deflection

Greek

δ_{A}	rad	axle steer angle
δ_{A0}	rad	axle static steer angle
δ_{BS}	rad	single-wheel bump steer angle
δ_{DBS}	rad	axle double-bump steer angle
δ_{RS}	rad	axle roll steer angle
δ_{RT}	rad	axle roll toe-out angle
δ_{RU}	rad	axle roll understeer angle
δ_{S}	rad	driver-controlled steer angle
δ_{T}	rad	axle toe-out angle
δ_{T0}	rad	static toe-out angle
ε	various	suspension geometric coefficients (various units)
ε_{BS1}	rad/m	single-wheel linear bump steer coefficient
ε_{BS2}	rad/m^2	single-wheel quadratic bump steer coefficient
ε_{DBS1}	rad/m	axle linear double-bump steer coefficient
ε_{DBS2}	rad/m^2	axle quadratic double-bump steer coefficient
ε_{RS1}	—	axle linear roll steer coefficient (rad/rad)
ε_{RS2}	rad^{-1}	axle quadratic roll steer coefficient (rad/rad^2)
ε_{RT1}	—	linear axle roll toe coefficient (rad/rad)
ε_{RT2}	rad^{-1}	axle quadratic roll toe coefficient(rad/rad^2)
ε_{RU1}	—	axle linear roll understeer coefficient (rad/rad)
ε_{RU2}	rad^{-1}	axle quadratic roll understeer coefficient (rad/rad^2)
ε_{VHU1}	rad/m	vehicle linear heave understeer coefficient
ε_{VHU2}	rad/m^2	vehicle quadratic heave understeer coefficient
ε_{VPU1}	—	vehicle linear pitch understeer coefficient (rad./rad)
ε_{VPU2}	rad^{-1}	vehicle quadratic pitch understeer coefficient (rad/rad^2)
ε_{VRU1}	—	vehicle linear roll understeer coefficient (rad/rad)
ε_{VRU2}	rad^{-1}	vehicle quadratic roll understeer coefficient (rad/rad^2)
θ_{S}	rad	suspension pitch angle
ρ_{A}	rad	axle location-axis inclination angle
ϕ_{S}	rad	suspension roll angle

Subscripts

f	front
L	left
r	rear
R	right

Chapter 7 — Camber and Scrub

c_{y}	—	swing centre lateral movement coefficient
c_{z}	—	swing centre vertical movement coefficient
$k_{\phi A \phi S}$	—	variation of axle roll angle with suspension roll
T	m	track (tread)
V_{CH}	m/s	contact patch horizontal velocity
V_{S}	m/s	suspension bump velocity
y	m	swing centre lateral coordinate
z	m	swing centre vertical coordinate
z_{B}	m	body vertical ride position

z_S	m	suspension bump
z_W	m	wheel vertical ride position

Greek

α	rad	slip angle
β	rad	body attitude angle
γ	rad	camber angle
γ_A	rad	axle camber angle
γ_{BC}	rad	bump camber angle
γ_d	rad	semi-difference of wheel camber angles ($= \Gamma_A$)
γ_L	rad	left wheel camber angle
γ_m	rad	mean of wheel camber angles ($= \gamma_A$)
γ_R	rad	right wheel camber angle
γ_0	rad	static camber angle
Γ	rad	inclination angle
Γ_A	rad	axle mean inclination angle
Γ_{ARI}	rad	axle roll inclination angle
Γ_L	rad	inclination angle of left wheel
Γ_P	rad	inclination angle of path (road) surface
Γ_R	rad	inclination angle of right wheel
$\Gamma_{W,L}$	rad	inclination angle of left wheel relative to the local road surface
$\Gamma_{W,R}$	rad	inclination angle of right wheel relative to the local road surface
ε_{ARC1}	—	axle linear roll camber coefficient
ε_{ARC2}	—	axle quadratic roll camber coefficient
ε_{ARI}	—	axle roll inclination coefficient (rad/rad)
ε_{ARI1}	—	axle first roll inclination coefficient (rad/rad)
ε_{ARI2}	rad^{-1}	axle second roll inclination coefficient
ε_{BC}	rad/m	bump camber coefficient
ε_{BC1}	rad/m	linear bump camber coefficient
ε_{BC2}	rad/m^2	quadratic bump camber coefficient
ε_{BSc}	—	local bump scrub coefficient ds/dz_S
ε_{BSc1}	—	linear bump scrub coefficient
ε_{BSc2}	m^{-1}	quadratic bump scrub coefficient
ε_{BScd}	—	local bump scrub rate coefficient
ε_{BScd0}	—	initial bump scrub rate coefficient
ε_{BScd1}	m^{-1}	linear bump scrub rate coefficient
ε_{BScd2}	m^{-2}	quadratic bump scrub rate coefficient
θ_{SA}	rad	swing arm angle
ϕ_A	rad	axle roll angle
ϕ_S	rad	suspension roll angle

Subscripts

L	left
R	right
0	static

Chapter 8 — Roll Centres

B	m	lateral position of GRC in vehicle coordinates
f_{LT}	—	load transfer factor

F_{LJ}	N	link jacking force
F_{LT}	N	suspension normal force lateral transfer ('load transfer')
F_Y	N	lateral force in suspension links
F_1	N	suspension force reacted by springs
F_2	N	suspension force reacted by links
g	m/s^2	gravitational field strength
H	m	height of GRC in vehicle coordinates
H_{CM}	m	height of body centre of mass in vehicle coordinates
H_{FRC}	m	height of FRC in vehicle coordinates
H_{KRC}	m	height of KRC in vehicle coordinates
J_{AR}	—	anti-roll factor
k_{B1}	s^2	linear coefficient in $B(A_Y)$ [m/(m s^{-2})]
k_{B2}	s^4/m	quadratic coefficient in $B(A_Y)$ [m/(m s^{-2})^2]
k_{H1}	s^2	linear coefficient in $H(A_Y)$ [m/(m s^{-2})]
k_{H2}	s^4/m	quadratic coefficient in $H(A_Y)$ [m/(m s^{-2})^2]
k_{ZA}	s^2	linear coefficient in $z_A(A_Y)$ [m/(m s^{-2})]
$k_{\phi S}$	rad.s^2/m	linear coefficient in $\phi_S(A_Y)$
m_B	kg	mass of body (sprung mass)
M_{FYL}	Nm	moment of link force
M_{YCM}	Nm	moment of body weight force
T	m	track (tread)
Y	m	lateral position of GRC in road coordinates
Y_{CM}	m	lateral position of body centre of mass in roll
Y_P	m	lateral position of body pivot point in roll
z_A	m	axle suspension double bump (mean of two wheels)
z_B	m	body heave
z_L	m	suspension left bump
z_R	m	suspension right bump
Z	m	height of GRC in road coordinates
$Z_{B\phi}$	m	body heave after roll

Greek

ε	var.	suspension geometric coefficient
ε_{BSc}	—	bump scrub coefficient in vehicle coordinates
ε_{BScR}	—	bump scrub coefficient in road coordinates
ε_{BSc1}	m^{-1}	linear bump scrub coefficient
ε_{BSc2}	m^{-2}	quadratic bump scrub coefficient
ε_{Bscd0}	—	bump scrub rate at zero bump
ε_{BScd1}	m^{-1}	linear bump scrub rate variation coefficient
ε_{BScd2}	m^{-2}	quadratic bump scrub rate variation coefficient
$\varepsilon_{B1,DB}$	—	linear variation of B in double bump
$\varepsilon_{B1,Roll}$	m/rad	linear variation of B in roll, with respect to suspension roll angle
$\varepsilon_{B1,Roll,ZB}$	—	linear variation of B in roll, with respect to wheel bump Z_R (m/m)
$\varepsilon_{H1,DB}$	—	linear variation rate of GRC height in double bump, in vehicle body coordinates
$\varepsilon_{H1,Roll}$	m/rad	linear variation of GRC height in roll
$\varepsilon_{H2,Roll}$	m/rad^2	quadratic variation of GRC height in roll
$\varepsilon_{Z1,DB}$	—	linear variation rate of GRC height in double bump, in road coordinates
θ_{SA}	rad	angle of swing arm
ϕ_S	rad	suspension roll angle

Subscripts

L	left
R	right
L + R	sum of left and right parts
L − R	left part minus right part
L × R	product of left and right parts
0	initial position, constant
1	linear, or spring part of force
2	quadratic, or link part of force

Chapter 9 — Compliance Steer

$\boldsymbol{C}$	various	compliance matrix
$C_{X,FX}$	m/N	compliance coefficient for X displacement due to F_X
$C_{\gamma,FY}$	rad/N	compliance camber coefficient for F_Y
$C_{\gamma,MZ}$	rad/Nm	compliance camber coefficient for M_Z
$C_{\delta,FY}$	rad/N	compliance steer coefficient for F_Y
$C_{\delta,MZ}$	rad/Nm	compliance steer coefficient for M_Z
$C_{\delta,FX,WCH}$	rad/N	compliance steer coefficient for F_X at wheel centre height
$\boldsymbol{F}$	N, Nm	force–moment vector
F_X	N	longitudinal force
F_Y	N	lateral force
F_Z	N	normal force
$k_{\delta,C}$	rad/m.s^{-2}	compliance steer gradient
$k_{\delta,FY}$	rad/m.s^{-2}	compliance steer gradient due to F_Y
$k_{\delta,MZ}$	rad/m.s^{-2}	compliance steer gradient due to M_Z
K_S	N/m	suspension vertical stiffness ('wheel rate')
K_{ZT}	N/m	tyre vertical stiffness
m_{Bf}	kg	front body mass (front sprung mass)
m_{Br}	kg	rear body mass (rear sprung mass)
M_D	Nm	driveshaft moment (torque)
M_X	Nm	overturning moment
M_Y	Nm	pitch moment
M_Z	Nm	yaw moment (aligning moment)
R_L	m	wheel loaded radius
t	m	tyre pneumatic trail
$\boldsymbol{X}$	m, rad	compliance linear and angular displacement vector
X, Y, Z	m	axis coordinates
X_C, Y_C, Z_C	m	compliance linear deflections

Greek

γ	rad	road wheel inclination angle, camber angle
γ_C	rad	road wheel compliance inclination angle
γ_G	rad	road wheel geometric inclination angle
δ	rad	road wheel steer angle
δ_C	rad	road wheel compliance steer angle
δ_G	rad	road wheel geometric steer angle
θ	rad	wheel hub pitch angle
θ_C	rad	wheel hub compliance pitch angle
θ_G	rad	wheel hub geometric pitch angle

Subscripts

C	compliance
D	driveshaft
G	geometric
X, Y, Z	axis directions

Chapter 10 — Pitch Geometry

a	m	front axle to centre of mass
A_B	m/s^2	braking deceleration ($-A_x$)
A_x	m/s^2	longitudinal acceleration
b	m	rear axle to centre of mass
F	N	force
F_S	N	effective spring force
F_{TX}	N	longitudinal load transfer
F_V	N	axle vertical force
H	m	height of vehicle centre of mass
H_S	m	height of sprung centre of mass
J_{AD}	—	anti-dive coefficient
J_{AL}	—	anti-lift coefficient
J_{AR}	—	anti-rise coefficient
J_{AS}	—	anti-squat coefficient
k_{SP}	Nm/rad	pitch stiffness on suspension
k_{TP}	Nm/rad	pitch stiffness on tyre vertical stiffness
K_S	N/m	suspension vertical stiffness (one wheel)
K_t	N/m	tyre vertical stiffness (one wheel)
L_{AD}	m	anti-dive pitch arm radius
L_{AE}	m	anti-dive pitch arm length
L_W	m	wheelbase
m	kg	vehicle mass
m_S	kg	sprung mass
M	Nm	moment
M_{TXS}	Nm	suspension pitch moment
M_{TXT}	Nm	total longitudinal pitch moment
p	—	fraction of braking force at the front axle
t	—	fraction of tractive force at the front axle
W	N	weight force (mg)
z_{Sf}	m	front suspension bump

Greek

ε_{BCas}	rad/m	bump caster coefficient
θ_B	rad	pitch angle of body
θ_{BCas}	rad	bump caster angle
θ_{cf}	rad	angle of front line to wheel centre
θ_{cr}	rad	angle of rear line to centre
θ_C	rad	caster angle
θ_{C0}	rad	static caster angle
θ_{gf}	rad	angle of front line to ground
θ_{gr}	rad	angle of rear line to ground

θ_S	rad	suspension pitch angle
θ_U	rad	pitch angle of unsprung mass

Subscripts

BCas	bump caster
c	to wheel centre
C	caster
f	front
g	to ground
i	ideal
r	rear
S	suspension

Chapter 11 — Single-Arm Suspensions

C_α	N/rad	tyre cornering coefficient $dF_Y/d\alpha$
C_γ	N/rad	tyre camber coefficient $dF_Y/d\gamma$
F_Y	N	tyre lateral force
H_{Ax}	m	height of pivot axis
H_P	m	height of pitch centre
H_S	m	height of swing centre
l	—	*X* direction cosine
L	m	a length
L_A	m	arm length pivot axis to wheel centre
L_C	m	pivot axis to contact patch/point
L_P	m	pitch arm length
L_S	m	swing arm length
m	—	*Y* direction cosine
n	—	*Z* direction cosine
R_A	m	plan view arm length
R_C	m	plan view length to contact point ($\approx R_A$)
R_L	m	wheel loaded radius
R_P	m	pitch arm length, plan view
R_S	m	swing arm length, plan view
z_{CP}	m	contact patch bump
Z_{CP}	m	contact patch height above datum
z_S	m	suspension bump
z_W	m	wheel bump
Z_W	m	wheel centre height above datum

Greek

γ	rad	road wheel camber angle
δ	rad	road wheel steer angle
ε_{BC1}	rad/m	linear bump camber coefficient
ε_{BC2}	rad/m^2	quadratic bump camber coefficient
ε_{BKC1}	rad/m	linear bump kingpin caster angle coefficient
ε_{BKC2}	rad/m^2	quadratic bump kingpin caster angle coefficient
ε_{BKI1}	rad/m	linear bump kingpin inclination angle coefficient
ε_{BKI2}	rad/m^2	quadratic bump kingpin inclination angle coefficient
$\varepsilon_{BScd0,T}$	—	total bump scrub rate coefficient

$\varepsilon_{\mathrm{BScd0,X}}$	—	bump scrub rate coefficient, X component
$\varepsilon_{\mathrm{BScd0,Y}}$	—	bump scrub rate coefficient, Y component
$\varepsilon_{\mathrm{BS1}}$	rad/m	linear bump steer coefficient
$\varepsilon_{\mathrm{BS2}}$	rad/m^2	quadratic bump steer coefficient
θ_{A}	rad	arm angle about axis, wheel end up
θ_{Ax}	rad	pivot axis yaw angle ($\pi/2 - \psi_{\mathrm{Ax}}$)
θ_{C}	rad	angle of line pivot axis to contact patch/point
θ_{K}	rad	kingpin total angle from vertical
θ_{KC}	rad	kingpin inclination angle
ϕ_{Ax}	rad	pivot axis angle (out of horizontal)
ψ_{Ax}	rad	pivot axis sweep angle
ψ_{K}	rad	kingpin sweep angle

Subscripts

0	initial, zero bump, conditions

Superscripts

$\hat{\boldsymbol{u}}, \hat{\boldsymbol{v}}$	circumflex cap indicates unit vectors

Chapter 12 — Double-Arm Suspensions

f_{BJH}	—	$H_{\mathrm{BJS}}/H_{\mathrm{BJD}}$, sum/difference height factor
H_{BJD}	m	$H_{\mathrm{BJU}} - H_{\mathrm{BJL}}$, height difference
H_{BJL}	m	static height of lower outer ball joint
H_{BJM}	m	$H_{\mathrm{BJS}}/2 = (H_{\mathrm{BJU}} + H_{\mathrm{BJL}})/2$, mean height
H_{BJS}	m	$H_{\mathrm{BJU}} + H_{\mathrm{BJL}}$, height sum
H_{BJU}	m	static height of upper outer ball joint
H_{P}	m	height of pitch centre
H_{S}	m	height of swing centre
L	m	length
L_{P}	m	length of pitch arm
L_{S}	m	length of swing arm
L_{XL}	m	length of equivalent lower longitudinal arm
L_{XU}	m	length of equivalent upper longitudinal arm
L_{YL}	m	length of equivalent lower lateral arm
L_{YU}	m	length of equivalent upper lateral arm
R	m	arc radius
R_{P}	m	pitch arm radius
R_{S}	m	swing arm radius
s	m	scrub
S	m^{-1}	shortness, reciprocal of length
S_{XD}	m^{-1}	longitudinal shortness difference $S_{\mathrm{XU}} - S_{\mathrm{XL}}$
S_{XL}	m^{-1}	shortness of equivalent lower longitudinal arm
S_{XS}	m^{-1}	longitudinal shortness sum $S_{\mathrm{XU}} + S_{\mathrm{XL}}$
S_{XU}	m^{-1}	shortness of equivalent upper longitudinal arm
S_{YD}	m^{-1}	lateral shortness difference $S_{\mathrm{YU}} - S_{\mathrm{YL}}$
S_{YL}	m^{-1}	shortness of equivalent lower lateral arm
S_{YS}	m^{-1}	lateral shortness sum $S_{\mathrm{YU}} + S_{\mathrm{YL}}$
S_{YU}	m^{-1}	shortness of equivalent upper lateral arm
u	m	local coordinate

v	m	local coordinate
V	m/s	velocity
w	m	local coordinate
y	m	lateral coordinate
y_L	m	lateral displacement of lower ball joint
y_U	m	lateral displacement of upper ball joint
z	m	vertical coordinate
z_S	m	suspension bump

Greek

γ_B	rad	bump camber angle
ε_{Bcas1}	rad/m	linear bump caster coefficient
ε_{Bcas2}	rad/m^2	quadratic bump caster coefficient
ε_{BC1}	rad/m	linear bump camber coefficient
ε_{BC2}	rad/m^2	quadratic bump camber coefficient
ε_{BScd}	—	local bump scrub rate coefficient (at z_S) (m/m)
$\varepsilon_{BScd0,X}$	—	longitudinal bump scrub rate static (m/m)
$\varepsilon_{BScd1,X}$	m^{-1}	longitudinal bump scrub rate variation (m/m^2)
$\varepsilon_{BScd0,Y}$	—	lateral bump scrub rate static (m/m)
$\varepsilon_{BScd1,Y}$	m^{-1}	lateral bump scrub rate variation (m/m^2)
ε_{BSc1}	—	linear lateral bump scrub coefficient (m/m)
ε_{BSc2}	m^{-1}	quadratic lateral bump scrub coefficient (m/m^2)
θ	rad	angle
θ_{SL}	rad	angle of strut slider from vertical
θ_{XD}	rad	$\theta_{XU} - \theta_{XL}$
θ_{XL}	rad	angle of equivalent lower longitudinal arm
θ_{XS}	rad	$\theta_{XU} + \theta_{XL}$
θ_{XSL}	rad	side-view angle of strut slider from vertical
θ_{XU}	rad	angle of equivalent upper longitudinal arm
θ_{YD}	rad	$\theta_{YU} - \theta_{YL}$
θ_{YL}	rad	angle of equivalent lower lateral arm
θ_{YS}	rad	$\theta_{YU} + \theta_{YL}$
θ_{YSL}	rad	front-view angle of strut slider from vertical
θ_{YU}	rad	angle of equivalent upper lateral arm
ϕ	rad	angle
ψ	rad	angle
ω	rad/s	angular velocity
Ω	rad/s	angular velocity

Subscripts

0 Static value (zero suspension bump)

Chapter 13 — Rigid Axles

$C_{I,J}$	var	link space sensitivity matrix
E	m	link space length errors
H	m	roll centre height
H_A	m	axle vertical spacing of links
l, m, n	—	direction cosines (e.g. of a link)
L_{Ax}	m	axle length between wheel centres
L_{T13PV}	m	length in plan view between triangle points 1 and 3

R	m	a radius
T	m	axle track (tread)
x, y, z	m	coordinate system
x	m	longitudinal coordinate (forwards)
X	m	surge, and a general variable
y	m	lateral coordinate (to the left)
Y	m	sway
z	m	vertical coordinate (upwards)
Z	m	heave
z_{AS}	m	axle suspension heave (symmetrical double bump)
z_{SBL}	m	suspension bump left
z_{SBR}	m	suspension bump right

Greek

δ	rad	steer angle (axle yaw angle)
$\varepsilon_{\mathrm{AHP}}$	rad/m	axle heave pitch coefficient
$\varepsilon_{\mathrm{AHS}}$	rad/m	axle heave steer coefficient
$\varepsilon_{\mathrm{AHSc,X}}$	—	axle heave scrub coefficient, longitudinal
$\varepsilon_{\mathrm{AHSc,Y}}$	—	axle heave scrub coefficient, lateral
$\varepsilon_{\mathrm{AHX}}$	—	axle heave surge coefficient
$\varepsilon_{\mathrm{AHY}}$	—	axle heave sway coefficient
$\varepsilon_{\mathrm{ARP}}$	—	axle roll pitch coefficient
$\varepsilon_{\mathrm{ARS}}$	—	axle roll steer coefficient
$\varepsilon_{\mathrm{ARSc,X}}$	m/rad	axle roll scrub coefficient, longitudinal
$\varepsilon_{\mathrm{ARSc,Y}}$	m/rad	axle roll scrub coefficient, lateral
$\varepsilon_{\mathrm{ARX}}$	m/rad	axle roll surge coefficient
$\varepsilon_{\mathrm{ARY}}$	m/rad	axle roll sway coefficient
θ	rad	pitch/caster angle, angle generally
ϕ	rad	roll angle
ϕ_{A}	rad	axle roll angle (on tyres)
ϕ_{B}	rad	body roll angle
ϕ_{S}	rad	axle suspension roll

Chapter 14 — Installation Ratios

C_{D}	Ns/m	damper coefficient
C_{SD}	Ns/m	effective damper coefficient seen at suspension
e	m	offset
F	N	force
F_{D}	N	force at damper
F_{K}	N	compression force at spring
F_{m}	N	force at point mass
f_{R}	-	rising rate factor
F_{SK}	N	spring force active at suspension
K_{K}	N/m	spring stiffness
K_{SK}	N/m	effective spring stiffness seen at suspension
K_{W}	N/m	total spring/bush wheel rate (mainly K_{SK})
l	m	rocker arm length
l	—	X-direction cosine
l_{WP}	m	length from wheel centre to pivot axis (in plan)
L	m	length

L_K	m	spring length
m	—	Y-direction cosine
n	—	damper characteristic exponent
n	—	Z-direction cosine
R	—	motion ratio
R'	m^{-1}	motion ratio variation dR/dz_S
$R_{A/B}$	—	motion ratio of item A relative to item B
R_{APH}	—	motion ratio for pitch in heave
R_D	—	damper motion ratio dx_D/dz_S
R_{DC}	—	damper coefficient ratio
$R_{D\phi}$	m/rad	damper velocity ratio in roll
R_K	—	spring motion ratio dx_K/dz_S
R_m	—	motion ratio of a point mass
R_R	—	rocker motion ratio
R_{RL}	—	rocker arm length motion ratio
$R_{R\psi}$	—	rod angle motion ratio
$R_{R\psi 0}$	—	rod angle motion ratio at $\theta = 0$
t	s	time
T	m	axle track
u	m/s	X -velocity component
v	m/s	Y-velocity component
V	m/s	velocity
V_A	m/s	velocity of point A
(V_X, V_Y, V_Z)	m/s	velocity coordinates
V_D	m/s	damper compression velocity
V_K	m/s	spring compression velocity
V_P	m/s	pushrod velocity
V_S	m/s	suspension bump velocity
w	m/s	Z-velocity component
x	m	displacement
x_D	m	damper compression from static
x_K	m	spring compression from static
x_P	m	pushrod displacement from static
(X, Y, Z)	m	coordinates
z_B	m	body displacement
z_D	m	damper compression
z_S	m	suspension displacement in bump
z_W	m	wheel displacement

Greek

α_D	rad	damper out-of-plane angle
α_K	rad	spring out-of-plane angle
γ	rad	wheel camber angle
ε_{APH}	rad/m	axle pitch/heave coefficient
ε_{BC}	rad/m	bump camber coefficient
ε_{BScd}	-	local bump scrub rate coefficient
ε_{DBScd0}	-	linear bump scrub rate coefficient (m/m)
ε_{DBScd1}	m^{-1}	quadratic bump scrub rate coefficient (m/m^2)
θ_R	rad	rocker position angle
θ_{RD}	rad	rocker deviation angle
ϕ_R	rad	rocker included angle

ψ	rad	rocker rod offset angle from tangent
ψ_Z	rad	ψ at static ride height (zero bump)
ω	rad/s	angular speed
ω_R	rad/s	rocker angular velocity

Subscripts

D	damper
K	spring
m	mass (point mass)
P	pushrod
S	suspension
W	wheel
z	static (zero) ride height
1	input
2	output

Chapter 15 — Computational Geometry in Three Dimensions

In some cases the units may vary. This depends on the type of problem being solved (e.g. a position problem or a velocity diagram problem).

Variables are in italics, vectors are in bold italics, and magnitudes of vectors are in non-bold italics.

$a_1,\ldots,a_5$	various	various constant values
$\boldsymbol{A}$	m	position or displacement vector
A	m	magnitude of $\boldsymbol{A}$
a, b, c	m	side lengths of triangle
A, B, C	rad	angles in a triangle
b_1, b_2	various	constant values
$\boldsymbol{B}$	m	position or displacement vector
B	m	magnitude of $\boldsymbol{B}$
$c_1,\ldots,c_5$	various	various constant values
d	various	determinant value, divisor
$\boldsymbol{D}$	m	position or displacement vector
D	m	magnitude of $\boldsymbol{D}$
f	—	position factor (e.g. on a line)
f_1, f_2, f_3	—	position factors in a triangle
$\hat{\boldsymbol{i}}, \hat{\boldsymbol{j}}, \hat{\boldsymbol{k}}$	—	unit vectors along (x, y, z)
L	m	length
l, m, n	—	direction cosines
L, M, N	—	direction numbers
$\boldsymbol{N}$	m	vector normal to a plane
N	m	magnitude of $\boldsymbol{N}$
p	m	constant in plane equation
$\boldsymbol{P}$	various	vector cross product
P	various	vector dot product, or cross product magnitude
R	m	radius of circle or sphere
s	m	displacement along a line
u, v, w	m	local coordinate position
u, v, w	m/s	velocity components
$\hat{\boldsymbol{u}}, \hat{\boldsymbol{v}}, \hat{\boldsymbol{w}}$	—	unit vectors in local coordinates
$\boldsymbol{V}$	m/s	velocity vector

V	m/s	velocity magnitude
V_N	m/s	velocity component normal to a plane
V_T	m/s	velocity component tangential
x, y, z	m	coordinate position
z_{HP}	m	height of horizontal plane

Greek

α, β, γ	rad	direction angles
θ	rad	angle (e.g. between two vectors)
Ω	rad/s	angular velocity

Chapter 16 — Programming Considerations

f	—	interpolation factor
l, m, n	—	direction cosines or numbers
p	m	a constant
u, v, w	m	coordinates
x, y, z	m	coordinates

Chapter 17 — Iteration

a	—	$1 - r$
a, b, c, d	—	Muller's method coefficients
A	—	asymptotic error constant
C	—	a constant
d_K	—	difference between successive changes
$\mathrm{d}x$	—	a small change in x
D	—	simultaneous equation divisor
e	—	an error value
e_K	—	error at Kth iteration
f	—	a function $f(x)$
f'	—	first derivative of f
f''	—	second derivative of f
g	—	$ff''/2f'^2$
m	—	gradient (e.g. numerical)
p	—	root multiplicity factor
P	—	order of convergence
r	—	sequential error ratio
R	—	remainder
w	—	iteration 'acceleration' factor
x	—	a variable
x_{ITER}	—	basic proposed new value by iteration
x_K	—	value of x after Kth iteration
y	—	a variable $y(x)$
z	—	possibly complex variable
Z_{LBJ}	m	height of lower ball joint
Z_{SB}	m	suspension bump position

Greek

Δ	—	a change in a variable

———— // ————

Appendix B

Units

This book is, of course, in SI units throughout, but variations of Imperial ('English') units are still in everyday engineering use in the USA, so the conversion factors in Table B.1 may be useful. Note that lbm denotes pound mass, lbf denotes pound force (the weight force that acts on one pound mass due to standard gravity), kg denotes kilogram mass of course, whilst kgf denotes a 'kilogram force', the weight of 1 kg mass in standard gravity. The kgf is sometimes denoted the kp, the kilopond, which is in common use in Continental Europe. This must not be confused with the kip, a US unit of 1000 lb. The force units kgf and kp are not true SI, in which forces are always in newtons. The pound mass is 0.453592370 kg (exact by definition).

In basic geometry, the only measures are those of length and angle. Even here, true world standardisation has not been achieved, with the inch (25.400 mm exactly by definition in the UK and US since the 1960s) still being used in the US. The international spelling of the length unit is *metre*, accepted in the UK and the rest of the world except for the US and Germany, where it is spelt *meter*. In international English, a *meter* is a measuring device. The number 25.4 is only moderately round, and one can look forward, once the now only 'US' inch eventually dies out, to the introduction of a convenient and useful metric inch of exactly 25 mm.

In the context of suspensions, the length unit of the decimetre is useful, denoted dm, this being exactly 0.1 m by definition, 100 mm and 10 cm, approximately 4 US inches. Coefficients such as linear bump steer are conveniently expressed in degrees per decimetre (°/dm), with $°/\text{dm}^2$ for quadratic bump steer. One decimetre is a realistic large displacement for a suspension, so in these units the relative magnitude of the coefficients is significant.

In angular measure, the fundamental unit is the radian (about 57.29578 degrees, exactly $180/\pi$ degrees). This angular unit is a great mystery to non-mathematicians, but makes a great deal of sense once one appreciates its role in geometry and computation, or, indeed, in the relationship between engine torque and energy per revolution, or power. Physically, it is the angle for which the arc length equals the radius. For human consumption, the degree is generally preferred, or the complete revolution. In France, as an effect of the political revolution of *circa* 1795, the gradian is often used, this being by definition one hundredth of a right angle, being exactly 0.9 degrees. The sexagesimal base (the very ancient base 60) system of angular measure, with 60 angular seconds in one angular minute and 60 angular minutes in one degree is still used by astronomers, but in general science and engineering it is now common, and certainly easier, to use decimal degrees.

The main angular conversions needed are therefore between radians and degrees, and some computer languages provide this facility by intrinsic routines, e.g. rads(angdeg) and degs(angrad).

Suspension Geometry and Computation J. C. Dixon

Table B.1 Conversion factors

Length (SI m, metre):
1 inch = 0.025400 m (exact by definition)
1 foot = 0.304800 m (exact)
1 m = 39.37008 inch (approximate)
Angle (SI rad, radian)
1 rad = $1/2\pi$ revolutions (exact)
1 rad = $180/\pi$ degrees (exact)
1 rad = 57.29578 deg
1 deg = 17.4533 mrad
1 rev = 360.000 deg (exact)
1 rev = 2π rad (exact)
1 grad = 1/400 rev (exact)
1 deg = 60 min (angular minutes, exact)
1 min = 60 sec (angular seconds, exact)
Mass (SI kg):
1 lbm = 0.453592370 kg (exact, definition)
1 kg = 2.204622622 lbm (approximate)

Basic usually has RAD and DEG. Of course, these are only multiplications by the appropriate constant of $180/\pi$ or $\pi/180$.

The correct use of capital letters in units is essential. Consider, for example, the possible errors in incorrect use of kg, Kg, KG and kG, which would correctly mean kilogram, kelvin gram, kelvin gauss and kilogauss, although the gauss is not SI.

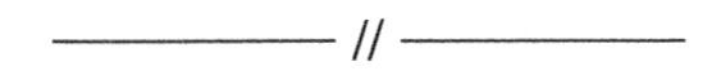

Appendix C

Greek Alphabet

Lower case

Name	Letter	Main applications
alpha	α	slip angle
beta	β	attitude angle
gamma	γ	camber angle
delta	δ	steer angles
epsilon	ε	geometric coefficients
zeta	ζ	—
eta	η	compliance coefficients
theta	θ	pitch angle, angles
iota	ι	—
kappa	κ	path curvature
lambda	λ	angles, Langensperger angle
mu	μ	—
nu	ν	path angle
xi	ξ	—
omicron	o	—
pi	π	geometric constant
rho	ρ	density, angles
sigma	σ	—
tau	τ	path turn-in
upsilon	υ	—
phi	ϕ	roll angles
chi	χ	—
psi	ψ	heading angle
omega	ω	angular velocity

(continued)

Upper case

Name	Letter	Main applications
alpha	A	—
beta	B	—
gamma	Γ	inclination angle
delta	Δ	non-small increments
epsilon	E	—
zeta	Z	—
eta	H	—
theta	Θ	angles
iota	I	—
kappa	K	—
lambda	Λ	angles
mu	M	—
nu	N	—
xi	Ξ	—
omicron	O	—
pi	Π	—
rho	P	—
sigma	Σ	mathematical summation
tau	T	—
upsilon	Y	—
phi	Φ	angles
chi	X	—
psi	Ψ	angles
omega	Ω	angular velocity

Many of the upper case letters are the same as the usual Latin alphabet of English, so do not arise in mathematical symbols.

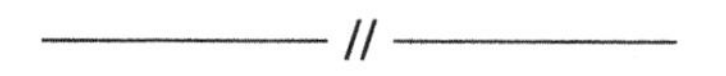

Appendix D

Quaternions for Engineers

D.1 Introduction

A quaternion is a group of four numbers (hence the name *quat-*), for example (a, b, c, d). Early work on the properties of such four-number groups was done by Euler, but the real invention of quaternions was by the prominent Irish mathematician Sir William R. Hamilton in 1843, who established the rotational properties and the three imaginary numbers involved.

In particular, a quaternion is a scalar number a with a three-dimensional vector (b, c, d). There are simple rules for addition, multiplication and so on. Quaternions can be used for efficient calculations of the position of a rotated solid body in space, for example aircraft and space vehicles. Multiplication of quaternions corresponds to successive rotations about possibly different axes, so they are intimately related to the nature of three-dimensional space. This involves the use of three distinct square roots of -1, so quaternions are an important generalisation of the usual two-dimensional complex numbers.

Quaternion addition and subtraction are simply performed term by term. Multiplication is more complicated, being performed across all combinations of terms, with rules for the various vector products. Division is by multiplication of a reciprocal.

The theory of quaternions was described by Hamilton as the theory of lines in space, and as the theory of vector quotients. A quaternion converts one vector (line) into another by multiplication. Division and reciprocation of quaternions is also required.

D.2 History

There is a saying that 'Anything is more fun if you shouldn't be doing it'. We are fascinated by the taboo. The Catholic church used to have a long list of forbidden books, a list that became unofficial in the 1950s. Many religious people would still never consider owning a book for fear of reprisals. There are religious people who would destroy all books totally, perhaps short of the Bible in Latin. Haters of mathematics have a special fear and loathing of complex numbers, and of quaternions in particular. These mathematical devices are used, *inter alia*, to calculate rotations, which is the source of the fear.

Six hundred years ago, the official religious dogma was that the Earth was fixed, and was the centre of the universe, and the world was largely a theocracy. Science showed that this was false – the Earth is in orbit around the Sun (giving the year) and revolves on its own axis (the day), despite fierce opposition from the church (persecution of Galileo: 'It moves nonetheless'). The Earth has combined rotations. The Sun is a non-special star in a non-special galaxy. This severely undermined claims that human beings are central to the universe. Hence, the hatred of 'rotations'. Also, the theocracy was threatened by Science, so there was fear of 'revolutions', which would deny religious power, particularly a fear of combined revolutions.

The hatred of Science and books in some religions contrasts interestingly with the respect accorded to these in others, such as Islam, Hinduism, Buddhism and, of course, by non-believers.

The application of quaternions to solid body rotation is alone enough to justify its inclusion in undergraduate mechanics texts. Its absence therefrom, and the general dearth of material, is very interesting itself. The suppression of quaternions has made them a *cause célebre* with the inevitable attraction of the 'forbidden'. Quaternions are also applied in subatomic physics. The correspondence between the quaternion and four-dimensional space-time is itself fascinating.

In view of the above historical vignette, and bearing in mind the fact that the Spanish mathematician Valmes was burned to death at the stake by the Inquisition for solving a quartic equation, let us, with a frisson of fear and excitement, consider these dangerous quaternions, these 'agents of revolution'. In comparison with the apparent dramatic political importance attending them, the simple mathematical reality is mundane indeed. It is hard to imagine anything more innocuous.

D.3 Position in Space

The general position of a body in three-dimensional space requires specification of one particular point of the body, usually the centre of mass, in fixed coordinates (X, Y, Z), and the body's angular position. The rotational position is specified by rotations in a coordinate system centred at the body reference point. Often the Euler angles are used, which are three successive rotations about, in turn, the vertical axis (giving a heading angle), the body lateral axis (giving a pitch angle), and finally the body longitudinal axis (giving a roll angle). However, Euler's theorem of rotation states that any angular position of a body, however actually reached, may be reached by a single rotation about some axis, generally inclined. This means that the rotation can be given an axis-and-angle representation, with the axis defined by direction cosines. This is a representation in terms of four numbers, these being one scalar, the angle θ, and a three-dimensional axis vector: (θ, l, m, n). Quaternions are not far from this, but modified for convenient calculation.

For an object with an initial position represented by a quaternion $\boldsymbol{P}_1$, when it is rotated to a new position that new position is given by $\boldsymbol{P}_2 = \boldsymbol{Q}^{\mathrm{C}}\boldsymbol{P}_1\boldsymbol{Q}$, that is, premultiplication by the conjugate $\boldsymbol{Q}^{\mathrm{C}}$ and post-multiplication by the quaternion $\boldsymbol{Q}$.

D.4 Quaternion Representations

A quaternion is a four-part number, being one part scalar and one part three-dimensional vector. The vector part is also known as the imaginary part. According to the job in hand, the best representation may vary. As a four part number it is simply

$$\boldsymbol{Q} = (a, b, c, d)$$

If the vector is, as usual, directed in normal space, then the (x, y, z) axes have corresponding unit vectors $(\hat{\boldsymbol{i}}, \hat{\boldsymbol{j}}, \hat{\boldsymbol{k}})$ and the quaternion is conveniently written as

$$\boldsymbol{Q} = q_0 + q_{\mathrm{x}}\hat{\boldsymbol{i}} + q_{\mathrm{y}}\hat{\boldsymbol{j}} + q_{\mathrm{z}}\hat{\boldsymbol{k}}$$

The Euclidean length is

$$L_{\mathrm{Q}} = \sqrt{q_0^2 + q_{\mathrm{x}}^2 + q_{\mathrm{y}}^2 + q_{\mathrm{z}}^2}$$

A unit quaternion, having a total Euclidean length of 1, may be written in terms of its unit direction vector as

$$\boldsymbol{Q} = \cos\phi + \hat{\boldsymbol{u}}\sin\phi$$

More generally, for a quaternion of length L_{Q},

$$\boldsymbol{Q} = L_{\mathrm{Q}}(\cos\phi + \hat{\boldsymbol{u}}\sin\phi)$$

D.5 Conjugates

The conjugate of a quaternion (like the usual diernion complex number) requires the negation of the vector (imaginary) part, so for

$$\boldsymbol{Q} = (a, b, c, d)$$

the conjugate is simply

$$\boldsymbol{Q}^{\mathrm{C}} = (a, -b, -c, -d)$$

Some other useful conjugate relationships are

$$(\boldsymbol{A}^{\mathrm{C}})^{\mathrm{C}} = \boldsymbol{A}$$
$$(\boldsymbol{A} + \boldsymbol{B})^{\mathrm{C}} = \boldsymbol{A}^{\mathrm{C}} + \boldsymbol{B}^{\mathrm{C}}$$
$$(\boldsymbol{AB})^{\mathrm{C}} = \boldsymbol{B}^{\mathrm{C}}\boldsymbol{A}^{\mathrm{C}}$$

with a change of order occurring in the last equation. This last relationship is used in combined rotations.

D.6 Addition and Subtraction

Addition and subtraction of quaternions is simply performed by separate addition of the four component parts.

D.7 Multiplication

Quaternion multiplication is associative, which means that

$$\boldsymbol{Q}_1\boldsymbol{Q}_2\boldsymbol{Q}_3 = (\boldsymbol{Q}_1\boldsymbol{Q}_2)\boldsymbol{Q}_3 = \boldsymbol{Q}_1(\boldsymbol{Q}_2\boldsymbol{Q}_3)$$

so the rotated position quaternion $\boldsymbol{P}_2 = \boldsymbol{Q}^{\mathrm{C}}\boldsymbol{P}_1\boldsymbol{Q}$ can be calculated as $(\boldsymbol{Q}^{\mathrm{C}}\boldsymbol{P}_1)\boldsymbol{Q}$ or as $\boldsymbol{Q}^{\mathrm{C}}(\boldsymbol{P}_1\boldsymbol{Q})$ with correct results. Quaternion multiplication is also distributive, so

$$\boldsymbol{Q}_1(\boldsymbol{Q}_2 + \boldsymbol{Q}_3) = \boldsymbol{Q}_1\boldsymbol{Q}_2 + \boldsymbol{Q}_1\boldsymbol{Q}_3$$

The actual quaternion multiplication is performed term by term. In general,

$$\boldsymbol{P} = p + p_{\mathrm{x}}\hat{\boldsymbol{i}} + p_{\mathrm{y}}\hat{\boldsymbol{j}} + p_{\mathrm{z}}\hat{\boldsymbol{k}} = p + \boldsymbol{P}_{\mathrm{v}}$$

$$\boldsymbol{Q} = q + q_{\mathrm{x}}\hat{\boldsymbol{i}} + q_{\mathrm{y}}\hat{\boldsymbol{j}} + q_{\mathrm{z}}\hat{\boldsymbol{k}} = q + \boldsymbol{Q}_{\mathrm{v}}$$

By reducing the products, such as $\hat{\boldsymbol{i}}\hat{\boldsymbol{j}}$ to $\hat{\boldsymbol{k}}$, and $\hat{\boldsymbol{i}}^2$ to -1 (detailed later), and collecting terms, then $\boldsymbol{R} = \boldsymbol{PQ}$ may be expressed in its components as

$$\boldsymbol{R} = r + r_{\mathrm{x}}\hat{\boldsymbol{i}} + r_{\mathrm{y}}\hat{\boldsymbol{j}} + r_{\mathrm{z}}\hat{\boldsymbol{k}}$$

with

$$r = pq - (p_{\mathrm{x}}q_{\mathrm{x}} + p_{\mathrm{y}}q_{\mathrm{y}} + p_{\mathrm{z}}q_{\mathrm{z}})$$
$$r_{\mathrm{x}} = pq_{\mathrm{x}} + qp_{\mathrm{x}} + p_{\mathrm{y}}q_{\mathrm{z}} - p_{\mathrm{z}}q_{\mathrm{y}}$$
$$r_{\mathrm{y}} = pq_{\mathrm{y}} + qp_{\mathrm{y}} - p_{\mathrm{x}}q_{\mathrm{z}} + p_{\mathrm{z}}q_{\mathrm{x}}$$
$$r_{\mathrm{z}} = pq_{\mathrm{z}} + qp_{\mathrm{z}} + p_{\mathrm{x}}q_{\mathrm{y}} - p_{\mathrm{y}}q_{\mathrm{x}}$$

This may be expressed conveniently in terms of the scalar and vector parts as

$$\boldsymbol{PQ} = (pq - \boldsymbol{P}_\text{v} \cdot \boldsymbol{Q}_\text{v}) + p\boldsymbol{Q}_\text{v} + q\boldsymbol{P}_\text{v} + \boldsymbol{P}_\text{v} \times \boldsymbol{Q}_\text{v}$$

where the vector parts $\boldsymbol{P}_\text{v}$ and $\boldsymbol{Q}_\text{v}$ behave as conventional vectors, with a dot product on the left and a cross product on the right. This type of quaternion product is known as the Grassman product. There are other types of product, not relevant to rotations. The quaternion product is, then, effectively the cross product minus the dot product.

The final vector cross product is the standard determinant

$$\boldsymbol{P} \times \boldsymbol{Q} = \begin{vmatrix} \hat{\boldsymbol{i}} & \hat{\boldsymbol{j}} & \hat{\boldsymbol{k}} \\ p_\text{x} & p_\text{y} & p_\text{z} \\ q_\text{x} & q_\text{y} & q_\text{z} \end{vmatrix} = \hat{\boldsymbol{i}}(p_\text{y}q_\text{z} - p_\text{z}q_\text{y}) - \hat{\boldsymbol{j}}(p_\text{x}q_\text{z} - p_\text{z}q_\text{x}) + \hat{\boldsymbol{k}}(p_\text{x}q_\text{y} - p_\text{y}q_\text{x})$$

The point position quaternion $\boldsymbol{P}_2$ resulting from the calculation should be a true position quaternion, with zero as its scalar part. This can be demonstrated to be algebraically correct, but may be numerically imperfect.

To obtain the quaternion multiplication result requires consideration of multiplication of the component parts with their direction unit vectors, which has a somewhat mysterious quality (see Table D.1), as with planar complex numbers.

Table D.1 Multiplication tables

For *ab* use *a* in the left column and *b* in the top row. Pre- or postmultiplication by 1 always gives the same vector.

Unit vectors, dot product

	$\hat{\boldsymbol{i}}$	$\hat{\boldsymbol{j}}$	$\hat{\boldsymbol{k}}$
$\hat{\boldsymbol{i}}$	1	0	0
$\hat{\boldsymbol{j}}$	0	1	0
$\hat{\boldsymbol{k}}$	0	0	1

Unit vectors, cross product

	$\hat{\boldsymbol{i}}$	$\hat{\boldsymbol{j}}$	$\hat{\boldsymbol{k}}$
$\hat{\boldsymbol{i}}$	0	$\hat{\boldsymbol{k}}$	$-\hat{\boldsymbol{j}}$
$\hat{\boldsymbol{j}}$	$-\hat{\boldsymbol{k}}$	0	$\hat{\boldsymbol{i}}$
$\hat{\boldsymbol{k}}$	$\hat{\boldsymbol{j}}$	$-\hat{\boldsymbol{i}}$	0

Planar complex numbers

	i
i	-1

Quaternions

	$\hat{\boldsymbol{i}}$	$\hat{\boldsymbol{j}}$	$\hat{\boldsymbol{k}}$
$\hat{\boldsymbol{i}}$	-1	$\hat{\boldsymbol{k}}$	$-\hat{\boldsymbol{j}}$
$\hat{\boldsymbol{j}}$	$-\hat{\boldsymbol{k}}$	-1	$\hat{\boldsymbol{i}}$
$\hat{\boldsymbol{k}}$	$\hat{\boldsymbol{j}}$	$-\hat{\boldsymbol{i}}$	-1

This is the vector cross product minus the vector dot product. For quaternions, also note that

$$\hat{\boldsymbol{i}}^2 = \hat{\boldsymbol{j}}^2 = \hat{\boldsymbol{k}}^2 = \hat{\boldsymbol{i}}\hat{\boldsymbol{j}}\hat{\boldsymbol{k}} = -1$$
$$\hat{\boldsymbol{i}}\hat{\boldsymbol{j}} = -\hat{\boldsymbol{j}}\hat{\boldsymbol{i}} = \hat{\boldsymbol{k}}$$
$$\hat{\boldsymbol{j}}\hat{\boldsymbol{k}} = -\hat{\boldsymbol{k}}\hat{\boldsymbol{j}} = \hat{\boldsymbol{i}}$$
$$\hat{\boldsymbol{k}}\hat{\boldsymbol{i}} = -\hat{\boldsymbol{i}}\hat{\boldsymbol{k}} = \hat{\boldsymbol{j}}$$

Here, the mystery is that in considering the direction of the axis of rotation, and for the positions of points, the $\hat{\boldsymbol{i}}, \hat{\boldsymbol{j}}$ and $\hat{\boldsymbol{k}}$ are taken to be unit vectors. However, in the multiplication table they behave as three distinct square roots of minus one. The vector cross product $\hat{\boldsymbol{i}} \times \hat{\boldsymbol{i}}$ is zero, but, for quaternion multiplication as for planar complex numbers, $\hat{\boldsymbol{i}} \times \hat{\boldsymbol{i}}$ is -1. This means that the negated dot product appears in the scalar part of the result quaternion. There are therefore three multiplication results for self-products of unit vectors, the dot product 1, the vector cross product zero and the quaternion product -1. This applies to any general unit vector, not just to axis-aligned unit vectors.

The quaternion multiplication property that $\hat{\boldsymbol{i}}^2 = -1, \hat{\boldsymbol{j}}^2 = -1$ and $\hat{\boldsymbol{k}}^2 = -1$ shows that a quaternion is a generalised form of a complex number in which there are three distinct square roots of -1, namely i, j and k, about independent axes. This is directly related to the fact that in three-dimensional space there are three independent rotations, that is, three ways to select a pair of the three axes for a rotation.

D.8 Reciprocals

The reciprocal or inverse of a quaternion is given by

$$\boldsymbol{Q}^{-1} = \frac{\boldsymbol{Q}^{\mathrm{C}}}{L_{\mathrm{Q}}^2}$$

where L_{Q} is the Euclidean length of $\boldsymbol{Q}$, defined earlier, with

$$\boldsymbol{Q}\boldsymbol{Q}^{-1} = \boldsymbol{Q}^{-1}\boldsymbol{Q} = (1, 0, 0, 0)$$

This is also the accepted method for obtaining the reciprocal of a complex number. In this special case of the reciprocals the multiplication is commutative, $\boldsymbol{Q}\boldsymbol{Q}^{-1} = \boldsymbol{Q}^{-1}\boldsymbol{Q}$.

D.9 Division

Quaternion division is, like multiplication, non-commutative, and is performed by multiplication by a reciprocal. The expression

$$\boldsymbol{Q} = \frac{\boldsymbol{Q}_1}{\boldsymbol{Q}_2}$$

is ambiguous, and could mean

$$\boldsymbol{Q} = \boldsymbol{Q}_1 \frac{1}{\boldsymbol{Q}_2} \quad \text{or} \quad \boldsymbol{Q} = \frac{1}{\boldsymbol{Q}_2} \boldsymbol{Q}_1$$

which will not, in general, have the same value. The correct one to use depends on the problem in hand. The reciprocal may be found as shown earlier, so the quotient is either

$$\boldsymbol{Q} = \boldsymbol{Q}_1 \frac{1}{\boldsymbol{Q}_2} = \frac{\boldsymbol{Q}_1 \boldsymbol{Q}_2^{\mathrm{C}}}{L_{\mathrm{Q2}}^2}$$

or

$$\boldsymbol{Q} = \frac{1}{\boldsymbol{Q}_2} \boldsymbol{Q}_1 = \frac{\boldsymbol{Q}_2^{\mathrm{C}} \boldsymbol{Q}_1}{L_{\mathrm{Q2}}^2}$$

D.10 Powers and Roots

Using the axis unit vector notation, the square of $\boldsymbol{Q}$ is

$$\boldsymbol{QQ} = L^2(\cos^2\phi + 2\hat{\boldsymbol{u}}\sin\phi\,\cos\phi + (-1)\sin^2\phi)$$

where $\hat{\boldsymbol{u}}\hat{\boldsymbol{u}} = -1$, showing how the quaternion self-multiplication of a unit vector must be $\hat{\boldsymbol{u}}^2 = -1$ to give the $\cos 2\phi = \cos^2\phi - \sin^2\phi$:

$$\begin{aligned}\boldsymbol{Q}^2 &= L^2(\cos 2\phi - \sin 2\phi + \hat{\boldsymbol{u}}(2\sin\phi\,\cos\phi))\\ &= L^2(\cos 2\phi + \hat{\boldsymbol{u}}\sin 2\phi)\end{aligned}$$

which is squaring of the length and doubling of the argument (angle), exactly as for a diernion complex number, and correspondingly for the nth power:

$$\boldsymbol{Q}^N = L^N(\cos \mathrm{N}\phi + \hat{\boldsymbol{u}}\sin \mathrm{N}\phi)$$

Note that the direction of the axis vector $\hat{\boldsymbol{u}}$ is unchanged.

Correspondingly, the square root of the quaternion $\boldsymbol{Q}$ is

$$\sqrt{\boldsymbol{Q}} = \sqrt{L}\left(\cos\tfrac{1}{2}\phi + \hat{\boldsymbol{u}}\sin\tfrac{1}{2}\phi\right)$$

and correspondingly for the nth root:

$$\sqrt[N]{\boldsymbol{Q}} = \sqrt[N]{L}\,(\cos(\phi/N) + \hat{\boldsymbol{u}}\sin(\phi/N))$$

However, multiplication is not conveniently achieved by this means because in general $\hat{\boldsymbol{u}}_1 \neq \hat{\boldsymbol{u}}_2$ (unlike complex numbers).

D.11 Interpolation

In the simulation of solid body motions, as used in flight training simulators and computer games, it is often desired to interpolate smoothly between two angular positions of a body, If these angular positions are specified by $\boldsymbol{Q}_1$ and $\boldsymbol{Q}_2$, it is required to find the intermediate transform quaternion $\boldsymbol{Q}$ where

$$\boldsymbol{QQ}_1 = \boldsymbol{Q}_2$$

so

$$\boldsymbol{Q} = \boldsymbol{Q}_2\boldsymbol{Q}_1^{-1}$$

$$\boldsymbol{Q} = \frac{\boldsymbol{Q}_2\boldsymbol{Q}_1^{\mathrm{C}}}{L_{\mathrm{Q1}}^2}$$

It is then easy to interpolate for intermediate positions by fractioning the angle of the quaternion $\boldsymbol{Q}$.

D.12 Point Position Quaternions (Vector Quaternions)

For the purposes of basic physics and engineering, it is convenient to think in terms of two types of quaternion. After the body reference point is specified, a local set of axes centred there is considered, parallel to the main Earth-fixed axes. Any point in the body has a position in the local axes, represented by a position vector and a corresponding local point position quaternion. The other particular type of quaternion is the rotation quaternion itself. These two types are just special cases. For a point position vector in local axes (x, y, z) the corresponding point position quaternion is

$$\boldsymbol{P} = (0, x, y, z)$$

All point position quaternions have a zero first part. This also illustrates how a quaternion is a combination of a scalar part, the first number, and a three-dimensional vector part, the last three numbers, analogous to time and space. Using the usual unit vectors $\hat{\boldsymbol{i}}$, $\hat{\boldsymbol{j}}$, $\hat{\boldsymbol{k}}$, along axes x, y, and z, the point position quaternion may be expressed as

$$\boldsymbol{P} = 0 + x\hat{\boldsymbol{i}} + y\hat{\boldsymbol{j}} + z\hat{\boldsymbol{k}}$$

This use of a point position quaternion touches upon the use of homogeneous coordinates, which in three dimensions are (a, x, y, z) or (x, y, z, a), where the extra member a is used to scale the basic point position (x, y, z) to handle points at infinity without loss of direction.

D.13 Standard Rotation Quaternions

Consider now rotation by angle θ (radians) with, for convenience, $\phi = \theta/2$, positive right-handed about an axis through the origin defined by direction cosines (l, m, n). The corresponding standard rotation quaternion is

$$\boldsymbol{Q} = (\cos\phi, l\sin\phi, m\sin\phi, n\sin\phi)$$

also expressible in terms of unit vectors as

$$\boldsymbol{Q} = \cos\phi + (l\sin\phi)\hat{\boldsymbol{i}} + (m\sin\phi)\hat{\boldsymbol{j}} + (n\sin\phi)\hat{\boldsymbol{k}}$$

The vector part of the standard rotation quaternion points along the axis of rotation, and has a length of sin $\phi = \sin(\theta/2)$.

The Euclidean length of a standard rotation quaternion is

$$L_{\mathrm{Q}} = \sqrt{(\cos^2\phi + \sin^2\phi(l^2 + m^2 + n^2))} = \sqrt{(\cos^2\phi + \sin^2\phi)} = 1$$

This unit Euclidean length is a useful feature of a standard rotation quaternion, unlike a position quaternion which can be any length but must have a zero first (scalar) part. Hence, the scalar part of a standard rotation quaternion bulks up the Euclidean length to unity. This eliminates any scaling effect.

D.14 Angular Position

Given an initial position, then the angular position of a body at any other time may be represented by the appropriate rotation quaternion from that reference state. This angular position quaternion is the product of all the angular movement quaternions in the intervening time, taken in the correct sequence.

For a rotation through angle $\theta = 2\phi$, the rotation quaternion and its conjugate are

$$\begin{aligned}\boldsymbol{Q} &= (\cos\phi, l\sin\phi, m\sin\phi, n\sin\phi)\\ \boldsymbol{Q}^{\mathrm{C}} &= (\cos\phi, -l\sin\phi, -m\sin\phi, -n\sin\phi)\end{aligned}$$

which is the same as negating the angle.

When the rotation of the body about the axis occurs, any point P_1 with angular position quaternion $\boldsymbol{P}_1$ moves to position P_2, with angular position quaternion

$$\boldsymbol{P}_2 = \boldsymbol{Q}^{\mathrm{C}}\boldsymbol{P}_1\boldsymbol{Q}$$

giving the desired values for the new position. Because quaternion multiplication is not commutative, it is essential that the premultiplication be by the conjugate, not the other way round. If they are reversed, it is as if the angle were negated. This would be equivalent to rotating the coordinate system with the body stationary, rather than moving the body itself.

In the above equation is a clue to the use of only half of the physical rotation angle in each quaternion. It is useful to think that postmultiplication moves the body by half of the angle, but with some side effects. Premultiplication by $\boldsymbol{Q}$ would move it in the other direction, so the angle is negated by using the conjugate. The combination of pre- and postmultiplication eliminates the side effects, leaving only the required angular movement.

The angular position calculation requires two quaternion multiplications. Multiplication of quaternions is non-commutative, meaning that $\boldsymbol{Q}_1\boldsymbol{Q}_2$ does not in general have the same value as $\boldsymbol{Q}_2\boldsymbol{Q}_1$. This aspect of some operators is well understood nowadays, but was highly 'revolutionary' in 1843. In effect it is a mathematical consequence of the physical fact that two rotations about different axes have different final results if done in a different order.

D.15 Reversed Axis

In three-dimensional geometry, if the direction of the rotation axis is reversed and the angle is negated then the physical result of the rotation must be the same. For a quaternion, this is

$$\begin{aligned}\boldsymbol{Q}_1 &= \cos\phi + u\sin\phi \\ \boldsymbol{Q}_2 &= \cos\phi + (-u)\sin(-\phi) \\ &= \cos\phi + (-u)(-\sin\phi) \\ &= \cos\phi + \sin\phi\end{aligned}$$

so the quaternion is the same, as it should be.

D.16 Special Rotation Cases

A rotation quaternion with a rotation angle of 90° has cos 45° and sin 45°, and does not look special. At 180° rotation, it has cos 90° and sin 90°, and the scalar part becomes zero, $\boldsymbol{Q}$ being simply

$$\boldsymbol{Q} = (0, l, m, n)$$

A rotation of $-180°$, which is 180° in the other direction, is

$$\boldsymbol{Q} = (0, -l, -m, -n)$$

which has the same final result.

A zero rotation, leaving the object in its original position, has cos 0° $(=+1)$ and sin 0° $(=0)$ with rotation quaternion

$$\boldsymbol{Q} = (1, 0, 0, 0)$$

A complete 360° rotation, or $-360°$, leaving the object in its original position, has cos 180° $(=-1)$ and sine 180° $(=0)$ with rotation quaternion

$$\boldsymbol{Q} = (-1, 0, 0, 0)$$

If this is applied twice, 720° rotation, then the original position is achieved once again. The product of the two 360° rotation quaternions is $(1, 0, 0, 0)$, the other rotation quaternion that leaves the body in its original position.

The need for two 360° rotations to restore the quaternion to its original state is significant. It has been held to imply something about the nature of space-time, and is important in the description of some subatomic particles which seem to have this same property.

D.17 Negation

If a rotation quaternion is fully negated, $\boldsymbol{Q}_2 = -\boldsymbol{Q}_1$, its effect is unchanged. Solving for the angle ϕ from the quaternion components shows that it is changed by 180°. The actual rotation angle is changed by twice that, 360°, a complete rotation, so the results are the same. This means that there are two quaternions representing any one rotation. When a negative scalar part occurs, the entire quaternion may therefore be negated, to produce a 'reduced' quaternion of the same rotational properties but with a positive scalar part.

D.18 Non-Unit Rotations

The foregoing explanation of rotation was couched in terms of standard rotation quaternions, also known as unit quaternions, with Euclidean length 1. Strictly, this is not necessary. If $\boldsymbol{P}_1$ is rotated by unit quaternion $\boldsymbol{Q}$ to $\boldsymbol{P}_2$ then

$$\boldsymbol{P}_2 = \boldsymbol{Q}^{\mathrm{C}} \boldsymbol{P}_1 \boldsymbol{Q}$$

If $\boldsymbol{Q}$ is multiplied by a scalar constant A, then $\boldsymbol{R} = A\boldsymbol{Q}$ does not have unit Euclidean length, and the above equation no longer holds – a scaling with factor A^2 occurs. However, it is still correct that

$$\boldsymbol{P}_2 = \boldsymbol{R}^{-1} \boldsymbol{P}_1 \boldsymbol{R} = (A\boldsymbol{Q})^{-1} \boldsymbol{P}_1 (A\boldsymbol{Q})$$

using premultiplication by the reciprocal instead of the conjugate. The factor A then increases the intermediate product but this scaling is corrected by the second multiplication. The precalculation of $\boldsymbol{R}^{-1}$ requires

$$\boldsymbol{R}^{-1} = \frac{\boldsymbol{R}^{\mathrm{C}}}{L_{\mathrm{R}}^2}$$

In the special case of unit quaternions, with $L = 1$, $\boldsymbol{R}^{-1} = \boldsymbol{R}^{\mathrm{C}}$, the reciprocal equals the conjugate, making for a more efficient calculation. For this reason, rotations are, in practice, represented by normalised (unit) quaternions.

D.19 Sequential Rotations

Two rotations in succession, $\boldsymbol{Q}_1$ and $\boldsymbol{Q}_2$ in that order, have a total result $\boldsymbol{Q}$. Point position P_1 goes via $\boldsymbol{Q}_1$ into P_2 and then via $\boldsymbol{Q}_2$ into point position P_3 with quaternion $\boldsymbol{P}_3$:

$$\boldsymbol{P}_3 = \boldsymbol{Q}_2^{\mathrm{C}} (\boldsymbol{Q}_1^{\mathrm{C}} \boldsymbol{P}_1 \boldsymbol{Q}_1) \boldsymbol{Q}_2$$

The multiplication is associative, so

$$\boldsymbol{P}_3 = (\boldsymbol{Q}_2^{\mathrm{C}} \boldsymbol{Q}_1^{\mathrm{C}}) \boldsymbol{P}_1 (\boldsymbol{Q}_1 \boldsymbol{Q}_2)$$

It is a property of the conjugates that

$$\boldsymbol{Q}_2^{\mathrm{C}} \boldsymbol{Q}_1^{\mathrm{C}} = (\boldsymbol{Q}_1 \boldsymbol{Q}_2)^{\mathrm{C}}$$

with the order changing, so

$$P_3 = (Q_1 Q_2)^C P_1 (Q_1 Q_2)$$

and the total effect is given simply by

$$Q = Q_1 Q_2$$

taken in the order in which they occur. This applies to any succession of rotations:

$$Q = Q_1 Q_2 \cdots Q_N$$

A long sequence of such calculations can introduce cumulative numerical errors. With a matrix representation of rotation this can be difficult to correct, because the rotation matrix has a complex relationship between its parts. A rotation quaternion is easily renormalised because the one specification to be met by a standard rotation quaternion is that it have a Euclidean length of exactly 1, which is a helpful feature in numerical work.

D.20 Relationship with Complex Numbers

In two dimensions, in the complex plane, rotation through an angle θ may be performed by multiplying by the complex number $z = (\cos\theta + i\sin\theta)$. In two dimensions, there is only one possible rotation, which is an interchange between x and y without change of length. In three dimensions, three possible independent rotations are possible. Considering axis-aligned rotations, these are exchanging x with y, y with z or z with x, so a quaternion has three independent rotations, and so three roots of -1 for three-dimensional rotations instead of the diernion complex number single root i for two-dimensional rotations.

In two-dimensional geometry, multiplying by $i = \sqrt{-1}$ is a 90° turn to the left, about the virtual z axis. Two such turns, which is a multiplication by $i^2 = -1$, gives a 180° turn, which is a full reversal ($+1$ into -1). Another square root of -1 is $-i$. This is a 90° turn in the opposite direction, to the right. Two such turns equals $-180°$, which is the same position as $+180°$.

In complex numbers, rotation (e.g. multiplication) is supposed to be about a virtual axis perpendicular to the plane $(1, i)$, but uses equations for rotation about an i axis. However, in three dimensions, rotations about the i axis are in the (j, k) plane. It happens that $i \times j = k$ and $i \times k = -j$. In two dimensions, we use $i \times 1 = i$ and $i \times i = -1$, which is two-dimensional complex numbers are really in the (j, k) plane, not the $(1, i)$ plane, but the renaming of j as 1 and k as i happens to work.

D.21 Infinitesimal Rotations

For an infinitesimal rotation, the angles θ and ϕ themselves are infinitesimals and all higher orders such as ϕ^2 can be neglected. Then $\sin\phi \approx \phi$ and $\cos\phi \approx 1$, so

$$Q = (1, l\phi, m\phi, n\phi)$$

Table D.2 Comparison of i in two and three dimensions

2D	3D
(j, k) plane	$(1, i)$ plane
$i \times j = k$	$i \times 1 = i$
$i \times k = -j$	$i \times i = -1$
$i^2 = \frac{-j}{j} = -1$	$i^2 = \frac{-1}{1} = -1$

Considering two such infinitesimal rotations, Q_1 and Q_2, the total effect is $Q = Q_1 Q_2$ which, by multiplying out and eliminating higher-order infinitesimals, gives

$$Q = (1, l(\phi_1 + \phi_2), m(\phi_1 + \phi_2), n(\phi_1 + \phi_2))$$

which is why angular velocities, unlike angular positions, behave as simple vectors.

D.22 Vector Quaternions

For pure vector quaternions (zero scalar part)

$$Q_{V1} Q_{V2} = -Q_{V1} \cdot Q_{V2} + Q_{V1} \times Q_{V2}$$

$$Q_{V2} Q_{V1} = -Q_{V1} \cdot Q_{V2} - Q_{V1} \times Q_{V2}$$

where the first term is a scalar dot product, and the second is a vector.

Two vectors are perpendicular (orthogonal) if their dot product is zero. For pure vector quaternions (zero scalar part) this can be expressed as

$$Q_{V1} Q_{V2} + Q_{V2} Q_{V1} = 0$$

If the cross product is zero then two vectors are parallel. For vector quaternions this is

$$Q_{V1} Q_{V2} - Q_{V2} Q_{V1} = 0$$

Three vectors are coplanar in general if the cross product of any two is perpendicular to the third (zero dot product). This is expressed by the vector triple product, for example

$$(V_1 \times V_2) \cdot V_3 = 0 \text{ (coplanar)}$$

Co-planarity of the vector part of complete quaternions is correspondingly specified as requiring

$$Q_3(Q_1 Q_2 - Q_2 Q_1) + (Q_1 Q_2 - Q_2 Q_1) Q_3 = 4\,\mathrm{Re}(Q_3 \mathrm{Im}(Q_1 Q_2)) = 0$$

These conditions are useful for hand analytical work, but for numerical work, particularly in computers, one works with the component parts anyway, so the vector conditions can be applied directly.

Another useful form of expression of a quaternion is in terms of the angle ϕ and a unit vector $\hat{u}$ along the axis of rotation:

$$\hat{u} = p_x \hat{i} + p_y \hat{j} + p_z \hat{k}$$

giving

$$Q = L(\cos\phi + \hat{u}\sin\phi)$$

D.23 Finale

In mathematics, quaternions are an important part of basic complex number theory, generalising the simple two-dimensional complex plane. In engineering, the single application of quaternions is in the representation of physical body rotational position, particularly for aircraft and spacecraft. Euler angles are good for human understanding of angular position. Matrices can do the calculations. However, quaternions really *are* the theory of rotations, in a way that Euler angles and matrices are not, and they are computationally efficient whilst also avoiding the singularities of Euler angles. They have a valuable part to play in mechanical engineering.

It is not known who said: 'God created complex numbers, and his greatest work was quaternions.' It is rumoured that there is a mathematical proof of the existence of a god using quaternions, possibly due to W. K. Clifford (1845–1879), since lost or destroyed. Newton may have worked on some foundations of this, involving number quadruples, numerology and the tetragrammaton. Euler worked on quadruple numbers, but apparently with no religious connection.

References

Katz, A. (1997) Computational Rigid Vehicle Dynamics, Krieger publishing Company, Inc., Florida, ISBN 1-57524-016-5.

——————— // ———————

Appendix E

Frenet, Serret and Darboux

In the analysis of a smooth line in space, the Frenet–Serret equations are important. They are described briefly here as an introduction to path analysis, and to explain why they are not readily applied to the road vehicle.

Considering a space curve which is smooth, that is, having a suitable number of continuous derivatives, any point, P, with vector position $\boldsymbol{P}$, may be chosen for scrutiny. The path length to this point from a reference point on the curve is s. At this point the curve lies locally in a plane, the local plane of the curve.

The unit tangent vector to the curve,

$$\boldsymbol{T} = \frac{\mathrm{d}\boldsymbol{P}}{\mathrm{d}s}$$

is in the direction of the changing s, and lies in the local plane. The curvature of the path is κ (lower case Greek kappa). The particular unit normal vector $\boldsymbol{N}$ points towards the centre of curvature, and is also in the local plane. Finally, the binormal unit vector $\boldsymbol{B}$ is perpendicular to the local plane, and is the cross product

$$\boldsymbol{B} = \boldsymbol{T} \times \boldsymbol{N}$$

The three unit vectors ($\boldsymbol{T}$, $\boldsymbol{N}$, $\boldsymbol{B}$) form a local right-handed coordinate system, useful for general path analysis. The variations of these vectors with path length are given by

$$\frac{\mathrm{d}\boldsymbol{T}}{\mathrm{d}s} = \kappa \boldsymbol{N}$$

$$\frac{\mathrm{d}\boldsymbol{N}}{\mathrm{d}s} = -\kappa \boldsymbol{T} + \tau \boldsymbol{B}$$

$$\frac{\mathrm{d}\boldsymbol{B}}{\mathrm{d}s} = -\tau \boldsymbol{N}$$

where κ is the path curvature and τ (lower case Greek tau) is here the curve torsion. The torsion is effectively the rate at which the curve tries to move out of the local plane. The above are known as the Frenet–Serret equations (after the French mathematicians J.F. Frenet and J.A. Serret, both active in the mid-nineteenth century).

The plane of $\boldsymbol{T}$ and $\boldsymbol{N}$ is called the osculating plane. The plane of $\boldsymbol{N}$ and $\boldsymbol{B}$ is called the normal plane. The plane of $\boldsymbol{T}$ and $\boldsymbol{B}$ is called the rectifying plane.

Additionally, the unit normal vector $\boldsymbol{N}$ is in the direction of the spatial rate of change of the tangent vector, so

$$\boldsymbol{N} = \frac{\mathrm{d}\boldsymbol{T}/\mathrm{d}s}{\mathrm{abs}(\mathrm{d}\boldsymbol{T}/\mathrm{d}s)}$$

Also,

$$\kappa = \frac{\mathrm{d}\boldsymbol{T}/\mathrm{d}s}{\boldsymbol{N}}; \qquad \tau = -\frac{\mathrm{d}\boldsymbol{B}/\mathrm{d}s}{\boldsymbol{N}}$$

The Frenet–Serret axis system, moving with the point, has an angular velocity. Dividing this by the (signed) point speed, that is, taking the derivative of the angular position of the axis system with respect to the path position, gives the Darboux vector, $\boldsymbol{\omega}$ (lower case Greek omega) which is given in value by

$$\boldsymbol{\omega} = \tau\boldsymbol{T} + \kappa\boldsymbol{B}$$

giving

$$\boldsymbol{T}' = \boldsymbol{\omega} \times \boldsymbol{T}$$
$$\boldsymbol{N}' = \boldsymbol{\omega} \times \boldsymbol{N}$$
$$\boldsymbol{B}' = \boldsymbol{\omega} \times \boldsymbol{B}$$

The lack of application of these pleasing equations to vehicles on roads is a result of the fact that the road is not a single path, but has a lateral slope angle of great importance, that is, the road is a ribbon, and the preferred coordinate system for vehicle analysis is always based on the local surface alignment of the ribbon, not on the local plane of path curvature, although these systems coincide on a flat road.

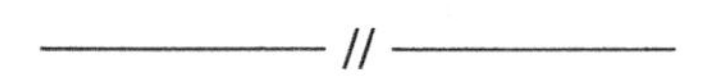

Appendix F

The Fourier Transform

F.1 Introduction

The Fourier transform is a representation of a periodic waveform by harmonic components, the method having been invented by the French engineer and mathematician, J.P. Fourier (1768–1830), for analysis of heat flux and temperature distributions. The method has subsequently found wide application throughout the sciences (Newland, 1984; IEEE, 1967).

There are various types of transform, varying in the exact job and the method. The main acronyms to know are:

DFT	discrete Fourier transform
EFT	engineering Fourier transform
FFT	fast Fourier transform
FT	Fourier transform
IDFT	inverse discrete Fourier transform
IFT	inverse Fourier transform
MFT	mathematical Fourier transform
SFT	slow Fourier transform

The discrete Fourier transform is the discrete transform of a discrete data set sampled from a continuous periodic wave. Naturally, this lends itself to digital computation.

The fast Fourier transform is an efficient method of calculation of a discrete Fourier transform. It is clearly defined mathematically, and would appear to be simple to use. In practice there are some difficulties, not least because there are several extant programs that give differing results. This needs to be explained. The first FFT algorithm was by C.F. Gauss in 1805 (published posthumously in 1866). Theoretical analysis followed by Runge and Koenig in 1924, and a good algorithm by Danielson and Lanczos in 1942.

In practical application, Fourier analysis is the representation of a cyclic (periodic) waveform by a series of sine and cosine terms of various frequencies which sum to the original wave. For a continuous original wave, the number of transformed frequencies is infinite. For a discrete (sampled) original data set, as normally used nowadays, the discrete Fourier transform gives a number of frequencies basically equal to half the number of points, that is, N data points (where N is even) give $N/2$ frequencies. Each of these has a sine and cosine component, the amplitudes of which are calculated by the transform. Actually N points give a top frequency $N/2$. The first frequency is actually zero, so the number of frequencies is a total of $N/2+1$. This is two too many coefficients to determine from the N points. Two of the coefficients are always zero. For the first and last frequencies (0 and $N/2$) the coefficients of the sine terms are indeterminate, and therefore considered to be zero, having no significance. The zero cosine term corresponds to the mean value. Because the frequencies begin at zero, it is convenient to number the frequency array from zero. In practical computation of the DFT, it is efficient to use a data length N equal to 2 raised to an integer power, $N = 2^M$. Then a data array $(1:N)$ has $N/2$ frequencies numbered 0 to $N/2$.

For any one frequency, these two amplitudes can be combined into a total amplitude and a phase angle. These amplitudes represent the spectral density. The wavelengths of the transform are given by the original total wavelength of the data divided by successive integers. The frequencies are therefore in linear progression.

The equations are well conditioned, although subtractive cancellation is a possible problem. Numerical accuracy is generally good. To perform a Fourier transform is to solve the N simulataneous equations for the coefficents. The advantage of the Fourier transform is that the waves are orthogonal, so the results can be obtained much more efficiently than for normal simultaneous equations. The fast Fourier transform is normally used, but if only a few frequencies are required then a slow transform is acceptable, and is also useful for checking purposes. The transform can be inverted to recover the data from the spectral distribution.

This sounds straightforward, but in practice there are complications. There are various programs available from several books, giving different results, which can be very confusing. Also the method of use varies.

F.2 Types of Fourier Transform

There are several variations on the basic FT theme, and these need to be understood for correct use of a DFT. The various programs can be classified in the following ways:

(1) Mathematical transform or engineering transform. The mathematical transform is the underlying process. The engineering transform does a little more work, and actually produces the sine and cosine coefficients, including the zero frequency terms (including the mean) which are handled differently, or starts with the sine and cosine coefficients and produces a data array.
(2) Dividing by N. If the bare mathematical process is applied, and then reapplied for an inverted transform, the original data set is not recovered; instead the result is the original data set multiplied by N. There is a variety of opinion on what to do about this. Some programs do nothing, some divide by N at the forward transform stage, some divide by N at the inverse transform stage, and some make the transform as symmetrical as possible by dividing by $\sqrt{N}$ at both stages. To obtain agreement between various published programs for the mathematical transform it is therefore often necessary to adjust the output up or down by N or $\sqrt{N}$. In the engineering transform, the division must definitely be performed exclusively in the forward direction to give the correct coefficients.
(3) Type of transform. There are two methods of transform, called Cooley–Tukey and Sande–Tukey, In the FFT, there are two phases of operation. One is a bit rearrangement involving a two-deep loop. The other is the basic transform itself involving a three-deep loop. In the Cooley–Tukey transform the bit shifting is done first. In the Sande–Tukey transformation, the bit shifting is done last. Otherwise, they seem to be considered equivalent, simply alternative methods that should produce the same results.
(4) Various published programs seem to give different numerical results when applied to a complex input. If applied to a purely real input then they agree on the complex output. If applied to an imaginary input, they have opposite signs for the complex output, which agrees in magnitude. If applied to a complex input, they combine the complex parts to give not just sign disagreement but numerical value disagreement. In some cases, it seems that the order of the coefficients is reversed. There is a unique correct solution to the DFT of a given input data set, so there seems to be something amiss with some published programs.
(5) Argument passing. In practical computation, the arguments are passed to subroutines in various configurations. Programs for the mathematical transform usually take a complex argument, that is, an

array of complex numbers (available as an intrinsic data type in Fortran). In languages without the intrinsic complex data type, two real arrays may be passed, one with the real parts, one with the imaginary parts, or a single pseudo-complex real array of alternating real and imaginary parts.

(6) Type of result. A basic engineering transform will typically accept a single real array argument and return two arrays, the sine and cosine coefficients (or possibly the complete amplitude and phase values). The MFT returns a complex array result.

(7) Number of input data sets. A good multiple EFT will transform simultaneously two valid independent real inputs (which correspond to the real and imaginary parts of a complex input in an MFT), so a suitable engineering transform can accept two independent real arrays and simultaneously produce the two real transforms (two independent sets of sine and cosine coefficients) in an efficient manner.

(8) In place or not. The transformation may be 'in place' or not, that is, the returned results may replace the incoming argument array or be returned as separate output arrays. It is usual for an MFT to take a complex argument and to transform it in place, whereas an EFT will usually take a data set and return the transform in separate arrays without altering the data. The 'in-place' method was once a necessary memory saver, less necessary nowadays unless the arrays are very large, and not really good practice.

(9) There is more than one type of FFT, and the speeds vary considerably. One claimed FFT was found by the author to be a factor of 10 slower than the best program tried, although still faster than a slow transform. Also, the SFT may be speeded up by various means, so there are SFT, fast SFT, slow FFT, several varieties of FFT, and even very fast VFFT.

Considering the above alternative methods, configurations, argument passing conventions and performances, the alternative Cooley–Tukey and Sande–Tukey transforms, and division by N or $\sqrt{N}$, or not, the potential for confusion is apparent.

F.3 Engineering Fourier Transform

The EFT is firmly grounded in reality. The transform is between the data points in terms of definite displacements from zero (e.g. the height of points along the road), and the amplitudes of the sine and cosine waves that correspond to, and can recreate, the positions. There is no room for discussion about the signs, or about whether or not to divide by N or $\sqrt{N}$ at various stages. The forward transform produces the circular function (sine and cosine) amplitudes from the displacements, and the reverse transform produces the displacements from the amplitudes. There is no ambiguity. To an engineer, the forward transform is going from the data to the harmonic analysis (sine and cosine coefficients), and an inverse transform is going from the frequency spectrum coefficients to a displacement data set. Therefore, as may be required in practice, when starting with a frequency spectrum and producing a displacement array, the first thing that an engineer may do is an inverse transform.

F.4 Mathematical Fourier Transform

In contrast, the MFT is less clearly defined, particularly in terms of when to do any divisions by the data length. Here there seems to be a conceptual problem. The MFT is not grounded in reality as is an EFT, so the MFT is whatever it is defined to be, and there seems to be no obvious way to check it against reality without further processing, that is, to change it into an EFT. This leaves room for varying definitions, and introduces the possibility of confusion. The mathematical definition typically given is

$$H_n \equiv \sum_{k=0}^{N-1} h_k \exp S(i2\pi kn/N)$$

where S is a sign value, $+1$ or -1. Changing the sign of S changes the direction of the transform. A process with one sign followed by a process with the opposite sign is said to recover the data, but this is not entirely true. An increase by a factor of N occurs, requiring a compensating division by N at some stage, but it is difficult to say exactly where this happens and at what point the correction should be introduced. Mathematicians simply call it a Fourier transform whichever way it is used first, and then an inverse transform is going back the other way. This contrasts sharply with the engineer's idea of forward and inverse transforms, in which the data in the two forms have different physical meanings – for example, in one case a series of point heights along a road, in the other a series of amplitudes of sines and cosine waves.

F.5 Mathematical Definitions

Press et al. (1992) give the forward transform definition as

$$H_n \equiv \sum_{k=0}^{N-1} h_k \exp\left(i2\pi kn/N\right)$$

where the data set is $h(0:N-1)$, with n, the index of the various frequencies, running from 0 to $N/2$. The inverse transform is given as

$$h_k \equiv \frac{1}{N}\sum_{n=0}^{N-1} H_n \exp\left(-i2\pi kn/N\right)$$

with a division by N clearly stated to be in the inverse transform only. The forward transform has no negative sign in the exponent, which appears in the inverse transform only.

In contrast, Newland (1984) gives the formal definition of the 'forward' transform as

$$X_k = \frac{1}{N}\sum_{r=0}^{N-1} x_r \exp\left(-i2\pi kr/N\right); \quad k = 0, 1, 2, \ldots, N-1$$

and the 'reverse' transform as

$$x_r = \sum_{k=0}^{N-1} X_k \exp\left(i2\pi kr/N\right); \quad r = 0, 1, 2, \ldots, N-1$$

exhibiting not only a difference in when to divide by N, which now appears in the 'forward' transform only, but also changing of the signs in the exponents, with the negative exponential sign now occurring in the forward transform. In effect, then, the definitions of forward and reverse transforms have been interchanged.

In Harris and Stocker (1998) the forward transform is expressed (with slightly changed notation) as

$$f_r = \frac{1}{N}\sum_{k=0}^{N-1} c_k \exp(i2\pi kr/N)$$

agreeing with Press et al. There is no explicit statement of any reverse transform.

F.6 Equations

The data set is $Y(J)$ with $J = 1:N$. The number N is preferably, for convenience and efficiency in an FFT, a whole power of 2. In any case, the equations vary slightly for an odd number of points, so an even number is assumed, even for the SFT. The independent variable X is equally spaced at ΔX. The first wavelength is $\lambda = N\Delta X$. The notional point number $N+1$ is the first point of the next wave, and has the same Y value as the first point of the data set, $Y(N+1)=Y(1)$. This can be used as a check in the regeneration of the data. The independent variable for point J is $X(J)=(J-1)\times\Delta X$ with the first point at $X(1)=0$. The last point is at $X(N)=(N-1)\times\Delta X$. The first point of the next wave is at $X(N)=N\times\Delta X$.

The lowest frequency is zero. The top frequency is $M=N/2$. The total number of frequencies is $M+1=N/2+1$. The data set (points $J=1,\ldots, N$) is produced from the trigonometric function coefficients by the slow inverse-transform equation

$$Y(J)=\frac{1}{2}C_0+\sum_{K=1}^{M-1}\left[C_K\cos\left(\frac{2\pi JK}{N}\right)+S_K\sin\left(\frac{2\pi JK}{N}\right)\right]+\frac{1}{2}C_M\cos(\pi J)$$

Here it is apparent that the first and last frequencies ($K=0$ and $K=M=N/2$) have no sine term (no S_K coefficient).

To obtain the trigonometric coefficients from the data, the slow forward-transform equations are

$$C_K=\frac{2}{N}\sum_{J=1}^{N}Y(J)\cos\left(\frac{2\pi JK}{N}\right);\quad K=0,1,\ldots,N/2$$

$$S_K=\frac{2}{N}\sum_{J=1}^{N}Y(J)\sin\left(\frac{2\pi JK}{N}\right);\quad K=1,2,\ldots,N/2-1$$

where J is the data index and K is the frequency index. For the sine coefficients, it is apparent here that for $K=0$ the sine argument would be zero, so the value of S_0 is zero. For $K=N/2$, the sine argument is a multiple of π and would give $S_N=0$.

Considering the complex coefficient

$$Z_K=C_K+iS_K$$

then the transform into trigonometric coefficients may be expressed as

$$Z_K=\frac{2}{N}\sum_{J=1}^{N}Y(J)\left\{\cos\left(\frac{2\pi JK}{N}\right)+i\sin\left(\frac{2\pi JK}{N}\right)\right\}$$

$$=\frac{2}{N}\sum_{J=1}^{N}Y(J)\exp\left(\frac{i2\pi JK}{N}\right);\quad K=0,1,\ldots,N/2$$

The fact that the product JK occurs with integer values ranging from 1 to N and from 0 to $N/2$ respectively gives an integer product range from 0 to $N^2/2$. This makes possible a more efficient calculation of the slow transform, by precalculating the necessary sine and cosine values – an FSFT. Also, the slow transform may often be applied for a limited frequency range, or for a limited number of frequencies that are of interest. The FFT of necessity does all frequencies, many of which may not be of much engineering interest. The practical speed gain of the FFT, whilst very valuable, is therefore a little exaggerated.

Comparing the equations here, for the SFT, with the earlier definitions for the MFT, it is apparent that the SFT has an additional factor 2, and divides by N in the forward direction.

F.7 Published Programs

The FORK program is an MFT by V. Herbert (1962) (pre-dating Cooley and Tukey (1965), who were, however, the first to publish in recent times), modified by Brenner, and further modified and published by

Claerbout (1975). It works correctly in all respects. It divides by $\sqrt{N}$ in both directions, so it is symmetrical, and it has a direction switch argument to provide forward and reverse transforms. It has bit reversal first (Cooley–Tukey).

The MFT program Four1 in Press et al. (1992) is stated to based on 'a program by Brenner'. It does not do a division in either direction. It produces the correct signs. There is a direction switch. Bit reversal is first (Cooley–Tukey). With adjustment for the division, it is in good agreement with FORK.

The MFT program in Newland (1984) has no switch, being intended for forward transforms only. It includes division by N. An appendix explains how to reverse it, including, *inter alia*, removal of the division. Bit reversal is first. The signs for the imaginary part of the argument appear to be wrong, so it works for the real part only. Newland has a usefully different way of handling some of the details.

The MFT program in Harris and Stocker (1998) is in pseudo code. It is easily expressed in Fortran. It is stated to be Sande–Tukey, and does indeed have the bit reversals last. It accepts two real argument arrays, but apparently produces the wrong signs for the second one, so the output appears correct only for a single real array. It is intended for forward transforms only, and there is no explanation about how to reverse it.

Barrett and Mackay (1987) provide a program in Basic which they describe as based on the 'successive doubling' method. This converts readily to Fortran, but seems relatively slow (by FFT standards) and works for a single real array only, giving incorrect signs for the second input array.

Various other published programs have been investigated with a view to use, but have generally been found wanting in one way or another. Typically they are either incorrect or very slow by good FFT standards.

F.8 Dividing by N

For the MFT, it is desirable for successive forward and reverse transforms (or *vice versa*) to recover the original data, which end up multiplied by N in the bare process, so a compensating division by N should occur somewhere around the two-way cycle. For symmetry, this can be a division by $\sqrt{N}$ in each process. However, it is more efficient to do a single division at one stage.

For the EFT, the forward transform requires a division by N to obtain the correct coefficients. If the MFT in use does a division by $\sqrt{N}$ then the EFT must do a further division, which is not ideal. For the reverse transform, no overall division is needed, so if the MFT does a division then the EFT wrapper must do a compensating multiplication.

Therefore, for efficient operation of the EFT, it is better if the MFT in use does no N or $\sqrt{N}$ division at all, or does the division by N in the forward transform only. Because the EFT has an additional factor of 2 to deal with anyway, the most efficient computation is achieved if the factoring is left entirely to the EFT stage, with no divisions in the basic MFT.

The best solution, then, is probably to have an underlying service routine that does no divisions. The MFT will use this and add its own divisions by $\sqrt{N}$ in both directions, keeping it symmetrical and providing reversibility. The EFT will also use the underlying routine, and can do so in its own way efficiently.

F.9 Sign Discrepancies

When the data are placed in the imaginary part of the complex argument of an MFT, or if two data sets are processed simultaneously, or if the transform of a complex data set is required, there is disagreement on the results from the various published programs (not just those listed above). If only one real data set is used, placed in the real part of the complex argument of the MFT, there is agreement on the result. The fact that the programs agree in this simple case has increased the confusion. When the MFT is used in the service of an EFT, as it generally is for practical applications, the sine and cosine amplitudes are deduced from the complex coefficients found by the MFT. This process involves taking the symmetric and antisymmetric aspects of the output, adding and subtracting corresponding terms in various ways. With the output from an

MFT that has faulty signs, the error can be corrected by changing (fiddling) the signs used in the additional EFT processing. Therefore the final result of the EFT may still be correct, because of compensating errors. This seems to have disguised the fact that there is inconsistency in the output of published MFT programs. However, two wrongs do not make a right, and it seems that some of the published MFT programs must be incorrect in their signs. If two programs differ in the magnitude or sign of their output, one of them must be wrong, unless there is ambiguity in the definition of the FT.

F.10 Speed

There are significant differences in the speed of various FFTs. There are some slow ones around that are a factor of 10 worse (successive doubling method) than a good standard one. There are various ways to improve the performance of even a good one, for example by using a base-4 transform, which can be 20% faster than a conventional FFT, promotion of unnecessary calculations out of inner loops, careful organisation of exit conditions or use of determinate loops. When only one real array is to be processed, it may be expected that it would run faster than a double array (or complex array) because the zero imaginary parts will process more rapidly, but this seems to be a small effect in practice. A further improvement may still be possible by using a folded single array in a complex half-length array; see the Press et al routine 'realft'.

F.11 Equation Check Program

The following program is purely for checking the FT equations for consistency.

```
!
!
! TEST DFT EQUATIONS
!
!
      Program Test_FT_Eqns
      Implicit none
      Integer J,K,KD,KF,N,ND,NF,M,IT
      Real*8 d(1024),dr(1024),de(1024),a(0:512),b(0:512),RandH
      Real*8 pi,s,sa,sb,t,er
      IT=0
      Print*
      Print*,'Numerical check of DFT equations by slow FT:'
      Print*
      pi=acos(-1d0)
      ND=1024
      Print '(A,I0)',' ND = ',ND
      Do KD=1,ND
            d(KD)=RandH()
      enddo

      If(IT>0)Print*,'Start analysis:'
      NF=ND/2
      Do KF=1,NF-1
            sa=0
            sb=0
```

```
          Do KD=1,ND
                t=(2*pi*KF*KD)/ND
                sa=sa+d(KD)*cos(t)
                sb=sb+d(KD)*sin(t)
          enddo
          a(KF)=2*sa/ND
          b(KF)=2*sb/ND
          If(IT>0.and.NF<9)Print*,KF,a(KF),b(KF)
enddo
! Extra cosine terms:
a(0)=2*sum(d)/ND
sa=0
Do KD=1,ND
          sa=sa+d(KD)*cos(KD*pi)
enddo
a(NF)=2*sa/ND

If(IT>0)Pause 'Start data recovery:'
Do KD=1,ND
          s=0
          Do KF=1,NF-1
                t=(2*pi*KF*KD)/ND
                s=s+a(KF)*cos(t)+b(KF)*sin(t)
          end do
          dr(KD)=a(0)/2 + s + a(NF)/2*cos(pi*KD)
enddo
er=sum(abs(d-dr))
Print '(A,ES9.2)',' Checksum = ',er
Print*
Pause 'RTF:'
End
```

References

Barrett, A.N. and Mackay, A.L. (1987) *Spatial Structure and the Microcomputer*. Basingstoke: Macmillan. ISBN 0333392841.

Claerbout J. F. (1975) *Fundamentals of Geophysical Data Processing*. London: McGraw-Hill. Also published by Blackwell in 1985: ISBN 0865423059 (Program FORK is on p 12).

IEEE (1967) Special Issue on Fast Fourier Transform. *Transactions on Audio and Electroacoustics*, **15**(2).

Harris J. W. and Stocker H. (1998) *Handbook of Mathematics and Computational Science*. New York: Springer-Verlag. ISBN 0387947469.

Newland D. E. (1984) *An Introduction to Random Vibration and Spectral Analysis*. Harlow: Longman. ISBN 0582305306.

Press, W.H., Teukolsky, S.A., Vetterling, W.T. and Flannery, B.P. (1992) *Numerical Recipes in Fortran: The Art of Scientific Computing*. Cambridge: Cambridge Uuniversity Press, ISBN 052143064X.

Programs

Owners of this book can download a suite of FFT programs by the author of this book from the publisher's website. These include engineering EFTs that transform road heights directly into sine and cosine coefficients and vice versa. The URL is wiley.com/go/Dixon.

——————— // ———————

References and Bibliography

References

Adams, D.A. (1967) A stopping criterion for polynomial root finding. *Communications of the ACM*, **10**(10), 655–658.

Bolaski M. P. (1967) Front Vehicle Suspension with Automatic Camber Adjustment, US Patent 3,497,233, filed 3 Nov 1967, patented 24 Feb 1970.

Cebon D. and Newland D. E. (1984) The Artificial Generation of Road Surface Topography by the Inverse FFT Method, in *8th IAVSD Conference on Dynamics of Vehicles on Roads and Tracks*, pp 29–42. Lisse: Swets and Zeitlinger.

Conte S. D. and De Boor C. (1972) *Elementary Numerical Analysis: An Algorithmic Approach*, 2nd edn. New York: McGraw-Hill. (A classic, well worth seeking out second hand.)

Dixon J. C. (1991) *Tyres, Suspension & Handling*, 1st edition, Cambridge University Press.

Dixon J. C. (1996) *Tires, Suspension and Handling*, 2nd edition, London: Edward Arnold, ISBN 0340677961 (In US SAE, ISBN 1560918314).

Dixon J. C. (2007) *The Shock Absorber Handbook*, Chichester: J. Wiley & Sons Ltd, ISBN 9780470510209.

Drechsel A. (1956) Radführungsanordnung für Fahrzeuge, German Patent 12-14-100, filed 18 Dec 1956, patented 10 Nov 1966.

Durstine, J. W. (1973) *The Truck Steering System from Hand Wheel to Road Wheel*, SAE 730039, The 19th Ray Buckendale Lecture, January 1973, SAE SP-374, New York: Society of Automotive Engineers.

Ford H. (1937), *Motor Vehicle*, US Patent 2,214,456, filed 1 Nov 1937, patented 10 Sep 1940.

Hamy N. (1968) The Trebron DRC Racing Chassis, *Motor*, Dec. 21, 1968, pp. 20–24.

Heldt, P. M. (1948) *The Automotive Chassis*, 2nd edition, Published by P. M. Heldt, 600pp.

Hurley L. G. (1939) Automobile Chassis, US Patent 2,279,120, filed 5 Jun 1939, patented 7 April 1942.

Jeandupeux P. A. (1971) Suspension pour Véhicules Automobiles, Swiss Patent 537818, filed 4 May 1971, patented 31 July 1973.

Jenkins, M.A. (1975) Zeros of a real polynomial (Algorithm 493), *ACM Transactions on Mathematical Software*, **1**(2), 178–189.

Jenkins, M.A. and Traub, J.F. (1972) Zeros of a complex polynomial, *Communications of the ACM*, **15**(2), 97–99.

Katz A. (1997) *Computational Rigid Vehicle Dynamics*, Malabar, Florida: Krieger Publishing Company, Inc., ISBN 1575240165. (Discussion of calculation and use of quaternions.)

Lapidus L. (1962) *Digital Computation for Chemical Engineers*, New York: McGraw-Hill. (Hard to find. Good explanations of general numerical methods. The chemical content is limited to the end-of-chapter examples.)

MacPherson E. S. (1949) Wheel Suspension System for Motor Vehicles, US Patent 2,660,449, filed 27 Jan 1949, patented 24 Nov 1953.

Madsen, K. (1973) A root-finding algorithm based on Newton's method. *BIT*, **13**, 71–75. (Includes a program in ALGOL.)

M.I.R.A. 1965/1 *Definition of Handling Terms*.

Morrison J. L. M. and Crossland B. (1970) *An Introduction to the Mechanics of Machines*, 2nd edn in SI units, London: (Longman. ISBN 0582447291 and 0582447313. Contains material on velocity and acceleration diagrams.)

Suspension Geometry and Computation J. C. Dixon

Muller, D.E. (1956) A method for solving algebraic equations using an automatic computer, *Mathematical Tables and Other Aids to Computation*, **10**, 208–215.

Nemeth J. (1989) Optimization on steering mechanisms for high speed coaches, In *Second International Conference on New Developments in Powertrain and Chassis Engineering*, pp 325–330, Bury St Edmunds, Mechanical Engineering Publications, ISBN: 0852986890.

Open University, Course T235 *Engineering Mechanics – Solids* contains detailed material on drawing velocity diagrams. The entire course (on basic mechanical engineering) is available free at http://intranet.open.ac.uk/open-content/. Click on 'Visit the OpenLearn website', enter 'LabSpace', click on topics 'Technology', go to page 3, scroll down to 'Engineering Mechanics: Solids' and click on 'Browse'. See Block 3 for velocity diagrams.

Ostrowski, A.M. (1973) *Solution of Equations in Euclidean and Banach Space*, New York: Academic Press.

Parsons C. F. D. (1969) Vehicle Wheel Suspension System, US Patent 3,598,385, filed 27 Mar 1969, patented 10 Aug 1971.

Pellerin L. (1997) Vehicle Suspension System, UK Patent GB2309206, priority date 19 Jan 1996, filed 17 Jan 1997, patented 11 Mar 1998.

Phillippe M. P. (1975) Vehicle Suspension Systems, UK Patent 1526970, filed 30 Dec 1975, patented 4 Oct 1978.

Press, W.H., Teukolsky, S.A., Vetterling, W.T. and Flannery, B.P. (1992) *Numerical Recipes in Fortran: The Art of Scientific Computing*, Cambridge: Cambridge University Press, ISBN 052143064X. (Also available in C and Basic. Well known reference, more-or-less essential, many programs, although no help on geometry.)

Rall, L.B. (1966) Convergence of the Newton process to multiple solutions. *Numerische Mathematik*, **9**, 23–27.

Robson J. D. (1979) Road Surface Description and Vehicle Response, *International Journal of Vehicle Design*, **1**(1), 25–35, Inderscience.

SAE (1976) *Vehicle Dynamics Terminology*, SAE J670e, Warrendale, PA: Society of Automotive Engineers (included as an appendix in Dixon, 1996, above).

Schröder, E. (1870) Über unendlich viele Algorithmen zur Auflösung der Gleichungen. *Mathematische Annalen*, **2**, 317–365.

Shoup T. E. (1979) *A Practical Guide to Computer Methods for Engineers*, Englewood Cliffs, NJ: Prentice Hall, ISBN 0136906516. (Good explanations.)

Shoup, T.E. (1984) *Applied Numerical Methods for the Microcomputer*. Englewood Cliffs, NJ: Prentice Hall. ISBN 0130414182. (This is a development of Shoup, 1979.)

Sloan A. P. (1963) *My Years with General Motors*, Doubleday.

Spiegel M. R. (1968) *Mathematical Handbook of Formulas and Tables*, New York: McGraw-Hill, ISBN 0070382034.

Sternberg E. R. (1976) Heavy Duty Truck Suspensions, S.A.E. Paper 760369.

Verschoore R., Duquesme F. and Kermis L. (1996) Determination of the Vibration Comfort of Vehicles by Means of Simulation and Measurement, *European Journal of Mechanical Engineering*, **41**(3), 137–143.

Walker P. (2008) Camber Nectar, US patent 7407173, filed 29 May 2003, patented 5 Aug 2008.

Weiss W. (1997) Wheel Suspension with Automatic Camber Adjustment, US Patent 6267387, filed 12 Sep 1997, patented 31 July 2001.

Bibliography

For a qualitative discussion of many example suspensions, see Norbye (1980). A more quantitative analysis with many practical examples is given in Campbell (1981) and Bastow et al (2004). For more examples specifically on front-wheel drive see Norbye (1979). Sternberg (1976) above is a useful reference on the suspension of heavy commercial vehicles. Many details of suspension design from old, previously unpublished, General Motors reports by Olley have been gathered together with added explanation by Milliken and Milliken in Olley et al (2002), which also gives historical insight. Matschinsky (2000) gives a useful analysis of suspensions in some depth. See also Dixon (1996) and (2007) above.

Bastow D., Howard G. and Whitehead J. P. (2004) *Car Suspension and Handling*, 4th Edn, Professional Engineering Publishing, ISBN 186058439X. (In US, SAE, ISBN 0768008727.)

Campbell D. (1981) *Automobile Suspensions*, Chapman and Hall Ltd, ISBN 0412164205. (A general introduction to vehicle suspensions.)

Matschinsky W. (2000) *Road Vehicle Suspensions*, Bury St Edmunds: Professional Engineering Publishing Ltd, ISBN 1860582028.

Milliken W. F. and Milliken D.L. (1995) *Race Car Vehicle Dynamics*, New York: Society of Automotive Engineers, ISBN 1560915269.
Norbye J. P. (1979) *The Complete Handbook of Front Wheel Drive Cars*, Tab Books, ISBN 0-8306-2052-4.
Norbye J. P. (1980) *The Car and Its Wheels – A Guide to Modern Suspension Systems*, Tab Books, ISBN 0-8306-2058-3.
Olley M., Milliken W. F. and Milliken D. L. (2002) *Chassis Design Principles and Analysis*, New York: SAE, ISBN 186058389X.

——— // ———

Index

After the page number, a letter e means a significant equation, f means a figure, p means an example program fragment or program output, and t means a table.

Printed and bound by CPI Group (UK) Ltd, Croydon, CR0 4YY

16/04/2025

14658472-0004